Enzyme-Based Organic Synthesis

Enzyme-Based Organic Synthesis

Cheanyeh Cheng
Department of Chemistry
Chung Yuan Christian University
Chungli
Taiwan

WILEY

This edition first published 2022

Registered Office
John Wiley & Sons, Inc., 111 River Street, Hoboken, NJ 07030, USA

Editorial Office
111 River Street, Hoboken, NJ 07030, USA

For details of our global editorial offices, customer services, and more information about Wiley products visit us at www.wiley.com.

Wiley also publishes its books in a variety of electronic formats and by print-on-demand. Some content that appears in standard print versions of this book may not be available in other formats.

Library of Congress Cataloging-in-Publication Data
Names: Cheng, Cheanyeh, author.
Title: Enzyme-based organic synthesis / Cheanyeh Cheng, Department of Chemistry, Chung Yuan Christian University, Chungli, Taiwan.
Description: Hoboken, NJ : Wiley, 2022. | Includes bibliographical references and index.
Identifiers: LCCN 2021031980 (print) | LCCN 2021031981 (ebook) | ISBN 9781118027943 (hardback) | ISBN 9781118995143 (adobe pdf) | ISBN 9781118995150 (epub)
Subjects: LCSH: Enzymes–Synthesis. | Organic compounds–Synthesis. | Biocatalysis. | Enzymes–Biotechnology.
Classification: LCC TP248.E5 C44 2021 (print) | LCC TP248.E5 (ebook) | DDC 661/.805–dc23
LC record available at https://lccn.loc.gov/2021031980
LC ebook record available at https://lccn.loc.gov/2021031981

Cover Design: Wiley
Cover Image: Courtesy of author

Set in 9.5/12.5pt STIXTwoText by Straive, Pondicherry, India

10 9 8 7 6 5 4 3 2 1

MIX
Paper from responsible sources
FSC® C013604

Contents

Preface

The first time I was introduced to the field of microbiology was during the year 1982–1987, when I studied in the United States in the Department of Chemical & Biochemical Engineering of the Graduate and Professional School of Rutgers University to pursue my PhD degree. I was surprised that bacteria can live in an environment without air and at a temperature much higher than room temperature as well as ferment glucose to acetic acid. This study ignited my interest in the research of microorganisms and enzymes. As soon as I finished my PhD study in 1987, I came back to my alma mater, Chung Yuan Christian University, and worked as an associate professor in the Department of Chemistry, the place where I obtained my BS degree in 1974. I decided to continue my PhD research work to study the enzymatic cellulose hydrolysis for producing glucose using raw materials such as waste paper, dead tree branch, or waste bamboo chopsticks and the enantioselective bioreduction of ketones catalyzed by whole yeast cells for producing chiral secondary alcohols. I also taught a course called bioorganic chemistry, which focuses on the chemo-, stereo-, and regioselective enzyme or whole microbial cell catalyzed organic synthesis.

Ten years ago, I received an invitation from Wiley to write this book. At that moment I did not realize it is a big challenge for me to write a comprehensive book concerning enzyme-catalyzed organic synthesis using six classes of enzymes. With the kind of courage that "the newborn calf is not afraid of tigers," I accepted this invitation and wrote a book writing proposal. Fortunately, my proposal was approved by reviewers and I immediately started writing this book. Then I found I cannot concentrate my mind on writing this book and it takes me a long time to finish a chapter due to my teaching loads, my research works, student and family affairs, and many other trivial things. In fact, I could put all my time and mind on writing this book only after my retirement from school three and half years ago. I really appreciate the tolerant heart of Wiley editor to allow me to finish this book in such a long time. I learned a lot from writing this book, which also opened a new vision for me in the field of microbes and enzyme-catalyzed organic synthesis. I also deeply understood the meaning of the Chinese proverb "Live and Learn." What I did and what knowledge I acquired in my 30 years of academic career is only a small part of the field, just like a drop in the ocean. However, I sincerely hope that through this book more people will be interested in the field of enzyme-catalyzed organic synthesis.

Life originated from single-cell microorganisms, and microorganisms that cannot be seen by the human eye have existed on Earth since prehistoric times. Enzymes catalyze diverse chemical reactions in microbial cells from time to time and silently participate in the progress of life. The life phenomena presented by the variety of chemistry involved in the microbial cells is like a solemn and brisk music suite of life. No one would have expected that the relationship between enzymes and the tiny universe of microorganisms is so close and inseparable. Microorganisms are also taken as a cell factory by scientists due to their ability to produce various kinds of useful chemicals for human. However, as a result of the division of labor in science today, chemists, biochemists, biologists, biomedical scientists, biochemical engineers, etc., each use their own specialized scientific expertise to explore this life community, which has led to the difficulty in communication and the inefficient integration among different academic disciplines. Therefore, one of the goals of this book is to enable researchers from different disciplines to communicate and gain consensus to achieve integration.

The difference between enzyme-based organic synthesis and traditional organic synthesis is that it uses a highly selective biocatalyst (enzyme), and the enzyme selectivity includes reaction substrate specificity, stereospecificity, and regiospecificity. The selectivity of enzyme also makes the enzyme-based organic synthesis, particularly the asymmetric synthesis, more easy, convenient, and efficient to produce specialty chemicals. Because the enzyme-based reaction is usually performed in aqueous solution under mild conditions and in many cases using sustainable renewable substrates, which demonstrates environmentally friendly, enzyme-based organic synthesis fulfils the requirements of green chemistry. The development of enzymatic biotransformation or microbial fermentation has been over 50 years and has been implemented in numerous industrial applications. The recent advances in enzyme technology, such as protein engineering, site-specific evolution, metabolic engineering, and enzyme immobilization, have made enzyme-based organic synthesis more and more competitive with organic synthesis derived from fossil fuels.

This book contains eight chapters. Chapter 1 is an introduction to enzyme, coenzyme, enzyme specificity, and the green chemistry. Chapter 2 is about organic syntheses and their applications using class I oxidoreductases. Chapter 3 focuses on the transamination, glycosyl-transfer, phosphorylation, and acetyl-group transfer reactions using class II transferases and their applications. Chapter 4 is about class III hydrolases-based organic syntheses including hydrolysis reactions of ester bond, amide bond, phosphate esters, epoxides, hydantoins, glycosidic bonds with natural polysaccharides, and their applications. Chapter 5 contains organic syntheses and applications using class IV lyases and concentrates on carbon-carbon bond formation, carbon-oxygen bond formation, carbon-nitrogen bond formation, carbon-sulfur bond formation, and carbon-halide bond formation. Chapter 6 describes organic syntheses using class V isomerases including racemases and epimerases, *cis–trans* isomerase, intramolecular oxidoreductases, intramolecular transferases, intramolecular lyases, and their applications. Chapter 7 presents class VI ligases-based organic syntheses and their applications focusing on carbon-oxygen bond formation, carbon-sulfur bond formation, carbon-nitrogen bond formation, and carbon-carbon bond formation reactions. The final Chapter 8 shows two major techniques that could assist the advancements of enzyme-based organic syntheses in the future: one is the combinatorial chemistry and the other is artificial intelligence.

The level of contents of this book is medium to high suitable for readers having some basic knowledge of chemistry, organic chemistry, biochemistry, biology, microbiology, and chemical engineering. This book would be a very good reference book for academic researchers and industrial experts working on research and development. This book could also be used as a textbook for one semester course in senior class or graduate school class. Finally, I hope that this book will be able to "throw a brick to attract jade" to elicit truly outstanding books by experts and scholars in this field.

Cheanyeh Cheng
Chungli, January 2021

Acknowledgements

I would like to thank my parents, Mr. Yet-Sen Cheng and Mrs. Yen-In Chen Cheng, for working hard to raise me and for providing me the necessary fees for higher education.

I would like to thank my former teachers and professors for the education they gave me and for inspiring me so that I continue in the academic field and have a good performance.

I would like to thank God for giving me the opportunity to write this book.

1 Introduction

1.1 Discovery and Nature of Enzyme

Although the historical discovery of enzyme can be sourced back to Spallanzani as early as in 1783 with his noting to the liquefied meat by gastric juice of hawks [1], the discovery of enzyme is in general ascribed to the first "isolation" of an enzyme by two chemists, Anselme Payen and Jean-François Persoz, who worked at a sugar factory in Paris. In 1833, they obtained a substance from the malt extract called *diastase* (now known as amylase) that can hydrolyze starch to soluble sugar. Next year, Schwann succeeded in extracting the first enzyme from animal source, pepsin, which digests meat from stomach wall [2]. Later, he also identified trypsin, a peptidase in digestive fluids. By 1837, Jön Berzelius made a remarkable foresight for the catalytic nature of all these biological diastases. In the 1950s, Louis Pasteur acknowledged that sugar fermentation by yeast to produce alcohol is catalyzed by "ferments." Then, in 1860, Berthelot obtained an alcohol precipitate from yeast that can convert sucrose to glucose and fructose and concluded that there was much such ferment in yeast. Not until 1878, the name *enzyme*, which means "in yeast," was proposed by Frederick W. Kühne for these biological catalysts. The catalytic activity of enzyme was proved by Eduard Bücher in 1987 using yeast extract for catalytic alcohol fermentation. One year later, Duclaux proposed that all enzymes should give the suffix "ase" for an easily recognition [3].

The intensive studies of enzymes and proteins were both performed by biochemists in the 1800s. However, not until 1926 the protein nature of enzyme was seriously considered by biochemists that the jack bean urease that was recognized as a protein was first crystallized and recrystallized by James Sumner showing the catalytic ability for hydrolysis of urea to CO_2 and NH_3 [4, 5]. However, the crystal structure of urease which in essence is a nickel-containing enzyme as known nowadays was obtained by Andrew Karplus from *Klebsiella* aerogenes [6, 7]

Enzyme-Based Organic Synthesis, First Edition. Cheanyeh Cheng.

almost 70 years later after Sumner's work. Sumner's conclusion was widely accepted in the 1930s, after John Northrop and Moses Kunitz crystallized pepsin, trypsin, and other digestive enzymes and found to be proteins. Due to the prosperous development of separation and purification technology and corresponding instrumentation, hundreds of enzymes had been purified and discovered in the middle of nineteenth century. Nevertheless, the sequencing of proteins and enzymes was not until the work of William H. Stein et al [8]. who first complete the sequence of ribonuclease A (an enzyme with only 124 amino acids) in 1960. Later, in 1972, William H. Stein and Stanford Moore shared the Nobel Prize in chemistry.

1.2 Enzyme Structure and Catalytic Function

1.2.1 Enzyme Structure

With the exception of a small group of catalytic RNA molecules, all enzymes are protein. The protein nature of enzyme has been elucidated about a century ago that has led fast and broad progress in chemistry, biochemistry, and biology, in addition, led the development of many new fields such as enzymology, bioorganic chemistry, and molecular biology. In order to understand enzyme and how its function as a catalyst, one must know the enzyme structure first. Since enzyme is a kind of protein, its structure follows the four-level structure of protein, namely, the primary structure, the secondary structure, the tertiary structure, and the quaternary structure.

Protein is a polymer of amino acids that is referred to as peptides or proteins. Peptides are chains of amino acids joined through a substituted amide linkage, termed a peptide bond, which is formed by dehydration to remove the elements of water from the α-carboxyl group of one amino acid and the α-amino group of another. When a few amino acids (usually, less than 10) are joined in this fashion, the structure is called an oligopeptide. When many amino acids are joined, the product is called a polypeptide. Proteins that may have thousands of amino acid residues are polypeptides. Therefore, "protein" and "polypeptide" are sometimes used interchangeably. However, molecules with molecular weight below 10 000 are generally, referred to as polypeptides. In a peptide, the amino acid residue at one end with a free α-amino group is the amino-terminal (or *N*-terminal) residue, while at the other end, the residue with a free carboxyl group is the carboxyl-terminal (C-terminal) residue [9].

Since proteins are large macromolecules, the complexity of their 3D structure cannot be described easily like small molecules. Therefore, four levels of structure are used to define the complete 3D structure of protein. The primary structure is the amino acid sequence of a polypeptide chain that describes the order of all covalent

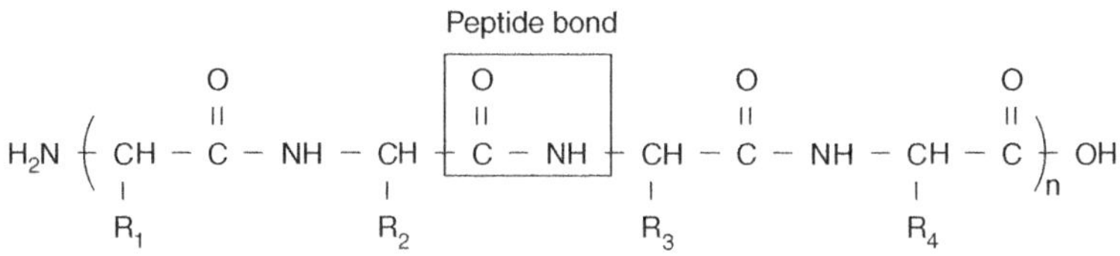

Scheme 1.1 The primary structure of a polypeptide chain linked by the peptide bond shows the sequence of amino acids.

bonds, mainly peptide bonds and disulfide bonds, linking amino acid residue in the polypeptide as illustrated in Scheme 1.1. The importance of primary structure is in determining the secondary, tertiary, quaternary structures of proteins, and thus their biological functions, which can be demonstrated by the hereditary disease sickle-cell anemia of human.

Part of the very long chain polypeptide can be coiled or folded into units by amino acid residues within a short distance to form recurring structural patterns of secondary structure such as the α-helix of α-keratin. The helix is a part of the tertiary structure that is the overall 3D arrangement or folding of a polypeptide. An example of the tertiary structure is myoglobin, a globular protein with 153 amino acid residues. The secondary structure refers to the spatial arrangement of amino acid residues that are adjacent in the primary structure, whereas tertiary structure includes longer-range aspects of the primary structure. When a protein has two or more polypeptide subunits that are associated with each other or one another, their arrangement in space is referred to as quaternary structure. Hemoglobin consisting of four polypeptide subunits is the most well-known protein with a complex quaternary structure.

1.2.2 Catalytic Function

The folding of long polypeptide chain to form tertiary or quaternary structure is caused by chemical or physical forces such as disulfide linkage, hydrogen bonding, acid–base interaction (salt bridge), and hydrophobic interaction. Folding of polypeptides into two or more stable globular units is called domains. Different domains often play distinct functions, such as binding molecules or interaction with other proteins. A molecule bound reversibly by a protein is called ligand. The site on the protein that binds the ligand is called the binding site. When a protein binds a ligand, the 3D structure of protein is often caused by a conformational change to permit a tighter binding to the ligand. This kind of binding with structural adaption between protein and ligand is called induced fit mechanism. Enzymes have catalytic function that binds and chemically rapidly transforms other molecules. For enzyme-catalyzed reaction, the molecule bound and acted by the enzyme is called substrate; the binding site is called the active site or catalytic site.

Enzyme is an efficient catalyst and is responsible for thousands integrated chemical reactions of the biological process occurred in the living system. Just like the usual inorganic catalyst, enzymes catalyze a reaction by lowering the transition state energy (the activation energy) of the activated complex and by raising the ground state energy. On the other hand, the catalysis of enzymes, not like simple inorganic catalysts, proceeds by forming several transition states and each with low activation energy instead of one activated complex of greater activation energy. The rate of enzyme-catalyzed reaction for a simple one enzyme, one substrate and one product system with the following mechanism was studied by Michaelis and Menten in 1913 (Scheme 1.2). In this mechanism, enzyme (E) binds the substrate (S) to form an enzyme–substrate (ES) complex and subsequently the ES breaks down to the product (P).

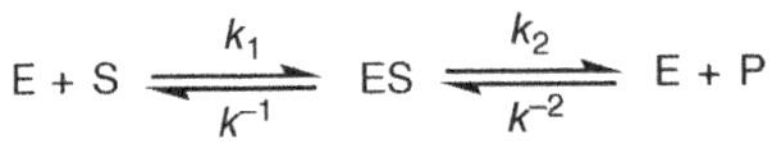

Scheme 1.2 The proposed enzyme reaction mechanism by Michaelis and Menten.

According to this mechanism, the enzyme-catalyzed reaction rate equation called Michaelis–Menten equation (Eq. 1.1) was derived by Michaelis and Menten with the second step as the rate-limiting step and derived by Briggs and Haldane using steady-state assumption. The term V_0 is the initial rate, V_{max} is the maximum reaction rate, $[S]$ is the substrate concentration, and K_m is the Michaelis constant.

$$V_0 = \frac{V_{max}[S]}{K_m + [S]} \tag{1.1}$$

For a more common case, enzyme product complexes (EP) release, EP → E + P, is the rate-limiting step, which is described as the reaction Scheme 1.3.

Therefore, a more general rate constant called the turnover number, k_{cat}, is defined to describe the limiting rate of any enzyme-catalyzed reaction at saturation, that is, $V_{max} = k_{cat}[E_t]$. In this situation, the Michaelis–Menten equation becomes Eq. 1.2

$$V_0 = \frac{k_{cat}[E_t][S]}{K_m + [S]} \tag{1.2}$$

The turnover numbers of several enzymes are given in Table 1.1.

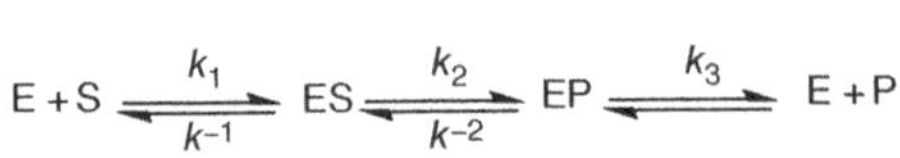

Scheme 1.3 A generalized enzyme-catalyzed reaction.

Table 1.1 Turnover numbers for some enzymes.

Enzyme	Turnover number k_{cat} (s^{-1})
Catalase	40 000 000
Carbonic anhydrase	400 000
Acetylcholinesterase	140 000
β-Lactamase	2000
Fumarase	800
β-Galactosidase	208
Phosphoglucomutase	21
Tryptophan synthetase	2
RecA protein (an ATPase)	0.4

1.3 Cofactors and Coenzymes

Enzymes have protein nature and molecular weights ranging from about 12 000 to over 1 million. The large molecule of enzymes is flexible for binding natural and unnatural substrates at their active site. The active site contains moieties consisted with amino acid residues. Although the activity of some enzymes requires no chemical groups other than their amino acid residues, others require an additional chemical component called cofactor. A cofactor, also called a coenzyme, is either one or more inorganic ions, such as Fe^{2+}, Mg^{2+}, Mn^{2+}, or Zn^{2+} (Table 1.2) [9], or an organic or metallo-organic molecule. Coenzyme are often derived from vitamins and organic nutrients required in small amounts in the diet (Table 1.3) [9]. The cofactor binds to the active site, in some cases covalently and

Table 1.2 Some inorganic metal ions as cofactor of enzymes.

Fe^{2+} or Fe^{3+}	Cytochrome oxidase, catalase, peroxidase
K^+	Pyruvate kinase
Mg^{2+}	Hexokinase, pyruvate kinase, enolase
Mn^{2+}	Arginase, ribonucleotide reductase
Ni^{2+}	Urease
Zn^{2+}	Carbonic anhydrase, alcohol dehydrogenase, carboxypeptidases A and B

Source: Based on Nelson and Cox [9].

Table 1.3 Some coenzymes as transient carriers of specific atoms or functional groups.

Coenzyme	Chemical groups transferred	Dietary precursor in mammals
Biocytin	CO_2	Biotin
Coenzyme A	Acyl group	Pantothenic acid and other compounds
5'-Deoxyadenosylcobalamin (coenzyme B_{12})	H atoms and alkyl groups	Vitamin B_{12}
Flavin adenine dinucleotide	Electrons	Riboflavin (vitamin B_2)
Lipoate	Electrons and Acyl groups	Not required in diet
Nicotinamide adenine Dinucleotide	Hydride ion (:H^-)	Nicotinic acid (niacin)
Pyridoxal phosphate	Amino groups	Pyridoxine (vitamin B_6)
Tetrahydrofolate	One-carbon groups	Folate
Thiamine pyrophosphate	Aldehydes	Thiamine (vitamin B_1)

Source: Based on Nelson and Cox [9].

in others noncovalently, which serves as transient carriers of redox equivalents, such as NAD(P)H or chemical energy (ATP) and is essential for the catalytic action of those enzymes that require cofactors.

For some enzymes, a coenzyme is required for their activity. A coenzyme or metal ion that is bound to the enzyme protein at the active site is called a prosthetic group. The protein part of such an enzyme is called the apoenzyme or apoprotein, and the entire enzyme is called a holoenzyme. Most of these cofactors are relatively unstable molecules. We will consider various coenzymes throughout the text in more detail for those related enzyme-catalyzed reactions.

1.4 Molecular Recognition and Enzyme Specificity

The science by which molecules interact via geometric orientations of atoms is called molecular recognition. The specific geometry of a molecule controls how it will react with other substances. Nearly, all enzymes are made up of more than 100 amino acid residues. However, an enzyme binds a substrate molecule at the catalytic active site that is just a small pocket or cleft region of the enzyme. The 3D structure of the active site is surrounded by amino acid side chains that come from different parts of the linear amino acid sequence. Water is usually excluded unless it is a reactant. The nonpolar character of much of the cleft enhances the binding of substrate. However, the cleft may also contain some polar residues that create a microenvironment essential for catalysis and extraordinary enzyme specificity. The specificity of binding of atoms in a substrate to the active site leads the proposed lock-and-key model of interaction between the substrate and enzyme by Emil Fischer in 1894 that points out the binding of a substrate to an enzyme just like the relationship of a key to a lock. The lock-and-key model is now modified to the induced fit model by Daniel Koshland in 1958 with the evident that the enzyme and substrate must adjust to fit one another to take up a configuration to stabilize the transition state [10, 11].

However, to assure the necessary geometric accuracy of the substrate binding and the orientation of catalytic functional group for enzyme interaction, the number of specific binding sites or points needed between the substrate and the active site of enzyme depends on the size of a molecule. For a large molecule such as glycyl tyrosine, a dipeptide, can bind carboxypeptidase A through total of five points at the active site [10, 12]: the electrostatic force, two hydrogen bonds, the hydrophobic interaction, and the coordination bonding to perform the hydrolysis of peptide bond of glycyl tyrosine. For small molecule such as carbon dioxide and water can coordinative bind carbonic anhydrase with only one point through the cofactor zinc ion and react to form bicarbonate or the reversible reaction [10]. Another example for small molecule is the binding of a covalent adduct formed between pyruvate and nicotinamide adenine dinucleotide (NAD^+) to lactate dehydrogenase (LDH) to produce lactate in which only two binding points, namely, the carbonyl group and the carboxylate group, of pyruvate are used to bind with the LDH [12]. The number of binding sites of an enzyme to the substrate is important in determining the type of enzyme specificity. The molecular recognition for enzyme specificity has been categorized into three major types of specificities: substrate specificity, regiospecificity, and stereospecificity [13].

1.4.1 Substrate Specificity

As a catalyst, the most distinguished property of enzyme is its substrate specificity. This kind of specificity determines its unique biological function. Many enzymes catalyze only one particular biological substrate. Enzymes with this kind of specificity are, therefore, called having absolute substrate specificity. For example, glucose-6-phosphatase binds only glucose-6-phosphate (G6P) to hydrolyze catalytically the phosphate moiety from the G6P. Therefore, the substrate specificity of glucose-6-phosphatase is absolute substrate specificity. Other enzymes possess a broader substrate specificity called relative group specificity that takes a group of biological molecules of similar chemical structure as substrate. As an example, acid and alkaline phosphatase can form phosphate ester bond for a variety of substrates. Therefore, acid and alkaline phosphatase presents wider relative group specificity.

Whether an enzyme has absolute substrate specificity or relative group specificity depends on the number of binding sites of the enzyme–substrate complex. Figure 1.1 illustrates the number of binding sites of an enzyme to the substrate makes the enzyme absolute substrate specificity or relative group substrate specificity. There is only one specific molecule that can attach the enzyme with four binding sites as shown in Figure 1.1a that gives absolute substrate specificity for the enzyme. The number of biding sites for the enzyme with substrate shown in Figure 1.1a–c is three, two, and one, respectively, that makes a variety of different groups available for the unbound group in the substrate to produce a relative group specificity for the enzyme. The substrate specificity of enzyme brings the advantage of reducing the possible side reactions, thus eliminating the laborious purification and separation works.

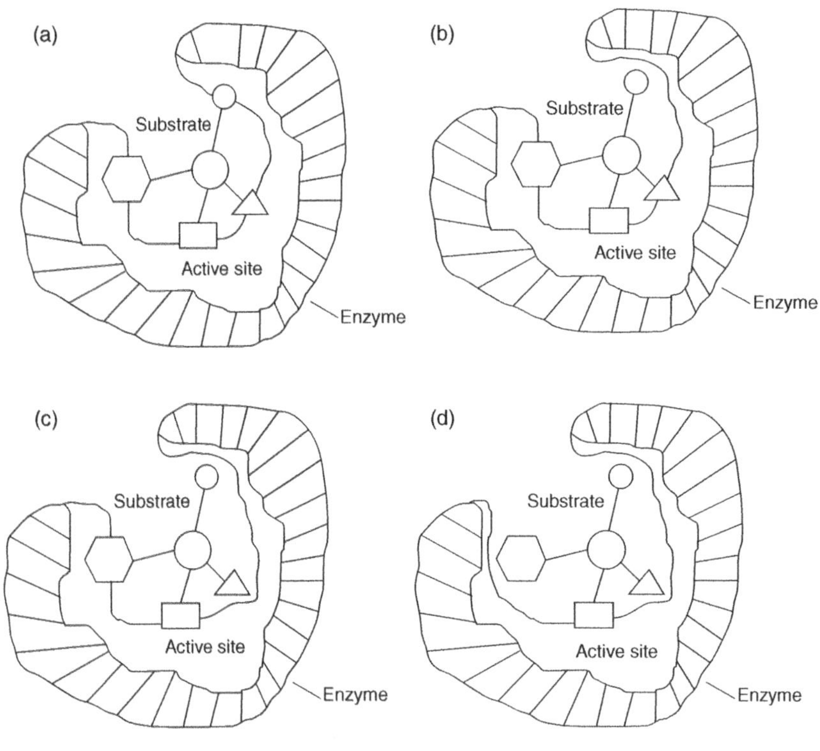

Figure 1.1 The number of binding sites of a substrate with enzyme determines the type of enzyme specificity: (a) Absolute substrate specificity of enzyme, (b–d) relative group specificity of enzyme.

1.4.2 Regiospecificity

Some of the enzymes may selectively catalyze one functional group at certain region of the molecule among several similar functional groups located at different regions of the same molecule. This kind of substrate specificity is called regiospecificity or diastereospecificity [13]. Example of regiospecificity in organic synthesis has been found since 1986 that the regioselective deacylation of methyl 2,3,4,6-tetra-*O*-acyl-D-hexopyranosides gives the 6-OH derivatives in high yields using the lipase from *Candida cylindracea* [14]. The enzyme-catalyzed regioselective *O*-acylation of ribo-, arabino-, xylo-, rhamnopyranosides, and aryl pyranosides is reviewed, and the methodology is applied to the total synthesis of the naturally occurring rhamnopyranoside by Bashir et al [15]. Regioselective biotransformation of dinitrile compounds 2-, 3-, and 4-(cyanomethyl) benzonitrile can be performed by whole bacterium cell *Rhodococcus rhodochrous* to the corresponding 2-(cyanophenyl) acetic acid, 3- or 4-(cyanomethyl) benzoic acid with high yield [16]. Whole cell bacterium *Bacillus cereus* has also been used for regioselectively converting 2-phenylenediamine to 2-aminoacetanilide with a76% molar yield [17]. Other than bacterium, monooxygenase in the phytopathogenic fungi *Colletotrichum gloeosporioides* and *Botrytis cinerea* has been found having the ability of regioselective hydroxylation of the C-H bonds to yield the corresponding diols [18]. Enzyme in cultured plant cells of *Phytolacca Americana* can reduce, and regioselectively hydroxylate and glucosylate, raspberry ketone and zingerone to their β-glycosides [19]. *Ginkgo biloba* cell suspension cultures were used to regio- and stereoselectively convert sinenxan A, 2α,5α,10β,14β-tetra-acetoxy-4(20), 11-taxadiene, a taxoid isolated from callus tissue cultures of Taxus spp., in Taxol® synthesis [20]. The regioselective oxidation of (−)-verbenone, an important component of the essential oil from rosemary, to (−)-10-hydroxyberbenone with human liver microsomes has been investigated by Miyazawa et al [21]. Although regioselective oxidation of terpenoids is difficult by chemical methods, regioselective oxidation of (+)- and (−)-citronellene was recently performed with *Spodoptera litura*, a larvae of common cutworm, to (2*R*,3*S*)-3,7-dimethyl-6-octene-1,2-diol (yield: 89.7%) and (2*S*,3*R*)-3,7-dimethyl-6-octene-1,2-diol (yield: 56.3%) [22]. These examples of enzyme-catalyzed regiospecificity clearly elucidates that the specific enzyme–substrate binding configuration at the active site allows only one of several similar function groups in different regions of the substrate molecule reacts to produce the product. The number of binding points of enzyme–substrate complex for regioselective molecular recognition must be a “multi-point” binding case to match the complexity of the substrate molecule.

1.4.3 Stereospecificity

The most magnificent specificity of enzyme is its distinguishable ability for only one enantiomeric structure of racemic substrate molecules. The molecular recognition of an enzyme for enantiomeric molecules is called enantiospecificity or stereospecificity. The stereospecificity is an intrinsic property of enzyme which is due to the chirality of active site of the enzyme. Except for a few cases, all enzymes are chiral catalysts because they all made from L-amino acids, thus the binding of asymmetric substrate at the active site is stereoselective. Since the stereospecificity of enzyme involves enzyme–substrate complex formation with only one enantiomer of a racemate, only one product is formed from the enzyme-catalyzed reaction. Therefore, enzyme-catalyzed reactions are in great favor of the organic asymmetric synthesis [23].

The theoretical explanation for the stereospecificity of enzyme was based on the rationale of three-point attachment rule [24, 25]. This rule suggests that at least three different binding points should occur between enzyme and substrate at the active site to make recognition for the correct stereostructure of substrate molecule as illustrated in Figure 1.2. However, the three-point attachment is not strict for the stereospecificity of enzyme to asymmetric molecules. As shown in Figure 1.3, the recognition of the stereostructure of asymmetric molecules with enzyme can be accessed by two-point binding of enzyme–substrate complex, under the situations that the stability of the enzyme–substrate complex is greatly influenced by the formation of two different 3D tetrahedral structure of enzyme–substrate complex or the ability of interaction for the two unbinding groups of the substrate with enzyme at the active site is obviously different. With only two-point binding of the enzyme–substrate complex, one of the two substrate enantiomers possessing greater stability of the enzyme–substrate complex will have enough time to react to form product, but this situation will not happen for the other substrate enantiomer.

The preparation of enantiopure compounds is highly demanded by industries, particularly, in pharmaceutical and agrochemical applications because of the different biological activities of drug enantiomers. As a result of advances in asymmetric synthesis and separation technologies, there were many cases in using the stereospecificity of enzyme for the production of single enantiomer to allow the study of its pharmacodynamics and pharmacokinetic properties. For example, in the case of total synthesis of D-biotin, the novel enantioselective synthesis of the optically active (3*aS*,6*aR*)-lactone (the key D-biotin intermediate) was through kinetic resolution by inexpensive microbial lipase instead of pig liver esterase [26]. In Scheme 1.4, optically active (3*aS*,6*aR*)-lactone **1** was enantioselectively produced with high enantiomeric excess.

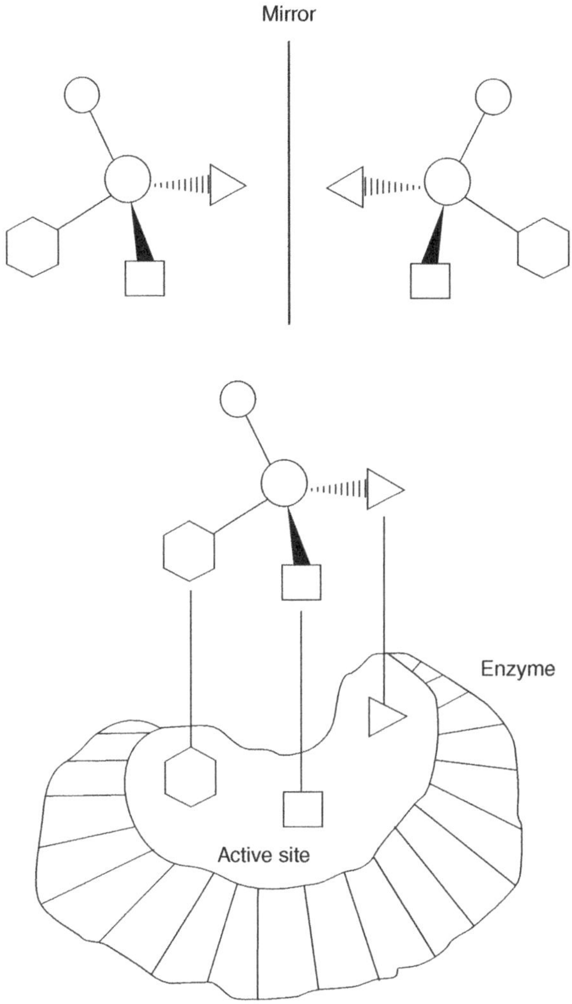

Figure 1.2 Three-point attachment rule shows that only one enantiomer of the asymmetric molecule can successfully bind with enzyme at the active site to produce the stereospecificity of enzyme.

(ee > 98%) and conversion ratio (≥ 40%) by dry microbial cells of *Aspergillus oryzae* WZ007 on racemic acid **1** (1,3-dibenzyl-5-(hydeoxymethyl)-2-oxo-4-imidazolidineca rboxylic acid) that was obtained via chemical hydrolysis of racemic lactone **1**.

Recently, an indirect strategy was used for the synthesis of (*R*)-phenylephrine (an α_1-adrenergic receptor agonist) that is widely used in over-the-counter drugs to treat the common cold. An amino alcohol dehydrogenase gene isolated from

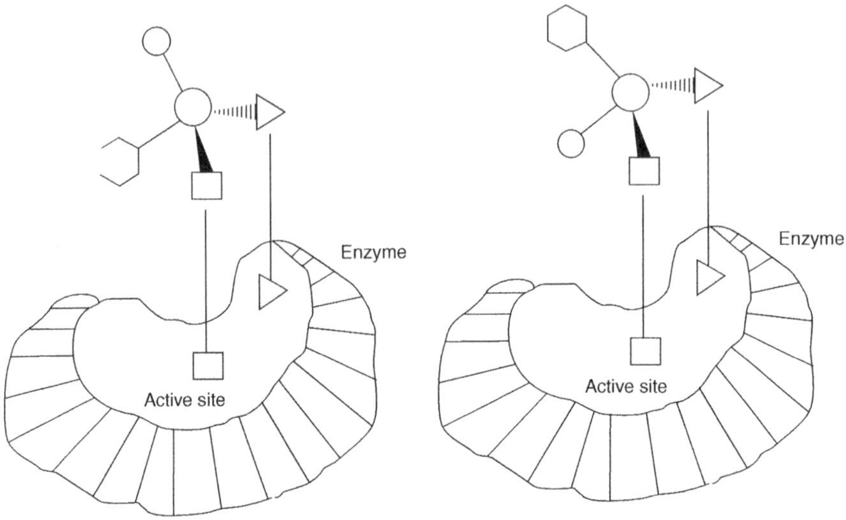

Figure 1.3 The stereospecificity of enzyme for two substrate enantiomers by two-point binding.

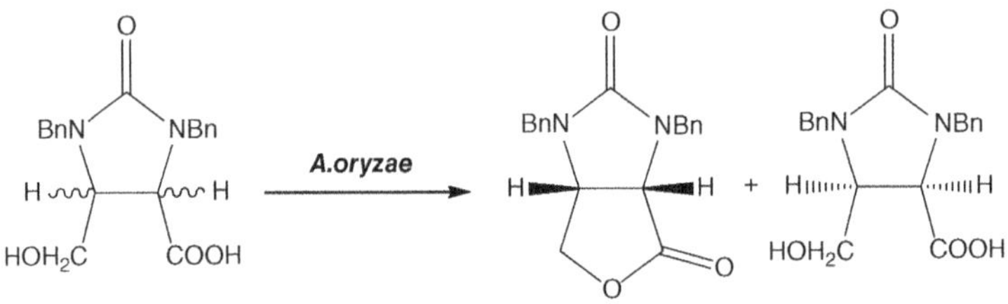

Scheme 1.4 Enantioselective synthesis of (3*aS*,6*aR*)-lactone.

Rhodococcus erythropolis BCRC 10909 was expressed in *Escherichia coli* NovaBlue, which is able to convert 1-(3-hydroxyphenyl)-2-(methylamino) ethanone (HPMAE) to (*S*)-phenylephrine with more than 99% enantiomeric excess (ee) and 78% yield as shown in Scheme 1.5 [27]. The (*S*)-phenylephrine was subsequently converted to (*R*)-phenylephrine by Walden inversion reaction [28]. Since enantiopure vicinal diols are useful and valuable intermediate for pharmaceutical production, a simple and green method for preparing several enantiopure 1,2-diols was developed via regio- and stereoselective concurrent oxidations of the racemates with microbial cell *Sphingomonas* sp. HXN-200.

As shown in Scheme 1.6, concurrent biooxidations of racemic 3-*O*-benylglycerol **1** with resting cells gave (*S*)-**1** in 99.2% enantiomeric excess (ee) and 32% yield. Similar biooxidations of racemic 1-(4-chlorophenyl)-1,2-ethanediol **2**, 1-(4-methylp

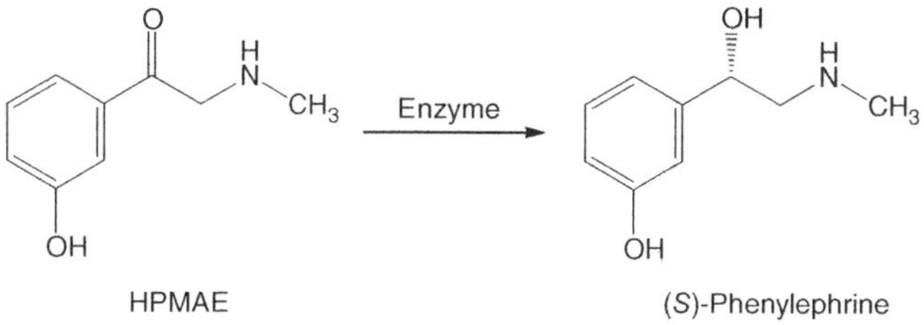

Scheme 1.5 The enantioselective conversion of HPMAE to (*S*)-phenylephrine.
Source: Lin et al. [27].

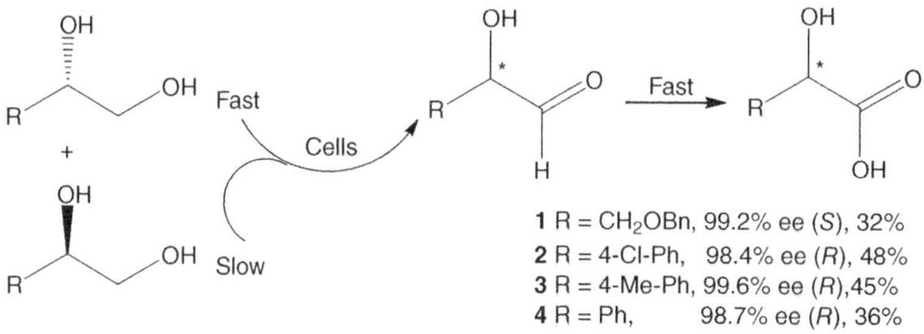

Scheme 1.6 Regio- and stereoselective concurrent oxidations of racemic vicinal diols to enantiopure 1,2-diols.

henyl)-1,2-ethanediol **3**, and phenyl-1,2-ethanediol **4** gave (*R*)-**2** in 98.4% ee and 48% yield, (*R*)-**3** in 99.6% ee and 45% yield, and (*R*)-**4** in 98.7% ee and 36% yield, respectively [29].

1.5 Enzyme Classes and Nomenclature

Since Duclaux proposed that all enzymes should give the suffix "ase" for an easy recognition [3], the suffix "ase" has been added to the name of many enzymes according to their substrate or to a word or phrase for describing the activity. For example, glucose oxidase catalyzes the oxidation of glucose to produce gluconolactone, and cellulase catalyzes the hydrolysis of cellulose to form glucose. However, enzymes such as pepsin and trypsin have names that do not relate with their substrates or functions. Because more and more enzymes are discovered accompanied with the progress of scientific researches, the name of new enzyme may have two or more names, or two different enzymes may be given the same name. To avoid the ambiguity for naming enzymes, a systematic method for naming and classifying enzymes should be used and agreed globally.

In 1960s, the Commission on Enzyme Nomenclature was formed by International Union of Biochemistry (IUB) to classify enzymes into six major classes according

Table 1.4 Six major classes of enzyme.

No.	Class	Catalytic function or reaction
1	Oxidoreductases	Transfer of electrons (hydride ions or H atoms), e.g. $H_3CCH_2OH \longrightarrow H_3CC(=O)H + 2H\cdot$
2	Transferases	Group-transfer reactions, e.g. $R{-}CH(NH_3^+){-}COO^- + R'{-}C(=O){-}COO^- \rightleftharpoons R{-}C(=O){-}COO^- + R'{-}CH(NH_3^+){-}COO^-$
3	Hydrolases	Hydrolysis reactions (transfer of functional groups to water), e.g. $R{-}O{-}P(=O)(O^-){-}O^- + HOH \longrightarrow R{-}OH + HPO_4^{2-}$
4	Lyases	Addition of groups to double bonds, or formation of double bonds by removal of groups, e.g. $R{-}CH_2CH(NH_2)R' \rightleftharpoons RCH{=}CHR' + NH_3$
5	Isomerases	Transfer of groups within molecules to yield isomeric forms, e.g. L-Alanine $\rightleftharpoons$ D-Alanine
6	Ligases	Formation of C–C, C–S, C–O, and C–N bonds by condensation reactions coupled to ATP cleavage, e.g. $CH_3C(=O)COO^- + CO_2 + H_2O + ATP \rightleftharpoons {}^-OOCCH_2C(=O)COO^- + ADP + P_i$

Source: Based on Armstrong [2]; Nelson and Cox [9]; Kula [30].

to the type of reaction catalyzed as indicated in Table 1.4 [2, 9, 30]. Each of the six major classes is further divided into subclasses and subgroups.

By international agreement, the catalytic reaction is assigned and identified by a group of four-digit number according to the enzyme classification system. For example, the enzyme catalyzes the transfer of a phosphoryl group from ATP to D-glucose is named as ATP:glucose phosphotransferase. The enzyme is classified as

Transferase	Main class 2
Phosphotransferase	Subclass 7
Using a hydroxyl group as acceptor	Subgroup 1
D-Glucose as the phosphoryl-group acceptor	The serial number 1

The serial number of the last digit of an enzyme is identified by the first three entrees. Therefore, the Enzyme Commission number (E.C. number) of this enzyme is 2.7.11 denoted as E.C. 2.7.1.1. However, a trivial name, hexokinase, is more commonly used for this enzyme.

1.6 Enzyme and Green Chemistry

Green chemistry, also known as sustainable chemistry, is an emerging field in chemistry and is highly advocated by global researchers recently. Green chemistry emphasizes the design of products and chemical processes that reduce or eliminate the use or production of hazardous substances [31]. Whereas sustainability has been defined as "meeting the needs of the present generation without compromising the ability of future generations to meet their needs." Therefore, green chemistry can also be thought as a critical tool in attaining sustainability by developing new technologies in all kinds of applications such as food and drink, medicine, energy, biofuels, plastics, and nanotechnology. Also, the long-term entanglement of the "three *E*'s" problems – *Energy*, *Economy*, and *Environment* – could be solved by applying the 12 principles of green chemistry to assure a sustainable society in the future.

The 12 principles of green chemistry listed in Table 1.5 [31, 32] were proposed by Paul T. Anastas and John C. Warner in 1998 that has become the index to implement the green/sustainable chemistry. How chemical synthesis catalyzed by enzyme is related with green chemistry? As mentioned before, the substrate specificity of enzyme greatly reduces the formation of byproducts, thus greatly increases the atom economy of the reaction. Enzymes are biodegradable natural material that does not threat human health and make environmental pollution. The regioselectivity, particularly, the stereoselectivity of enzymes can be used to design and produce safe drugs of no toxicity and without side-effects by raising the enantiomeric excess (%ee) value or yield. The reaction conditions for enzyme-catalyzed reaction are mild, usually at low temperature and under atmospheric pressure, and the solvent used is innocuous water that makes an energy efficient and clean synthetic process. The substrate for enzyme can be either renewable or nonrenewable source that gives the enzyme reaction wide and flexible applications. Various specificities of enzyme show great advantage in avoiding the use of blocking groups, protection/deprotection, or temporary modification of physical/chemical processes, thus reduce the synthetic steps for a final target product to a minimum that saves a lot of chemical reagents and generates minimal wastes. Immobilized enzyme-catalyzed reaction is most feasible to couple an online analytical technique to perform real-time monitoring and control of hazardous substances [33, 34].

Table 1.5 Twelve principles of green chemistry.

1) Prevention: It is better to prevent waste than to treat or clean up waste after it has been created.
2) Atom Economy: Synthetic methods should be designed to maximize the incorporation of all materials used in the process into the final product.
3) Less Hazardous Chemical Syntheses: Wherever practicable, synthetic methods should be designed to use and generate substances that possess little or no toxicity to human health and the environment.
4) Designing Safer Chemicals: Chemical products should be designed to affect their desired function while minimizing their toxicity.
5) Safer Solvents and Auxiliaries: The use of auxiliary substances (e.g. solvents, separation agents) should be made unnecessary wherever possible and innocuous when used.
6) Design for Energy Efficiency: Energy requirements of chemical processes should be recognized for their environmental and economic impacts and should be minimized. If possible, synthetic methods should be conducted at ambient temperature and pressure.
7) Use of Renewable Feedstocks: A raw material or feedstock should be renewable rather than depleting whenever technically and economically practicable.
8) Reduce Derivatives: Unnecessary derivatization (use of blocking groups, protection/deprotection, and temporary modification of physical/chemical processes) should be minimized or avoided, if possible, because such steps require additional reagents and can generate waste.
9) Catalysis: Catalytic reagents (as selective as possible) are superior to stoichiometric reagents.
10) Design for Degradation: Chemical products should be designed so that at the end of their function they break down into innocuous degradation products and do not persist in the environment.
11) Real-time Analysis for Pollution Prevention: Analytical methodologies need to be further developed to allow for real-time, in-process monitoring, and control prior to the formation of hazardous substances.
12) Inherently Safer Chemistry for Accident Prevention: Substances and the form of a substance used in a chemical process should be chosen to minimize the potential for chemical accidents, including releases, explosions, and fires.

Source: Based on Anstas and Warner [31]; Hill et al. [32].

Probably, the greatest disadvantage of free-enzyme catalyzed reaction for green chemistry is only one time usage of the expensive purified enzyme due to the difficulty of enzyme recovery. To overcome this drawback, the enzyme-catalyzed reaction can be performed by using whole microbial cell. However, the use of whole cell catalyzed reaction can be considered convenient against the free enzyme reaction under three situations: (i) as the enzyme is intracellular; (ii) as the enzyme needs a cofactor to carry out the catalysis; and (iii) the formation of product through a multienzymatic processes. The other strategy is to perform the enzyme reaction with immobilized enzyme that can be recovered and reused many times. However, the maintenance of activity for the recovered enzyme is a big challenge in present technology. Nevertheless, this is probably the fundamental approach for enzyme-catalyzed reaction to fulfill green chemistry.

To encourage and propel the implementation of green chemistry, the Presidential Green Chemistry Challenge Award was established by the United States government in 1996. The following are several award winners that can be used as models in green chemistry by using enzyme and related microbe for production of chemicals [35].

The winner of year 2011: Production of Basic Chemicals from Renewable Feedstocks at Lower Cost

Genomatica has developed a microbe using sophisticated genetic engineering to make 1,4-butanediol (BDO) (a high-volume chemical building block used to make many common polymers, such as spandex) by fermenting sugars. When produced at commercial scale, Genomatica's Bio-BDO will be less expensive, require about 60% less energy, and produce 70% less carbon dioxide emissions than BDO made from natural gas.

1.7 The Winner of Year 2010: Greener Manufacturing of Sitagliptin Enabled by an Evolved Transaminase

Merck and Codexis have developed a second-generation green synthesis of sitagliptin, the active ingredient in Januvia, a treatment for type 2 diabetes. This collaboration has led to an enzymatic process with transaminase that reduces waste, improves yield and safety, and eliminates the need for a metal catalyst. Early research suggests that the new biocatalysts will be useful in manufacturing other drugs as well.

1.8 The Winner of Year 2009: A Solvent-Free Biocatalytic Process for Cosmetic and Personal Care Ingredients

Esters are an important class of ingredients in cosmetics and personal care products. Usually, they are manufactured by harsh chemical methods that use strong acids and potentially hazardous solvents; these methods also require a great deal of energy. Eastman's new method uses immobilized enzymes to make esters, saving energy, and avoiding both strong acids and organic solvents. This method is so gentle that Eastman can use delicate, natural raw materials to make esters never before available.

References

1 Segal, I.H. (**1975**). *Enzyme Kinetics. Behavior and Analysis of Rapid Equilibrium and Steady-State Enzyme Systems*. New York: Wiley.
2 Armstrong, F.B. (**1989**). *Biochemistry*. Oxford: Oxford University Press.
3 Silverman, R.B. (**2002**). *The Organic Chemistry of Enzyme-Catalyzed Reactions*. Amsterdam: Academic Press.
4 Sumner, J.B. (**1926**). *J. Biol. Chem. 69*: 435–441.
5 Sumner, J.B.J. (**1926**). *Boil. Chem. 70*: 97–98.
6 Jabri, E., Carr, M.B., Hausinger, R.P., and Karplus, P.A. (**1995**). *Science 268*: 998–1004.
7 Jabri, E. and Karplus, P.A. (**1996**). *Biochemistry 35*: 10616–10626.
8 Hirs, C.H.W., Moore, S., and Stein, W.H. (**1960**). *J. Biol. Chem. 235*: 633–647.
9 Nelson, D.L. and Cox, M.M. (**2000**). *Lehninger Principles of Biochemistry*. New York: Worth Publishers.
10 Lubert, S. (**1988**). *Biochemistry*, Thirde. New York: W. H. Freeman and Company.
11 Mathews, C.K., van Holde, K.E., and Ahern, K.G. (**2000**). *Biochemistry*, Thirde. San Francisco: Addison Wesley Longman, Inc.
12 Zubay, G. (**1988**). *Biochemistry*, Seconde. New York: Macmillan Publishing Company.
13 Faber, K. (**2004**). *Biotransformations in Organic Chemistry, A Textbook*, Fifthe. Berlin: Springer-Verlag Berlin Heidelberg.
14 Sweers, H.M. and Wong, C.-H. (**1986**). *J. Am. Chem. Soc. 108*: 6421–6422.
15 Bashir, N.B., Phythian, S.J., Reason, A.J., and Roberts, S.M.J. (**1995**). *Chem. Soc., Perkin Trans. 1: Org. Bio-org. Chem 18*: 2203–2222.
16 Dadd, M.R., Claridge, T.D.W., Walton, R. et al. (**2001**). *Enz. Microbial Technol. 29*: 20–27.
17 Takenaka, S., Mulyono, Sasano, Y. et al. (**2006**). *J. Biosci. Bioeng. 102*: 21–27.
18 Bustillo, A.J., García-Pajón, C.M., Aleu, J. et al. (**2003**). *Tetrahedron.: Asymmetry 14*: 3755–3760.
19 Shimoda, K., Harada, T., Hamada, H. et al. (**2007**). *Phtochem. 68*: 487–492.
20 Dai, J., Ye, M., Guo, H. et al. (**2002**). *Tetrahedron 58*: 5659–5668.
21 Miyazawa, M., Sugie, A., and Shindo, M. (**2002**). *Biosci. Biotechnol. Ciochem. 66*: 2458–2460.
22 Miyazawa, M., Marumoto, S., Masuda, A. et al. (**2009**). *J. Agric. Food Chem. 57*: 7800–7804.
23 Sih, C.J. and Wu, S.-H. (**1989**). *Topics Stereochem. 19*: 63–125.
24 Ogston, A.G. (**1948**). *Nature* 162: 963.
25 Jones, J.B. (**1976**). *Biochemical Systems in Organic Chemistry: Concepts, Principles and Opportunities*. In: *Applications of Biochemical Systems in Organic Chemistry* (ed. J.B. Jones, C.J. Sih and D. Perlman). New York: Part I. Wiley.

26 Zheng, J.-Y., Wang, Z., Zhu, Q. et al. (**2009**). *Mol. Catal. B 56*: 20–23.
27 Lin, W.-D., Chen, C.-Y., Chen, H.-C., and Hsu, W.-H. (**2010**). *Process Biochem. 45*: 1529–1536.
28 Dorokhova, M.I., Somolina, N.E., Tikhonova, Y.O., and Mikhalev, V.A. (**1974**). *Pharm. Chem. J. 8*: 209–211.
29 Jia, X.; Xu, Y.; Li, Z. (**2011**) ACS Catal. *1*, 591-596.
30 Kula, M.-R. (**1995**). *Introduction*. In: *Enzyme Catalysis in Organic Synthesis: A Comprehensive Handbook, Volume I* (ed. K. Drauz and H. Waldmann). New York: VCH Publishers, Inc.
31 Anstas, P. and Warner, J. (**1998**). *Green Chemistry: Theory and Practice*. New York: Oxford University Press.
32 Hill, J.W.; McCreary, T.W.; Kolb, D.K. (**2010**) *Chemistry for Changing Times*, Twelfth Edition, Pearson Education, Inc., New Jersey: Upper Saddle River.
33 Cheng, C., Chen, C.-S., and Hsieh, P.-H. (**2010**). *J. Chromatogr. A 1217*: 2104–2110.
34 Cheng, C. and Chang, K.-C. (**2007**). *Anal. Scis 23*: 305–310.
35 FedCenter.gov. Presidential Green Chemistry Challenge Awards. https://www.fedcenter.gov/Bookmarks/index.cfm?id=1205&printable=1.

2

Organic Synthesis with Oxidoreductases

2.1 Oxidation Reactions

Enzymatic and microbial oxidations can be dated back to 2000 BCE with vinegar production that is based on the oxidation of ethanol by acetic acid bacteria. Enzymes involved in the biocatalyzed oxidations are dehydrogenases and oxidases. The more common coenzymes associated with dehydrogenases are nicotinamide adenine dinucleotide/nicotinamide adenine dinucleotide hydrogen (NAD^+/NADH), nicotinamide adenine dinucleotide phosphate/nicotinamide adenine dinucleotide phosphate hydrogen ($NADP^+$/NADPH), and flavin adenine dinucleotide/flavin adenine dinucleotide hydrogen (FAD/$FADH_2$), whereas oxidases are usually assisted by flavoproteins for transferring electrons to molecular oxygen. Dehydrogenases can be found in aerobic and anaerobic organisms or microorganisms; however, oxidases are not present in strictly anaerobic species. Nowadays, dehydrogenases and oxidases have been extensively used for selective oxidations and as an alternative synthetic strategy for conventional oxidations with the advantages of being environmentally friendly and highly chemo-, regio-, and stereoselective. Nevertheless, due to poor stability of the two enzymes at high substrate/product/organic solvent concentration and temperature, large-scale bio-oxidation processes were few. The functional groups involved in those bio-oxidations include hydroxyls of primary and secondary alcohols, carbonyls of aldehydes, saturated C–C bonds, C–N bond of amino acids, amines, nitroalkanes, and thiols [1].

2.1.1 Oxidation of Alcohols and Aldehydes

Since primary alcohols can be oxidized to aldehydes and further to carboxylic acids, which are versatile building blocks in organic synthesis, the selective oxidation of primary alcohols using enzymes to produce corresponding aldehydes is important for both fundamental and industrial research. Benzaldehyde, which

Enzyme-Based Organic Synthesis, First Edition. Cheanyeh Cheng.

is important and widely applied in cosmetic, flavor, and pharmaceutical industries [2, 3], is generally prepared by oxidation of toluene or hydrolysis of benzyl chloride. Due to environmental concern, scientific reports have shown that *Gluconobacter oxydans* can be used for the selective oxidation of benzyl alcohol to obtain benzaldehyde in an organic/aqueous biphasic system to substitute for these environmentally unfavorable processes [4, 5]. Optimization of the immobilization parameters for *G. oxydans* not only improves the bioconversion of the selective oxidative process but also increases the stability of immobilized cells so that the immobilized cells can be used repeatedly for 10 cycles and a 53.2% of the oxidative activity of the immobilized cells compared with that of free cells was retained [6].

Fourteen commercial alcohol oxidases, 33 selected commercial alcohol dehydrogenases (ADHs), and 218 microorganisms were tested for the oxidation of benzyl alcohol for the production of benzaldehyde via different types of hydrogen transfer [7]. If alcohol oxidases were used for the oxidation, the reaction just required molecular oxygen as oxidant and hydrogen peroxide was produced as a side product, which subsequently is decomposed disproportionately by a catalase to form water and molecular oxygen (Scheme 2.1). If isolated ADHs or lyophilized microorganisms were employed for the oxidation of benzyl alcohol, the hydrogen acceptor of the hydrogen transfer reaction could be acetaldehyde (Scheme 2.2), chloroacetone, or acetone.

Direct oxidation of simple primary alcohols such as methanol and ethanol to corresponding aldehydes using either free whole cell or immobilized whole cell was also reported in literature. For methanol oxidation to formaldehyde, methanol oxidase (MOX) in the yeast *Hansenula polymorpha* was involved for the biotransformation [8]; for ethanol oxidation to acetaldehyde, ADH in *Saccharomyces cerevisiae* was employed for the biotransformation. In these studies, the immobilization

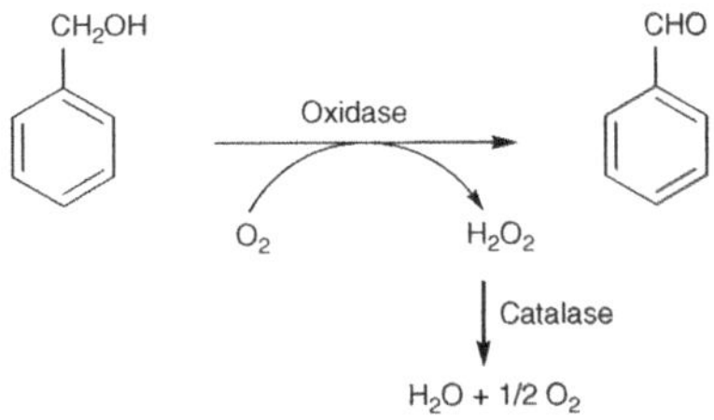

Scheme 2.1 Oxidation of benzyl alcohol using an oxidase and a catalase.

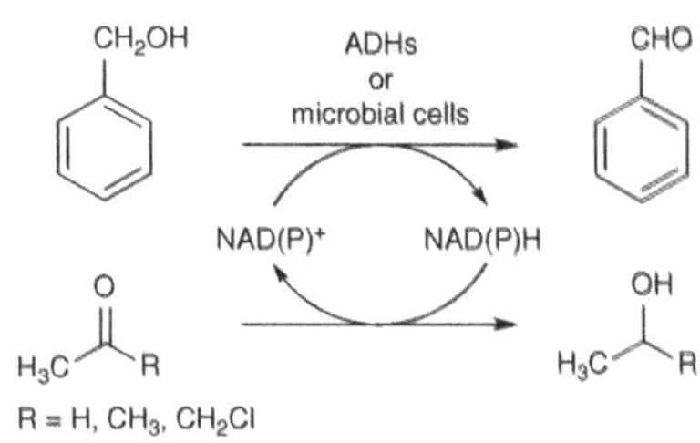

Scheme 2.2 Oxidation of benzyl alcohol via alcohol dehydrogenases or microbial cells.

of microbial cells was found to be of great help for both the cell activity and stability [9].

The stereoselective oxidation of secondary alcohols to produce ketones is of greatest interest in organic synthesis for its applications in pharmaceutical industries. Simple *sec*-alcohol, 2-butanol, has been oxidized to butanone by the immobilized yeast *S. cerevisiae* with a 45% yield [9]. Three screened yeast, *Williopsis californica*, *Williopsis saturnus*, and *Pachysolen tannophilus*, have been used for the oxidation of six cycloalkanols with different ring size including cyclobutanol, cyclopentanol, cyclohexanol, cycloheptanol, cyclooctanol, and cyclododecanol. The results show that *W. californica* and *P. tannophilus* are active against all six cycloalkanols and can be thought as nonselective, while *W. saturnus* is active against cycloalkanols of four, five, and six carbon atoms and is selective for small cyclohexanols [10]. These three selected strains have also been employed for exploring the stereoselectivity of several *sec*-alcohols such as (1*R*)-(2-furyl)-ethanol, (1*S*)-(2-furyl)-ethanol, (1*R*)-phenylethanol, (1*S*)-phenolethanol, (1*R*)-tetrahydronapthol, (1*S*)-tetrahydronapthol, (−)-neo-menthol ((1*R*,2*R*,5*S*)-2-isopropyl-5-methyl-cyclohexanol), (+)-menthol ((1*S*,2*R*,5*S*)-2-isopropyl-5-methyl-cyclohexanol), and iso-menthol ((1*S*,2*R*,5*R*)-2-isopropyl-5-methyl-cyclohexanol). The results indicate that all the strains are stereoselective toward the *S*-enantiomer [10].

The application of primary alcohol oxidation is proved to be valuable for the preparation of enantiopure 1,2-diols [11]. Several enantiopure 1,2-diols were produced via a simple and green regio- and stereoselective concurrent oxidation of the racemates using *Sphingomonas* sp. HXN-200 (Scheme 2.3). For the four tested 1,2-diols, this concurrent oxidation process demonstrates better enantioselectivity than any other known biocatalytic processes.

Another method used for the preparation of enantiopure (*S*)-1-phenyl-1,2-ethanediol from the racemic (*R*,*S*)-1-phenyl-1,2-ethanediol was performed with the whole-cell yeast *Candida parapsilosis*. In this method,

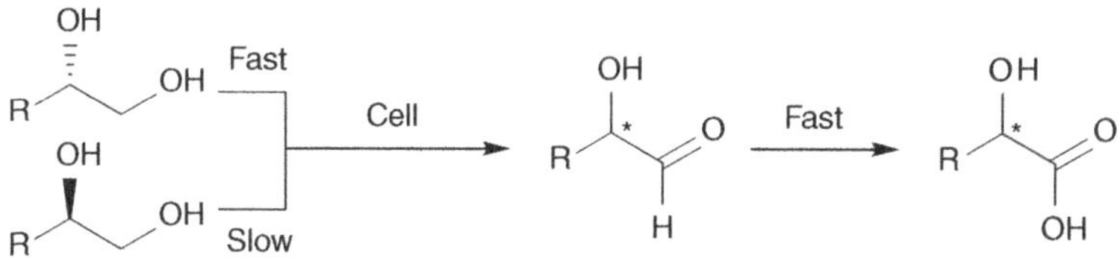

1. R = CH_2OBn, 99.2% ee (*S*), 32%
2. R = 4-Cl-Ph, 98.4% ee (*R*), 48%
3. R = 4-Me-Ph, 99.6% ee (*R*), 45%
4. R = Ph, 98.7% ee (*R*), 36%

Scheme 2.3 Regio- and stereoselective concurrent oxidations of (±)-1,2 diols.

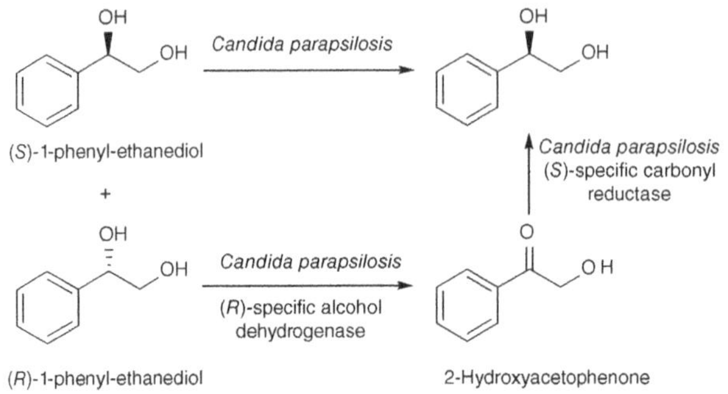

Scheme 2.4 Deracemization of racemic 1-phenyl-1,2-ethanediol by *C. parapsilosis* through an oxidation–reduction sequence.

(*R*)-1-phenyl-1,2-ethanediol is first stereoselectively oxidized to 2-hydroxyacetophenone by a (*R*)-specific ADH, and subsequently the intermediate is stereoselectively reduced by an (*S*)-specific carbonyl reductase to form (*S*)-1-phenyl-1,2-ethanediol (Scheme 2.4). The catalytic activities of the two key enzymes in *C. parapsilosis* responsible for the deracemization were significantly enhanced by the increase of dissolved oxygen concentration with higher agitation speed [12].

Enantiopure *sec*-alcohols and α-hydroxyketones (acyloins) are popular and important synthons for the asymmetric synthesis of many bioactive compounds [13–15]. A biocatalytic racemization is proved to be a useful tool for the preparation of enantiopure simple alcohols and acyloins via an enzymatic oxidation–reduction process with the formation of a nonchiral ketone or α-diketone intermediate, respectively, using whole lyophilized cells of various bacteria, fungi, and one yeast [16]. The kinetic resolution process was applied for three selected *sec*-alcohols (2-octanol, 6-methyl-5-heptene-2-ol, and 1-phenylethanol) and three selected acyloins (benzoin, phenyl acetylcarbinol, and 2-hydroxy-1-indanone) in which phenyl acetylcarbinol and derivatives are key intermediates for industrially manufacturing (−)-(pseudo)ephedrine, antidepressants, and smoking cessation agents [17–20], and 2-hydroxy-1-indanone is used for the synthesis of several human immunodeficiency virus (HIV)-protease inhibitors [21, 22]. *Sec*-alcohol dehydrogenases and α-diketone reductases are responsible for the equilibrium-controlled enzymatic oxidation–reduction sequence, respectively, and the racemization protocol gives (*S*)-enantiomers (Scheme 2.5).

Glucose dehydrogenases, glucose oxidases, and gluconate dehydrogenases have been exploited for the oxidation of aldoses on large-scale applications and biosensors [1]. For instance, glucose oxidase (GO_X) has been widely employed in enzymatic kits for fast glucose determination and for the industrial production

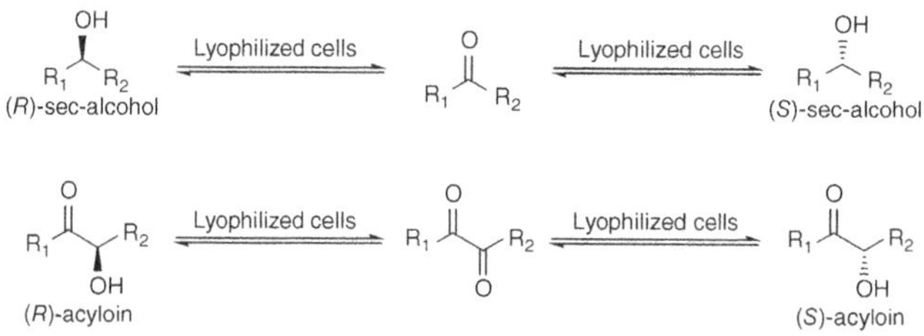

Scheme 2.5 Biocatalytic racemization of *sec*-alcohols and acyloins using lyophilized microbial cells.

of gluconic acid from glucose, which was carried out using whole cells of *Aspergillus niger*. The industrial production of gluconic acid from glucose can be also performed by a membrane-bound dehydrogenase from *G. oxydans*. A more recent industrial application of GO_X is the direct oxidation of D-glucosamine to D-glucosaminic acid with the combined use of catalase to degrade the H_2O_2 produce.

Direct oxidation of heterocyclic and aromatic aldehydes to the corresponding carboxylic acids can be accomplished by *Acetobacter rancens* IFO3297, *Acetobacter pasteurianus* IFO13753, and *Serratia liquefaciens* LF14 [1, 23]. For instance, oxidation of furfural by *A. rancens* IFO3297 can produce $110\,g\,L^{-1}$ of 2-furoic acid with a 95% yield. 5-Hydroxymethyl-2-furancarboxylic acid obtained from corresponding aldehydecanbeobtainedbywholecellsLF14.Isophthalaldehyde,2,5-furandicarbaldehyde, 2,5-thiophenedicarbaldehyde, and 2,2′-biphenyldicarbaldehyde can be converted to the corresponding formylcarboxylic acid with 86–91% yields by both IFO13753 and LF14. The aromatic carboxylic acids such as vanillic acid, *p*-hydroxybenzoic acid, and syringic acid can be produced by the oxidation of corresponding aromatic aldehydes using whole-cell *Burkholderia cepacia* TM1 [1, 24].

The oxidation of alcohols and aldehydes to form corresponding aldehydes, ketones, and carboxylic acids has also been used by cells to produce important intermediates or precursors during the biosynthetic pathways. In the biosynthesis of depside atranorin from acetate using immobilized lichen cells *Evernia prunastri*, the most probable precursor, an aldehyde-substituted phenolic aldehyde, haematommic acid, is produced by oxidation of an alcohol intermediate as shown in Scheme 2.6 [25]. In another example, resting cells of *Nocardia iowensis* DSM 45197 have been selected for the synthesis of vanillin and vanillic acid from the starting material isoeugenol. Three possible pathways used by *N. iowensis* have been suggested for the conversion of isoeugenol to vanillic acid and vanillin. ^{18}O-labeling studies showed that most likely route appears to be the initial side-chain olefin epoxidation of isoeugenol, epoxide hydrolysis to a vicinal diol, followed by diol cleavage to vanillin and subsequently oxidation of the aldehyde group of vanillin to vanillic acid by an aldehyde oxidase (Scheme 2.7) [26].

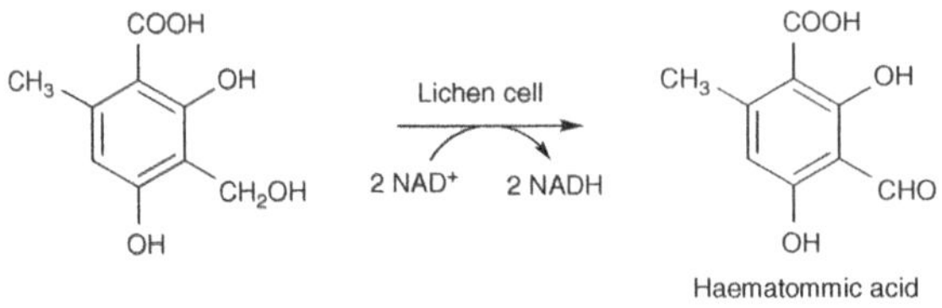

Scheme 2.6 Oxidation of an alcohol intermediate to the precursor of atranorin in the biosynthetic pathway of lichen cells.

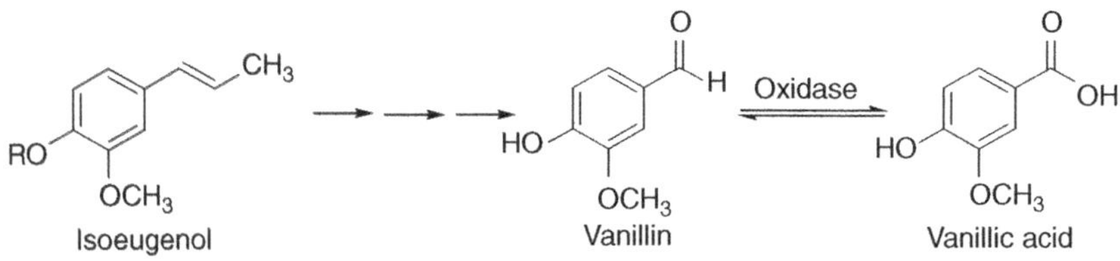

Scheme 2.7 Bioconversion of isoeugenol to vanillin and vanillic acid by *N. iowensis*.

2.1.2 Hydroxylation of Alkanes

Alkanes are saturated hydrocarbons that constitute about 20–50% of crude oil, and living organisms, such as bacteria, plants, and some animals, also produce them. They are chemically quite inert, low value, and usually burned as energy source to produce carbon oxides. Thus, there are two main reasons to carry out the catalytic hydroxylation of inert C–H bonds in alkanes for chemical industry applications. The first one is the providing of high-value compounds from the low-value oil refinery products such as the manufacturing of solvents, plasticizers, and surfactants. The second one is the removal of pollutants from the environment [27, 28].

Because carbon and hydrogen atoms have almost equal electronegativity, the activation of alkanes by the hydroxylation process, especially at the terminal positions, remains a challenging topic in synthetic chemistry. Despite the use of high-temperature heterogeneous catalysts or environmentally unfriendly organometallic catalysis for the selective hydroxylation of alkanes, biocatalysts provide an alternative approach that has the advantages of high selectivity, mild reaction conditions, and low by-product formation [29]. Nature has demonstrated the ability of biocatalysts for the hydroxylation of alkanes that many microorganisms have evolved enzymes for activating alkanes by oxidation of one of the terminal methyl groups to generate the corresponding primary alcohol in the presence of oxygen. The formation of primary alcohol with oxidative enzymes can be further oxidized to fatty acid by dehydrogenases and further metabolized. Oxidative enzymes are called oxygenases that can catalyze the selective insertion of oxygen atoms into a wide range of organic compounds. Oxygenases such as methane monooxygenase, alkane hydroxylase, and cytochrome P450 are the common biocatalysts used for the oxidation of a variety of alkanes [30].

The diversity of alkane oxygenases evolved in prokaryotes and eukaryotes has been strategically used in synthetic chemistry to catalyze the chemo-, regio-, and stereoselective oxygenation of alkanes and many other compounds for the production of useful alcohols, aldehydes, epoxides, and carboxylic acids [31]. In particular, cytochromes P450 are external monooxygenases that convert a broad variety of substrates and catalyze many interesting chemical reactions including the hydroxylation of hydrocarbons [32, 33]. As an example, cytochrome P450 BM3(CYP102A1) isolated from *Bacillus megaterium* catalyzes the hydroxylation of gaseous alkanes, such as methane, ethane, propane, and butane, and cyclohexane to corresponding primary alcohols and 2-propanol and 2-butanol together with a dummy substrate perfluorocarboxylic acids of different alkyl chain length (8–14 carbon atoms) [34]. The screening of the mutant library revealed that cytochrome P450 BM3 F87a/A328V mutant has the ability to effectively hydroxylate cyclooctane, cyclodecane, and cyclododecane to corresponding alcohols. Cytochrome P450 BM3 F87V/A328F mutant demonstrated the ability to hydroxylate acylic *n*-octane to 2-(*R*)-octane with 46% e.e. and high regioselectivity of 92% [35]. For cytochrome P450 catalysis, the nicotinamide adenine dinucleotide cofactor in its reduced form (NADH or NADPH) is needed to provide the electrons necessary for the catalytic cycle. However, the transfer of electrons cannot be made directly and a cytochrome P450 reductase must be present to shuttle the electrons from NAD(P)H to the hydrolase domain (Figure 2.1). It was discovered that cytochrome P450 BM3 was the first natural self-sufficient P450, which still makes it one of the most efficient cytochromes to date [36].

Medium-chain alkanes (C_5–C_{16}) are usually oxidized by heme-iron-containing cytochrome P450 monooxygenases (P450s or CYPs) or by integral-membrane nonheme diiron monooxygenases (Alk B) [37, 38]. Alkane hydrolases of the CYP153A from *Mycobacterium marinum* (CYP153A16) and *Polaromona* sp. (CYP153A *P.* sp.) were examples to catalyze regioselectively ω-hydrolation of C_6–C_{11} alkanes, alkenes, cycloalkanes, and alicyclic compounds [39, 40]. The cloned and expressed CYP153A16 and CYP153A *P.* sp. monooxygenases were capable of performing the ω-hydroxylation *in vitro* with C_5–C_{12} alkanes and C_6–C_{12} primary alcohols into their corresponding primary alcohols and α,ω-diols [37]. However, a soluble

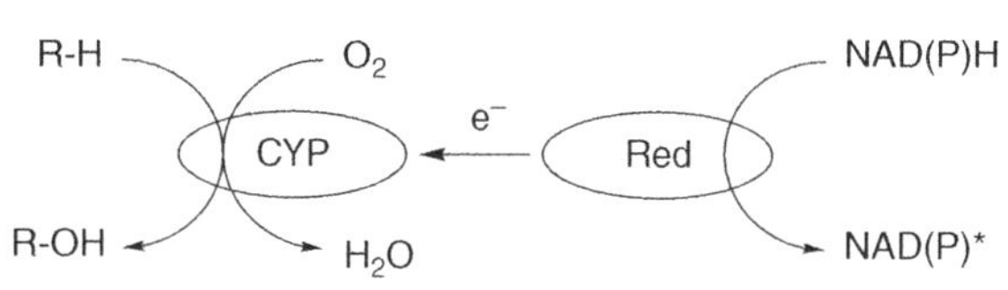

Figure 2.1 Hydroxylation of alkanes by cytochrome P450 monooxygenase (CYP).

three-component diiron monooxygenase, butane monooxygenase (sBMO), has been purified from the gram-negative β-proteobacterium *Pseudomonas butanovora* (ATCC 43655), which was capable of hydroxylating C_3–C_6 linear and branched aliphatic alkanes (propane, butane, pentane, hexane, isobutene, and isopentane) at the terminal carbon atom mostly (≥80% regiospecificity) to produce primary alcohols with only small fractions of secondary and tertiary alcohols [41].

Since long-chain alkanes are more persistent in the environment than shorter alkanes, the degradation of long-chain alkane is an important topic in synthetic chemistry for dealing with the crude oil contamination in environment. Enzymes used for the activation of alkanes with long chain length (≥C_{16}) by hydroxylation have been found including yeast P450s [42], integral-membrane monooxygenases (AlkM) [43], flavin-containing alkane monooxygenases (LadA) [44], and dioxygenases [45]. It was acknowledged that thermophilic bacillus *Geobacillus thermodenitrificans* NG80-2 degrades the long-chain alkane (C_{15}–C_{36}) by utilizing a terminal oxidation pathway to convert long-chain alkanes to their corresponding primary alcohols (Scheme 2.8). It is also found that LadA is the key initiating enzyme in the terminal oxidation pathway of *G. thermodenitrificans* NG80-2 [46]. However, studies on the bacterial strain *Acinetobacter venetianus* 6A2 show that this bacterial strain can utilize *n*-alkanes with chain lengths ranging from decane ($C_{10}H_{22}$) to tetracontane ($C_{40}H_{82}$) [47]. Two genes of the AlkB-type alkane hydroxylase homologous, *alkMa* and *alkMb*, isolated from *A. venetianus* 6A2 were involved in the utilization of alkanes with chain lengths ranging from C10 to C18 implying that other enzyme(s) should be required for the utilization of C20-C40 alkanes.

In addition to the notorious cytochrome P450 monooxygenase for alkane hydroxylation, a group of novel extracellular heme-thiolate peroxygenases secreted by fungal *Agrocybe aegerita* have been demonstrated to catalyze efficiently the H_2O_2-dependent hydroxylation of a variety of alkanes, which include linear, branched, and cyclic saturated hydrocarbons. As shown in Scheme 2.9, linear C3-C16 alkanes can be hydroxylated at either 2- or 3-positon of the carbon chain to yield corresponding secondary alcohols. For example, *n*-propane and *n*-butane were hydroxylated regioselectively to produce 2-propanol and 2-butanol, whereas *n*-heptane and *n*-octane were hydroxylated both stereoselectively and regioselectively to give 99.9% e.e. (*R*)-enantiomer and the corresponding 3-alcohol. Branched alkanes such as 2,3-dimethylbutane were hydroxylated regioselectively to 2,3-dimethylbutane-2-ol and isobutene was oxidized to 2-methyl-propan-2-ol. While cyclic alkanes of C5–C8 gave monohydroxylated products [48], there are also non-heme enzymes that are capable of catalyzing the hydroxylation of alkanes [49]. Studies showed that

$$RCH_3 + O_2 \xrightarrow{\text{LadA}} RCH_2OH$$

$$\text{R: } CH_3\text{-}(CH_2)_n^-, \qquad (13 \le n \le 34)$$

Scheme 2.8 Terminal hydroxylation of long-chain alkanes by LadA.

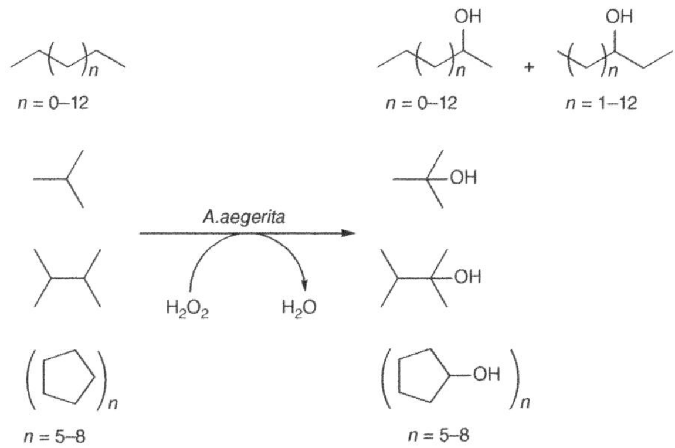

Scheme 2.9 Hydroxylation of alkanes by fungal peroxygenase.

non-heme alkane hydroxylase involves high-valent oxoiron moiety to catalyze the initial H abstraction from alkane followed by the rebound mechanism of FeOH/ radical species. Examples are the enzyme taurine:α-ketoglutarase dioxygenase (TauD) toward the cyclohexanol and cyclopentanol from cyclohexane and cyclopentane, respectively [50], and the diiron alkane monooxygenase of *Pseudomonas oleovorans* (AlkB) hydroxylates 2-methyl-1-phenylcyclopropane to produce only 1-phenyl-3-buten-1-ol and norcarane to produce approximately 85% *cis*- and *trans*-2-norcaranol [51].

2.1.3 Hydroxylation of Aromatic Compounds

Oxidative enzymes performing the introduction of one or two oxygen atoms on aromatic compounds is important in industrial applications [52, 53]. The four classes of enzyme: oxygenases, hydroxylases, peroxidases, and laccases are the most representative enzymes to perform the hydroxylation of aromatic compounds. The four classes of oxidative enzymes differ in many aspects, such as the metal present in the active site, the number of electrons transferred for the reaction, and the kind of reductive cofactor [54]. The most common hydroxylases are the cytochrome P-450-dependent hydroxylases and the flavoprotein phenol hydroxylases [55]. Since hydroxylations introduce the hydroxyl group into organic compounds primarily via the substitution of functional groups or hydrogen atoms, it is a great challenge to organic chemists to perform the direct and selective introduction of the hydroxyl group into aromatic ring.

One of the well-known industrial applications with hydroxylase for the hydroxylation of aromatic ring is the catalytic production of tyrosine from phenylalanine by phenylalanine hydroxylase and the subsequent hydroxylation of tyrosine yielding the catecholamine neurotransmitter L-dihydroxyphenylalanine (L-DOPA), as in Scheme 2.10 [53, 55].

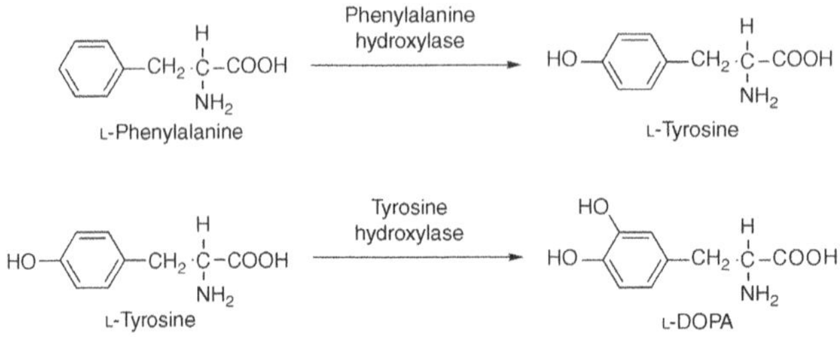

Scheme 2.10 The catalytic hydroxylation of L-phenylalanine and L-tyrosine by phenyl hydroxylase and tyrosine hydroxylase, respectively, to produce corresponding L-tyrosine and L-DOPA.

Recently, a microsomal P450 designated as PcCYP65a2 that was derived from the white rot fungus *Phanerochaete chrysosporium* can catalyze 3′-hydroxylation of naringenin to yield eriodictyol in the culture of the recombinant *S. cerevisiae* AH22/pG65a2 cells (Scheme 2.11) [56].

Isoliquiritigenin (2′,4′,4-trihydroxychalcone) is a chalcone found in licorice root and other plants, which has shown potent antioxidant, anti-inflammatory, phytoestrogenic, tyrosinase inhibitory, and antitumor activity *in vitro*. However, when prepared *in vivo*, aromatic hydroxylation of isoliquiritigenin on the A or B ring was found by the metabolism of human liver microsomes to produce 2′,4′4′,5′-tetrahydroxychalcone or butein, respectively (Scheme 2.12) [57]. The enzyme involved in the formation of the hydroxylated metabolites of isoliquiritigenin was cytochrome P450.

Amphetamine was once extensively used for weight reduction, and it has been employed in treating mild depression and narcolepsy as well. The administration of amphetamine may cause excitability, restlessness, tremors, insomnia, dilated pupils, increased pulse rate and blood pressure, hallucinations, and psychoses. Since amphetamine is inexpensive, it has been a drug of abuse. The biotransformation of *d*-amphetamine into *p*-hydroxyamphetamine (HA) by cytochrome P450 occurs in

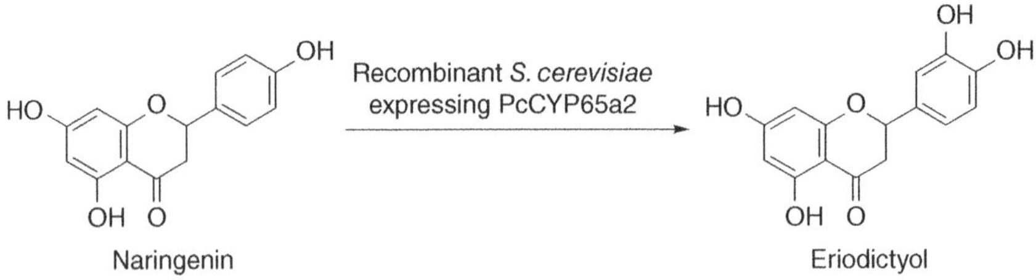

Scheme 2.11 Hydroxylation of naringenin in the culture of the recombinant *S. cerevisiae* expressing *P. chrysosporium* PcCYP65a2 to produce eriodictyol.

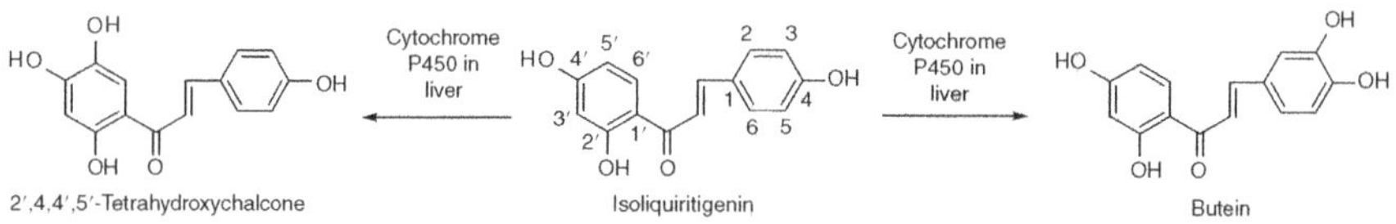

Scheme 2.12 Hydroxylation of isoliquiritigenin in human liver by cytochrome P450.

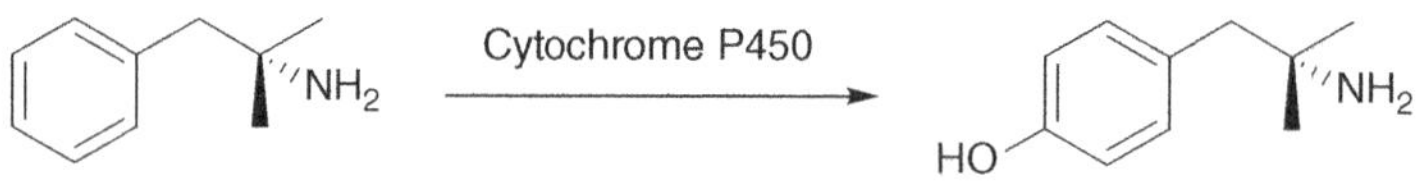

d-Amphetamine *p*-Hydroxyamphetamine

Scheme 2.13 Hydroxylation of *d*-amphetamine by cytochrome P450 to give *p*-hydroxyamphetamine.

human and mouse (Scheme 2.13) [58]. The urinary excretion of HA not only is a biomarker of exposure but also gives an insight into *d*-amphetamine reactivity.

2.1.4 Dihydroxylation of Aromatic Compounds

The initial degradation of aromatic compounds by bacteria often involves the *cis*-dihydroxylation, which is catalyzed by Rieske non-heme iron dioxygenases to yield *cis*-dihydrodiol derivatives. This type of reaction leads to a permanent disruption of aromaticity and offers a strategy for regio- and stereoselectivity in organic syntheses to a variety of useful natural and unnatural compounds [59]. The first report in literature for the arene *cis*-dihydroxylation to produce *cis*-dihydrodiol metabolite was with the use of bacterium *Pseudomonas putida* F1 and the substrate was benzene [60]. Later developments of this synthetic strategy extend the use of monosubstituted arenes such as toluene, chlorobenzene, phenol, etc.; polycyclic arene such as naphthalene; and biphenyl as the starting material to produce various kinds of *cis*-1,2-dihydrodiols that in turn have been used as chiral synthons for the generation of valuable biologically active products through sequential chemoenzymatic steps. The following examples show the wide applications of this synthetic strategy.

It is noted that the hybrid toluene/biphenyl dioxygenase (TDO/BPDO) in *Escherichia coli* encoded by the *todC1* gene of *P. putida* F1 and the *bphA2A3A4* genes of *Pseudomonas pseudoalcaligenes* KF707 was able to biotransform monocyclic aromatic compounds including benzene, toluene, styrene, *p*-xylene, acetophenone, propiophenone, butyrophenone, and trifluoroacetophenone to their corresponding *cis*-dihydrodiols (Scheme 2.14). Subsequently, these *cis*-dihydrodiols were bioconverted by the *bphB* (dihydrodiol dehydrogenase)

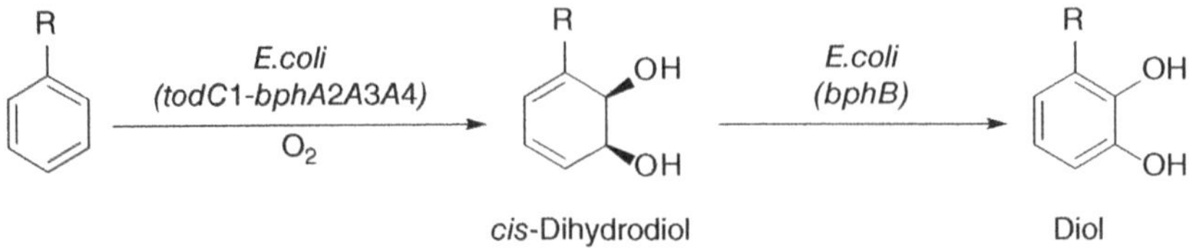

R = H, CH_3, $CH_2 = CH_2$, 1,4-CH_3, CH_3CO, CH_3CH_2CO, $CH_3CH_2CH_2CO$, CF_3CO

Scheme 2.14 Catabolic pathways of monosubstituted benzene to diol via *cis*-dihydrodiol.

expressed in *E. coli* cells in addition to *todC1-bphA2A3A4* to produce their monocyclic arene-diols [61].

Alkaloid pancratistatin is tested as anticancer, antiviral, and antiparasitic agents. The *cis*-dihydroxylation of bromobenzene by TDO from the bacterium strain *P. putida* 39/D to produce *cis*-bromocyclohexadienediol was utilized to chemically synthesize the key intermediate bromoazide (Scheme 2.15) for the preparation of pancratistatin analogs [62]. (−)-Conduramine C-4 was also synthesized in six steps from the whole-cell fermentation of bromobenzene with *P. putida* 39/D in 23% overall yield [63].

The same chemoenzymatic method has been used for the synthesis of certain alkaloids and terpenoids by starting with the whole-cell biotransformation of a variety of monosubstituted benzene to form the corresponding *cis*-1,2-dihydrocatechol using TDO expressing *P. putida* 39-D or *E. coli* JM109 (pDTG601) [64, 65]. *E. coli* JM109 (pDTG601) was also used for the dihydroxylation of methyl 2-iodobenzoate to give two *cis*-dihydrodiol metabolites in a molar ratio of 4:1 (Scheme 2.16) [66]. The minor product in the diol mixture was further chemically transformed to an alcohol intermediate for the asymmetric preparation of kibdelone C and its congeners.

Ten 1,4-disubstituted benzene substrates using *P. putida* UV4 as a source of TDO have been biotransformed to yield the corresponding *cis*-dihydrodiol metabolites [67]. The yields of these *cis*-dihydrodiols were generally lower than those obtained from the corresponding monosubstituted benzene substrates. The

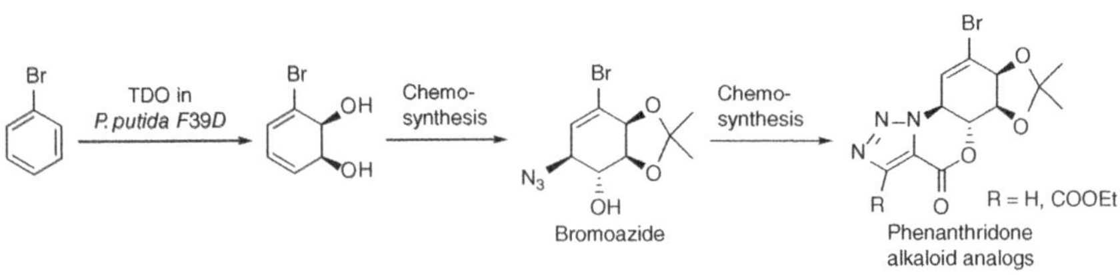

Scheme 2.15 Chemoenzymatic preparation of pancratistatin analogs.

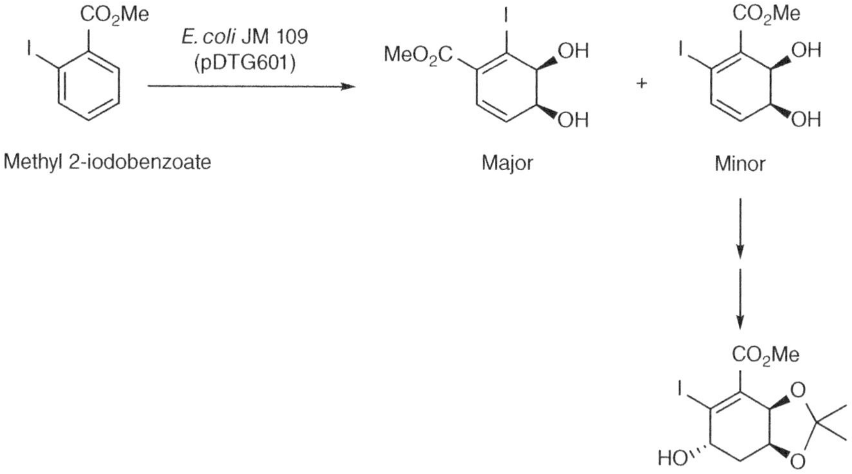

Scheme 2.16 Whole-cell fermentation of methyl 2-iodobenzoate for organic synthesis.

enantiomeric excess (*ee*) values were determined by a combination of methods to show their enantiopurity. The *cis*-dihydroxylation of *meta*-substituted phenols (*m*-phenols) catalyzed by TDO, an arene dioxygenase from mutant strain of *P. putida* UV4, yields the corresponding cyclohexenone *cis*-diol metabolites and several of their cyclohexene and cyclohexane *cis*-triol derivatives [68]. Using the same bacterium strain and *p*-xylene as the substrate, the major biotransformation product was the corresponding *cis*-1,2-dihydrodiol (95% relative yield) along with the two minor products, catechol (1% relative yield) and cyclohexenone *cis*-diol (4% relative yield) via the intermediate *p*-xylenol (Scheme 2.17). Another metabolite of *p*-xylenol with TDO present in *P. putida* UV4 was *o*-quinol dimmer (Scheme 2.17) [69]. However, *p*-xylene was first catalyzed to *cis*-*p*-xylene dihydrodiol and further undergone dehydration to give 2,5-dimethylhydroquinone by a novel *o*-xylene dioxygenase from *Rhodococcus* sp. strain DK17 [70]. *cis*-Dihydroxylation of aromatic carboxylic acid such as benzoic acid with mutants of *Alcaligenes eutrophus* B9 or *P. putida* U103 produces a diol containing a tertiary alcohol moiety is another interesting case for enantioselective synthesis, which demonstrates two stereogenic centers can be introduced simultaneously during organic synthesis [71]. A series of benzoate esters have been studied for the enzymatic dihydroxylation by whole-cell fermentation with *E. coli* JM109 (pDTG601) to produce their corresponding *cis*-cyclohexadienediols except for *n*-butyl and *t*-butyl benzoates [72]. The diols obtained from the enzymatic dihydroxylation of benzoate esters serve as intermediates for the synthesis of pseudo-sugars, amino cyclitols, and bicyclic ring systems.

Naphthalene dioxygenase (NDO), a Rieske dioxygenase [73, 74], from *Pseudomonas* strain NCIB9816-4 catalyzes the *cis*-hydroxylation of

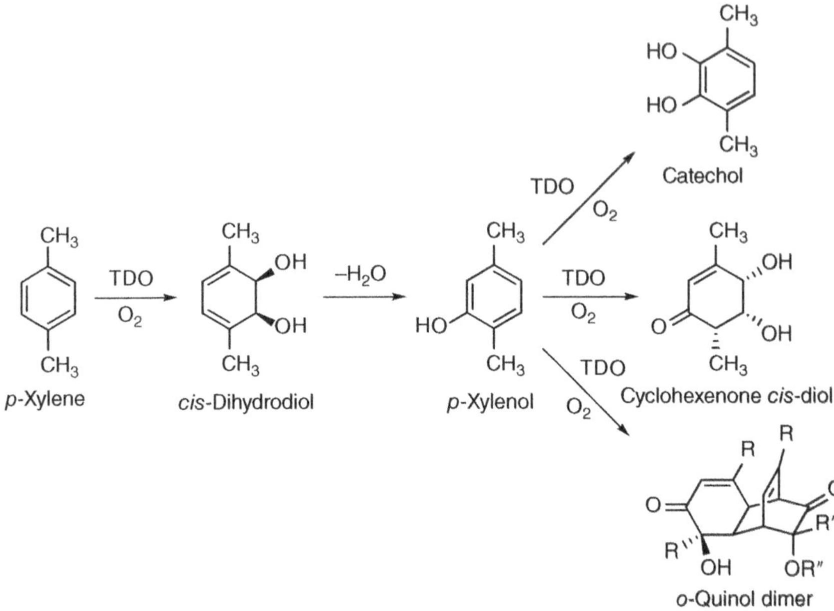

Scheme 2.17 Toluene dioxygenase catalyzed *cis*-dihydroxylation of phenols toward catechol, cyclohexenone *cis*-diol, and *o*-quinol dimmer metabolites.

naphthalene to *cis*-(1*R*,2*S*)-dihydroxy-1,2-dihydronaphthalene (naphthalene *cis*-dihydrodiol) (Scheme 2.18) [75, 76]. Incubation of naphthalene with *E. coli* BL21(DE3) harbored with a recombinant expression plasmid of the *Rhodococcus* sp. strain DK17 *o*-xylene dioxygenase also produces *cis*-1,2-naphthalene dihydrodiol [70]. In the same manner, the recombinant *E. coli* JM109(DE3) (pDTG141) expressing the NDO from *Pseudomonas* sp. NCIB9616-4 has been used for the whole-cell biotransformation of a series of azaarene compounds. The study showed that several bicyclic azaarenes can be catalyzed well by NDO to give *cis*-dihydrodiol derivatives [77]. In addition, the DK17 *o*-xylene dioxygenase catalyzes the *cis*-dihydroxylation of biphenyl to form *cis*-2,3-biphenyl dihydrodiol (Scheme 2.19) [70]. BPDOs from mutant bacterial strains that lack the corresponding diol dehydrogenase enzymes can be utilized to transform polycyclic arenas such as anthracene and benz[*a*]anthracene to *cis*-hydrodiols

Scheme 2.18 Naphthalene dioxygenase catalyzed *cis*-dihydroxylation of naphthalene.

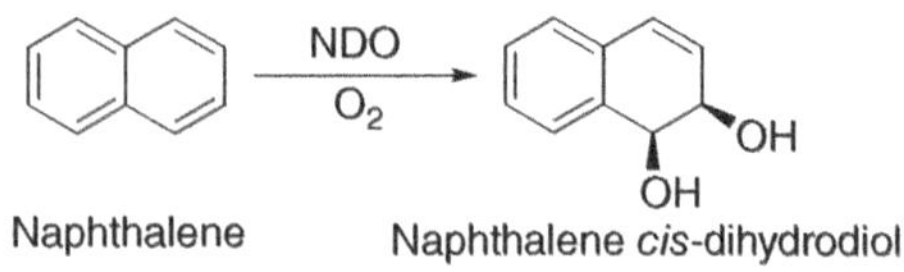

Scheme 2.19 Regioselective oxidation of biphenyl by the *Rhodococcus* sp. DK17 *o*-xylene dioxygenase.

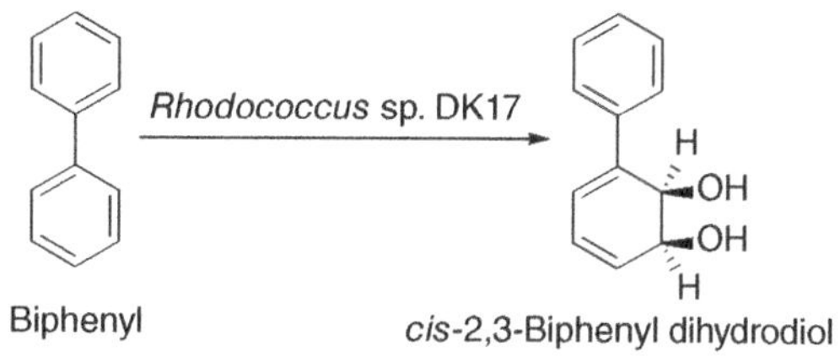

and acetonide derivatives of the *cis*-dihydrodiols of biphenyl and phenanthrene to enantiopure bis(*cis*-dihydrodiol) metabolites [78]. The produced *cis*-diols in turn have been widely used for organic synthesis.

2.1.5 Epoxidation

The cytochromes P450 (CYPs) catalyzed a wide variety of reactions including the oxidation of alkenes to form reactive epoxides. The oxidations of propene, cyclohexene, and styrene for producing their corresponding epoxides have been reported by P450cam from *P. putida* and P450 BM-3 variant 139-3 [79, 80]. A mutant, T252A P450cam, prepared from the water-soluble camphor-hydroxylating P450cam of *P. putida* has been used to catalyze the epoxidation of more complex 1*R*-camphor [81]. CYP153 enzymes expressed in *P. putida* GPo12 also showed the ability of epoxidation for styrene, octene, and cyclohexene in a large scale [40]. Except for one CYP153 produced (*R*)-epoxide, others yielded (*S*)-epoxide with up to 80% e.e. The CYP450 was further used for the epoxidation of tea tree oil ingredient *p*-cymene to form thymol by a postulated enzymatic reaction mechanism that involves the epoxidation of one of the π-double bonds in the benzene moiety followed by an National Institutes of Health (NIH) shift as shown in Scheme 2.20 [82]. Dihydrochalcomycin (DHC) is a 16-membered macrolide antibiotic that is clinically applied in the treatment of bacterial infections against gram-positive as well as gram-negative bacteria except against *E. coli*. Two CYP450 enzymes, *gerPI* and *gerPII*, were characterized for the hydroxylation at the C_8 position by the *ger*PII P450 followed by the epoxidation reaction by

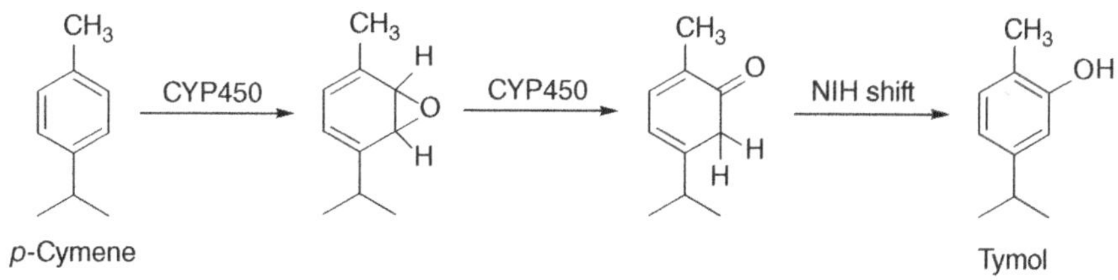

Scheme 2.20 Postulated reaction mechanism for the formation of thymol from *p*-cymene through arene epoxidation.

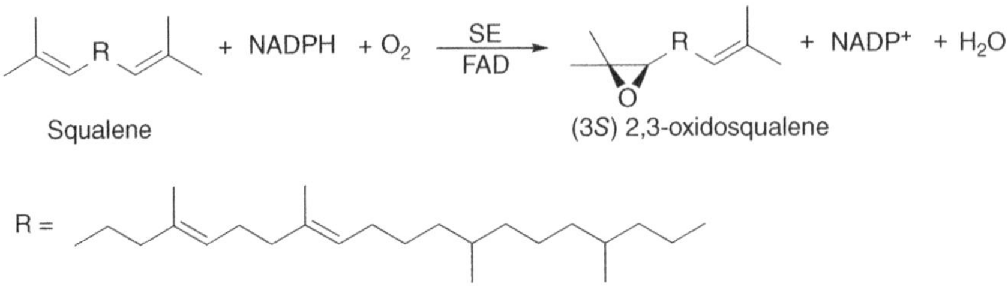

Scheme 2.21 The epoxidation reaction catalyzed by squalene epoxidase (SE).

*ger*PI P450 at the C_{12}–C_{13} position in the biosynthetic pathway for DHC from *Streptomyces* sp. KCTC 0041BP [83].

Squalene monooxygenase (squalene epoxidase, SE, E.C. 1.14.99.7) converts the squalene by regio- and stereospecifically forming an epoxide on the C–C double bond to yield (3*S*)2,3-oxidosqualene that requires molecular oxygen, FAD, and depending on the organism, either NADH or NADPH (Scheme 2.21) [84]. This reaction is a key step in eukaryotes for sterol biosynthesis. Since SEs are essential for the synthesis of cholesterol in mammals and ergosterol in fungi, SE is an important target to lower cholesterol levels, thus, for hypercholesterolemia and antifungal therapies [85]. Study of SE in the squalene epoxidation shows that it is involved in the biosynthesis of both the antitumor clavaric acid, an inhibitor of the human Ras-farnesyl transferase, and sterols in the basidiomycetes *Hypholoma sublateritium* [86]. Sterol biosynthesis by prokaryotes is very rare. However, through functional expression of both SE and lanosterol synthase, which are from the methanotrophic bacterium *Methylococcus capsulatus*, in *E. coli*, lanosterol can be formed from squalene [87].

The one-pot epoxidation of cyclohexene and *N*-benzyloxycarbonyl-3,4-dihydroxy-pyrrolidine by Novozym 435® gave cyclohexene oxide and (3*R*,4*R*)-*N*-benzyloxycarbonyl-3,4-epoxy-pyrrolidine, respectively, which were subsequently *trans*-dihydrolyzed by resting cells of *Sphingomonas* sp. to produce corresponding *trans*-vicinal diols in high enantiomeric excess values and high yields [88]. The silicone composites of Novozym 435 (silicoat-NZ435) have been used in the solvent-free three-phase system chemoenzymatic epoxidation of 1-dodecene to demonstrate the ability of the most difficult terminal alkene epoxidation [89]. The reaction showed an 80% yield with almost 100% selectivity and is better than the native Novozym 435 (NZ435).

Growing cells and resting cells of recombinant *E. coli* containing the styrene monooxygenase StyAB were used for enantioselective styrene epoxidation to efficiently produce (*S*)-styrene oxide in an organic/aqueous two-liquid-phase system and batch or fed-batch reaction [90–93]. Styrene monooxygenase (StyAB) from *Pseudomonas* sp. VLB120 is composed of an FAD-dependent monooxygenase component (StyA) that catalyzes the epoxidation reaction and an NADH-dependent reductase component (StyB) that delivers the reducing equivalents from NADH to StyA via $FADH_2$ [94]. The enzyme StyAB catalyzes the specific (*S*)-epoxidation of a broad range of *m*- and *p*- as well as α- and β-substituted styrene derivatives [95]. Since the motif of enantiopure 1,2-amino alcohol is present in alkaloids, amino sugars, enzyme inhibitors, and antibiotics, various enantiopure 2-phenyl, 2-amino ethanols were prepared chemically through the nitrogen nucleophilic addition to the epoxy group of a number of styrene epoxide derivatives (Scheme 2.22) [96]. However, the formation of various styrene epoxide derivatives was performed enzymatically by using the recombinant *E. coli* to oxidize various styrene derivatives. The bioconversions were performed in an aqueous buffer using an organic phase to separate both the epoxide product and unreacted styrene derivatives. The enantiomeric excess values for epoxide products are quite high (>95% e.e.) except the epoxide derived from 1-phenyl, 2-methyl propene.

Two reasons cause monooxygenase catalyzed epoxidation reactions using whole cells rather than isolated enzyme more attractive, which are the very unstable monooxygenase *in vitro* and the easy regeneration of the cofactor for NAD(P) H-dependent monooxygenase in whole cell. The epoxidation of ally phenyl ether (APE) for producing chiral phenyl glycidyl ether (PGE) with an enantiomeric excess of 86% has been investigated by encapsulating whole-cell *Mycobacterium* M156 in water-in-oil reverse micelles [97] as in Scheme 2.23.

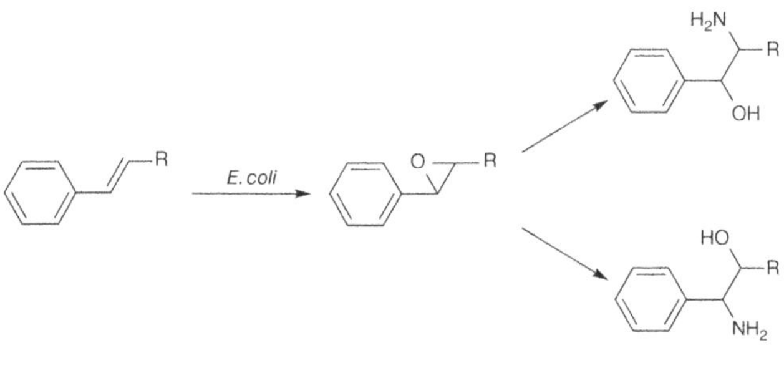

Scheme 2.22 The synthesis of enantiopure 2-amino-1-phenyl and 2-amino-2-phenyl ethanols through enantioselective enzymatic epoxidation of styrene derivatives.

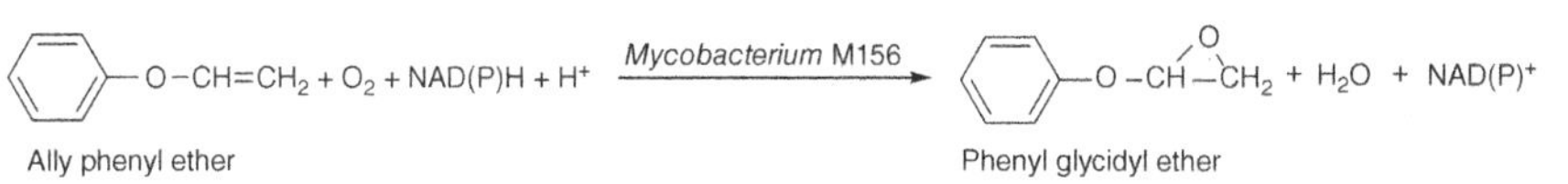

Scheme 2.23 Epoxidation of ally phenyl ether for producing chiral phenyl glycidyl ether.

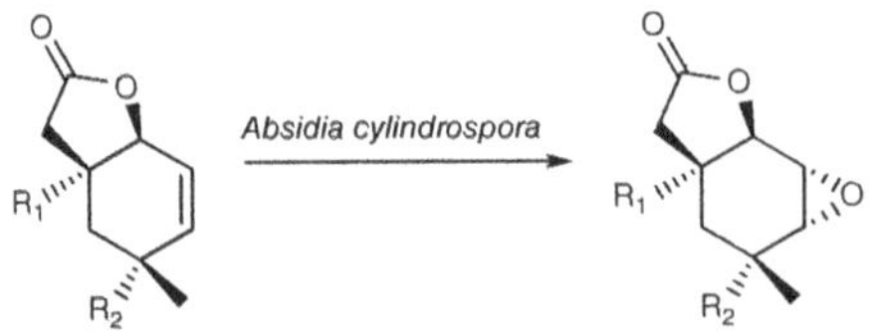

1a,b 2a,b

a: R_1, R_2 = CH_3
b: R_1 = CH_3, R_2 = H

Scheme 2.24 Stereoselective epoxidation of unsaturated bicyclic γ-lactones.

The epoxidation of *cis*-propenylphosphonic acid (cPPA) by bacterium *Bacillus simplex* strain S101 has been used for the preparation of fosfomycin that is more delicate, environmentally friendly, and has a higher conversion yield (81.3%) than the large-scale industrial process (<20%) [98]. The biologically active unsaturated bicyclic γ-lactones (4,4,6-trimethyl-9-oxabicyclo[4.3.0]non-2-en-8-one (**1***a*) and 4,4-dimethyl-9-oxabicyclo-[4.3.0]non-2-en-8-one (**1***b*)) were stereoselectively transformed into the corresponding *trans*-epoxylactones (2,3-epoxy-4,4,6-trimethyl-9-oxabicyclo[4.3.0]nonan-8-one (*2a*) and 2,3-epoxy-4,6-dimethyl-9-oxabicyclo[4.3.0]nonan-8-one (*2b*)) by the strain *Absidia cylindrospora* as shown in Scheme 2.24 [99]. These epoxylactones can be further converted to yield hydroxylactones with the secondary hydroxy group. Oleic acid in the lipophilic extractives can be oxidized with *Pycnoporus cinnabarinus* laccase in the presence of 1-hydroxybenzotriazole (HBT) to form an epoxy oleic acid at the C_9 and C_{10} double bond. The conversion was 88% after a two-hour treatment [100].

2.1.6 Sulfoxidation Reactions

There are natural organosulfur compounds such as sulfur containing amino acids (cysteine, methionine, and cystine), allicin, lipoic acid, and unnatural organosulfur compounds such as dibenzothiophene in petroleum products or penicillin in pharmaceutical products. Among the variety of organosulfur compounds, chiral organic sulfoxides (COSs) are useful chiral building blocks or stereodirecting groups in asymmetric synthesis of important pharmaceuticals that contain a functional sulfinyl group attached to the alkyl moieties [101, 102]. However, the preparation of COSs also can be obtained through sulfoxidation by the high regioselectivity and stereoselectivity of enzymes [103]. For example, the purified catalase-peroxidase (KatG) characterized as a heme-containing protein from the bacterium *Bacillus pumillis* was employed for stereoselective oxidation of β-lactams, represented by penicillin-G, penicillin-V, and cephalosporin-G to their *R*-sulfoxides [104].

The use of co-expression system that is the co-expression of formate dehydrogenase from gene originated from *Candida boidinii* and the cyclohexanone monooxygenase (CHMO) gene cloned from *Acinetobacter calcoaceticus* NCIMB 9871 in *E. coli* BL21 has been used as the whole-cell biocatalyst to selectively synthesize chiral *R*-phenyl methyl sulfoxide (*R*-PMSO) from thioanisole. In this reaction system, NADPH has also been regenerated to improve the catalytic efficiency [105]. Another strategy was utilized to selectively synthesize corresponding *S*-sulfoxide from *p*-chlorothioanisole as shown by Scheme 2.25 [106]. In this investigation, the asymmetric oxygenation of sulfide to *S*-sulfoxide was with co-expressed *E. coli* that contains the P450SMO gene from *Rhodococcus* sp. and the glucose dehydrogenase gene from *Bacillus subtilis*. In this study, NADPH was efficiently regenerated when glucose was supplied to the reaction.

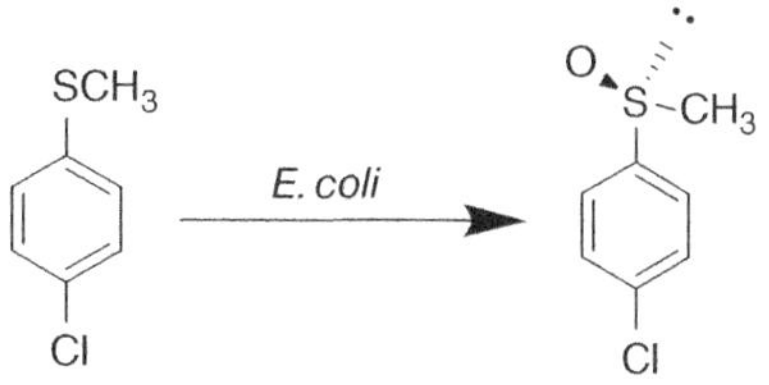

Scheme 2.25 Synthesis of optically pure *S*-sulfoxide by co-expressed *E. coli*.

The enantioselective sulfoxidation of benzyl methyl sulfide to its corresponding *S*-sulfoxide was performed by phenylacetone monooxygenase (PAMO) from *Thermobifida fusca* in a nonconventional Tris-HCl buffer medium containing hydrophilic organic solvents such as polyethylene glycol (PEG), methanol (MeOH), acetonitrile, iso-propanol, and alcohol with high conversion rate and moderate e.e.% [107]. The reaction also used glucose 6-phosphate/glucose-6-phosphate dehydrogenase (G6P/G6PDH) as secondary ancillary system for regenerating the NADPH cofactor. An alternative method for producing enantiopure sulfoxides by direct asymmetric oxidation of prochiral sulfides was the optical resolution of racemic sulfoxides [108]. Therefore, *S*-phenyl methyl sulfoxide (*S*-PMSO) accompanied by a by-product sulfone was formed at 93.7% *ee*(*S*) in a fed-batch reaction with the use of bacterium *Rhodococcus* sp. ECU0066. For other substrates such as *para*-substituted (methyl and chloro) PMSOs and ethyl phenyl sulfoxide, an *S*-enantioselectivity (*ee*(*S*)) larger than 99.0% was also obtained.

2.1.7 Baeyer–Villiger Reactions

Baeyer–Villiger (BV) oxidation of ketones is the insertion of a molecular oxygen into the Baeyer–Villiger monooxygenases (BVMOs) [109–111]. BVMO catalyzes the carbonyl group in a compound to form esters, or lactones, from ketones, often with great enantioselectivity [112]. It has been used as an intermediate step for the production of 11-hydroxyundec-9-enoic acid, a precursor for synthesizing Nylon-11, from ricinoleic acid via recombinant *E. coli* (Scheme 2.26) [113]. A variety of secondary metabolites of plants, oxygenated unsaturated carboxylic acids, which are difficult to synthesize were also explored by designing an artificial biotransformation pathway consisting of fatty acid double-bond hydratases, ADHs, BVMOs, and esterases. In this case, γ-linolenic acid that contains three double bonds in the carbon skeleton has been used as the substrate for the

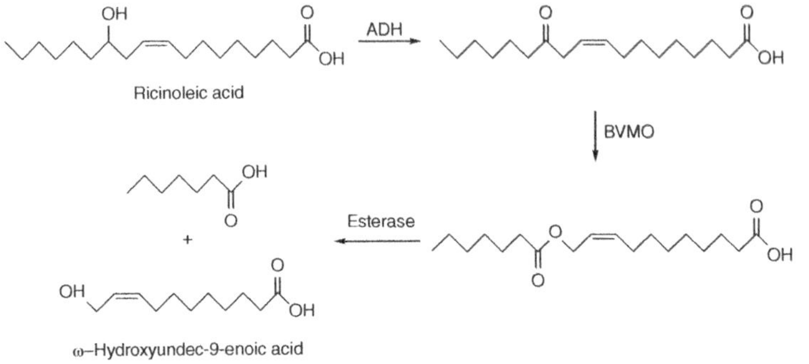

Scheme 2.26 The multiple enzyme biosynthesis of ω-hydroxyundec-9-enoic acid from ricinoleic acid via Baeyer–Villiger oxidation.

recombinant *E. coli* whole-cell biocatalysis to efficiently produce the target products (6*Z*,9*Z*)-12-hydroxydodeca-6,9-dienoic acid, (*Z*)-9-hydroxynon-6-enoic acid, (*Z*)-dec-4-enedioic acid, and (6*Z*,9*Z*)-13-hydroxyoctdeca-6,9-dienoic acid [114].

Recombinant *E. coli* expressing CHMO has been extensively used to investigate BV oxidation of cyclohexanone to ε-caprolactone (Scheme 2.27), which can be influenced by not only the efficient regeneration of NADPH but also a sufficient supply of oxygen [115]. The study of the crystal structure of CHMO reported that two crystal structures are found to bind ε-caprolactone: the $CHMO_{Tight}$ and $CHMO_{Loose}$ structures [116]. The $CHMO_{Tight}$ structure determines the substrate acceptance and stereospecificity, and the $CHMO_{Loose}$ is the first structure where the product is solvent accessible. Three regiodivergent BVMOs expressed in *E. coli* strains have been applied to enantioselectively oxidize a series of cyclic α,β-unsaturated ketones to produce either chiral enol-lactones or ene-lactones, which broadens the scope of BVMO activities [117].

The *S*-selective altered CHMO mutant 1-K2-F5 (Phe432Ser) from *Acinetobacter* sp. NCIMB has been tested in the BV oxidative desymmetrization of a number of structurally different ketones with excellent enantioselectivity [118]. An asymmetric BV bio-oxidation reaction catalyzed by *E. coli* has been scaled up to a pilot plant (kg) scale using a new "resin-based *in situ* substrate feeding and product removal (SFPR)" methodology with nearly enantiopure form (ee > 98%) and good yield [119]. In this application, racemic bicycle[3.2.0]hept-2-en-6-one was regiodivergent parallel kinetic resolution to two regioisomeric lactones that the

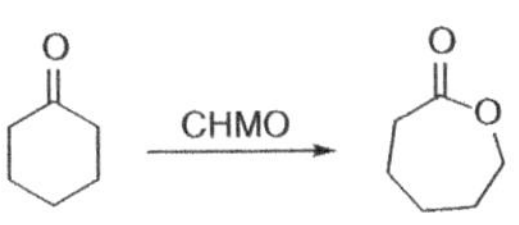

Scheme 2.27 The Baeyer–Villiger oxidation of cyclohexanone to ε-caprolactone by recombinant *E. coli* expressing cyclohexanone monooxygenase (CHMO).

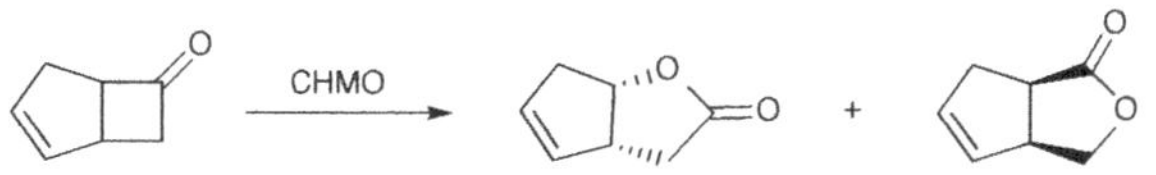

(+/–)-Bicyclo[3.2.0] hept-2-en-6-one (–)-(1*S*,5*R*)-lactone (ee > 98%) (–)-(1*R*,5*S*)-lactone (ee > 99%)

Scheme 2.28 Enantiopure asymmetric microbial Baeyer–Villiger oxidation of rac-bicyclo[3.2.0]hept-2-en-6-one.

(+)-bicycle[3.2.0]hept-2-en-6-one enantiomer is converted into the "expected" (−)-(1*S*,5*R*) lactone and (−)-bicycle[3.2.0]hept-2-en-6-one is converted into the "unexpected" (−)-(1*R*,5*S*) lactone as indicated by Scheme 2.28. The formation of "expected" lactone means that the oxygen is inserted into the more substituted carbon-carbon bond of the substrate ketone.

2.1.8 Peroxidation Reactions

In mammalian cells, hydrogen peroxide and organic hydroperoxides are synthesized continuously during aerobic metabolism. Peroxides can damage the cell components by their formation of highly reactive hydroxyl radicals that can initiate lipid peroxidation, to oxidize amino acid side chains in proteins, and to cause DNA strand breaks and base modification [120]. Therefore, peroxides must be detoxified continuously to prevent oxidation of cellular components by peroxides or peroxide-derived reactive oxygen species (ROS). In addition, the generation of peroxides in cells consumes oxygen, which causes the disposal of peroxides particularly important for human brain because brain cells utilize 20% of the oxygen used by the body [121]. In cells, H_2O_2 is produced by the disproportionation of superoxide generated through the mitochondrial respiratory chain as a by-product with superoxide dismutases (SODs). Besides, H_2O_2 can also be produced by the reactions using oxidases such as monoamine oxidases [122]. Stereospecifically defined organic peroxides are generated in cells through the pathways of prostaglandins and leukotrienes by cyclooxygenases and lipoxygenases. Hydroperoxides are also formed by unspecific oxidation of polyunsaturated fatty acids in membranes by radical-mediated lipid peroxidation [123].

Although not too many enzymatic syntheses of peroxides can be found in literature because of their substrate-dependent processes, peroxides are of interest to synthetic chemists due to their potent antimalarial and antimicrobial properties [124–126]. Crude enzyme from marine green algae, *Ulva pertusa*, has been used for the enantioselective α-hydroxylation of long-chain saturated and unsaturated fatty acids to corresponding (*R*)-2-hydroperoxy acids (Scheme 2.29) with >99% e.e. enantiopurity. The product of the same hydroperoxylation performed

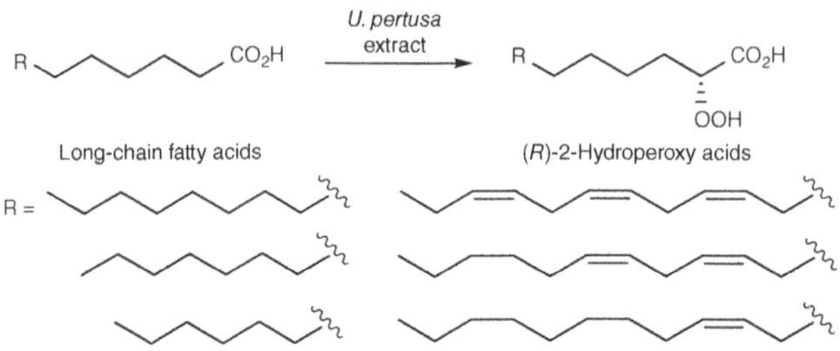

Scheme 2.29 Enantioselective 2-hydroperoxylation of saturated and unsaturated long-chain fatty acids with crude enzyme of *U. pertusa*.

with either brown or red algae on palmitic acid was also (*R*)-2-hydroperoxy acid in >99% e.e. enantiopure.

Lipid peroxidation has long been thought as a deleterious process. However, lipid peroxidation is a normal physiological process for eicosanoid synthesis or cell maturation in cell mediated reactions [123]. Furthermore, lipoxygenase-catalyzed lipid peroxidation has been found highly enantioselective. For instance, the 2,2′-azobis(2-amidinopropane) hydrochloride initiated peroxidation of linoleic acid in the presence of human serum albumin has been proved to be regio- and stereoselective with a large amount of 13(*S*)-hydroperoxy-(9*Z*,11*E*)-octadecadienoic acid [127]. Soybean lipoxygenase, the most studied of the lipoxygenase, has been used to catalyze the highly enantioselective and regioselective insertion of oxygen into the ω-6 position of polyunsaturated fatty acids to form diene hydroperoxides [128–132]. A regioselective and stereoselective synthesis of 1-stearoyl-2-[13′-(*S*)-hydroperoxy-(9′*Z*,11′*E*)-octadecadienoyl]-*sn*-glycero-3-phosphoch-oline was also performed without contamination by any regio- and stereochemical isomers by a combination of lipoxygenase-catalyzed peroxidation, lipase-catalyzed stearoylation, and dicyclohexyl carbodiimide-mediated esterification [133]. In this enzymatic peroxidation, linoleic acid is catalyzed by soybean lipoxygenase to give corresponding peroxide and is followed by treatment with diazomethane to produce methyl ester in 36% overall yield (Scheme 2.30).

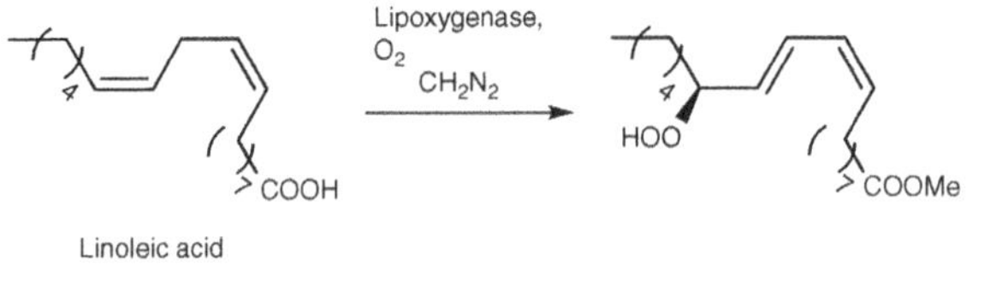

Scheme 2.30 Peroxidation of linoleic acid with soybean lipoxygenase and subsequent production of its methyl ester.

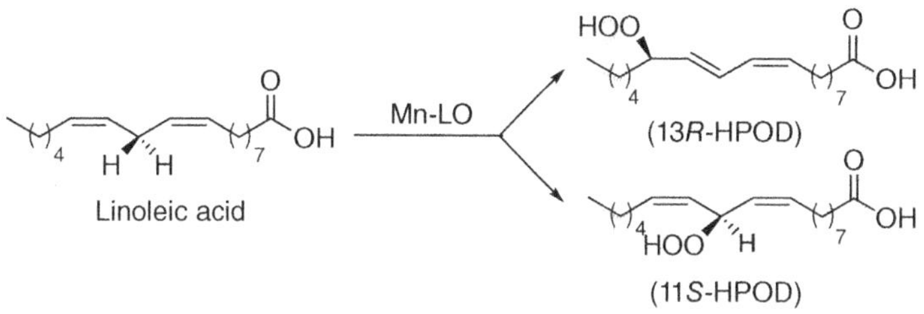

Scheme 2.31 Mn-LO mediated peroxidation of linoleic acid.

Manganese lipoxygenase (Mn-LO) from the fungus *Gäumannomyces graminis* has been reported for the peroxidation of linoleic acid to give a major product (13*R*)-hydro-peroxy-(9*Z*,11*E*)-octadecadienoic acid (13*R*-HPOD) and a new minor product (11*S*)-hydroperoxy-(9*Z*,12*Z*)-octadecadienoic acid(11*S*-HPOD) in a ratio of 1:5 (Scheme 2.31) [134]. An initial abstraction of the pro-*S* hydrogen from C-11 of linoleic acid is proceeded by Mn-LO to produce a linoleoyl radical. This radical is reversibly oxygenated at C-11 and further converted into (11*S*)-HPOD. Alternatively, this radical is irreversibly oxygenated at C-13 position to form (13*R*)-HPOD. Similar transformations with lipoxygenase enzymes have been described by other research groups [129, 130]. An iron mini-lipoxygenase from *Cyanobacterium Cyanthece* sp. (CspLOX2) was also reported by Feussner et al. to catalyze similar reactions [135].

Instead of the direct synthesis of enantiomerically pure hydroperoxides, enantioselective kinetic resolution using enzymes for racemic hydroperoxides is another effective method to reach the goal. The kinetic resolution of racemic hydroperoxides by horseradish peroxidase (HRP)-catalyzed reduction for aryl hydroperoxides shows that both the catalytic efficiency and the stereoselectivity of HRP are highly dependent on the structure of the hydroperoxides (Scheme 2.32) [124, 136]. Other

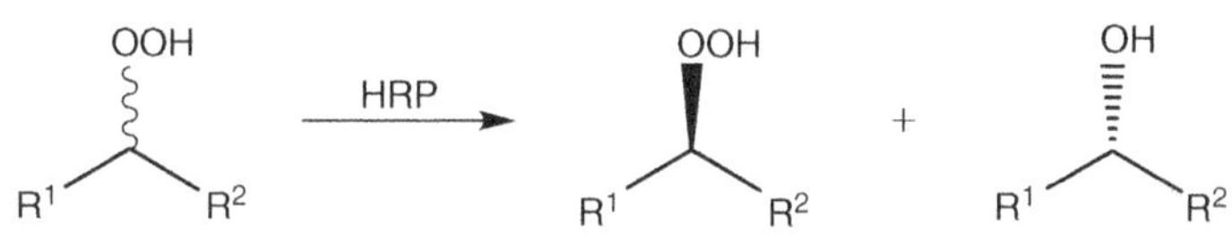

R^1	R^2
Ph (>99%, *S*- e.e.)	Me (>99%, *R*- e.e.)
Ph (93%, *S*- e.e.)	Et (95%, *R*- e.e.)
Ph (<5%, *S*- e.e.)	*n*-Pr (<5%, *R*- e.e.)
Ph (15%, *R*- e.e.)	*i*-Pr (14%, *S*- e.e.)
Ph (37%, *R*- e.e.)	*i*-Bu (48%, *S*- e.e.)
Et (97%, *R*- e.e.)	CO_2Me (97%, *S*- e.e.)
i-Pr (79%, *R*- e.e.)	CO_2Me (64%, *S*- e.e.)

Scheme 2.32 Kinetic resolution of aryl hydroperoxides by HRP.

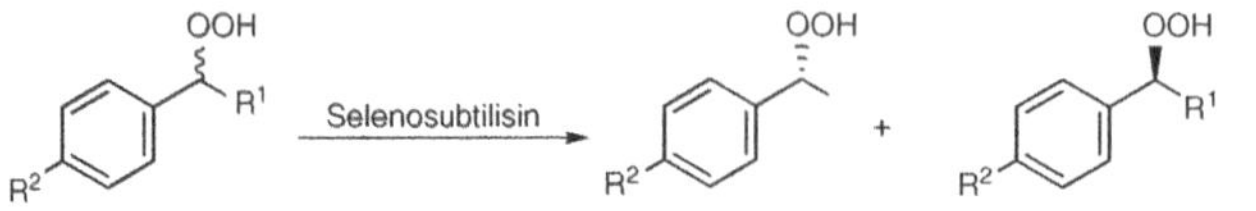

a. R^1 = Me,	R^2 = H	52%, *R*-e.e.	60%, *S*-e.e.
b. R^1 = Me,	R^2 = Cl	48%, *R*-e.e.	56%, *S*-e.e.
c. R^1 = CH_2OH,	R^2 = H	98%, *S*-e.e.	98%, *R*-e.e.
d. R^1 = $CH(CH_3)_2$,	R^2 = H	40%, *S*-e.e.	42%, *R*-e.e.

Scheme 2.33 Enantioselectivities of the selenosubtilisin-catalyzed kinetic resolution of hydroperoxides [137].

enzymes such as selenosubtilisin was also used for catalyzing the reduction of aryl hydroperoxides with thios via the catalytic cycle of glutathione peroxidase [137]. This reaction has also been applied for the kinetic resolution of racemic aryl hydroperoxides to form enantiopure hydroperoxides and alcohols [138, 139], and their corresponding enantioselectivity of enantiopure hydroperoxides and alcohols are shown in Scheme 2.33 [137].

2.2 Reduction Reactions

2.2.1 Reduction of Aldehydes and Ketones

Chiral alcohols are important building blocks and among the most valuable key intermediates for the production of pharmaceuticals and fine chemicals [140–143]. As an alternative to chemical processes, an efficient and powerful method to prepare enantiopure alcohols is the use of NAD(P)H-dependent ADHs to perform the asymmetric hydrogenation of prochiral ketones without protective groups that are common in traditional organic synthesis [144, 145]. Although asymmetric reduction of ketones to optically pure alcohols has become quite mature in asymmetric synthesis during the last decade, it is still the most interesting strategy in preparing single enantiomers of alcohols by recent advancement in genetic engineering, coupled enzyme reaction, reaction design, and the availability of a variety of ADH. At least 150 different ADHs are available from various commercial sources, which allow the most suitable ADH to be selected for a specific substrate to access the desired (*S*)- or (*R*)-enantiomer. The following are a concise introduction of the synthetic strategies used for the preparation of optically pure alcohol recently.

Since serine-threonine kinase, a polo-like kinase 1 (PLK1), is a key regulator of mitotic progression and cell division in eukaryotes and is highly expressed in tumor cells, it is considered a potential target for cancer therapy. To synthesize the most promising PLK1 inhibitor candidates, thiophene-benzimidazole and imidazopyridine, for chemotherapy, enantioselective reduction of *o*-chloroacetophenone is

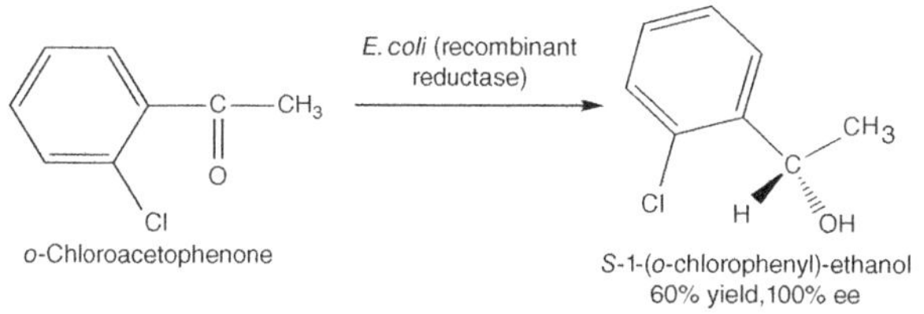

Scheme 2.34 Asymmetric reduction of ketone precursor *o*-chloroacetophenone with recombining microorganism toward chiral alcohol product.

preferred to produce the chiral key intermediate, (*S*)-1-(*o*-chlorophenyl)-ethanol, for their synthesis (Scheme 2.34) by whole-cell catalyst of recombining *E. coli* and recombining *S. cerevisiae* [146]. The recombinant xylose reductase was from *Candida tenuis*.

Aprepitant (Emend®), a trichiral compound, is a neurokinin-1 (NK-1) receptor antagonist that has been approved by the United States and Europe to treat moderately and highly emetogenic chemotherapy for the prevention of acute and delayed chemotherapy-induced nausea and vomiting. According to its structure (Figure 2.2), 3,5-bis(trifluoromethyl) acetophenone was used as the starting material to synthesize asymmetrically the key intermediate, chiral (1*R*)-[3,5-bis(trifluoromethyl)phenyl] ethanol, of aprepitant (Scheme 2.35) [147–149]. In the synthesis, a novel bacterial strain *Leifsonia xyli* HS0904 was isolated from soil that exhibits *R*-stereospecific carbonyl reductase. Under the optimal conditions, the best yield of 62% and an enantiomeric excess of 99.4% for the product was obtained with resting cells.

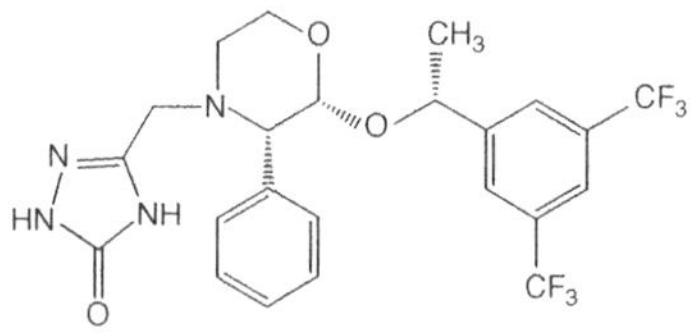

Figure 2.2 Structure of aprepitant.

An intermediate of amphetamine and amphetaminil, (*S*)-1-phenyl-2-propanol, has been prepared from the simple ketone, 1-phenyl-2-propanone, via bioreduction with 99% e.e. but low productivity through growing cells of *R. erythropolis*. The substrate inhibition of the biotransformation with growing

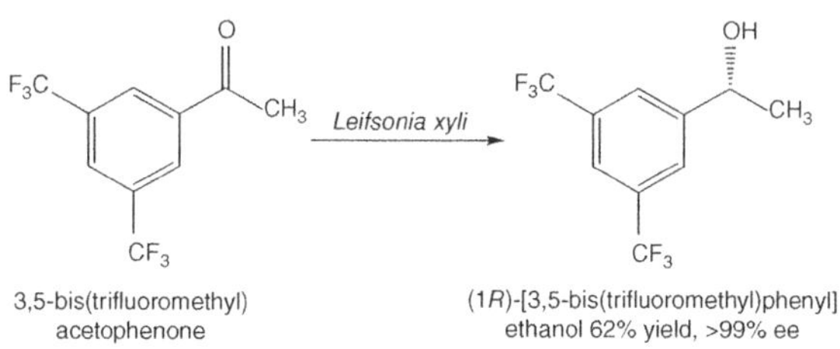

Scheme 2.35 The asymmetric synthesis of 3,5-bis(trifluoromethyl) acetophenone to (1*R*)-[3,5-bis(trifluoromethyl)phenyl] ethanol.

cells was solved by stepwise feeding, while product inhibition was solved by repeated removal of product using methods such as centrifugation, absorption with resin, and second phase. Higher productivity can be obtained for the reduction of 1-phenyl-2-propanone with resting cells through cofactor regeneration and recycling by the addition of glucose and permeabilized cells of *B. subtilis* [150].

The family of imidazole derivatives, including miconazole, econazole, isoconazole, ketoconazole, sertaconazole, and sulconazole, is well known for antifungal imidazolium compounds. Miconazole and econazole are usually employed in the treatment of vaginal diseases and several fungal infections in the skin of both human and animals by interfering with ergosterol biosynthesis of fungal organisms. Since the therapeutic efficacies of their enantiomers were usually different, the demand for synthesis of optically pure compounds rather than their racemates is required. However, the synthesis of miconazole and econazole single enantiomers with asymmetric chemical transformations [151–154] was rarely reported. Recently, a simple and novel bioreduction of prochiral ketones using ADHs has been applied for the synthesis of the precursors of miconazole and econazole single enantiomers. Then the target fungicides miconazole and econazole were produced from corresponding enantiomeric pure precursors by a series of chemical modifications. The best results were the synthesis of enantiopure precursor, (*R*)-2-chloro-1-(2,4-dichlorophenyl)ethanol, from 2-chloro-1-(2,4-dichlorophenyl)ethanone using screened ADHs under very mild conditions (Scheme 2.36) [155].

The use of *S. cerevisiae* (baker's yeast) for the reduction of ketones or aldehydes may result in mixtures of products due to the existence of a variety of reductases possessing overlapping substrate specificity and mixed stereoselectivity [156].

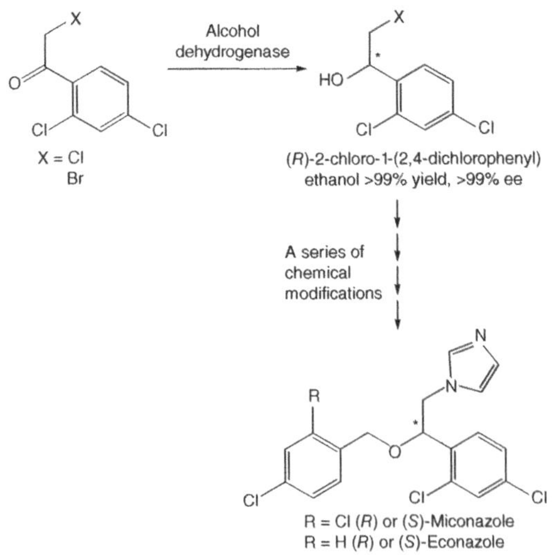

Scheme 2.36 Bioreduction of α-haloketones in aqueous medium using different alcohol dehydrogenase followed by a series of chemical modifications to optically pure miconazole or econazole.

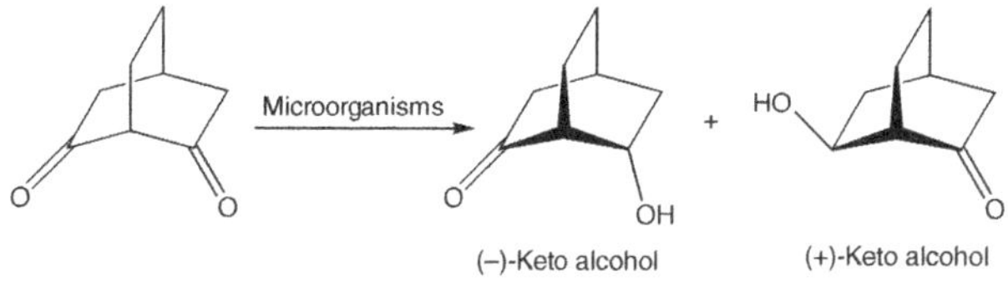

Scheme 2.37 *E. coli* or *S. cerevisiae* catalyzed reduction of bicycle[2.2.2]octane-2,6-dione to (−)-(1*R*,4*S*,6*S*)-6-hydroxy-bicyclo[2.2.2]octane-2-one and (+)-(1*S*,4*R*,6*S*)-6-hydroxy-bicyclo[2.2.2]octane-2-one.

Therefore, *S. cerevisiae* and the gram-negative bacterium *E. coli* were overexpressed with gene encoding the NADPH-dependent aldo-keto reductase YPR1 and compared for producing optically active alcohols through whole-cell bioreduction. The NADPH-dependent carbonyl reduction of bicycle[2.2.2]octane-2,6-dione to produce optically pure (−)-(1*R*,4*S*,6*S*)-6-hydroxy-bicyclo[2.2.2]octane-2-one was used as a model reaction (Scheme 2.37). High purity of the (−)-keto alcohol (>99% e.e., 97–98% de) was obtained for both engineered microorganisms. However, *E. coli* had higher initial rate but *S. cerevisiae* continued the reaction longer to give a higher substrate conversion (95%). *S. cerevisiae* also demonstrated higher viability during reaction than *E. coli* [157].

The asymmetric reduction of β-ketoesters mediated by microorganisms has become a standard method for the synthesis of chiral β-hydroxyesters [158, 159]. Recently, the immobilization of ADH from permeabilized brewer's yeast on derived attapulgite nanofibers through glutaraldehyde covalent binding for the bioreduction of ethyl 3-oxobutyrate (EOB) to ethyl (*S*)-3-hydroxybutyrate ((*S*)-EHB) was investigated. The effect of immobilization on ADH activity for the bioreduction shows that the immobilized ADH retained higher activity over wider ranges of pH and temperature than those of the free enzyme. The optimum temperature and pH of the immobilized ADH were 7.5 and 35 °C, respectively. Under the optimum conditions, the immobilized enzyme retained 58% of the original activity after 32 hours of incubation. The conversion of substrate (EOB) and the enantiomeric excess value of (*S*)-product reached 88 and 99.2%, respectively, within two hours. The immobilized ADH retained about 42% of the initial activity after eight cycles [160]. Also, although the yields and the enantioselectivity are low to moderate, the enantioselective bioreduction of ethyl benzoylacetate and their *p*-nitro and *p*-methoxy substituted derivatives to form corresponding chiral ethyl 3-hydroxy-3-phenylpropionate and substituted derivatives (Scheme 2.38) has long been of interest in pharmaceutical industry for synthesizing the key chiral building blocks of many compounds such as fluoxetine [161], chloramphenicol [162], and diltiazem [163]. However, the coupling of simple screening procedures and reaction engineering strategy can increase the (*S*)-enantioselectivity to 99% e.e. and shows a significant improvement

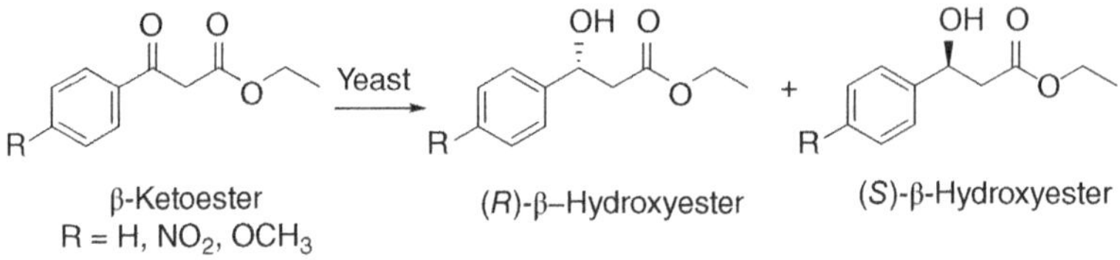

Scheme 2.38 Yeast mediated enantioselective reduction of ethyl benzoylacetate and substituted derivatives.

in the yields to around 85%. In this way, yeasts *Pichia kluyveri*, *Pichia stipites*, and *Candida utilis* were screened and better yields and e.e.'s for ethyl benzoylacetate, *p*-nitrobenzoylacetate, and *p*-methoxybenzoylacetate can be achieved by the addition of glucose, α-chloroacetophenone as inhibitor, and immobilization of the yeast in alginate beads, respectively. These processes can also be implemented on a preparative scale and still maintain the same yield and e.e. [164].

Although water is the first choice as solvent for biocatalysis, the low solubility of organic compounds in water, difficult product separation, and potential side reactions caused by other enzymes in the cell have led to alternative solvents being sought. Since optically pure (*S*)-1-(4′-methoxyphenyl) ethanol ((*S*)-MOPE) is a potential synthon for the preparation of cycloalkyl [b] indoles in clinical treatment of general allergic response, whole microbial cells have been used to synthesize for enantiopure (*S*)-MOPE from 4′-methoxyacetophenone (MOAP) in aqueous systems [165]. Recently, the bioreduction of MOAP to (*S*)-MOPE has been successfully performed in a hydrophilic ionic liquid (IL) containing cosolvent system using immobilized *Rhodotorula* sp. AS2.2241 cells. The novel IL 1-(2′-hydroxy)ethyl-3-methylimidazolium nitrate ($C_2OHMIM{\bullet}NO_3$) gave the best results. Under the optimized conditions of pH 8.5, a temperature of 25 °C, and an optimal $C_2OHMIM{\bullet}NO_3$ content of 5.0% (v/v), 12 mM, the initial reaction rate, the maximum yield, and the product e.e. were 9.8 μmol h^{-1} (g cell)$^{-1}$, 98.3%, and >99%, respectively. Also, the cells exhibited excellent operational stability with the cosolvent system. Moreover, the established system has been highly efficiently applied for the reduction of many other aryl ketones [166]. Immobilization of whole cell of *Geotrichum candidum* onto an ion exchange resin with polyallylamine was used for the enantioselective bioreduction of various ketones, such as acetophenone, *ortho*-fluoro and *para*-fluoro acetophenone, *ortho*-methyl, 2-phenylethyl methyl ketone, and phenyl trifluoromethane ketone, in aqueous and supercritical CO_2 ($scCO_2$) solvents. The immobilization of the cell improved not only the enantioselectivity but also the stability and enabled a continuous-flow reaction in aqueous solution. Recycling of the immobilized cell accompanying $scCO_2$

depressurization and continuous-flow reaction was also made when the reaction was performed in $scCO_2$ [167]. Hydrophobic ketone such as phenyl *n*-propyl ketone has been used as a model compound to survey ADH activity in *S. cerevisiae*. The enantioselectivity of yeast mediated reduction toward the product (*R*)-(+)- or (*S*)-(−)-1-phenyl-1-butanol was found to depend on the hexane-to-water volume percentage of biphasic cell culture and the cofactor zinc ion [167]. Without Zn^{2+} ion the biphasic cultures of middle to high hexane-to-water volume percentage possessed (*R*)-enantiomeric excess (54% to >99%) and (*S*)-enantiomeric excess (15–47%) for low hexane-to-water volume percentage. With Zn^{2+} ion in the biphasic cultures, the enantioselectivity was exclusively (*S*)-enantiomeric excess (28% to >99%). The bioreduction mediated with the yeast *C. utilis* of aqueous cultures showed an (*S*)-enantiomeric excess of 79–95% [144]. Glycerol is a nontoxic, biodegradable, and recyclable liquid and has a high boiling point and negligible vapor pressure that make it an ideal green reaction medium for many catalytic and noncatalytic organic syntheses. The high polarity of glycerol favorably facilitates the simple baker's yeast mediated reduction of benzaldehyde and ethyl acetoacetate because glycerol can dissolve glucose and ethyl acetoacetate and suspend baker's yeast. The use of glycerol as reaction medium for baker's yeast mediated reduction produced a high product yield and >99% enantioselectivity with either free or immobilized cells [168].

The ADH obtained from *Thermus* sp. ATN1 (TADH) is an NAD(H)-dependent enzyme, which shows a very broad substrate spectrum including aldehydes, aliphatic ketones, cyclic ketones, and double-ring systems and produces exclusively the (*S*)-enantiomer in high enantiomeric excess (>99%) for ketones. TADH can be used in the presence of 10% (v/v) water-miscible solvents like 2-propanol or acetone, which plays as sacrificial substrate in substrate-coupled cofactor regeneration reactions. TADH retained 80% of its activity when water-insoluble solvent like hexane or octane is used as cosolvent that forms an aqueous/organic biphasic reaction medium to allow the reaction of low-water-soluble substrates [169].

Asymmetric reduction of ketones to pure alcohols has been applied for material chemistry. For example, by carefully selecting the right ADH, the enantiomerically pure (*S*)- and the (*R*)-monomers were biosynthesized from the *p*-vinylacetophenone. Then, varied ratio of (*S*)- and (*R*)-monomers can be polymerized to prepare different polymers via free-radical polymerization. The polymers formed from the (*S*)- and (*R*)-monomer mixtures had a number-average molecular weight (*Mn*) of 5000–6000 $g\,mol^{-1}$ and a polydispersity of 1.7–2.1. The thermal properties of the polymer material (T_g) can be further fine-tuned by enantioselective grafting of the (*R*)-alcohol groups with vinylacetate by a lipase (Scheme 2.39). Therefore, a decrease of T_g for the acetylation modified polymers was shown by the differential scanning calorimetry (DSC) analysis [170].

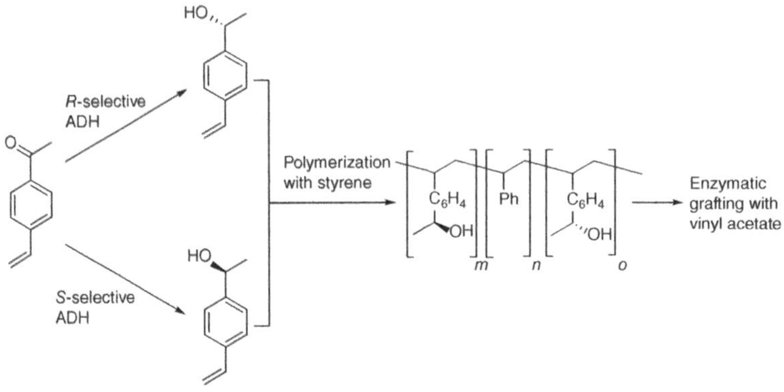

Scheme 2.39 Asymmetric reduction of ketones and polymerization of the optically pure monomers for application in material chemistry.

2.2.2 Reduction of C=C Bonds

The asymmetric bioreduction of activated C=C bonds with the generation of up to two stereogenic centers is one of the most widely employed synthetic strategies for producing chiral molecules [171, 172]. Enoate reductases (ERs), NAD(P)H-dependent flavin mononucleotide (FMN)-containing oxidoreductases from the "**Old Yellow Enzyme** (OYE)" family (E.C. 1.6.99.1) [171, 173], are widely distributed in microorganisms such as bacteria [174], yeast [175], and filamentous fungi [176] and in higher plants [177] and stereoselectively catalyze the reduction of activated C=C bonds containing electron-withdrawing groups (EWG) such as α,β-unsaturated ketones, aldehydes, nitroalkenes, carboxylic acids, and derivatives (ester, anhydride, lactone, and cyclic imide), thus, affording the production of fine chemicals, pharmaceuticals, and agrochemical intermediates. Note that EWGs such as halogen, oxime, sulfoxide, or sulfone usually enhance the reaction rates but are not sufficient to activate a C=C bond alone [178]. The overall stereoselective reduction of activated double bond catalyzed by the OYE family has been investigated in great detail. The reaction proceeds in a two-stage bi-bi ping-pong mechanism: the OYE flavin cofactor is first reduced at the expense of a nicotinamide cofactor NAD(P)H, which is followed by hydride transfer to the Cβ of the substrate (the reductive half reaction), whereas a tyrosine residue of the ER adds another proton to Cα of the double bond from the opposite side (the oxidative half reaction), with both reductive and oxidative substrates binding within the active site. This mechanism results in a *trans* addition of [2H] to the double bond of the substrate with absolute stereospecificity (Scheme 2.40).

In the past, the complexity of cofactor recycling has made the majority of the asymmetric reduction of activated C=C performed using whole fermenting cells, mostly baker's yeast and anaerobic bacteria. For example, whole cell of baker's

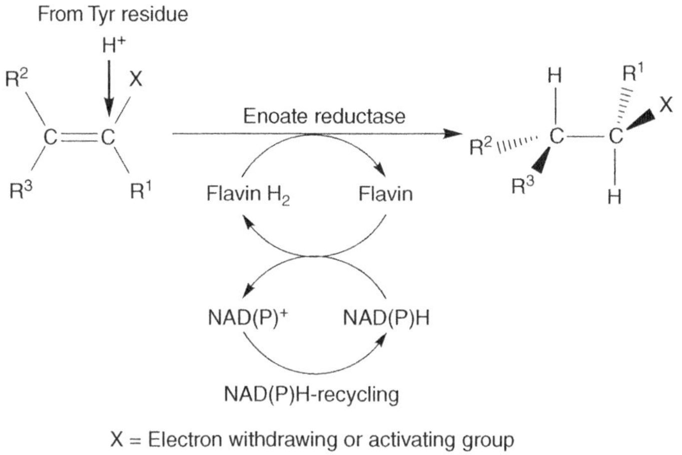

Scheme 2.40 Asymmetric reduction of activated alkene substrates catalyzed by OYE enzymes.

yeast (*S. cerevisiae*) has been broadly used for the stereospecific reduction of many compounds containing activated carbon-carbon double bond such as monoterpenes and sesquiterpenoids [179]. Since terpenes and terpenoids exhibit a wide variety of pleasant and floral scents, these properties make them extensively used in the perfumery and food industries. Monoterpenes and sesquiterpenoids, the chief components of the essential oils, not only act as precursors for new flavor derivatives but also are important building blocks for the preparation of new biologically active natural product drugs.

ERs used for asymmetric reduction of electronically activated C=C bonds, with a few exceptions, almost exclusively are isolated from *Saccharomyces* spp. yeasts, particularly, from the domesticated species *S. cerevisiae*. However, baker's yeast just represents a small fraction within the vast yeast kingdom. Recently, Buzzini et al. carried out a screening on 23 so-called **nonconventional yeasts** (NCYs) belonging to 21 species of the genera *Candida*, *Cryptococcus*, *Debaryomyces*, *Hanseniaspora*, *Kazachstania*, *Kluyveromyces*, *Lindnera*, *Nakaseomyces*, *Vanderwaltozyma*, and *Wickerhamomyces* for the whole-cell bioreduction of α,β-unsaturated ketones and aldehydes. Results show that extremely high yields (>90%) or even total bioconversion yields for the asymmetric reduction of the conjugated C=C bond of ketoisophorone (KIP) and 2-methyl-cyclopentanone (2-MCPO) were catalyzed by ERs in a few NCYs. The catalytic efficiency of these NCYs declined for aldehydes [(*S*)-(−)-perillaldehyde ((*S*)-PA) and α-methyl-cinnamaldehyde (MCA)]. The NCY whole cells were in lyophilized form and glucose was used for NAD(P)H cofactor recycling system [180].

Although the use of whole fermenting cells for the asymmetric reduction of activated alkenes can avoid the difficulty of cofactor recycling, the chemoselectivity for C=C was often low by the competing reduction of C=O with the carbonyl reductases in the cell. Therefore, purified ERs accompanied by the necessitating regeneration system for the nicotinamide cofactor [NAD(P)H] were employed to solve the problem of low chemoselectivity. The efficiency of the cofactor recycling is thus an important factor to determine both the yield and selectivity of the asymmetric bioreduction of activated C=C. A highly efficient, inexpensive, and robust ADH-coupled NADH-recycling system has been developed recently based on the concurrent oxidation of a sacrificial *sec*-alcohol 2-propanol to acetone by the solvent-stable ADH *Rhodococcus ruber* thereby providing a hydride via NADH to recycle the flavin cofactor of the ene-reduction catalyzed by ERs [181]. This highly NADH-recycling system has been applied to a variety of activated alkenes: α-methylmaleic acid dimethylester, 4-ketoisophorone, 2-methylcyclohexanone, 2-methylcyclopentenone, 2-methylpent-2-enal, and citral. For α-methylmaleic acid dimethylester, the conversion is quantitatively (>99%) and highly stereoselectively reduced to (*R*)-dimethyl-α-methyl-succinate with an excellent e.e. (>99%) by employing *Zymomonas mobilis* nicotinamide-dependent cyclohexenone reductase (NCR) reductase; the conversion is 90% and an e.e. > 99% is obtained by employing *B. subtilis* YqjM (Scheme 2.41).

It was reported that indirectly regenerating YqjM or NemA (*N*-ethylmaleimide reductase from *E. coli*) of reduced nicotinamide cofactors can be simplified by an efficient and convenient direct regeneration of catalytically active reduced flavins with the cell-free bioreductions of activated conjugated C=C double bonds. In this process, reducing equivalents are provided via photocatalytic oxidation of a lot of simple sacrificial electron donors such as phosphite, formate, ethylenediaminetetraacetate (EDTA) [182]. Even when using crude cell extracts, the chemoselectivity of the photoenzymatic reduction was exclusive, only C=C double bond reduction without altering the ketone or aldehyde groups was observed. Up to 65% rates of the NADH-driven reaction can be got while still preserving enantioselectivity.

Due to the production of 1 : 1 product mixture and unfavorable equilibria, disproportionation reactions generally are rarely used in organic synthesis. However, recent study shows that the nicotinamide-independent disproportionation of enones can be

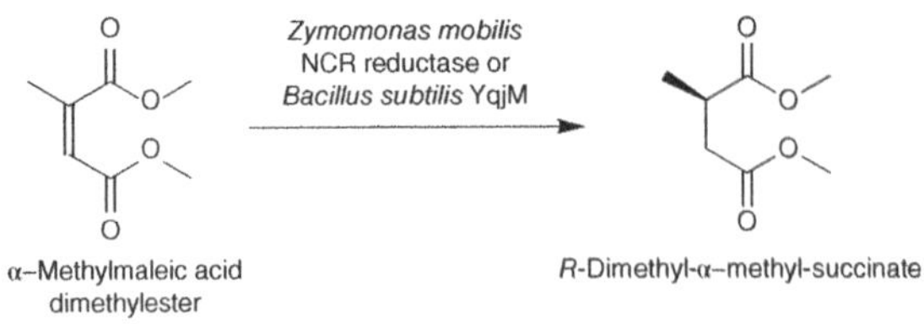

Scheme 2.41 Asymmetric reduction of α-methylmaleic acid dimethylester with *Z. mobilis* or *B. subtilis*.

used as a co-substrate for the asymmetric bioreduction of activated C=C bonds to eliminate the need for an NADH-recycling system. In this simplified system, the activated C=C bonds were catalyzed by a single flavoprotein and a sacrificial 2-enone or 1,4-dione was used as hydrogen donor for a direct hydrogen transfer without the requirement of nicotinamide cofactor. This strategy has been successively applied for the reduction of C=C bond in citraconic acid dimethylester to produce (*R*)-dimethyl-2-methylsuccinate with either 3-methylcyclohex-2-enone or cyclohexane-1,4-dione as co-substrate (Scheme 2.42) [183]. Other substrates such as 4-ketoisophorone and *N*-phenyl-2-methylmaleimide have also been successfully tested for the applicability to produce (*R*)-levodion and (*R*)-*N*-phenyl-2-methylsuccinimide, respectively.

A novel ER isolated from the bacterium *Z. mobilis* termed NCR reductase and OYEs 1–3 from yeast of *Saccharomyces* spp. were extended to use for the asymmetric reduction of a variety of activated C=C substrates. The activating groups in the investigated substrates include aldehyde, ketone, imide, nitro, carboxylic acid, and ester moieties. To control the stereospecificity of the reaction, strategies such as variation of the main substrate structure (cyclic vs. acylic), switching the (*E*/*Z*) configuration of the alkene moiety, modifying the substitution pattern (mono- vs. di-, α- vs. β-), or proper selection of the enzyme are utilized to allow the access to the opposite enantiomeric products. Results showed that reaction rates and stereoselectivities can be varied by different substrate type. The "substrate-based stereocontrol" is related with the ring size of cycloalkenones, position of substituents on the C=C bond, and its *E*/*Z* configuration. The "enzyme-based stereocontrol" was observed with nitroalkene in which the opposite enantiomeric products can be obtained by both NCR reductase and OYEs 1–3 [184].

Baker's yeast (BY) catalyzed enantioselective reduction of an EWG substituent activated double bond such as α,β-unsaturated aldehydes and ketones are generally very efficient with high enantioselectivities and conversions. However, it is

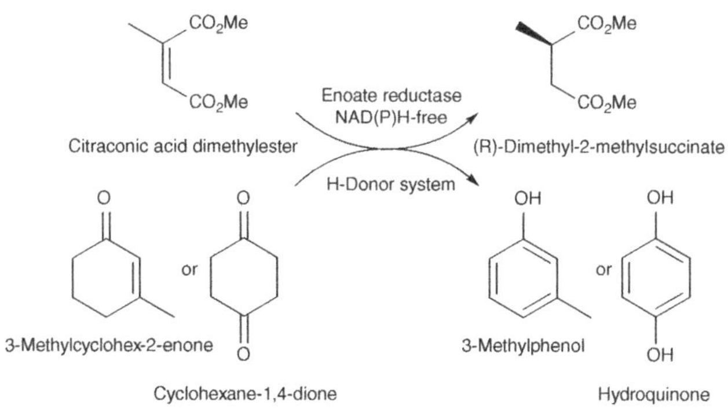

Scheme 2.42 Asymmetric bioreduction of citraconic acid dimethylester via a coupled-substrate system.

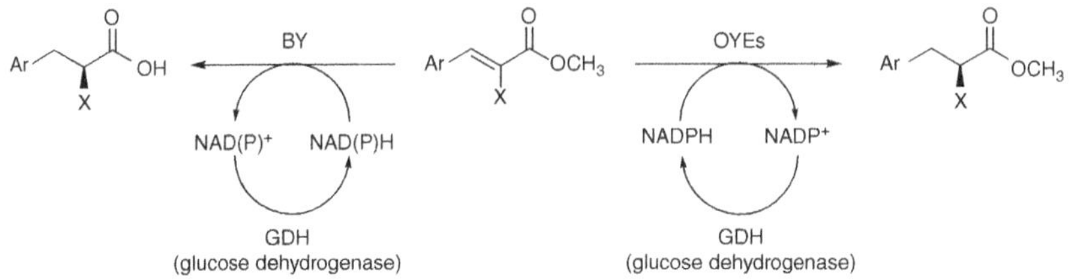

1. Ar = C_6H_5, X = Cl, 2. Ar = C_6H_5, X = Br, 3. Ar = 4-FC_6H_4, X = Cl, 4. Ar = 4-$CH_3OC_6H_4$, X = Cl,

5. Ar = 2-$NO_2C_6H_4$, X = Cl, 6. Ar = 4-CNC_6H_4, X=Cl

Scheme 2.43 BY fermentations and OYEs 1–3 mediated bioreductions of substrates 1–6.

necessary to explore the synthetic capability of the yeast-based ER for other activated substrates to make the enantioselective synthesis toolbox more fruitful and advantageous for implementation in industrial manufacturing processes. Therefore, the enantioselective reduction of the activated C=C substrates on some (*Z*)-methyl α-halo-β-arylacrylates was investigated and performed by baker's yeast fermentation or OYEs 1–3 mediated biotransformations to prepare (*S*)-α-halo-β-arylpropionic acid derivatives in high e.e. and with good yields. The fermentation medium promotes ester hydrolysis, but the isolated enzymes preserve the ester functionality (Scheme 2.43). Particularly, when aromatic ring was substituted by an EWG, high e.e. and conversion values were observed [185].

The reduction of the double bond associated with α,β-unsaturated ketones and catalyzed by microorganisms, animal, and plant cell cultures has long been studied and known by synthetic organic chemists. However, a strain of filamentous fungus identified as *Aspergillus versicolor* and named *A. versicolor* D-1 was isolated and used for the reduction of the γ,δ-double bond of the conjugated lactone in securinine to produce 14,15-dihydrosecurinine with high stereospecificity (Scheme 2.44) [186]. The NADPH-dependent securinine reductase found in *A. versicolor* D-1 is a cytosolic enzyme and is highly inducible in growing/resting cultures for securinine reduction. Moreover, it will not reduce the α,β-double bond in 14,15-dihydrosecurinine further. It was also found that the reductase is thermally unstable.

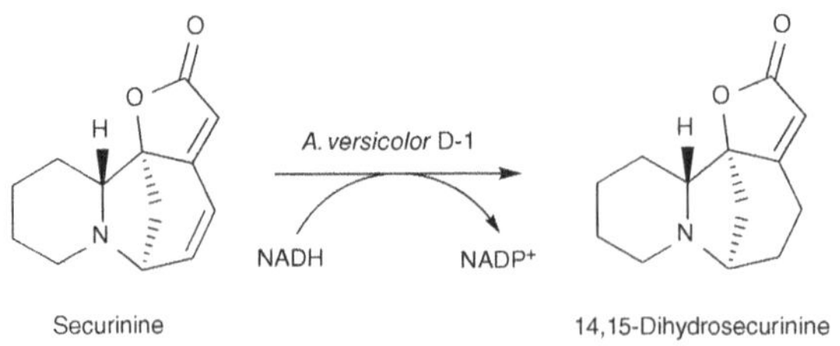

Scheme 2.44 The reduction of γ,δ-double bond of the conjugated lactone in securinine.

References

1 Molinari, F. (**2006**). *Curr. Org. Chem. 10*: 1247–1263.
2 Bouonomenna, M.G. and Drioli, E. (**2008**). *Org. Process Res. Dev. 12*: 982–988.
3 Jia, A.Z., Lou, L.L., Zhang, C. et al. (**2009**). *J. Mol. Catal. A 306*: 123–129.
4 Celik, D., Bayraktar, E., and Mehmetoglu, D. (**2004**). *Biochem. Eng. J. 17*: 5–13.
5 Molinari, F., Gandolfi, R., Aragozzini, F. et al. (**1999**). *Enzyme Microb. Technol. 25*: 729–735.
6 Wu, J., Wang, J.-L., Li, M.-H. et al. (**2010**). *Bioresour. Technol. 101*: 8936–8941.
7 Orbegozo, T., Lavandera, I., Fabian, W.M.F. et al. (**2009**). *Tetrahedron 65*: 6805–6809.
8 Sagiroglu, A. and Yavuz, M.O. (**2005**). *Artif. Cells Blood Substit. Biotechnol. 33*: 343–355.
9 Norouzian, D., Akbarzadeh, A., Inanlou, D.N. et al. (**2003**). *Enzyme Microb. Technol. 33*: 150–153.
10 Carballeira Rodríguez, J.D., García-Burgos, C., Quezada Alvarez, M.A. et al. (**2004**). *Biotechnol. Bioeng. 87*: 632–640.
11 Jia, X., Xu, Y., and Li, Z. (**2011**). *ACS Catal. 1*: 591–596.
12 Hilterhaus, L. and Liese, A. (**2007**). *Adv. Biochem. Eng. Biotechnol. 105*: 133–173.
13 Oedman, P., Wessjohann, L.A., and Bornscheuer, U.T. (**2005**). *J. Org. Chem. 70*: 9551–9555.
14 Hirscher, T., Gocke, D., Fernandez, M. et al. (**2005**). *Tetrahedron 61*: 7378–7383.
15 Scheid, G., Kuit, W., Ruijter, E. et al. (**2004**). *Eur. J. Org. Chem. 2004*: 1063–1074.
16 Nestl, B.M., Voss, C.V., Bodlenner, A. et al. (**2007**). *Appl. Microbiol. Biotechnol. 76*: 1001–1008.
17 Demir, A.S., Hamamci, H., Sesenoglu, O. et al. (**2001**). *Tetrahedr. Asymm. 12*: 1953–1956.
18 Smalleridge, A.J., Trewhalla, M.A., Maurice, A., and Wilkinson, A.K. (**2003**). Patent WO 2003018531 A1 20030306.
19 Rosche, B., Breuer, M., Hauer, B., and Rogers, P.L. (**2004**). *Biotechnol. Bioeng. 86*: 788–794.
20 Engel, S., Vyazmensky, M., Geresh, H. et al. (**2003**). *Biotechnol. Bioeng. 83*: 640–833.
21 Cheng, Y., Zhang, F., Rano, T.A. et al. (**2002**). *Bioorg. Med. Chem. Lett. 12*: 2419–2422.
22 Lunardi, I., Conceicao, G.J.A., Moran, P.J.S., and Rodrigues, J.A.R. (**2005**). *Tetrahedr. Asymm. 16*: 2515–2519.
23 Mitsukura, K., Sato, Y., Yoshida, T., and Nagasawa, T. (**2004**). *Biotechnol. Lett. 26*: 1643–1648.
24 Tanaka, M., Hirokane, Y., Mitsui, R., and Tsuno, T. (**2001**). *J. Biosci. Bioeng. 91*: 267–271.

25 Vicente, C., Fontaniella, B., Millanes, A.M. et al. (**2003**). *Int. J. Cosmet. Sci. 25*: 25–29.

26 Seshardri, R., Lamm, A.S., Khare, A., and Rosazza, J.P.N. (**2008**). *Enzyme Microb. Technol. 43*: 486–494.

27 Hamme, J.D., Singh, A., and Ward, O.P. (**2003**). *Microbiol. Mol. Boil. Rev. 67*: 503–549.

28 Beilen, J.B., Li, Z., Duetz, W.A. et al. (**2003**). *Oil Gas Sci. Technol.* 58: 427–440.

29 Labinger, J.A. (**2004**). *J. Mol. Catal. A 220*: 27–35.

30 Ayala, M. and Torres, E. (**2004**). *Appl. Catal. A 272*: 1–13.

31 van Beilen, J.B. and Funhoff, E.G. (**2005**). *Curr. Opin. Biotechnol. 16*: 308–314.

32 Bernhardt, R. (**2006**). *J. Bacteriol. 124*: 128–145.

33 Funhoff, E.G. and van Beilen, J.B. (**2007**). *Biocatal. Biotransformation 25*: 186–193.

34 Kawakami, N., Shoji, O., and Watanabe, Y. (**2011**). *Angew. Chem. Int. Ed. 50*: 5315–5318.

35 Weber, E., Seifert, A., Antonovici, M. et al. (**2011**). *Chem. Commun. 47*: 944–946.

36 Bordeaux, M., Galarneau, A., Fajula, F., and Drone, J. (**2011**). *Angew. Chem. Int. Ed. 50*: 2075–2079.

37 Scheps, D., Malca, S.H., Hoffmann, H. et al. (**2011**). *Org. Biomol. Chem. 9*: 6727–6733.

38 van Beilen, J.B., Funhoff, E.G., van Loon, A. et al. (**2006**). *Appl. Environ. Microbiol. 72*: 59–65.

39 Funhoff, E.G., Bauer, U., Garcia-Rubio, I. et al. (**2006**). *J. Bacteriol. 188*: 5220–5227.

40 Funhoff, E.G., Salzmann, J., Bauer, U. et al. (**2007**). *Enzyme Microb. Technol. 40*: 806–812.

41 Dubbels, B.L., Sayavedra-Soto, L.A., and Arp, D.J. (**2007**). *Microbiology 153*: 1808–1816.

42 Craft, D.L., Madduri, K.M., Eshoo, M., and Wilson, C.R. (**2003**). *Appl. Environ. Microbiol. 69*: 5983.

43 Tani, A., Ishige, T., Sakai, Y., and Kato, N. (**2001**). *J. Bacteriol. 183*: 1819.

44 Feng, L., Wang, W., Cheng, J. et al. (**2007**). *Proc. Natl. Acad. Sci. U. S. A. 104*: 5602.

45 Maeng, J.H., Sakai, Y., Tani, Y., and Kato, N. (**1996**). *J. Bacteriol. 178*: 3695.

46 Li, L., Liu, X., Yang, W. et al. (**2008**). *J. Mol. Biol. 376*: 453–465.

47 Throne-Holst, M., Markussen, S., Winnberg, A. et al. (**2006**). *Appl. Microbiol. Biotechnol. 72*: 353–360.

48 Peter, S., Kinne, M., Wang, X. et al. (**2011**). *FEBS J. 278*: 3667–3675.

49 Costas, M., Chen, K., and Que, L. Jr. (**2000**). *Coord. Chem. Rev. 200-202*: 517–544.

50 Usharani, D., Janardanan, D., and Shaik, S. (**2011**). *J. Am. Chem. Soc. 133*: 176–179.

51 Austin, R.N., Chang, H.-K., Zylstra, G.J., and Groves, J.T. (**2000**). *J. Am. Chem. Soc. 122*: 11747–11748.
52 Ullrich, R. and Hofrichter, M. (**2007**). *Cell. Mol. Life Sci. 64*: 271–293.
53 Hollmann, F., Arends, I.W.C.E., Buehler, K. et al. (**2011**). *Green Chem. 13*: 226–265.
54 Gennaro, P.D., Bargna, A., and Sello, G. (**2011**). *Appl. Microbiol. Biotechnol. 90*: 1817–1827.
55 Fitzpatrick, P.E. (**1999**). *Annu. Rev. Biochem. 68*: 355–381.
56 Kasai, N., Ikushiro, S.-I., Hirosue, S. et al. (**2009**). *Biochem. Biophys. Res. Commun. 387*: 103–108.
57 Guo, J., Liu, D., Nikolic, D. et al. (**2008**). *Drug Metab. Dispos. 36*: 461–468.
58 Carvalho, F., Soares, M.E., Fernandes, E. et al. (**2007**). *J. Health Sci. 57*: 371–377.
59 Boyd, D.R. and Bugg, T.D.H. (**2006**). *Org. Biomol. Chem. 4*: 181–192.
60 Hudlicky, T. and Reed, J.W. (**2009**). *Synlett 5*: 685–703.
61 Shindo, K., Nakamura, R., Osawa, A. et al. (**2005**). *J. Mol. Catal. B: Enzym. 35*: 134–141.
62 De la Sovera, V., Bellomo, A., and Gonzalez, D. (**2011**). *Tetrahedron Lett. 52*: 430–433.
63 Bellomo, A., Giacomini, C., Brena, B. et al. (**2007**). *Synth. Commun. 37*: 3509–3518.
64 Banwell, M.G., Lehmann, A.L., Menon, R.S., and Willis, A.C. (**2011**). *Pure Appl. Chem. 83*: 411–423.
65 Banwell, M.G., Kokas, O.J., and Willis, A.C. (**2007**). *Org. Lett. 9*: 3503–3506.
66 Endoma-Arias, M.A.A. and Hudlicky, T. (**2011**). *Tetrahedron Lett. 52*: 6632–6634.
67 Boyd, D.R., Sharma, N.D., Coen, G.P. et al. (**2007**). *Chem. A Eur. J. 13*: 5804–5811.
68 Boyd, D.R., Sharma, N.D., Stevenson, P.J. et al. (**2011**). *Org. Biomol. Chem. 9*: 1479–1490.
69 Boyd, D.R., Sharma, N.D., Malone, J.F., and Allen, C.C.R. (**2009**). *Chem. Commun.* 3633–3635.
70 Kim, D., Lee, J.S., Choi, K.Y. et al. (**2007**). *Enzyme Microb. Technol. 41*: 221–225.
71 Kourist, R., Dominguez de María, P., and Bornscheuer, U.T. (**2008**). *Chembiochem 9*: 491–498.
72 Fabris, F., Collins, J., Sullivan, B. et al. (**2009**). *Org. Biomol. Chem. 7*: 2619–2627.
73 Costas, M., Mehn, M.P., Jensen, M.P., and Que, L. Jr. (**2004**). *Chem. Rev. 104*: 939–986.
74 Solomon, E.I., Brunold, T.C., Davis, M.I. et al. (**2000**). *Chem. Rev. 100*: 235–349.
75 Lee, K. (**2006**). *FEMS Microbiol. Lett. 255*: 316–320.
76 Jeffrey, A.M., Yeh, H.J., Jerina, D.M. et al. (**1975**). *Biochem. 14*: 575–584.
77 Chopard, C., Bertho, G., and Prange, T. (**2012**). *RSC Adv. 2*: 605–615.
78 Boyd, D.R., Sharma, N.D., Belhocine, T. et al. (**2006**). *Chem. Commun.* 4934–4936.

79 Lonsdale, R., Harvey, J.N., and Mulholland, A.J. (**2010**). *J. Phys. Chem. B 114*: 1156–1162.

80 Farinas, E.T., Alcalde, M., and Arnold, F. (**2004**). *Tetrahedron 60*: 525–528.

81 Jin, S., Kakris, T.M., Bryson, T.A. et al. (**2003**). *J. Am. Chem. Soc. 125*: 3406–3407.

82 Meester, R.J.W., Duisken, M., and Hollender, J. (**2009**). *Xenobiotica 39*: 663–671.

83 Malla, S., Thuy, T.T.T., Oh, T.J., and Sohng, J.K. (**2011**). *Arch. Microbiol. 193*: 95–103.

84 Belter, A., Skupinska, M., Giel-Pietraszuk, M. et al. (**2011**). *Biol. Chem. 392*: 1053–1075.

85 Ruckenstuhl, C., Lang, S., Poschenel, A. et al. (**2007**). *Antimicrob. Agents Chemother. 51*: 275–284.

86 Godio, R.P., Fouces, R., and Martin, J.F. (**2007**). *Chem. Biol. 14*: 1334–1346.

87 Nakano, C., Motegi, A., Sato, T. et al. (**2007**). *Biosci. Biotechnol. Biochem. 71*: 2543–2550.

88 Xu, Y., Li, A., Jia, X., and Li, Z. (**2011**). *Green Chem. 13*: 2452–2458.

89 Prichanont, S., Leak, D.J., and Stuckey, D.C. (**2000**). *Enzyme Microb. Technol. 27*: 134–142.

90 Julsing, M.K., Kuhn, D., Schimd, A., and Bühler, B. (**2012**). *Biotechnol. Bioeng.* 109: 1109–1119.

91 Bae, J.-W., Doo, E.-H., Shin, S.-H. et al. (**2010**). *Process Biochem. 45*: 147–152.

92 Bühler, B., Park, J.-B., Blank, L.M., and Schmid, A. (**2008**). *Appl. Environ. Microbiol. 74*: 1436–1446.

93 Park, J.-B., Bühler, B., Habicher, T. et al. (**2006**). *Biotecnol. Bioeng. 95*: 501–512.

94 Hollmann, F., Lin, P.-C., Witholt, B., and Schimd, A. (**2003**). *J. Am. Chem. Soc. 125*: 8209–8217.

95 Schimd, A., Hofstetter, K., Feiten, H.-J. et al. (**2001**). *Adv. Synth. Catal. 343*: 732–737.

96 Sello, G., Orsini, F., Bernasconi, S., and Gennaro, P.D. (**2006**). *Tetrahedron Asymm. 17*: 372–376.

97 Bhattacharya, S., Drews, A., Lyagin, E. et al. (**2012**). *Chem. Eng. Technol. 35*: 1–9.

98 Xie, F., Chao, Y., Xue, Z. et al. (**2009**). *J. Ind. Microbiol. Biotechnol. 36*: 739–746.

99 Gladkowski, W., Grbarczyk, M., Wińska, K. et al. (**2007**). *J. Mol. Catal. B 49*: 79–87.

100 Molina, S., Rencoret, J., del Río, J.C. et al. (**2008**). *Appl. Microbiol. Biotechnol. 80*: 211–222.

101 Carreño, M.C. (**1995**). *Chem. Rev. 95*: 6129–6144.

102 Legros, J., Dehli, J.R., and Bolm, C. (**2005**). *Adv. Synth. Catal. 347*: 19–31.

103 Matsui, T.; Dekishima, Y.; Ueda, M. (**2014**). *Appl. Microbiol. Biotechnol.*. 98, 7699–7706. DOI https://doi.org/10.1007/s00253-014-5932-z.

104 Sangar, S., Pal, M., Moon, L.S., and Jolly, R.S. (**2012**). *Bioresour. Technol. 115*: 102–110.

105 Zhai, X.-H., Ma, Y.-H., Lai, D.-Y. et al. (**2013**). *J. Ind. Microbiol. Biotechnol. 40*: 797–803.

106 Zhang, J.-D., Li, A.-T., Yu, H.-L. et al. (**2011**). *J. Ind. Microbiol. Biotechnol. 38*: 633–641.

107 De Gonzalo, G., Rodríguez, C., Rioz-Martínez, A., and Gotor, V. (**2012**). *Enzyme Microb. Technol. 50*: 43–49.

108 Li, A.-T., Yu, H.-L., Pan, J. et al. (**2011**). *Bioresour. Technol. 102*: 1537–1542.

109 Franceschini, S., Van Beek, H.L., Pennetta, A. et al. (**2012**). *J. Biol. Chem. 287*: 22626–22634.

110 Orru, R., Dudek, H.M., Martinoli, C. et al. (**2011**). *J. Biol. Chem. 286*: 29284–29291.

111 Yachnin, B.J., Sprules, T., McEvoy, M.B. et al. (**2012**). *J. Am. Chem. Soc. 134*: 7788–7795.

112 Grogan, G. (**2012**). *Annu. Rep. Prog. Chem. Sect. B Org. Chem.* 108: 202–227.

113 Baek, A.-H., Jeon, E.-Y., Lee, S.-M., and Park, J.-B. (**2015**). *Biotechnol. Bioeng.* 112: 889–895.

114 Kim, S.-U., Kim, K.-R., Kim, J.-W. et al. (**2015**). *J. Agric. Food Chem. 63*: 2773–2781.

115 Lee, W.-H., Park, E.-H., and Kim, M.-D. (**2014**). *J. Micrbiol. Biotechnol. 24*: 1685–1689.

116 Yachnin, B.J., McEvoy, M.B., Maccuish, R.J.D. et al. (**2014**). *ACS Chem. Biol. 9*: 2843–2851.

117 Reignier, T., de Berardinis, V., Petit, J.L. et al. (**2014**). *Chem. Commun. 50*: 7793–7796.

118 Reetz, M.T. (**2009**). *J. Org. Chem. 74*: 5767–5778.

119 Hilker, I., Wohlgemuth, R., Alphand, V., and Furstoss, R. (**2005**). *Biotechnol. Bioeng. 92*: 702–710.

120 Halliwell, B. and Gutteridge, J.M.C. (**1999**). *Free Radicals in Biology and Medicine*. New York: Oxford University Press.

121 Dringen, R., Pawlowski, P.G., and Hirrlinger, J. (**2005**). *J. Neurosci. Res. 79*: 157–165.

122 Nicotra, A., Pierucci, F., Parvez, H., and Senatori, O. (**2004**). *Neurotoxicology 25*: 155–165.

123 Kuhn, H. and Borchert, A. (**2002**). *Free Radic. Biol. Med. 33*: 154–172.

124 Gandhi, H., O'Reilly, K., Gupta, M.K. et al. (**2017**). *RSC Adv. 7*: 19506–19556.

125 Dembitsky, V.M., Glorizova, T.A., and Poroikov, V.V. (**2007**). *Mini Rev. Med. Chem.* 571–589.

126 Dembitsky, V.M. (**2008**). *Eur. J. Med. Chem. 43*: 223–251.

127 Dufour, C. and Loonis, M. (**2005**). *Chem. Phys. Lipids 138*: 60–68.

128 Dussault, P.H., Anderson, T.A., Hayden, M.R. et al. (**1996**). *Tetrahedron 52*: 12381–12398.

129 Coffa, G., Imber, A.N., Maguire, B.C. et al. (**2005**). *J. Biol. Chem. 280*: 38756–38766.

130 Gao, B., Boeglin, W.E., and Brash, A.R. (**2010**). *Biochim. Biophys. Acta, Mol. Cell Biol. Lipids 1801*: 58–63.

131 Zheng, Y. and Brash, A.R. (**2010**). *J. Biol. Chem. 285*: 39876–39887.

132 Boeglin, W.E., Itoh, A., Zheng, Y. et al. (**2008**). *Lipids 43*: 979–987.

133 Baba, N., Yoneda, K., Tahara, S. et al. (**1990**). *J. Chem. Soc. Chem. Commun.* 1281–1282.

134 Hamberg, M., Su, C., and Oliw, E. (**1998**). *J. Biol. Chem. 273*: 13080–13088.

135 Andreou, A., Göbel, C., Hamberg, M., and Feussner, I. (**2010**). *J. Biol. Chem. 285*: 14178–14186.

136 Hoch, U., Adam, W., Fell, R. et al. (**1997**). *J. Mol. Catal. A Chem. 117*: 321–328.

137 Van de Velde, F., van Rantwijk, F., and Sheldon, R.A. (**2001**). *Trends Biotechnol. 19*: 73–80.

138 Haring, D., Herderich, M., Schüler, E. et al. (**1997**). *Tetrahedron Asymm. 8*: 853–856.

139 Haring, D., Schüler, E., and Schreier, P. (**1998**). *J. Mol. Catal. B: Enzym. 5*: 339–342.

140 Liese, A., Seelbach, K., and Wandrey, C. (**2006**). *Industrial Biotransformations*. New York: Wiley-VCH.

141 Patel, R.N. (**2006**). *Curr. Opin. Drug Discov. Devel. 9*: 741–764.

142 Genov, D.G. and Ager, D.J. (**2004**). *Angew. Chem. Int. Ed. 43*: 2816–2819.

143 Matsuda, T., Yamanaka, R., and Nakamura, K. (**2009**). *Tetrahedr. Asymm. 20*: 513–557.

144 Cheng, C. and Tsai, H.-R. (**2008**). *J. Chem. Technol. Biotechnol. 83*: 1479–1485.

145 Kratzer, R., Pukl, M., Egger, S. et al. (**2011**). *Biotechnol. Bioeng. 108*: 797–803.

146 Wang, P., Cai, J.-B., Ouyang, Q. et al. (**2011**). *Biotechnol. Products Proc. Eng. 90*: 1897–1904.

147 Hansen, K.B., Chilenski, J.R., Desmond, R., and Devine, P.N. (**2003**). *Tetrahedr. Asymm. 14*: 3581–3587.

148 Pollard, D., Truppo, M., Pollard, J. et al. (**2006**). *Tetrahedr. Asymm. 17*: 554–559.

149 Jin, J.-Z., Li, H., and Zhang, J. (**2010**). *Appl. Biochem. Biotechnol. 162*: 2075–2086.

150 Blay, G., Domingo, L.R., Hernandez-Olmos, V., and Pedro, J.R. (**2008**). *Chem. A Eur. J. 14*: 4725–4730.

151 Oh, K., Shimura, Y., Ishikawa, K. et al. (**2008**). *Bioorg. Med. Chem. 16*: 1090–1095.

152 Bisaha, S.N., Malley, M.F., Pudzianowski, A. et al. (**2005**). *Bioorg. Med. Chem. Lett. 15*: 2749–2751.

153 Lipshutz, B.H., Lower, A., Kucejko, R.J., and Noson, K. (**2006**). *Org. Lett. 8*: 2969–2972.

154 Mangas-Sánchez, J., Busto, E., Gotor-Fernández, V. et al. (**2011**). *J. Org. Chem.* *76*: 2115–2122.
155 Johanson, T., Katz, M., and Gorwa-Grauslund, M.F. (**2005**). *FEMS Yeast Res.* *5*: 513–525.
156 Parachin, N.S., Carlquist, M., and Gorwa-Grauslund, M.F. (**2009**). *Appl. Microbiol. Biotechnol.* *84*: 487–497.
157 Ch^enevert, R., Fortier, G., Rhalid, R.B., and Weuster-Botz, G.D. (**1992**). *Tetrahedron* *15*: 6769–6776.
158 Bhalerao, U.T., Chandraprakash, R., Babu, L., and Fadnavis, N.W. (**1993**). *Synth. Commun.* *23*: 1201–1208.
159 Zhao, Q., Hou, Y., Gong, G.-H. et al. (**2010**). *Appl. Biochem. Biotechnol.* *160*: 2287–2299.
160 Ribeiro, J.B., Ramos, M.C.K.V., de Aquino Neto, F.R.S. et al. (**2005**). *Catal. Commun.* *6*: 131–133.
161 Loncaric, C. and Wulff, W.D. (**2001**). *Org. Lett.* *3*: 3675–3678.
162 Anand, N., Kapoor, M., Koul, S. et al. (**2004**). *Tetrahedr. Asymm.* *15*: 3131–3138.
163 Milagre, C.D.F., Milagre, H.M.S., Moran, P.J.S., and Rodrigues, J.A.R. (**2009**). *J. Mol. Catal. B* *56*: 55–60.
164 Hillier, M.C., Marcoux, J.F., Zhao, D.L. et al. (**2005**). *J. Org. Chem.* *70*: 8385–8394.
165 Lou, W.-Y., Wang, W., Li, R.-F., and Zong, M.-H. (**2009**). *J. Biotechnol.* *143*: 190–197.
166 Matsuda, T., Marukado, R., Mukouyama, M. et al. (**2008**). *Tetrahedr. Asymm.* *19*: 2272–2275.
167 Wolfson, A. and Dlugy, C. (**2009**). *Org. Commun.* *2*: 34–41.
168 Höllrigl, V., Hollmann, F., Kleeb, A.C. et al. (**2008**). *Appl. Microbiol. Biotechnol.* *81*: 263–273.
169 Rajagopalan, A. and Kroutil, W. (**2011**). *Materials Today* *14*: 144–152.
170 Stuemer, R., Hauer, B., Hall, M., and Faber, K. (**2007**). *Curr. Opin. Chem. Biol.* *11*: 203–213.
171 Toogood, H.S., Fryszkowska, A., Hare, V. et al. (**2008**). *Adv. Synth. Catal.* *350*: 2789–2803.
172 Toogood, H.S., Gardiner, J.M., and Scrutton, N.S. (**2010**). *ChemCatChem* *2*: 892–914.
173 Blehert, D.S., Fox, B.G., and Chambliss, G.H. *J. Bacteriol.* *181*: 6254–6263.
174 Komduur, J.A., Leão, A.N., Monastyrska, I. et al. (**2002**). *Curr. Genet.* *41*: 401–406.
175 Quezada, M.A., Carballerira, J.D., and Sinisterra, J.V. (**2009**). *Bioresour. Technol.* *100*: 2018–2025.
176 Hall, M., Stueckler, C., Kroutil, W. et al. (**2007**). *Angew. Chem. Int. Ed.* *46*: 3934–3937.
177 Kawai, Y., Hayashi, M., Inaba, Y. et al. (**1998**). *Tetrahedron Lett.* *39*: 5225–5228.

178 Khor, G.K. and Uzir, M.H. (**2011**). *Yeast 28*: 93–107.

179 Goretti, M., Ponzoni, C., Caselli, E. et al. (**2011**). *Bioresour. Technol. 102*: 3993–3998.

180 Tauber, K., Hall, M., Krotil, W. et al. (**2011**). *Biotechnol. Bioeng. 108*: 1462–1467.

181 Grau, M.M., van der Toorn, J.C., Otten, L.G. et al. (**2009**). *Adv. Synth. Catal. 351*: 3279–3286.

182 Stueckler, C., Reiter, T.C., Baudendistel, N., and Faber, K. (**2010**). *Tetrahedron 66*: 663–667.

183 Hall, M., Stueckler, C., Hauer, B. et al. (**2008**). *Eur. J. Org. Chem.* 1511–1516.

184 Brenna, E., Gatti, F.G., Manfredi, A. et al. (**2011**). *Eur. J. Org. Chem.* 4015–4022.

185 Guan, H., You, S., Wang, X., and Yang, L. (**2010**). *Biocatal. Biotransform. 28*: 185–191.

186 Nie, Y., Xu, Y., Lv, T.F., and Xiao, R. (**2009**). *J. Chem. Technol. Biotechnol. 84*: 468–472.

3

Organic Synthesis with Transferases

3.1 Transamination with Aminotransferases

The use of transaminases, or aminotransferases, as a catalyst for the organic synthesis of chiral amino compounds from the corresponding keto acids, ketones, or aldehydes has been known a very powerful method since last decade. Their immense potential in the industrial applications can be due to not only their concise reaction, excellent enantioselectivity and environmental friendliness but also their easy combination with other existing enzymatic or chemical methods. The catalytic process for the transfer of an amino group from the amino donor to the amino group acceptor in general involves two steps: (i) the release of an amino group from the amino donor (deamination) and (ii) the addition of the amino group to the amino acceptor (amination). For this kind of transamination reaction, a vitamin B_6-based pyridoxal 5′-phosphate (PLP) is required as a cofactor for transaminase to act as an intermediate amine acceptor and electron sink. Since the cofactor releases the amino group and restores its initial state at the end of the transamination, there is no need to carry out additional cofactor regeneration for the enzyme [1, 2]. However, the use of transaminases in organic synthesis still suffers from the problem of unfavorable reversible reaction equilibrium. This problem is especially serious when an amino acid is used as amino donor in the asymmetric synthesis of amines.

To shift the reaction equilibrium to the product side of the transamination reactions catalyzed by transaminase, a conventional strategy that has been commonly used is to provide an excessive amount of the amino donor. Scheme 3.1 illustrates an example in which an excess amount of isopropyl amine was applied to push the transamination of acetophenone to 1-phenylethylamine toward complete conversion and an enantiomeric excess of >99% [3]. In addition to the method of

Enzyme-Based Organic Synthesis, First Edition. Cheanyeh Cheng.

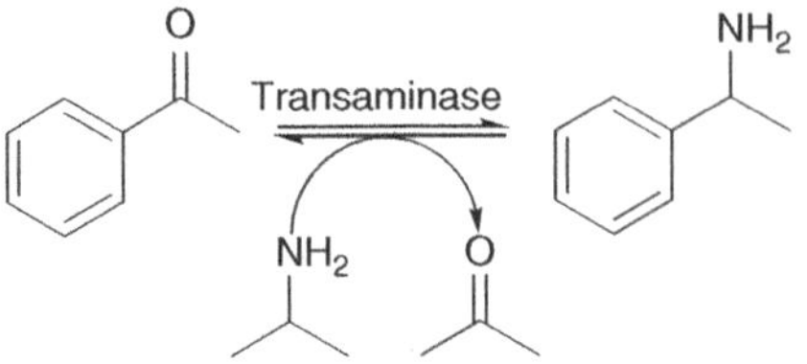

Scheme 3.1 The transamination between acetophenone and excessive isopropyl amine with transaminase. *Source:* Truppo et al. [3].

using excessive cosubstrate for shifting the equilibrium toward the product side, other methods such as removal of product or coproduct by vaporization, extraction, autodegrading smart coproducts, and biochemical cascade reactions were also utilized in many cases [1].

Another smart synthesis strategy is the combination of transaminase with other biocatalysts to form enzymatic cascade reactions that have been used to efficiently produce unnatural amino acids, optically pure amines, imines, secondary amines, and amides, or for the amination on the –OH or $-CH_3$ functional group, or for the amines with two stereocenters. The enzymatic cascade reactions in organic synthesis give the advantages of shortening reaction routes, avoiding unstable or toxic intermediate, increasing the atom efficiency, and reducing the amount of wastes [1]. Most importantly, the enzymatic cascade reactions involving transaminases show prominent potential for many industrial applications [2].

There have been many examples to demonstrate the use of ω-transaminase (ω-TA) for producing unnatural amino acids such as L-*tert*-leucine and L-phosphinothricine [4]. One of the many examples employing coupled enzyme reactions for synthesizing the unnatural amino acids was the production of the unnatural L-homoalanine from L-threonine by combining a threonine deaminase (TD) with the ω-TA and using an one-pot reaction process (Scheme 3.2) that resulted a 91% conversion [5]. The source of TD was from a cloned and overexpressed *Escherichia coli* and the (*S*)-specific ω-TA was from *Paracoccus denitrificans.* The formation of *N*-benzylacetamide from benzaldehyde through an efficient one-pot biocatalytic amine transaminase/acyl transferase cascade using methyl acetate as acyl donor is shown in Scheme 3.3 [6].

The conversion of the amide product in this enzymatic cascade reaction was up to 97% with an acyl transferase (AcT) from *Mycobacterium smegmatis* and an amine transaminase from *Silicibacter pomeroyi.*

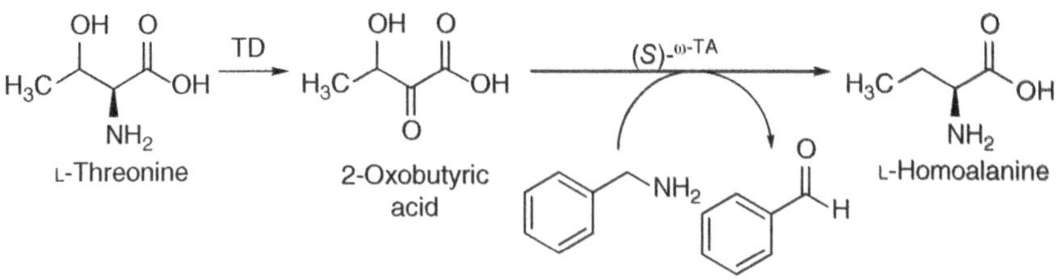

Scheme 3.2 Enzyme cascade reactions for the conversion of L-threonine to L-homoalanine using a one-pot reaction process. *Source:* Modified from Park et al. [5].

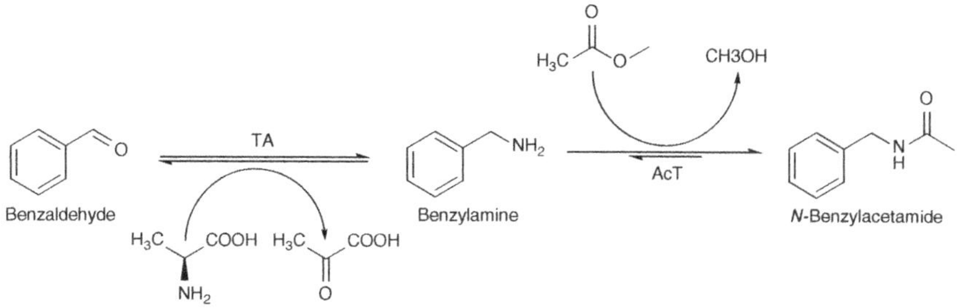

Scheme 3.3 The formation of *N*-benzylacetamide from benzaldehyde through an efficient one-pot biocatalytic amine transaminase/acyl transferase cascade. *Source:* Based on Land et al. [6].

Instead of compounds with one stereocenter, chiral amino alcohols with two stereocenters can be synthesized by the enzymatic cascade reactions without using protection step for the first stereocenter before the production of the second stereocenter. Scheme 3.4 shows an example of the enzymatic cascade route toward all four diastereomers of 4-amino-1-phenylpentane-2-ol started from a racemic 1,3-hydroxy ketone [7].

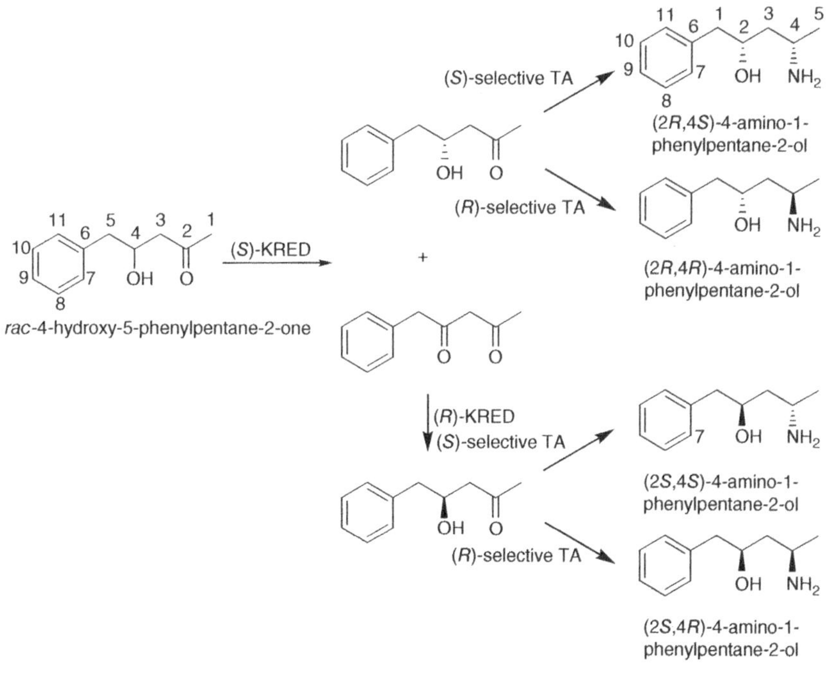

Scheme 3.4 The enzymatic cascade route toward all four diastereomers of 4-amino-1-phenylpentane-2-ol from a compound with two stereocenters. *Source:* Modified from Kohls et al. [7].

The first synthesis step of these four 1,3-amino alcohol diastereomers was through the kinetic resolution for a racemic 1,3-hydroxy ketone by applying using an (*S*)-selective keto reductase (KRED) to regioselectively provide the optically pure (*R*)-hydroxyl ketone (86% *e.e.*) and the corresponding diketone. Further transamination of the (*R*)-hydroxyl ketone using either an (*R*)- or an (*S*)-selective TA yields the (2*R*,4*R*)- and (2*R*,4*S*)-1,3-amino alcohol diastereomers.

The diketone was then used to prepare the remaining two diastereomers, (2*S*,4*S*)- and (2*S*,4*R*)-1,3- amino alcohol, by first converting the diketone to (*S*)-1,3-hydroxy ketone and subsequently using either an (*R*)- or an (*S*)-selective TA to stereoselective amination of the (*S*)-1,3-hydroxy ketone to (2*S*,4*R*)-1,3-amino alcohol and (2*S*,4*S*)-1,3-amino alcohol, respectively.

The industrial applications of enzymatic cascade reactions can be demonstrated by the *in vitro* biosynthesis of the nylon-6 monomer 6-aminohexanoic acid (hydrolyzed ε-caprolactam) from cyclohexanol as illustrated by Scheme 3.5 [8]. In addition to transaminase, this enzymatic cascade route additionally involves a Baeyer–Villiger monooxygenase (BVMO) and an esterase that all these biocatalysts can function independently in a one-pot synthesis under mild conditions. Two cofactor self-sufficient cascade modules are involved in this route. One is for the production of ε-caprolactone from cyclohexanol, the other is the subsequent production of 6-aminohexanoic acid. This biosynthesis route thus shows advantages over chemical approaches that require sophisticated transition-metal catalysts and high temperature and pressure.

Generally, the asymmetric amination of prochiral ketones with an amine donor employing ω-transaminases (ω-TAs) is performed in aqueous medium. Since most

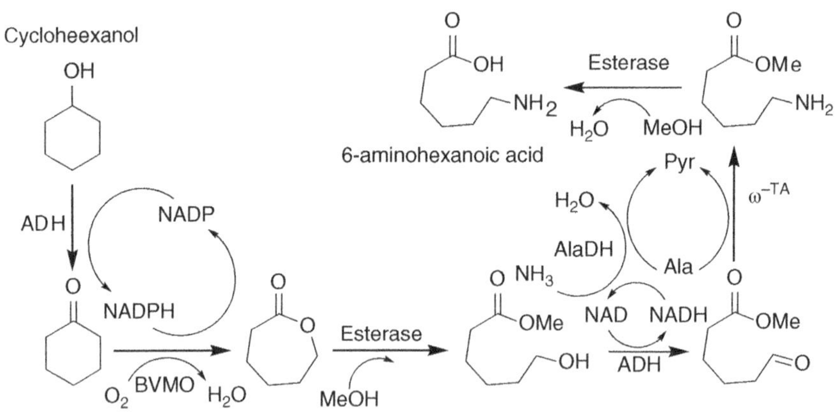

Scheme 3.5 The oxidation/transamination cascade reactions for the biosynthesis of 6-aminohexanoic acid from cyclohexanol. *Source:* Based on Sattler et al. [8].

ketones are lipophilic with only moderately soluble in aqueous solution, the transformations involving ω-TAs require the addition of an organic cosolvent, and enzyme engineering might be necessary to let it stand the harsher reaction conditions. In addition, the amination of prochiral ketones with ω-TAs is commonly employed at pH < 10; the product amines are generally protonated necessitating the basification of the reaction mixture and the extraction with organic solvents. As a result, it would be desirable to employ ω-TAs for the transamination exclusively in organic solvents. Mutti et al. employed nine different lyophilized crude cell-free extracts of (*R*)-selective and (*S*)-selective ω-TAs for the asymmetric amination of ketones in organic solvents with 2-propylamine as amine donor. The best enzyme activity for the transamination was found for methyl *tert*-butyl ether (MTBE) at a water activity of 0.6 that allowed excellent stereoselectivity of optically pure amines (*e.e.* > 99%) and a conversion rate > 99% [9].

The use of organic solvent MTBE has also been applied for the production of optically pure 1,2-amino-alcohols such as valinol from corresponding prochiral hydroxyl ketone using ω-TAs [9]. Chiral 1,2-amino-alcohols are common building blocks embedded in many synthetic and naturally occurring molecules having biological activity, and valinol is a typical example of the versatile vicinal amino alcohols. Thus, the reductive amination of isopropyl methyl alcohol ketone can be performed in MTBE using 2-propyl amine as amino donor to yield either (*R*)-valinol or (*S*)-valinol by the choice of (*R*)- and (*S*)-selective ω-TAs. The use of (*R*)-selective ω-TA purified from *Bacillus megaterium* afforded the (*R*)-valinol with an ideal optical purity (>99% *e.e.*), although a low conversion rate of 15%. The (*S*)-selective ω-TA originating from *Arthrobacter* sp. can produce the (S)-valinol with a much better conversion rate (95%) and a perfect stereoselectivity (>99% *e.e.*) [10].

A pharmaceutical industrial application of the ω-transaminase was the synthesis of antidiabetic compound sitagliptin. The synthesis protocol was the use of a stable (*R*)-selective amine transaminase and DMSO/water system to perform the transamination between prositagliptin ketone and isopropyl amine that gives an excellent stereoselectivity (99.95% *e.e.*) [11]. Later, the synthesis protocol was modified by immobilizing the transaminase on polymer-based resins, and the immobilized enzyme activity of the transamination was evaluated in neat organic solvent isopropyl acetate that results 91% sitagliptin production yield and >99% enantioselectivity (Scheme 3.6) [12].

The immobilized enzyme enables the use of flow reaction system to give high throughput, clean production, high enzyme stability, and excellent mass recovery. *E. coli* cells containing the overexpressed (*R*)-selective ω-TA from *Arthrobacter* and the cofactor pyridoxal 5′-phosphate (PLP) were immobilized on methacrylate polymeric resin beads that were used for continuous flow applications to produce

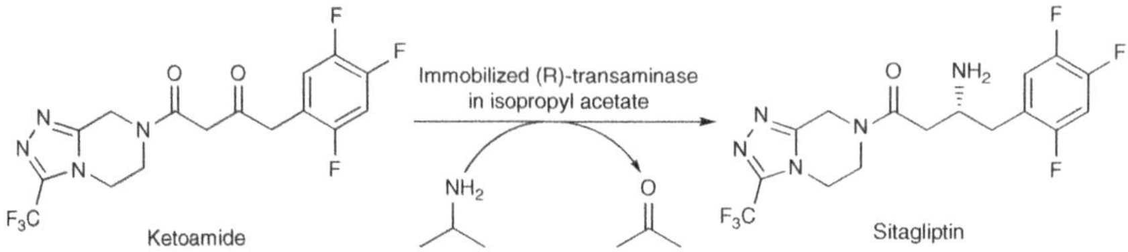

Scheme 3.6 Synthesis of sitagliptin from prositagliptin ketone using immobilized transaminase in organic solvent. *Source:* Truppo et al. [12].

chiral amines continuously by asymmetric transamination of ketones using a packed-bed reactor in organic solvent methy *tert*-butyl ether (MTBE). The use of organic solvent helps in the suppression of PLP leaching from the cells. Non-natural α-alkoxy- and α-aryl acetones were transformed under flow conditions using isopropyl amine as the amine donor with excellent enantioselectivity (>99% *e.e.*) [13].

3.2 Glycosyl-transfer with glycosyltransferase

The glycosylation reactions play a central role in the synthesis of well-defined carbohydrates and glycoconjugates and in the understanding of their roles and structure–function relationships in a variety of biological areas such as infections, signal transduction, cell–cell interactions, host–pathogen interactions, inflammation, immune recognition, targeting proteins to their correct destination, tumor propagation, and metastasis. The high information density coded in the sequence and linkage of carbohydrates on the molecular scale has led to an increased interest in the generation of new pharmacological agents as well [14]. Therefore, glycosylation is considered to be an important method for structural modification of compounds with useful biological activities in the pharmaceutical industries. Lipophilic compounds can be converted to hydrophilic ones through glycosylation to improve their pharmacokinetic properties. Sometimes, the attachment of a sugar moiety to a drug molecule can change its pharmacodynamics properties or obtain novel and more effective drug delivery systems (prodrugs). The applications of enzymes in sugar chemistry are indeed simple and the modification of carbohydrates by enzymes is also one of the most intensely exploited areas of enzyme applications. Especially, in the glycosylation of complex biologically active substances, enzymatic glycosylation methods are particular useful in comparison with chemical methods where generally harsh conditions or use of toxic (heavy metals) catalysts are undesirable. The enzymatic glycosylation is sometimes also superior to the synthetic chemistry for the synthesis of food additives [15].

The ubiquitous glycosyltransferases (GTs) are responsible for the synthesis of the diverse and complex array of oligosaccharides and glycoconjugates found in nature. The chemical diversity and complexity of glycoconjugates, reflecting the various chemical moieties, epimer at each chiral center, anomeric configuration, linkage position, and branching, require that the enzymes that catalyze their synthesis, degradation, and modification need to be highly specific [16]. The synthesis of most cell-surface glycoforms in mammalian systems is performed by Leloir-type glycosyltransferases. They usually show environmental conditions sensitive and often demand special buffers or detergents for solubilization [15, 17, 18]. A large number of eukaryotic Leloir glycosyltransferases have been cloned to date [14, 19, 20] to give highly regio- and stereospecific with respect to glycosidic linkage formation and provide products in high yield. In addition, these enzymes exhibit substrate specificity that transfers a given carbohydrate from the sugar nucleotide donor substrate to a specific hydroxyl group of the acceptor sugar.

The worldwide example of glycosylation is the β(1→4)-galactosylation by β-1,4-galactosyltransferase (β-1,4-GalT) as shown in Scheme 3.7 where the feedback product inhibition by the nucleoside diphosphates (NDP) has been solved by using phosphatase into the reaction to breakdown the NDP product [14, 15, 18]. The other problem associated with this glycosylation is the sugar nucleotide expense that can be a burden for large-scale production.

However, this problem was also solved by the *in situ* UDP-Gal regeneration from inexpensive starting sugar through multiple enzyme system (Scheme 3.8) [21].

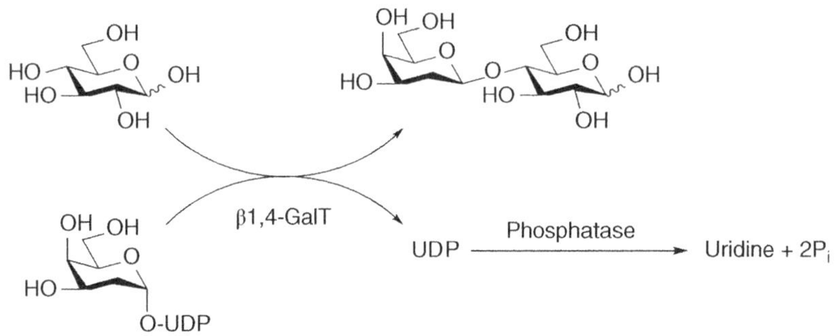

Scheme 3.7 Galactosyltransferase catalyzed glycosylation with UDP-2-d-Gal as donor. *Source:* Based on Wohlgemuth [14]; Křen and Thiem [15]; Koeller and Wong [18].

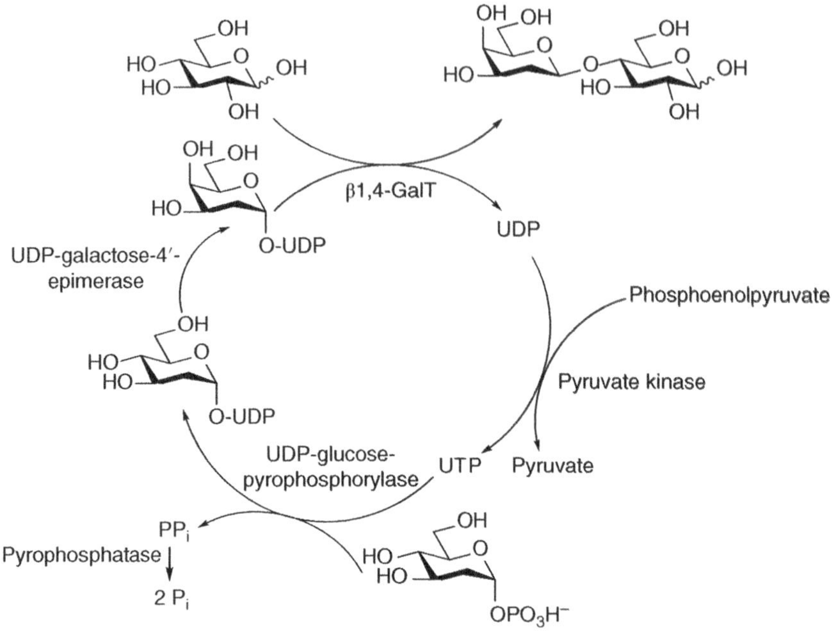

Scheme 3.8 Method for avoiding product inhibition in GalT-catalyzed glycosylation by *in situ* regenerating and recycling of sugar nucleotides. *Source:* Modified from Wong et al. [21].

Some natural products of nonsugar substances such as complex glycosides of ergot alkaloids required for immunological studies can also be prepared by glycosylation using GalTs. For example, bovine β-1,4-GalT was used for catalyzing UDP-Gal and elymoclavine 17-*O*-β-D-glucopyranoside (R = OH) or elymoclavine 17-*O*-(2-acetamido-2-deoxy-β-D-glucopyranoside) (R = NHAc) to produce the corresponding lactose and lactosamine derivatives (Scheme 3.9) [15, 22, 23]. In this reaction scheme, UDP-Gal was generated *in situ* from UDP-Glc by the use of UDP-Gal 4′-epimerase originated from yeast. It is interesting to note that glucose in the form of UDP-Glc can be concomitantly transferred to give in parallel β-1,4-D-glucopyranosyl elymoclavine 17-*O*-(2-acetamido-2-deoxy-β-D-glucopyranoside) by β-1,4-GalT [15, 22]. Following this protocol, other natural glycosides such as the sweetener stevioside and its congener steviolbioside can be transformed with good degree of conversion to their monogalactosyl derivatives with absolute regioselectivity and site-selectivity [23].

There are a group of non-Leloir glycosyltransferases such as cyclodextrin glucosyltransferase (CGTase). CGTases are cultivated and produced from

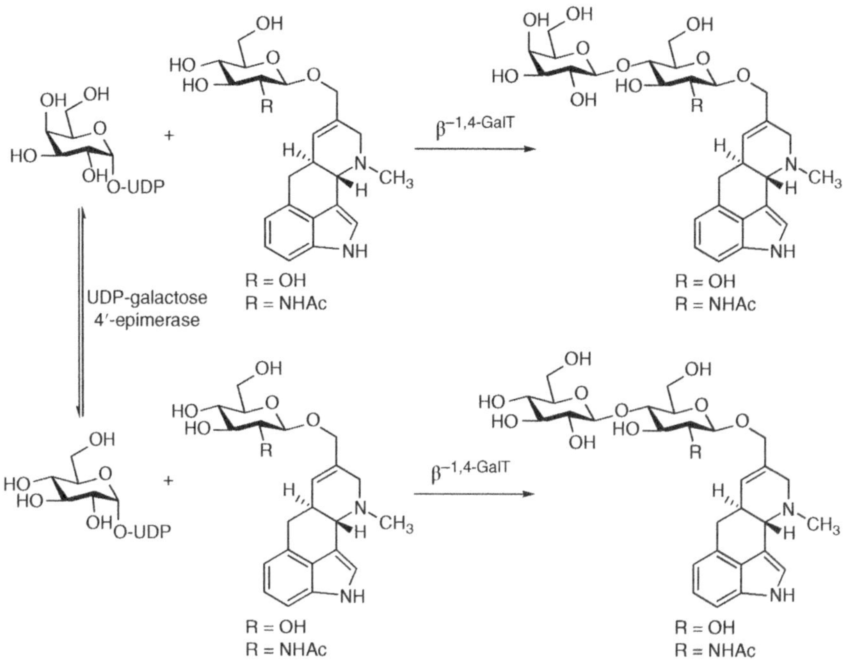

Scheme 3.9 β-1,4-GalT catalyzed galactosylation of natural glycosides and concomitant transfer of glucose. *Source:* Based on Křen and Thiem [15]; Křen [22]; Riva [23].

microorganisms [15, 24]. They catalyze four different kinds of reactions: cyclization, disproportionation, coupling, and hydrolysis so that cyclodextrins (CDs) can be formed from amylose or starch through the intramolecular circularization reaction; a linear malto-oligosaccharide is cleaved and one part is transferred to an acceptor sugar molecule in the disproportionation reaction; CD ring is opened and the resulting linear malto-oligosaccharide is transferred to a sugar molecule; and the glycosidic linkages in starch can be hydrolyzed to form oligosaccharides [25]. Since CDs of CD_6, CD_7, and CD_8 (α-CD, β-CD, and γ-CD) have been extensively used in the food, cosmetic, and pharmaceutical industries, the synthesis of CD by CGTases from starch is important. However, the formation of CD by native CGTases has the major advantage of lacking product specificity that they produce a mixture of CD_6, CD_7, CD_8, and larger-ring CD ($\geq CD_9$) and make the separation difficult. To solve the product specificity problem, CGTases from *Paenibacillus* sp. A11 and *Bacillus macerans* was crosslinked imprinted, thus the size of the CD products formed was shifted toward CD_8 and $\geq CD_9$, and the overall CD yield was increased. The crosslinked imprinted cyclodextrin glycosyltransferases also showed better stability in organic solvents and can be recycled several times [24].

Glycogen is a branched polysaccharide using glucose as monomer that contains a series of α-1,4-glycosidic linkages with branch points about every 10–13 glucose residues through the formation of α-1,6-glycosidic linkages. The synthesis of glycogen is as of other biological polymers a two-step process: the initiation stage and the elongation stage. The initiation stage of glycogen synthesis is catalyzed by glycogenin in an autocatalytic manner. The elongation stage is catalyzed by glycogen synthase in synergy with the branching enzyme. Glycogenin is a member of the glycosyltransferase family 8 (GT-8) that is a metal (Mn^{2+}) dependent retaining enzyme responsible for the synthesizing 6-10 α-1,4-linked glucose residues using UDP-Glc as the substrate donor during glycogen synthesis [26, 27]. The two-step reaction scheme is illustrated by Scheme 3.10 [27, 28]. In the first reaction step, the role for the aspartate residue at position 159 (Asp-159) of glycogenin serves in binding and activating the acceptor oligosaccharide chain that is covalently attached to the tyrosine residue at position 194 (Tyr-194), and the aspartate residue at position 162 (Asp-162) serves the role of chemical reaction of glucosyltransferase [27–29].

Protein glycosylation is one of the most common posttranslational modifications of proteins in eukaryotes that affect a wide range of protein functions, from folding and secretion to biomolecular recognition (targeting) and serum half-life (stability), and many other intercellular communication processes [30, 31].

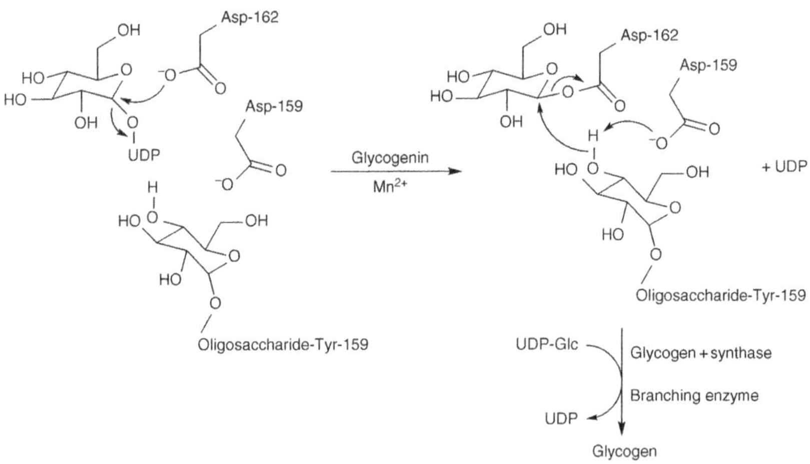

Scheme 3.10 Two-step synthesis of glycogen with glycogenin functioning as an autocatalytic initiator. *Source:* Based on Hurley et al. [27]; Smythe and Cohen [28].

However, owing to glycoprotein microheterogeneity [32], the specific covalently bound oligosaccharides to protein is difficult because protein glycosylation is not under direct genetic control. Therefore, glycoproteins are often produced as a mixture of glycoforms to make the isolation of individual glycoforms difficult and drive the development of new synthetic methods for producing well-defined oligosaccharides bound glycoproteins. The advancement in the field has provided some strategies to resolve this problem by synthesizing the glycoproteins *in vitro* that include the remodeling of recombinant glycoproteins with glycosidases and glycosyltransferases, the ligation of synthetic glycopeptides by enzymatic and chemical methods, the intein-mediated coupling of glycopeptides to larger proteins expressed as intein-fusion protein, the ligation of glycopeptides to larger proteins containing N-terminal cysteine expressed as TEV protease cleavable fusion proteins, *in vitro* translation, and the pathway reengineering in yeast systems to produce human-type *N*-linked glycoforms [18, 31, 33–35].

Besides *in vitro* synthesis of glycoproteins, *in vivo* synthesis method using suppressor tRNA has been described for the recombinant production of neoglycoproteins and glycoproteins [35]. The strategy to produce unique glycoforms in *E. coli* has been reported by evolving an orthogonal synthetase-tRNA pair that genetically encodes a glycosylated amino acid in responds to the amber stop codon (TAG). Further, a naturally occurring homogeneous glycoprotein can be produced in *E. coli* via the direct incorporation of the core glycosyl amino acids *N*-acetylglucosamine-β-serine and *N*-acetylgalactosamine-α-threonine [30, 31, 36, 37]. The sugar chains of these glycoproteins can be further elongated *in vitro* using glycosyltransferase.

Glycolipids are responsible for the organism's toxic and immunological properties. Like glycoproteins, glycolipids reside in cell membranes with their carbohydrate segments extending into the fluid surrounding the cells. In this location they function as receptors that are essential for recognizing chemical messengers, other cells, pathogens, and drugs [38]. The synthesis of glycolipid oligosaccharides is performed in the Golgi apparatus by a complex membrane-bound glycosyltransferases together with sugar nucleotide transporters and ceramide-bound accepter [39, 40]. The details are not discussed here for their fewer industrial applications.

3.3 Phosphorylation with Kinases

The fundamental roles played by kinases in cells are in different signal transduction pathways and in many cellular processes such as cell metabolism, division,

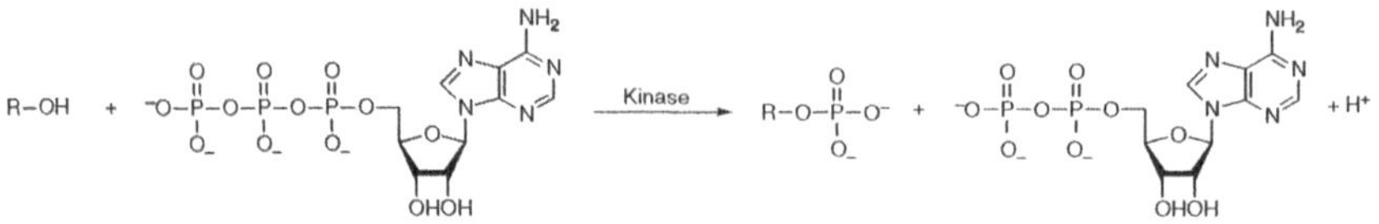

Scheme 3.11 The transfer of γ-phosphoryl group from ATP to hydroxyl group containing substrates by kinases. *Source:* Chapman and Wong [41].

differentiation, survival and apoptosis, as well as development and oncogenesis. They effectively catalyzed the transfer of the γ-phosphoryl group (PO_3^-) of ATP to a broad number of hydroxyl group containing substrates with the release of a proton (Scheme 3.11) [41]. Nowadays, the structural study of many different kinase classes indicates that all kinases present a similar nucleotide-binding sequence [42].

Protein crystallography in the solid state and ^{19}F NMR in solution have elucidated the formation of a trigonal bipyramidal transition state with phosphorus fully bonded to three equatorial oxygens and partially bonded to axial donor's oxygen and acceptor's oxygen [43]. The hydrolysis of ATP to form ADP and P_i is highly exergonic that is used to drive the synthesis of information-rich macromolecules, the transport of solute across membranes, and motion produced by muscle contraction.

Although most of the studies concerning about the phosphoryl group transfer using kinases are *in vivo* processes, there still some *in vitro* applications in organic synthesis. For instance, pyridoxal-5′-phosphate (PLP), a cofactor for enzymes catalyzing transaminations, decarboxylations, racemizations, and β- and γ-carbon elimination/replacements, and are mostly associated with amino acid metabolism, can be synthesized *in vitro* through the phosphorylation of toxic antimetabolite L-4-hydroxythreonine (4HT) into 4-phosphohydroxy-L-threonine (4PHT) by members of a novel kinase family DUF1537 (Scheme 3.12) [44].

Another example of the application of phosphorylation is about the bioinspired peptides containing 3 ser-ser-asp repeat motif (R-SSD3) that was designed by mimicking the highly phosphorylated protein, dentin phosphophoryn (DPP), found in dentin and alveolar bone. Then, the R-SSD3 peptides are sequentially

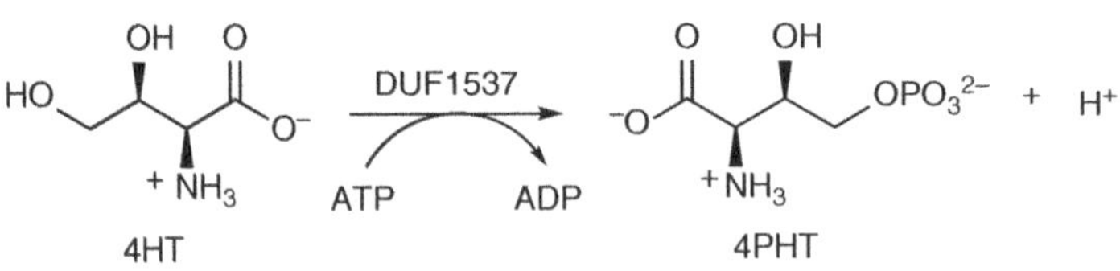

Scheme 3.12 Phosphorylation of 4HT to 4PHT by kinase DUF1537 for enzyme cofactor PLP biosynthesis. *Source:* Modified from Thiaville et al. [44].

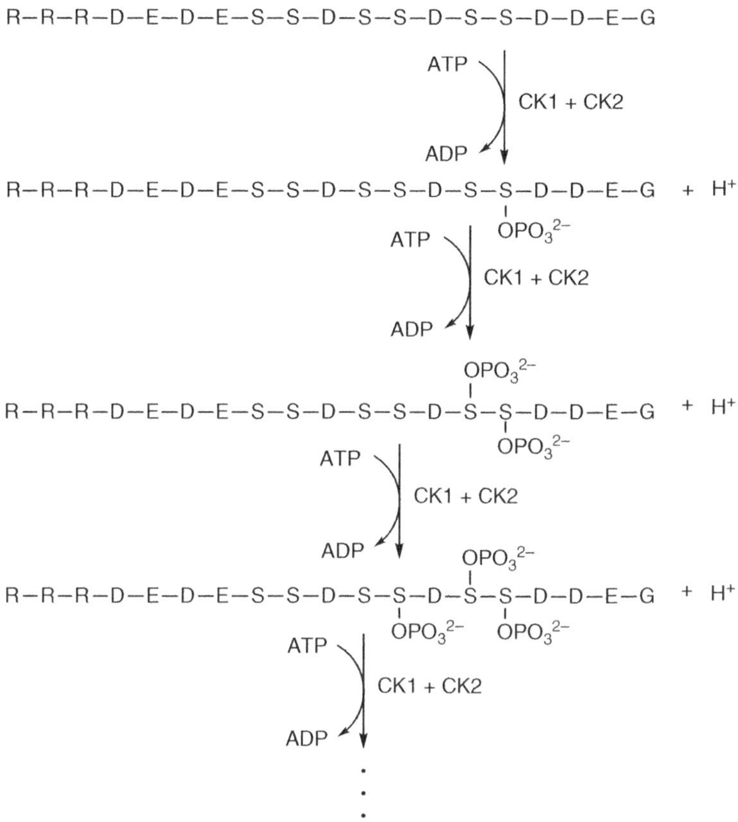

Scheme 3.13 Sequential phosphorylation of peptide R-SSD3 with casein kinases.

phosphorylated at multiple serine sites in (DSS)*n* sequence by casein kinases (CK1 and CK2) as shown in Scheme 3.13. These phosphorylated peptides were further used to induce biomimetic calcium phosphate mineralization of collagen fibrils that successfully produce biomimetic composite nanofibrils with integrated organic and inorganic phases [45].

3.4 Acetyl Group Transfer with Acetyltransferase

Acetyltransferases (ATs) catalyze the transfer of an acetyl group from acetyl coenzyme A (Acetyl-CoA) to a variety of small molecules including lipids, amino acids, drugs, and sugars as well as large molecules such as proteins. The acetyl group donor (acetyl-CoA) is generated in mitochondria from carbohydrate or amino acid catabolism by acetyl-CoA synthetase utilizing acetate, CoA, and ATP [46, 47]. The small molecule melatonin synthesis using AT was a typical example of application in organic synthesis concerning the acetyl group transfer. Melatonin has been found in almost all organisms from photosynthetic bacteria to human beings. Melatonin is a potent direct free radical scavenger and antioxidant that strongly protects unicellular organisms, plants, and animals from oxidative insults. Thus, it is important to human health, agriculture applications, and in food and beverage industries [48]. Based on the classic melatonin synthetic pathway in animals, serotonin is first acetylated to form *N*-acetylserotonin by arylalkylamine *N*-acetyltransferase (AANAT) or arylamine *N*-acetyltransferase (SNAT) and *N*-acetylserotonin is, subsequently, methylated to form melatonin by *N*-acetylserotonin *O*-methyltransferase (ASMT) (Scheme 3.14) [48–50]. However, evidences strongly support that the two steps illustrated by Scheme 3.14 could be reversed and may be more important in certain organisms and under certain conditions [48].

Physostigmine is a pyrroloindole alkaloid and is a parasympathomimetic drug that reversibly inhibits acetylcholinesterase. Physostigmine has been used clinically to treat a wide variety of disorders including Alzheimer's disease, glaucoma, delayed gastric emptying, and orthostatic hypertension. In addition, physostigmine can cross the blood–brain barrier, and this is used to counteract the effects

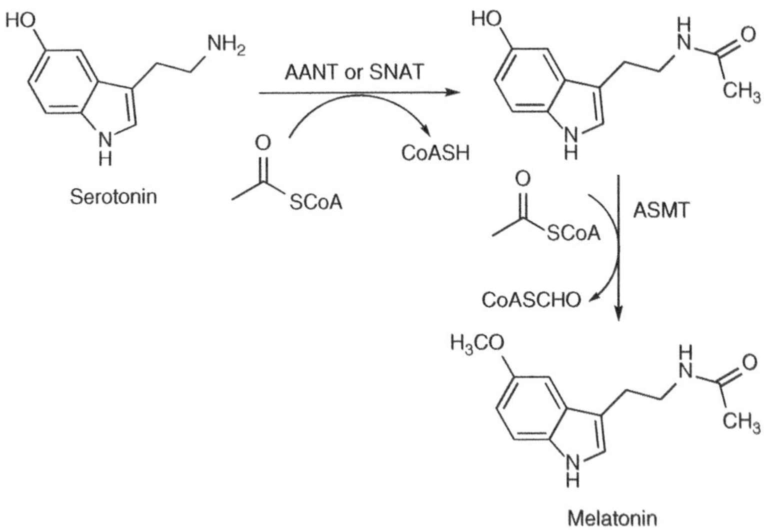

Scheme 3.14 Synthetic pathway of melatonin from serotonin by two kinds of acetyltransferases. *Source:* Based on Tan et al. [48]; Weissbach and Redfield [49]; Axelrod and Weissbach [50].

on the central nervous system of overdoses of atropine, scopolamine, and other anticholinergic drugs [51–53]. Just like melatonin, physostigmine is derived from tryptophan; however, the biosynthetic pathway for physostigmine involves more steps that are characterized by unusual reaction cascade consisting of highly coordinated methylation and acetylation/deacetylation reactions [51]. The *N*-acetylation of serotonin to produce *N*-acetyl-5-hydroxytryptamine catalyzed by *N*-acetyltransferase is the third reaction step of the reaction cascade that is the same step as the first step in Scheme 3.14 for melatonin synthesis.

The GCN5-related *N*-acetyltransferase (GNAT) superfamily encompasses more than 10 000 related enzymes in all kingdoms of life and human acetyl-CoA: glucosamine-6-phosphate *N*-acetyltransferase 1 (GNA1) belongs to a member of it [54, 55]. In the cytosol, GNA1 catalyzes the transfer of an acetyl group from the acetyl coenzyme A donor substrate to the acceptor substrate glucosamine-6-phosphate (GlcN6P) to form *N*-acetyl-glucosamine-6-phosphate (GlcNAc6P) that is used as an intermediate in the biosynthesis of UDP-GlcNAc. Since *N*-butyryl glucosamine (GlcNBu), an analog of GlcNAc, has been shown the healing properties of bone and articular cartilage in animal's arthritis, enzymatic synthesis of GlcNBu is important for biomedical applications [56–58]. Research results have shown that both acetyl and *n*-butyl groups were transferred to GlcN6P by GNA1 to form corresponding GlcNAc6P and GlcNBu6P, respectively [54]. Therefore, GlcNBu6P can be easily and subsequently converted to GlcNBu by alkaline phosphatase.

Instead of small molecules acetylation, reversible protein acetylation by corresponding acetyltransferase is a ubiquitous means for the rapid control of diverse cellular processes [59, 60]. The acetyl group on acetyl-CoA is transferred to the ε-amino group of lysine residues of the acceptor protein by acetyltransferase. This is just the case for the acetylation of histones by histone acetyltransferases (HATs) that has an important role in transcriptional regulation by remodeling chromatin structure. The acetylation of histones may result in altering interactions between protein–protein and protein–DNA complexes due to the neutralization of positive charged lysine in the amino termini of histone proteins [61, 62]. In higher organisms, aberrant acetylation of lysine residues in histone tails correlates with diseases such as cancers and developmental disorders and may contribute to modulation of cell life span [63, 64]. In mammalian mitochondrial matrix, endoplasmic reticulum lumen, and peroxisomes, carnitine acetyltransferase (CrAT) catalyzes the reversible transfer of acetyl groups between acetyl-CoA and L-carnitine (β-hydroxy-γ-trimethylammonium butyrate). The main function of carnitine is the transfer of long-chain fatty acids to mitochondria for subsequent β-oxidation. CrAT is homologous to other carnitine acyltransferases, particularly, to carnitine palmitoyltransferase 1 (CPT I) that serves the regulation of long-chain fatty acid metabolism. Therefore, the reversibly catalyzed reaction between

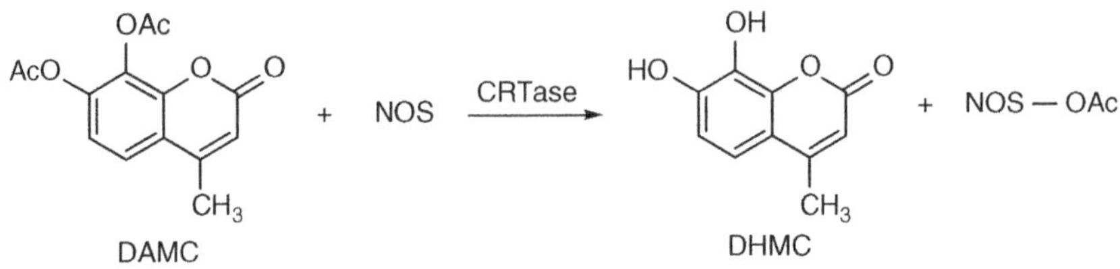

Scheme 3.15 Protein acetylation with CRTase and DAMC without involving acetyl-CoA. *Source:* Arora et al. [71].

acetyl-CoA and carnitine by CrAT makes CrAT a regulator for the cellular pool of CoA that, in turn, plays a role as a carrier of activated acetyl groups in the oxidation of energy metabolism substrates and in the synthesis of fatty acids and lipids. It is also known that the accumulation of fatty acyl-CoAs in heart may induce apoptosis and inflammation and acetyl-carnitine improves cognition in the brain [65–67].

The model acetoxy-coumarins (AC), 7,8-diacetoxy-4-methylcoumarin (DAMC), was shown to possess radical scavenger property by interacting with free radical to remove its acetyl group and give the acetyl cation (CH_3CO^+) and the phenoxyl radical [68]. The antioxidant action of DAMC is independent on the formation of parent 7,8-dihydroxy-4-methylcoumarin (DHMC). Calreticulin (CR) catalyzes the transfer of acetyl groups from AC to certain proteins [69, 70]; thus, CR was termed calreticulin transacetylase (CRTase). The enzymatic acetylation of protein by CRTase is unique and characterized as without involving acetyl-CoA. CRTase of rat tracheal smooth muscle cells (TSMC) was characterized the specificity of DAMC for acetylating and activating nitric oxide synthase (NOS) as illustrated by Scheme 3.15 [71]. Since the activated TSMC NOS will enhance NO in airway cells, and NO is believed to ameliorate the exacerbation of airway diseases such as asthma and coronary obstructive pulmonary diseases (COPD), AC may be expected to find therapeutic applications in respiratory diseases [71, 72].

References

1 Guo, F. and Berglund, P. (**2017**). *Green Chem. 19*: 333–360.

2 Fuchs, M., Farnberger, J.E., and Kroutil, W. (**2015**). *Eur. J. Org. Chem.* 6965–6962.

3 Truppo, M.D., Rozzell, J.D., Moore, J.C., and Turner, N.J. (**2009**). *Org. Biomol. Chem. 7*: 395–398.

4 Stewart, J.D. (**2001**). *Current Opin. Chem. Biol. 5*: 120–129.

5 Park, E., Kim, M., and Shin, J.-S. (**2010**). *Adv. Synth. Catal. 352*: 3391–3398.

6 Land, H., Hendil-Forssell, P., Martinelle, M., and Berglund, P. (**2016**). *Catal. Sci. Technol. 6*: 2897–2900.
7 Kohls, H., Anderson, M., Dickerhoff, J. et al. (**2015**). *Adv. Synth. Catal. 357*: 1808–1814.
8 Sattler, J.H., Fuchs, M., Mutti, F.G. et al. (**2014**). *Angew. Chem., Int. Ed. 53*: 14153–14157.
9 Mutti, F.G. and Kroutil, W. (**2012**). *Adv. Synth. Catal. 354*: 3409–3413.
10 Fuchs, C.S., Simon, R.C., Riethorst, W. et al. (**2014**). *Biorg. Med. Chem. 22*: 5558–5562.
11 Savile, C.K., Janey, J.M., Mundorff, E.C. et al. (**2010**). *Science* 329: 305–309.
12 Truppo, M.D., Strotman, H., and Hughes, G. (**2012**). *ChemCatChem 4*: 1071–1074.
13 Andrade, L.H., Kroutil, W., and Jamison, T.F. (**2014**). *Org. Lett. 16*: 6092–6095.
14 Wohlgemuth, R. (**2005**). *Chimia 59*: 735–740.
15 Křen, V. and Thiem, J. (**1997**). Chem. *Chem. Soc. Rev. 26*: 463–473.
16 Gloster, T.M. (**2014**). *Curr. Opin. Chem. Biol. 28*: 131–141.
17 Leloir, L.F. (**1971**). *Science* 172: 1299–1303.
18 Koeller, K.M. and Wong, C.-H. (**2000**). *Chem. Rev. 100*: 4465–4493.
19 Tsuji, S. (**1996**). *J. Biochem.* 120: 1–13.
20 Sears, P. and Wong, C.-H. (**1998**). *Cell. Mol. Life Sci. 54*: 223–252.
21 Wong, C.-H., Haynie, S.L., and Whitesides, G.M. (**1982**). *J Org. Chem. 47*: 5416–5418.
22 Křen, V. (**1997**). *Top. Curr. Chem. 186*: 45–64.
23 Riva, S.J. (**2002**). *Mol. Catal. B Enzym.* 19-20: 43–54.
24 Kaulpiboon, J., Pongsawasdi, P., and Zimmermann, W. (**2007**). *FEBS J. 274*: 1001–1010.
25 Van der Veen, B.A., van Alebeek, G.-J.W.M., Uitdehaag, J.C.M. et al. (**2000**). *Eur. J. Biochem. 267*: 658–665.
26 Pitcher, J., Smythe, C., and Cohen, P. (**1988**). *Eur. J. Biochem. 176*: 391–395.
27 Hurley, T.D., Stout, S., Miner, E. et al. (**2005**). *J. Biol. Chem. 280*: 23892–23899.
28 Smythe, C. and Cohen, P. (**1991**). *Eur. J. Biochem. 200*: 625–631.
29 Zeqiraj, E., Tang, X., Hunter, R.W. et al. (**2014**). *Proc. Natl. Acad. Sci. USA 111*: E2831–E2840.
30 Zhang, Z., Gildersleeve, J., Yang, Y.-Y. et al. (**2004**). *Science 303*: 371–373.
31 Wong, C.-H. (**2005**). *J. Org. Chem. 70*: 4219–4225.
32 Rush, R.S., Derby, P.L., Smith, D.M. et al. (**1995**). *Anal. Chem. 67*: 1442–1452.
33 Sears, P. and Wong, C.-H. (**2001**). *Science 291*: 2344–2350.
34 Macmillan, D. and Bertozzi, C.R.A. (**2004**). *Chem. Int, Ed. 43*: 1355–1359.
35 Wang, L. and Schult, P.G. (**2005**). *Angew. Chem. Int. Ed. 44*: 34–66.
36 Liu, H., Wang, L., Brock, A. et al. (**2003**). *J. Am. Chem. Soc. 125*: 1702–1703.
37 Xu, R., Hanson, S.R., Zhang, Z. et al. (**2004**). *J. Am. Chem. Soc. 126*: 15654–15655.

38 McMurry, J.; Castellion, M.E. (**1999**). *Fundamentals of General, Organic, and Biological Chemistry*, 3rd Ed., Prentice Hall, New Jersey: Upper Saddle River, pp. 673.

39 Uliana, A.S., Crespo, P.M., Martina, J.A. et al. (**2006**). *J. Biol. Chem. 281*: 32852–32860.

40 Giraudo, C.G. and Maccioni, H.J.F. (**2003**). *J. Biol. Chem. 278*: 40262–40271.

41 Chapman, E. and Wong, C.-H. (**2002**). *Bioorg. Med. Chem. 10*: 551–555.

42 Renzone, G., Salzano, A.M., Arena, S.D. et al. (**2006**). *J. Proteome Res. 5*: 2019–2024.

43 Jin, Y., Molt, R.W. Jr., and Blackburn, G.M. (**2017**). *Top. Curr. Chem.(Z) 375*: 36–66.

44 Thiaville, J.J., Flood, J., Yurgrl, S. et al. (**2016**). *ACS Chem. Biol. 11*: 2304–2311.

45 Sfeir, C., Fang, P.-A., Jayaraman, T. et al. (**2014**). *Acta Biomater. 10*: 2241–2249.

46 Rivière, L., Moreau, P., Allmann, S. et al. (**2009**). *PNAS 106*: 12694–12699.

47 Kuang, Y., Salem, N., Wang, F. et al. (**2007**). *J. Biochem. Biophys. Methods 70*: 649–655.

48 Tan, D.-X., Hardeland, R., Back, K. et al. (**2016**). *J. Pineal Res. 61*: 27–40.

49 Weissbach, H., Redfield, B.G., and Axelrod, J. (**1960**). *Biochim. Biophys. Acta 43*: 352–353.

50 Axelrod, J. and Weissbach, H. (**1960**). *Science 131*: 1312.

51 Liu, J., Ng, T., Rui, Z. et al. (**2014**). *Angew. Chem. Int. Ed. 53*: 136–139.

52 Witkop, B. (**1998**). *Heterocycles 49*: 9–27.

53 Granacher, R.P. and Baldessarini, R.J. (**1976**). *Clin. Neuropharmacol. 1*: 63–79.

54 Brockhausen, I., Nair, D.G., Chen, M. et al. (**2016**). *Biochem. Cell Biol. 94*: 197–204.

55 Vetting, M.W., de Carvalho, L.P.S., Yu, M. et al. (**2005**). *Arch. Biochem. Biophys. 433*: 212–226.

56 Wang, S.X., Cherian, A., Dumitriu, M. et al. (**2007**). *J. Rheumatol. 34*: 712–720.

57 Brockhausen, I. and Anastassiades, T.P. (**2008**). *Expert Rev. Clin. Immunol. 4*: 173–191.

58 Anastssiades, T., Rees-Milton, K., Xiao, H. et al. (**2013**). *Transl. Res. 162*: 93–101.

59 Walsh, C.T., Garneau-Tsodikova, S., and Gatto, G.J. Jr. (**2005**). *Angew. Chem. Int. Ed. Engl. 44*: 7342–7372.

60 Clark, R.S., Bayir, H., and Jenkins, L.W. (**2005**). *Crit. Care Med. 33*: S407–S409.

61 Kim, Y., Tanner, K.G., and Denu, J.M. (**2000**). *Anal. Biochem. 280*: 380–314.

62 Grunstein, M. (**1997**). *Nature 389*: 349–352.

63 Guarente, L. and Picard, F. (**2005**). *Cell 120*: 473–482.

64 Timmermann, S., Lehrmann, H., Polesskaya, A., and Harel-Bellan, A. (**2001**). Cell Mol. *Life Sci.* 58: 728–736.

65 Jaudzems, K., Kuka, J., Gutsaits, A. et al. (**2009**). *Enzym. Inhib. Med. Chem. 24*: 1269–1275.

66 Longo, N., Frigeni, M., and Pasquali, M. (**1863**). *Biochim. Biophys. Acta* **2016**: 2422–1325.

67 Ramsay, R.R. and Zammit, V.A. (**2004**). *Mol. Aspects Med. 25*: 475–493.

68 Raj, H.G., Parmer, V.S., Jain, S.C. et al. (**1999**). *Bioorg. Med. Chem. 7*: 2091–2094.

69 Raj, H.G., Parmer, V.S., Jain, S.C. et al. (**1999**). *Bioorg. Med. Chem. 7*: 369–373.

70 Raj, H.G., Parmer, V.S., Jain, S.C. et al. (**2000**). *Bioorg. Med. Chem. 8*: 1707–1712.

71 Arora, S., Vohra, P., Kumar, A. et al. (**2008**). *Biol. Pharm. Bull. 31*: 709–713.

72 Hogman, M., Frostell, C.G., Hedenstrom, H., and Hedenstierna, G. (**1993**). *Am. Rev. Respir. Dis. 148*: 1474–1478.

4

Organic Synthesis with Hydrolases

4.1 Hydrolysis of Ester Bond

4.1.1 Ester Hydrolysis with Esterases

Esterases (E.C. 3.1.1.1, carboxyl ester hydrolases) are one of the large versatile enzyme groups of hydrolases. They are interested in synthetic chemistry for their ability to catalyze the cleavage and formation of ester bonds. They have been described widely existing in animals, plants, and microorganisms as intra- or extracellular proteins. They have shown many advantages in organic synthesis such as a wide substrate tolerance, high regio- and stereospecificity, not requiring cofactors, and exceptionally robust catalysts of being able to act in the presence of organic solvents. Due to these reasons they have been included in the catalysts with the highest number of industrial applications and used for the production of optically pure fine chemicals in the areas of food and drinks, textile and leather, paper, and pharmaceuticals [1–3]. The three-dimensional structure of esterases belongs to the α/β-hydrolase superfamily, that is, a definite order of α-helices and β-sheets, and has a conserved catalytic triad composed of Ser-Asp-His and located in a highly conserved GXSXG sequence. The mechanism for ester hydrolysis or formation is composed of four steps: (i) the substrate is bound to Ser residue yielding a tetrahedral intermediate and stabilized by His and Asp residues, (ii) the alcohol is released and an acyl-complex is formed, (iii) the nucleophilic attack of the acyl-complex by either water in hydrolysis or alcohol or ester in (*trans-*)esterification forms again a tetrahedral intermediate, and (iv) the resolution of the intermediate yields the product (an acid or an ester) and free enzyme [2, 3]. The esterases catalyzed hydrolysis is classified in nonselective and selective transformations and is described separately as the following examples.

Wild-type thermophilic bacterium *Bacillus licheniformis* S-86 isolated from soil samples was used to produce organic-solvent-tolerant esterase and two different molecular weight esterases were found to possess ester hydrolysis activity. The

Enzyme-Based Organic Synthesis, First Edition. Cheanyeh Cheng.

substrate used for testing the esterase activity was *p*-nitrophenyl (*p*NP) acetate and the type of solvent was hydroxylic organic-solvents including propan-2-ol, butan-1-ol, and 3-methylbutan-1-ol. It was also observed that Type I esterase was inhibited by phenylmethylsulfonyl fluoride (PMSF) but no PMSF inhibitory effect was shown for Type II esterase [4]. Later, a novel solvent-activated esterase was isolated from a strain of bacterium, *Lysinibacillus sphaericus*, from soil. The solvent-activated esterase only hydrolyzes short chain 4-nitrophenyl (*p*NP) esters such as *p*NP acetate, *p*NP butyrate, *p*NP valerate, and *p*NP caproate in water-miscible organic solvents such as methanol, ethanol, and acetonitrile. The enzyme activity was the highest for *p*NP valerate which was enhanced by metal ions of Ca^{2+} and Mg^{2+} and inhibited by ethylenediaminetetraacetic acid (EDTA) and PMSF [5]. The hydrolysis of ethyl esters of phenylacetic and 2-phenylpropionic acids in phosphate buffer was performed separately by esterase from either lyophilized mycelia of *Aspergillus oryzae* or *Rhizopus oryzae*. The lyophilized mycelium of *A. oryzae* was more quicker and efficient catalyst than *R. oryzae* but both mycelia did not show valuable enantioselectivity for racemic ethyl 2-phenylpropionate. In addition, either biocatalyst showed a higher molecular conversion and starting rate for ethyl phenylacetate of hydrolysis than that with ethyl 2-phenylpropionate [6].

Esterase activity was found from heterotrophic bacteria isolated from neutral copper mine drainages. The esterase activity of the isolated bacteria was investigated by fluorescence-based high-throughput screening (HTS) assay using fluorogenic probes as indicated in Scheme 4.1. The enzymatic hydrolyzed fluorogenic products were monitored with fluorescence intensity at wavelength 460 nm (excitation wavelength at 390 nm) [7].

Pig liver esterase (PLE) is a serine-type esterase with wide substrate tolerant, thus has wide industrial applications. The kinetic parameters of PLE-catalyzed hydrolysis have been determined by fluorescence resonance energy transfer (FRET) based fluorescence probes. Two fluorescent probes, 1-naphthylacetic acid(*N*-metyl-*N*-dansyl-2amino)ethyl ester and 1-naphthylacetic acid(2-dansyloxy) ethyl ester, were designed and synthesized, respectively, and were used for the determination of PLE activity. 1-Naphthylacetic moiety of the fluorescent probes was selected as the fluorescence donor and the dansyl group was selected as the

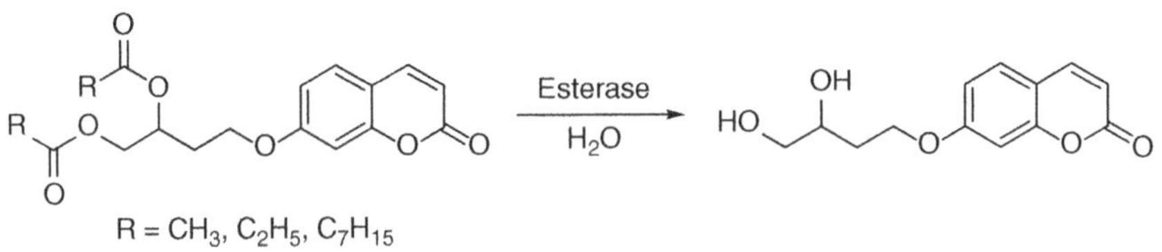

Scheme 4.1 Esterase activity with heterotrophic bacteria isolated from neutral copper mine drainages.

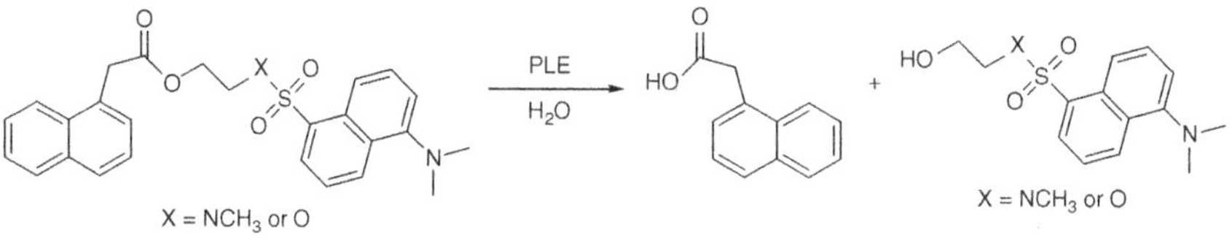

Scheme 4.2 FRET-based PLE catalyzed ester hydrolysis. *Source:* Yi et al. [8].

acceptor. The PLE-catalyzed hydrolysis of the two fluorescent probes is shown in Scheme 4.2 [8]. The real-time determination of PLE activity for ester hydrolysis is successful with the FRET assay and the two specifically designed fluorescent probes.

The enantioselective esterase-catalyzed hydrolysis has long been applied for the preparation or isolation of chiral starting materials such as alcohols, acids, and esters for the synthesis of useful pharmaceuticals and agrochemicals [9]. Continuous studies of the esterases activity for the enantioselective hydrolysis have been reported; for instance, an extracellular esterase produced by *Bacillus sphaericus* preferentially hydrolyzed the *R*-ester of ethyl 2-hydroxyalkanoates. The enantioselectivity was high larger than 100% *e.e.* for ethyl 2-hydroxyhexanoate and ethyl 2-hydroxypentanoate, and moderate for ethyl mandelate [10].

In order to develop new materials for photopolymerization systems or dental applications, a cyclopropanation reaction of allyl methacrylate with ethyl diazoacetate has been carried out to lead to the formation of a mixture of *cis*/*trans* isomers of 2-(2-methyl-acryloyloxymethyl)-cyclopropanecarboxylic ethyl ester in molar ratio 2:1. This cyclopropane-containing monomer was easily polymerized by free-radical initiators without worrying about ring-opening at relative high temperature. The method for the separation of *cis*/*trans*-ester isomers was the stereospecific hydrolysis of *cis*-ester with PLE (Scheme 4.3) that gives an optically active hydrolysis product

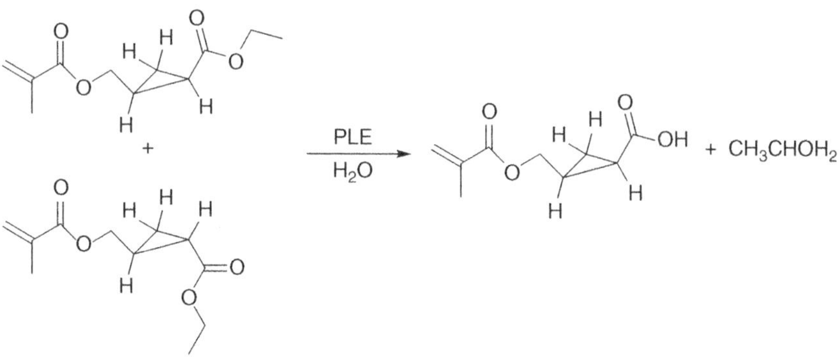

Scheme 4.3 *cis*-2-(2-Methyl-acryloyloxy-methyl)-cyclopropanecarboxylic acid ethyl ester stereoselectivity hydrolyzed by PLE. *Source:* Modified from Vretik and Ritter [11].

2-(2-methyl-acryloyloxy-methyl)-cyclopropanecarboxylic acid [11]. The PLE activity toward *cis*/*trans* isomers was also demonstrated by the stereospecific hydrolysis of *trans* chrysanthemic acid esters. Chrysanthemic acid esters are the essential building blocks for natural insecticide pyrethrin I synthesis. Since pyrethrin I is the ester of (+)-*trans*-chrysanthemic acid and a 4-hydroxy cyclopentenone derivative, two model ester substrates have been employed for PLE hydrolysis to obtain the corresponding *trans*-acid and 4-hydroxy cyclopentenone moiety (Scheme 4.4).

The hydrolysis of *cis*/*trans* isomer mixture of *rac*-cyclopentyl chrysanthemate shows highly diastereoselectivity (99% d.e.) to the product of (±)-*trans*-chrysanthemic acid with conversions from 62 to 71% but does not show any enantioselectivity.

For the cyclopentenone ester, the diastereoselectivity decreases from 99% d.e. to 80% d.e. at higher conversions; at lower conversions, PLE showed some enantioselectivity (19% *e.e.* for acid and 58% *e.e.* for alcohol) which indicates the effect of the alcohol structure of the chrysanthemic acid ester on the enantioselectivity [12].

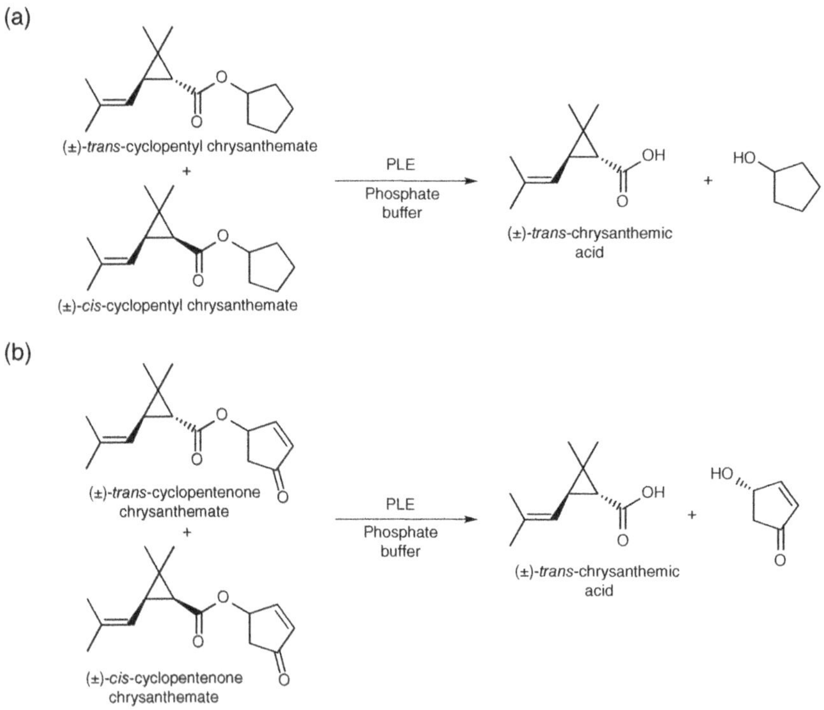

Scheme 4.4 PLE catalyzed hydrolysis of chrysanthemic acid esters: (a) *cis*/*trans* isomer mixture of *rac*-cyclopentyl chrysanthemate, (b) *cis*/*trans* isomer mixture of *rac*-cyclopentenone chrysanthemate.

Enantiopure tertiary alcohols (TAs) are versatile building blocks in organic synthesis for the preparation of high value pharmaceuticals [13]. The production of enantiopure TAs by enzymatic hydrolysis is highly interest by synthetic chemists for its mild reaction conditions and environmentally friendly. Esterase activity has been found for several bacterial strains for the enantioselective hydrolysis of bulky aliphatic, keto-derived, and pyridine substituted tertiary alcohol esters. In the hydrolysis of *para*-substituted pyridine tertiary alcohol ester using a strain of *Rhodococcus ruber*, the enantioselectivity can be up to 71% *e.e.* While in the hydrolysis of *meta*-substituted pyridine tertiary alcohol ester using strains of *Microbacterium* and *Alcaligenes* sp., an opposite enantioselectivity was found with 17% *e.e.* and 54% *e.e.*, respectively [14]. Since the direct organic synthesis of chiral TAs remains difficult and is still a great challenge in most cases for fine chemicals, kinetic resolution with the enantioselective esterases catalyzed hydrolysis of ester racemates can be an alternative method to obtain enantiopure TAs. Esterase EstBP7 obtained from *Bacillus* sp. BP-7 bearing rare GGG(A)X oxyanion hole motif that efficiently hydrolyzed esters of short-chain fatty acid esters such as *p*NP-butyrate and *p*NP-acetate was employed for the kinetic resolution of a wide range of TAs. The results showed that one of the mutants (EstBP7-AGA; Mut 5) produces excellent kinetic resolution toward a complex TA acetate (2-(4-pyridyl)3yn-2-yl acetate) at low reaction temperature (4 °C) (Scheme 4.5) [15].

The kinetic resolution of racemic 1,2-*O*-isopropylideneglycerol (IPG) esters for producing enantiopure IPG by esterase catalyzed enantioselective hydrolysis was another application example of the kinetic resolution. It has been found that (*S*)-IPG is an important chiral building block for the synthesis of β-blockers, prostaglandins, and leukotrienes [16]; therefore, esterase YbfF purified from *Escherichia coli* was used for the hydrolysis of IPG butyrate and IPG caprylate. There is high enantioselectivity for YbfF toward the *R*-enantiomer of IPG butyrate and IPG caprylate that produces (*S*)-IPG product with 72 to 94% *e.e.* (Scheme 4.6) [17]. In the kinetic resolution experiments for IPG caprylate, the enantioselectivity of YbfF could be significantly enhanced by the DMF or DMSO

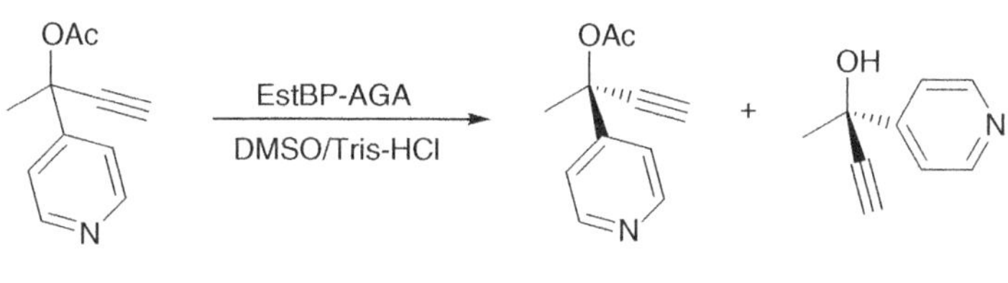

Scheme 4.5 Kinetic resolution of racemic ester for producing chiral TA with esterase. *Source:* Based on Fillat et al. [15].

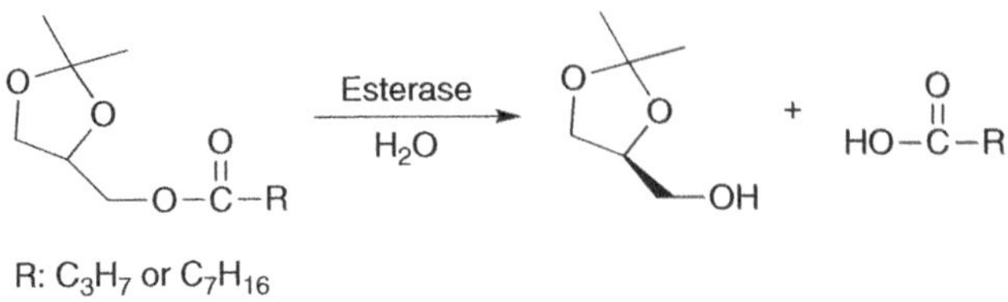

Scheme 4.6 Esterase-catalyzed kinetic resolution of racemic IPG esters. *Source:* Godinho et al. [17].

as co-solvent. Esterase YbfF also shows high enantioselectivity toward 1-phenylethyl acetate that produces (*R*)-1-phenylethanol with enantiomeric excess larger than 99%.

In addition to the use of free esterase for the enantioselective hydrolysis of esters, immobilized esterase has been applied for the enantioselective hydrolysis of esters so that esterase can be recovered and reused again and again. In one case, PLE was covalently bonded on silica through *p*-phenylenediamine derivatization and diazonium salt functionalization. The immobilized PLE was then employed to enantioselectively hydrolyze dimethyl 2-methyl-2-phenylmalonate to yield methyl hydrogen (+)-(*R*)-2-methyl-2-phenylmalonate with slightly higher enantiomeric excess (82–86% *e.e.*) than free PLE (Scheme 4.7). The immobilized PLE can be recovered by simple filtration and reused again without significant loss of enantioselectivity and enzyme activity. This study also demonstrated that the esterase-catalyzed hydrolysis can be used for the preparation of enantiopure acids [18]. In another case, a bacterial esterase called Esterase 30000 was immobilized in a liquid stationary phase and filled in a column while the substrate 2-cyano cyclopropyl-1,1-dicarboxylic acid dimethyl ester was pumped through the column with an apolar mobile phase 1,1,1-trichloroethane to perform the enzymatic resolution. This kind of countercurrent chromatography showed a *trans* hydrolytic selectivity as shown in Scheme 4.8 [19].

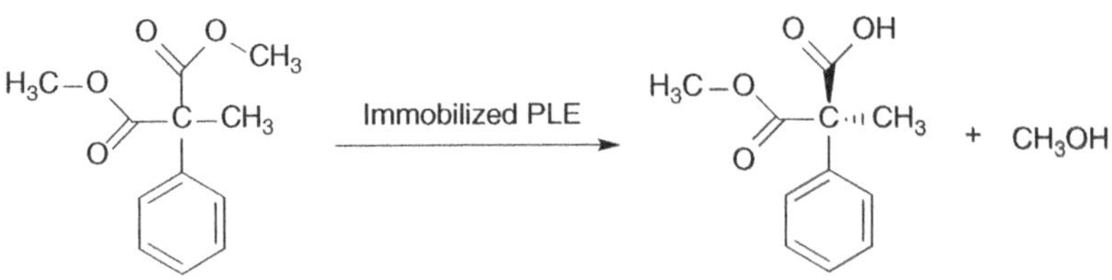

Scheme 4.7 Enantioselective hydrolysis of dimethyl 2-methyl-2-phenylmalonate using immobilized PLE.

Scheme 4.8 Esterase-catalyzed enantioselective hydrolysis of 2-cyano cyclopropyl-1,1-dicarboxylic acid dimethyl ester. *Source:* Modified from Bousquet et al. [19].

4.1.2 Ester Hydrolysis with Lipases

Lipases (E.C. 3.1.1.3), also known as triacylglycerol lipases, are distinguished from esterases by being able to catalyze the hydrolysis of acylglycerols with long acyl chains (≥10 carbon atoms) [20]. Esterases and lipases constitute the two major classes of hydrolases [21, 22] and just like esterases, lipases are a kind of serine hydrolases [3, 23]. The sources of lipases are broad including animals, plants, and natural or recombinant microorganisms which lead to many applications in the pharmaceutical and food industries [24]. Among these categories microbial lipases are more stable and more conveniently produced than their corresponding plant and animal enzymes; thus, they are particularly attractive for industrial applications such as the food industry, the paper, textile and leather industries, additives in detergents, synthesis in biopolymers, biodiesel production, synthesis of optically pure compounds and fine chemicals (antibiotics, anti-inflammatory drugs), cosmetic, agrochemical industries, and water treatment [25, 26]. In general, lipases are highly stable enzymes with essential roles in the digestion, transport, and processing of dietary lipids, that is, triglycerides, fats, and oils, by the hydrolysis/synthesis of carboxyl ester bonds of long-chain triacylglycerols [24, 27]. However, lipases-catalyzed hydrolysis of triglycerides to produce glycerol, free fatty acids, diglycerides, and/or mono glycerides under aqueous conditions (Scheme 4.9) that make them as an eco-friendly biocatalyst for the synthesis of pharmaceutical active intermediates and chemicals is the main theme in this section [28]. For the long-chain triacylglycerols hydrolysis, there are lipases unable to hydrolyze ester bonds at the secondary positions as the group of lipases hydrolyzes both primary and secondary ester.

There is still a group of lipases which shows fatty acid selectivity by cleaving ester bonds of particular types of fatty acids [25].

One of the most common excipients for oral lipid-based formulations is triglycerides (TG). In the lipolysis of triglycerides using porcine pancreatic lipase to

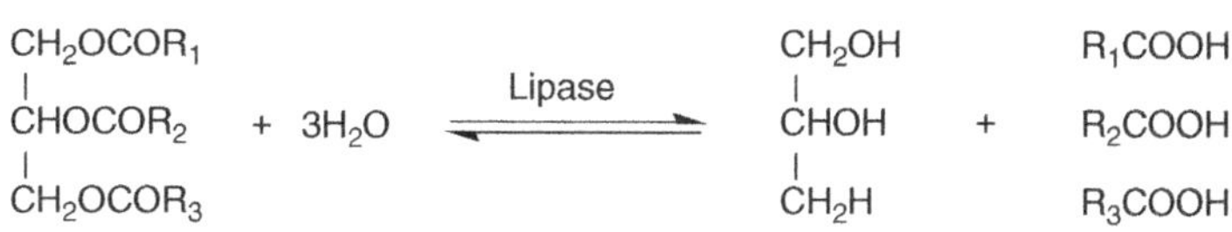

Scheme 4.9 Lipases-catalyzed hydrolysis of triacylglycerols. *Source:* Based on Sharma et al. [28].

understand the oral bioavailability of the co-administered drug, the chain length of fatty acids (FAs) in TG affects the activity of pancreatic lipase. The *in vitro* lipolysis study of TG with porcine pancreatic lipase under bio-relevant conditions at pH 6.80 showed that a specific chain length range (C2–C8) of the FAs in TG gives higher activity for porcine pancreatic lipase. It is proposed that the chain-length specificity should be due to the combination of physicochemical properties of TGs, 2-monoglycerides, and FAs, *viz.* the droplet size of the TGs, the solubility of 2-MGs within mixed micelles, and the relative stability of the FAs as leaving groups in the hydrolysis reaction [29].

Since substrates of the lipase are often insoluble or partially soluble in water, utilizing organic solvents or organic-aqueous solutions will be in favor of some hydrolysis reactions. There are advantages of using organic solvents in enzymatic reactions, mainly the increasing of substrates solubility, the ease of product recovery in organic phase, the reducing of substrate and/or product inhibition in organic solvent-water systems, and the shifting of reaction equilibrium toward the product by continuously removing the product in organic solvent-water systems. However, the use of organic solvents may make most microorganisms lose their activity or cease growing and denature and inactivate the enzymes. Therefore, the search of organic solvent tolerant microorganisms for producing organic solvent-stable lipase for TGs hydrolysis is important. Lipases hydrolyze ester bonds of TGs to glycerol and FAs at the oil–water interface and do not hydrolyze dissolved substrates in the bulk fluid [30]. Lipase produced from bacterium *B. sphaericus* 205y which was isolated from a total of 131 organic solvent tolerant strains originating from soil samples yielded the highest lipase activity of 0.42 U mL^{-1} min^{-1} in the hydrolysis of triolein with up to 75% (*v/v*) concentration of either *n*-hexane, toluene, ethyl benzene, or *p*-xylene organic solvent-water two phase systems. This lipase was also stable and can be activated by *n*-hexane and *p*-xylene [31].

Enantioselectivity of lipases used for producing optically pure enantiomers has been applied industrially for solving difficulties in organic synthesis methods [32]. Thus, lipase-catalyzed kinetic resolution for various racemates with lipases has been considered as a green method for separating and producing enantiomers using mild conditions [33, 34]. The asymmetric hydrolysis of (±)-1,2-diacetoxy-3-chloropropane (DACP), (±)-1,2-diacetoxy-3-bromoprpane (DABP), and (±)-1,2-diacetoxyethylbenzene (DAEB) was investigated by lipase in order to obtain their corresponding optically active esters or alcohols. The hydrolysis of (±)-1,2-diacetoxy-3-chloropropane with the lipoprotein lipase Amano 40 in sodium phosphate buffer gave (*S*)-1,2-diacetoxy-3-chloroprpane of 90% *e.e.* and 20% yield (Scheme 4.10). Then, the (*S*)-1,2-diacetoxy-3-chloroprpane enantiomer was chemically modified to form one of the β-adrenergic blocking agents, the (*S*)-isomer of propranolol [(*S*)-1-isoprpylamino-3-(1-naphthoxy)-2-propanol] [35].

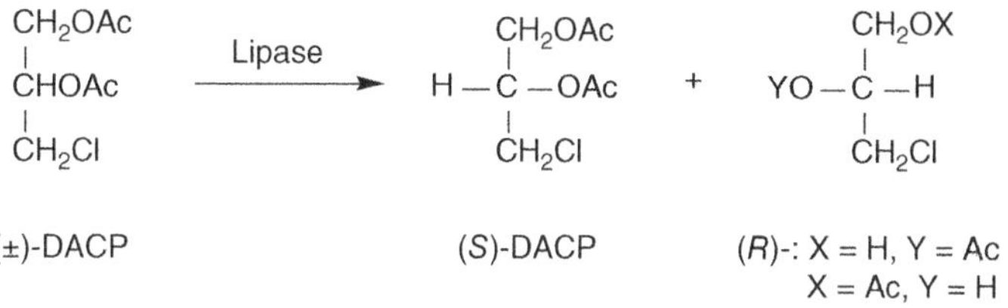

Scheme 4.10 Enantioselective hydrolysis of (±)-1,2-diacetoxy-3-chloropropane by lipase.

Chiral alcohols, including secondary and primary alcohols, are common chiral building blocks or chiral sources in synthetic chemistry. Optically pure secondary and primary alcohols play important roles in manufacturing drugs, cosmetic, and food. Therefore, chiral-recognition of alcohols by lipases through kinetic resolution has become a useful strategy recently in the preparation of chiral alcohols [25]. The kinetic resolution of racemic secondary alcohols via enantioselective acylation of lipases is efficient and widely used in organic synthesis. However, the lipase-catalyzed kinetic resolution through acylation or hydrolysis is more difficult for primary alcohols due to the lower enantioselectivity of lipases toward chiral primary alcohols or very limited for tertiary alcohols by the availability of commercial lipases for accepting tertiary alcohols [25, 36]. Researchers have found that lipase produced from *Burkholderia cepacia* (BCL) shows high enantioselectivity toward chiral primary alcohols, but this enantioselectivity is often unpredictable, especially for substrates containing an oxygen at the stereocenter. For instances, BCL resolves β-substituted-γ-acetyloxymethyl-γ-butyolactones (acetates of a primary alcohol) by hydrolysis of the acetate, but the enantioselectivity varies with the nature and orientation of the β-alkyl substituent (Scheme 4.11). BCL favors the (*R*)-primary alcohol when the β-alkyl substituent is hydrogen ($E = 30$) or *trans* methyl ($E = 38$), but the (*S*)-primary alcohol when it is *cis* methyl ($E = 145$) [37]. Citalopram, a highly selective inhibitor of antidepressant serotonin (5-HT) reuptake, resides its inhibitory activity in its *S*-enantiomer which is synthesized through a tertiary alcohol 3-[(acetoxy)methyl]-4-[4-(dimethylamino)-1-(4-fluorophenyl)-1-hydroxybutyl]-benzo nitrile (diol monoacetate) by Novozym 435 (*Candida antarctica* lipase B immobilized on acrylic resin) catalyzed asymmetric alcoholysis I organic media (Scheme 4.12) [38].

3-Amino-3-phenylpropionic acid (BPA) is a valuable β-amino acid used for the synthesis of the anticancer agent taxol and (*S*)-dapoxetine, (*S*)-(+)-*N*,*N*-dimethyl-α-[2-(1-naphthalenyoxy)ethyl]benzenemethanamine, which is a potent selective serotonin reuptake inhibitor used for the treatment of depression, bulimia, or anxiety. Since the *Carica papaya* lipase (CPL) catalyzed kinetic resolution of *N*-protected BPA conventional aliphatic esters showed moderate enantioselectivity (Scheme 4.13), an efficient synthesis of the (*S*)-BPA

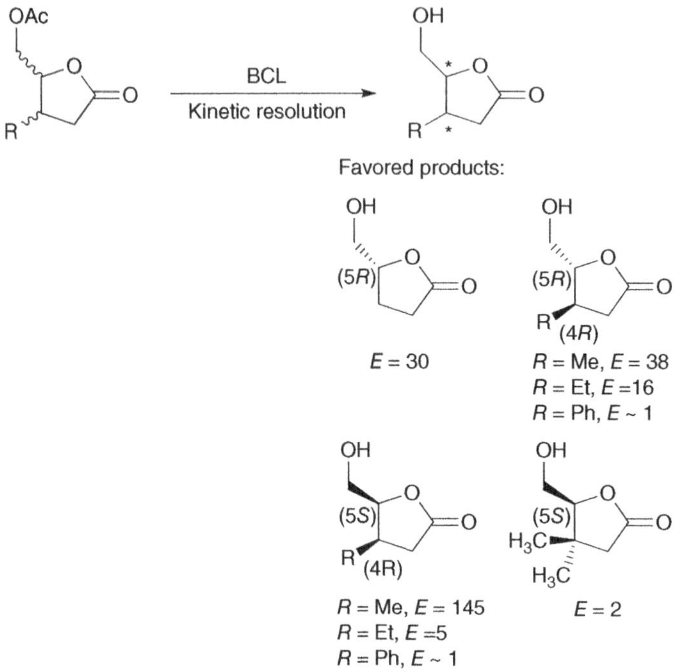

Scheme 4.11 BCL-catalyzed kinetic resolution of β-substituted-γ-acetyloxymethyl-γ-butyolactones by hydrolysis of the acetate.

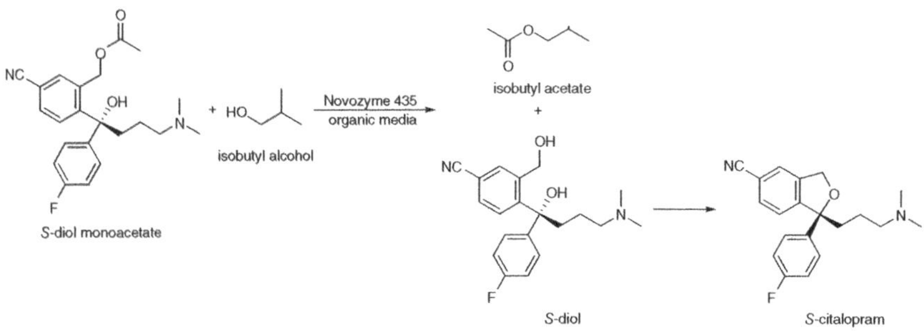

Scheme 4.12 *Candida antarctica* lipase B-catalyzed asymmetric alcoholysis of tertiary alcohol. *Source:* Based on Wang et al. [38].

enantiomer has been achieved by CPL-catalyzed enantioselective alcoholysis of the corresponding racemic *N*-protected 2,2,2-trifluoroethyl esters in an organic media [39].

Both enantiomeric β- and γ-amino acids have acquired considerable interest as potentially pharmaceutically active compounds, or as key intermediates for the

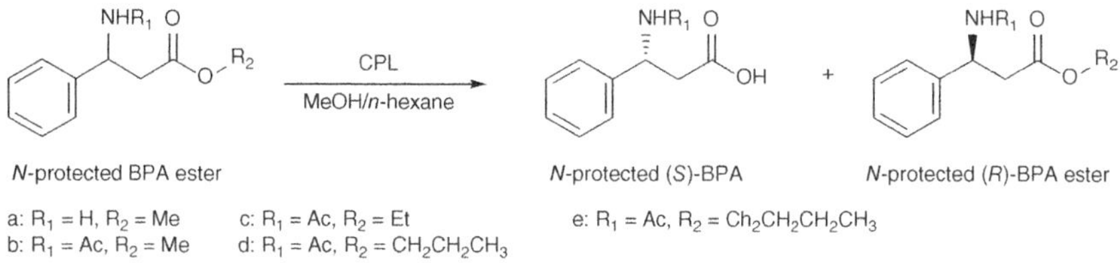

Scheme 4.13 CPL-catalyzed kinetic resolution of BPA alkyl esters (a–e) with different alkyl chain lengths. *Source:* You et al. [39].

synthesis of drugs. Thus, kinetic and sequential kinetic routes for the synthesis of enantiomeric β- and γ-amino acids through enantioselective *C. antarctica* lipase B- or *B. cepacia* lipase-catalyzed ring cleavage of the corresponding β- and γ-lactams in organic solvents or solvent-free systems or a supercritical CO_2 medium, and for the enantioselective hydrolysis of the corresponding amino esters in organic solvents have been adopted extensively [40].

In order to increase the activity, stability, and the repeated usage of lipase, porcine pancreatic lipase (PPL) has been covalently immobilized onto cross-linked polyvinyl alcohol (PVA) in organic media in the presence of a fatty acid substrate as the protective ligand. The effects of fatty acid additions to the immobilized media for the hydrolysis of tributyrin in water showed an enhanced activity with octanoic acid as the additive. After five repeated uses, the immobilized PPL activity was maintained at 63% of its original activity [41]. Immobilization of lipase for protecting from denaturation by the organic solvent has also been applied to the enantioselective hydrolysis of water-insoluble α-acyloxy esters. Lipases were immobilized with Celite or a synthetic prepolymer (ENTP-4000 or ENT-4000) and were used for the kinetic resolution of α-acyloxy esters (±)-**1** and (±)-**2** in a water-saturated organic solvent to produce chiral intermediates, (2S,3S)-**1** and (2S,3R)-**4**, for the synthesis of diltiazem hydrochloride and (−)-indolmycin, respectively (Scheme 4.14). The chiral intermediate (2S,3R)-**4** was also able to be formed through the enantioselective hydrolysis of (±)-**2** using lipid-lipase aggregates in water-saturated organic solvent [42].

Several very good properties of ionic liquids (ILs) make them used as reaction media in chemical reactions: less volatility, less flammability, low toxicity, and unique solubility for organic and inorganic materials. With these notable properties, ILs have been used as organic reaction media in lipase-catalyzed reactions. One application is the lipase-catalyzed synthesis of biodiesel oil (Scheme 4.15) which has gained strong interest from the standpoint of sustainable energy production and the results generally are excellent. Examples of lipase-catalyzed reaction using the IL solvent systems for biodiesel oil production have been reviewed [43].

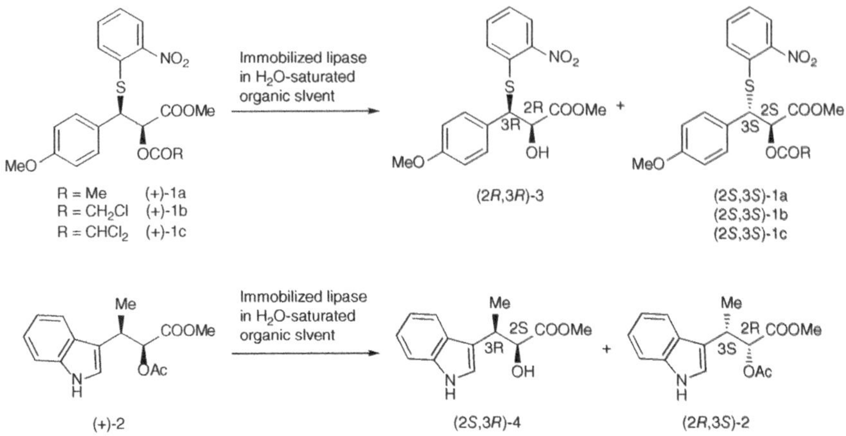

Scheme 4.14 Kinetic resolution of water-insoluble α-acyloxy esters with immobilized lipase.

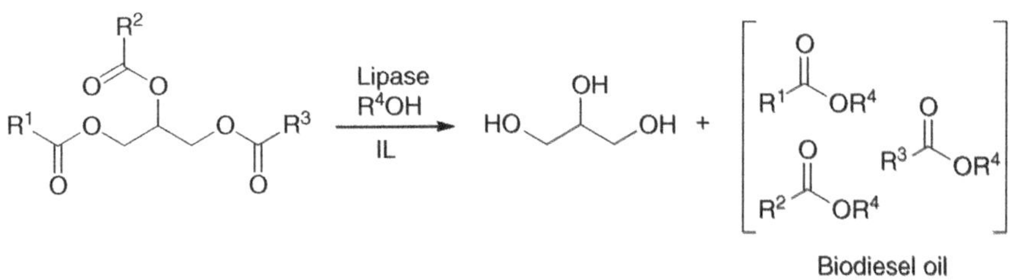

Scheme 4.15 Lipase-mediated biodiesel oil production using ionic liquids as media. *Source:* Itoh [43].

4.1.3 Ester Hydrolysis with Proteases

Generally, proteases (E.C. 3.4.21-24 and 99, also called as peptidases or proteinases) are defined as enzymes that perform the protein catabolism by hydrolysis of peptide bond in protein's peptide chain (proteolysis) with water and the peptide synthesis in organic solvents or in solvents of low water content. They can be found in animals, plants, microbes, and viruses. Proteases represent one of the three largest groups of industrial enzymes and are the most important enzymes produced commercially that are used in detergent, protein, brewing, meat, photographic, silver recovery, leather, waste treatment, and dairy industries [44–46].

Although proteases are intended for the hydrolysis of the amide bond of the protein molecules, many works have addressed to the hydrolysis of *N*-protected amino acid esters by serine and sulfhydryl proteases. Literature has shown that during the hydrolysis of *N*-acetyl-L-tyrosine ethyl ester or *N*-benzoyl-L-arginine ethyl ester by proteases such as chymotrypsin, subtilisin, trypsin, and papain, the substrate bond cleaved is the acyl–carbon–ethereal oxygen bond [47]. It also has been found that the methyl ester of *N*-acetyltyrosine is hydrolyzed much faster than the ethyl ester by subtilisin and the benzyl ester of *N*-acetyltyrosine is hydrolyzed even more rapidly than the ethyl ester [47, 48]. Instead of the *N*-protected amino acid ester, *N*-benzoyl-L-tyrosine thiobenzyl ester has been readily hydrolyzed by serine proteases such as α-chymotrypsin and subtilisin BPN'. In this kind of hydrolysis, indole was found a strict competitive inhibitor for the α-chymotrypsin-catalyzed hydrolysis but inole oppositely was observed to increase the maximal rate of hydrolysis of *p*-nitrophenyl acetate [49]. In addition to the hydrolysis of *N*-protected amino acid esters by proteases mentioned above, methyl β-phenylpropionate has also been hydrolyzed by chymotrypsin for elucidating the reaction mechanism some 60 years ago [50].

Valganciclovir hydrochloride is a mono-L-valyl ester prodrug of ganciclovir and is an antiviral active pharmaceutical ingredient used to treat cytomegalovirus infection which is associated with AIDS disease. One of the synthetic routes for the synthesis of Valganciclovir hydrochloride was through the synthesis of Valganciclovir hydrochloride intermediate, 2-[(2-amino-1,6-dihydro-6-oxo-9H-purin-9-yl)methoxyl]-3-hydropropyl-*N*-[(benzyloxy)carbonyl]-L-valinate (Intermediate 2) by the enzymatic desymmetrization of 2-[(2-amino-1,6-dihydro-6-oxo-9H-purin-9-yl)methoxyl]-1,3-propanediyl bis(L-valinate) (Intermediate 1) (Scheme 4.16). The effective hydrolysis enzymes used for cleaving one the ester bonds in ADOPMPBV were Protex 6L (a bacterial alkaline protease derived from a selected strain *B. licheniformis*) and Alcalase 2.4L (a *Subtilisin* protease) [51].

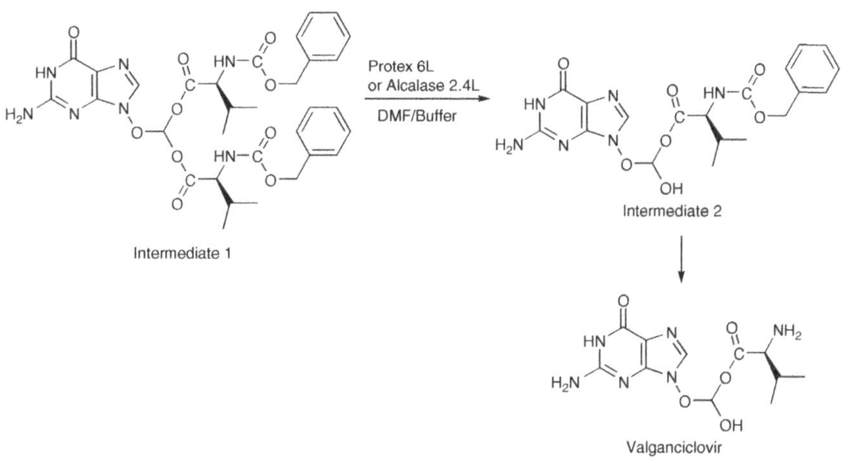

Scheme 4.16 Protease-catalyzed desymmetrization to form Valganciclovir hydrochloride intermediate.

Kinetic resolution of enantiomers employing protease-catalysis can be applied for reactions such as hydrolysis of carboxylic acid esters or amides, esterification, transesterification, and amide bond formation. For amino acid enantiomers synthesis, the resolution of racemic amino acids using the enantioselective hydrolysis of their esters by proteases is one of the most preferable strategies. It has been found that chymotrypsin, subtilisin, and protease from *A. oryzae* are by far the most widely used proteases for this purpose due to their broader substrate specificity. In general, these three proteases catalyzed the hydrolysis of L-amino acid ester derivative that leaves the corresponding D-amino acid ester unaffected. Besides this classical finding, many researches focused on the use of this approach toward the resolution of synthetic nonprotein amino acids, for examples, the resolution of a series of ring-substituted phenylalanine derivatives, γ,δ-unsaturated phenylnorvaline homologues, ring-size modified amino acids, α,β-disubstituted β-phenylalanine derivatives such as *threo*-methylphenidate, and the isoxazoline derivative of 2-amino-4-hydroxy-4-(4-methyloxyphenyl)-3-methylbutanoic acid methyl ester [52].

Homochiral nonprotein amino acids including those by synthesis are useful building blocks for the synthesis of analogs of biologically active peptides such as toxins, antibiotics, hormones, and enzyme inhibitors. In the *A. oryzae* protease-catalyzed ester hydrolysis, the kinetic resolution of a variety of nonprotein *N*-protected amino acid esters) resulted in a large enhancement of the hydrolysis rate, while the enantioselectivity deteriorated strikingly for the methyl esters. However, this difficulty has been overcome by using esters bearing a longer alkyl chain such as isobutyl ester. Therefore, isobutyl amino acid esters carrying an aromatic side chain were resolved with excellent enantioselectivities ($E = 50$ to >200). With isobutyl amino acid esters bearing an aliphatic side chain, good results in terms of the hydrolysis rate and enantioselectivity were also obtained. In addition, results showed that the enantioselectivity can be improved further by conducting the hydrolysis at low temperature for those substrates with not high enough enantioselectivity [53]. A further study using the obtained enhancement technique on the *N*-free nonprotein amino acid esters showed that the enantioselectivity enhancement is also attained by adjusting the pH of the reaction mixture to around 6.2 [54].

2-Aminosuberic acid, an aminoalkanedioic acid, is used to synthesize the deaminodicarba analogs of some peptide hormones such as [Asu^{1,6}]oxytocin and elcatonin which a cysteine residue is substituted by the L-α-aminosuberic acid (Asu). In order to obtain the enantiomeric pure L-Asu, fully *N*-protected racemic *N*-acetyl-α-aminosuberic acid diesters were first resolved with chymotrypsin to produce the optically pure L-amino acid monoesters in good yields and high optical purity. Subsequently, the obtained *N*-acetyl-α-aminosuberic acid monoesters were decarboxylated by an L-specific aminoacylase to the desired L-α-aminosuberic

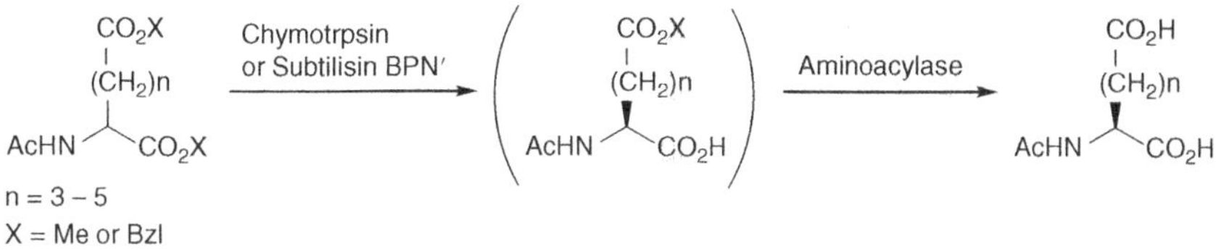

Scheme 4.17 Tandem enzymatic resolution of racemic α-aminoalkanedioic acid diester. *Source:* Based on Bordusa [52]; Nishino et al. [55].

acid (Scheme 4.17) [52, 55]. A later study showed that the resolution of Z-D,L-2-Asu methyl diester can be catalyzed regioselectively by papain (or subtilisin BPN') alone using an organic solvent with low water content to lead to Z-L-Asu(OMe)-OH and Z-D-Asu(OMe)-OMe [52, 56]. Then, Z-L-Asu-OH and Z-D-Asu-OH can be produced by saponifications.

Chiral δ-lactones are important compounds in the perfume, food, and pharmaceutical industries. A novel synthetic route to optically active saturated and unsaturated δ-lactones was based on enzymatic kinetic resolution and ring-closing metathesis reaction.

For example, besides lipases, protease K was used for the hydrolysis of racemic (*E*)-methyl 5-(phenyl)-5-acryloyloxypent-2-enoate to obtain enantiomerically pure (*E*)-methyl (5*R*)-hydroxy-5-phenylpent-2-enoate which in turn can be chemically transformed to (6*R*)-phenyltetrahydropyran-2-one. The enzymatic activity and enantioselectivity were substantially influenced by the organic co-solvent and surfactant type [57].

4.2 Hydrolysis of Amide Bond

Molecules containing amide bond are ubiquitous and are important as structural materials in both nature and technology. They have a variety of applications in our daily life such as food, drug, textile, transportation, and art. The most important biological compound with amide bond in nature is protein molecules which have a large polypeptide chain formed by the linkage of amino acid monomer using amide bond (also called peptide bond) and play different biological functions in organisms. The enzyme used for hydrolyzing the peptide bond of protein molecules is protease such as trypsin, papain, and α-chymotrypsin. In addition to the specificity on peptide amide hydrolysis of protease, different proteases show peptide amide hydrolysis specificity on substrate chain length. For instances, α-chymotrypsin hydrolyzes specific acetyl amino acid amides, while *Streptomyces griseus* protease 3 catalyzes the hydrolysis of peptide amide with the presence of up to four amino acid residues N-terminal to the scissile bond of the substrate [58].

Research results also have shown that the proteases catalyze the hydrolysis of amides belonging to serine proteases [59, 60]. The substrate specificity and kinetic mechanism of protease II (a cytoplasmic endopeptidase from *E. coli*) have been characterized to show trypsin-like specificity that cleaves essentially the carboxymethylated B chain of insulin at the Arg^{22}–Gly^{23} bond, but it is also able to cleave the Try^{16}–Leu^{17} bond with a prolonged incubation period. One of the two substrates used for the steady-state kinetic parameters estimation of protease II was *N*-benzoyl-L-arginine *p*-nitroanilide [61].

Instead of the hydrolysis for the amide bond of large polypeptide chain, proteases also have been used for the hydrolysis of low-molecular-mass oligopeptide. Thyrotropin-releasing hormone (TRH) is a tripeptide (pGlu-L-His-L-Pro-NH_2) discovered in 1969 which regulates the synthesis and secretion of thyrotropin from the anterior pituitary gland. Therefore, it is a potential drug for the treatment of various neurologic and neuropsychiatric disorders such as depression, shock, and schizophrenia. The initial degradation of TRH in the blood by the TRH-specific pyroglutamyl aminopeptidase serum enzyme is the hydrolysis of the pGlu–His bond to produce pyroglutamic acid and His-Pro-NH_2, while TRH goes deamination at the terminal proline amide residue to yield TRH-OH in organs such as the liver, the gut, and the kidney or in various brain tissues (Scheme 4.18) [62].

Protease was also used for low-molecular-mass amide molecule hydrolysis. Thermolysin (TL) is a thermostable Zn-dependent neutral protease from thermophilic strain *Bacillus thermoproteolyticus* that has been applied for the hydrolysis of 3-(2-furylacryloyl)-glycyl-L-leucine amide (FAGLA) to study its catalytic activity under extremes of pressure and temperature. The enzymatic hydrolysis of FAGLA using TL was monitored by the liberation of the hydrolysis product

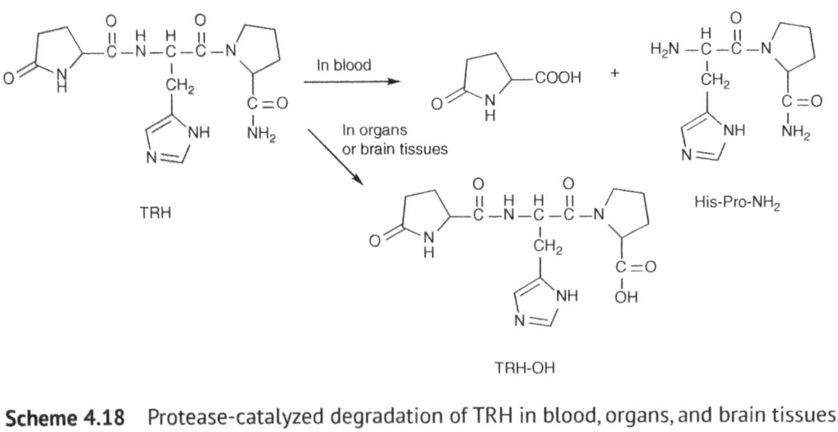

Scheme 4.18 Protease-catalyzed degradation of TRH in blood, organs, and brain tissues. *Source:* Møss and Bundgaard [62].

3-(2-furylacryloyl)-glycine on a spectrophotometer [63]. In another case, the hydrolysis ability on the of amide bond for the commercial protease Alcalase 2.4 L from *B. licheniformis* has been tested for the surface modification of polyamide (PA) 6.6 fabrics. The optimal conditions of the enzymatic hydrolysis reaction were investigated on adipic acid bishexyl-amide substrate, a PA model compound that mimics the surface structure of PA 6.6 fibers. After the enzymatic surface modification of PA textile, both the hydrophilicity and the dying ability of PA textiles were enhanced 2.7- and 1.24-fold, respectively [64].

In enzymology, the enzyme class used for the hydrolysis of the amide carbon–nitrogen bonds ($CO–NH_2$) other than peptide bonds is amidase (E.C. 3.5.1.4) which has the systematic name acylamide amidohydrolase. The two substrates of this enzyme are monocarboxylic acid amide and water which give products of monocarboxylic acid and ammonia, respectively. The enzymatic cleaving of amide bond has been an attractive tool in industries and leaded potential applications in commodity chemical synthesis, pharmaceuticals and agrochemicals production, and waste effluent treatments. Amidases widely occur in both eukaryotic and prokaryotic organisms [65]. According to their amino acid sequences and structural characteristics, amidases can be subdivided into two families viz. the amidase signature (AS) family and the nitrilase superfamily. The amino acid sequence of AS family amidases consists of a highly conserved stretch of approximately 130 serine- and glycine-rich amino acids with the catalytic triad of Lye-Ser-Ser which exhibits a broad substrate spectrum for aliphatic, aromatic, heterocyclic amides and their derivatives [65–68]. The nitrilase superfamily amidases, also known as aliphatic amidases, share significant structural homology in amino acid sequences with a Cys-Glu-Lys catalytic triad which shows substrate specificity toward only aliphatic amides [69–71].

2-Arylpropanoic acids such as naproxen, ibuprofen, ketoprofen, or flurbiprofen are important nonsteroidal anti-inflammatory drugs. For all this class of drugs, the corresponding (*S*)-enantiomers are more potent drugs than their racemic drugs. In this perspective, the enantiomeric pure (*S*)-naproxen was prepared by the enantioselective hydrolysis of racemic naproxen amide with immobilized resting cells of bacterial strain *Rhodococcus erythropolis* MP50 in organic solvents (Scheme 4.19). Under the optimal conditions, the amidase activity of *R.*

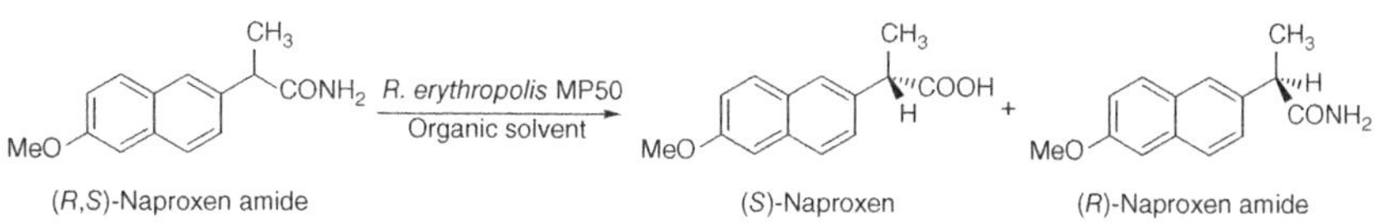

Scheme 4.19 Enantioselective hydrolysis of racemic naproxen amide with *R. erythropolis* in organic solvent.

erythropolis in the preparative scale application gave a conversion rate of 48%, the (*S*)-naproxen enantiomeric selectivity >99% *e.e.*, and an yield of 84% [72]. Since 2,5-disubstituted pyrrolidine derivatives are the core structure of natural products and synthetic pharmaceutical compounds, another amidase containing strain of *R. erythropolis* AJ270 has been used for the enantioselective hydrolysis of racemic and meso pyrrolidine-2,5-dicarboxamides.

During the whole cell catalyzed kinetic resolution, racemic *trans*-pyrrolidine-2,5-carboxamide was enantioselectively transformed into (2*S*,5*S*)-pyrrolidine-2,5-dicarboxamide and (2*R*)-5-carbamoylpyrrolidine-2-carboxylic acid in high yields and excellent enantioselectivity (Scheme 4.20A). While the whole cell catalyzed desymmetrization of meso *cis*-pyrrolidine-2,5-dicarboxamide gave (2*R*,5*S*)-5-carbamoylpyrrolidine-2-carboxylic acid in an almost quantitative yield (Scheme 4.20B) [73].

Instead of whole cell-catalyzed amide hydrolysis, an AS family amidase isolated and purified from *Pantoea* sp. (*Pa-Ami*) was exploited to catalyze the hydrolysis of 2-chloronicotinic amide to obtain 2-chloronicotinic acid because 2-chloronicotinic acid has been extensively used for the synthesis of two most efficient herbicides nicosulfuron and diflufenican. 2-Chloronicotinic acid is also an important building block for pharmaceuticals such as nevirapine, mirtazapine, pranoprofen, nixylic acid, and 2-aminonicotinic acids. A significant improvement in productivity for this amidase-catalyzed reaction can be acquired by using a fed-batch reactor at low reaction temperature which gave a substrate

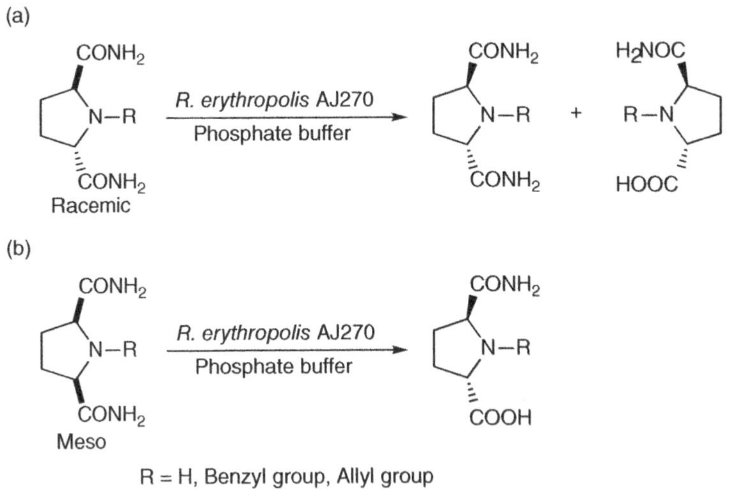

Scheme 4.20 Whole cell catalyzed enantioselective biotransformations of racemic and meso pyrrolidine-2,5-dicarboxamides: (a) kinetic resolution, (b) desymmetrization. *Source:* Chen et al. [73].

conversion of 94.2% [74]. In addition, an amidase purified from the bacterium *Pseudomonas chlororaphis* B23 was proved to have ability for the hydrolysis of a broad range of aliphatic and aromatic amides. This amidase also exhibited (*S*)-form enantioselectivity for three aromatic amides, 2-(4-chlorophenyl)-3-methylb utyramide, phenylalaninamide, and 2-phenylpropionamide, with *e.e.* value of 96, 55, and 100%, respectively [75].

Fatty acid amide hydrolase (FAAH, E.C. 3.5.1.99) is a membrane-bound serine hydrolase and belongs to the AS family. The enzyme is expressed in the central nervous system and in a variety of tissues such as the lung, the liver, the kidneys, the small intestine, the prostate, and the testis. FAAH hydrolyzes the amide bonds of various physiological active lipids, including anandamide, oleamide, *N*-oleoylethanolamide, and *N*-palmitoylethanolamide. The degradation ability of lipids at their amide bonds hydrolysis by FAAH has been applied to the development of a fluorometric assay for FAAH activity in which the substrates used were medium-chain fatty acid amides such as *N*-(4-nitrophenyl)octanamide, *N*-(4-nitrophenyl)decanamide, *N*-(6-methoxypyridin-3-yl)octanamide, and *N*-(6-methoxypyridin-3-yl)decanamide [76, 77].

A broad spectrum of carboxylic acids can be synthesized from nitriles through enzymatic hydrolysis using enzymes of nitrilase superfamily. For this kind of biological conversion, two hydrolysis pathways (Scheme 4.21) can be employed by nitriles to form their corresponding carboxylic acids. In one pathway, nitrilases (E.C. 3.5.5.1) are used to directly hydrolyze nitriles to the corresponding carboxylic acids and ammonia. The second pathway is a two-step process that nitriles are first hydrated and converted to their corresponding amides by nitrile hydratase (E.C. 4.2.1.84), then amides are hydrolyzed to the corresponding carboxylic acids by amidases (E.C. 3.5.1.4) [75, 78–80]. The enantioselectivity of this

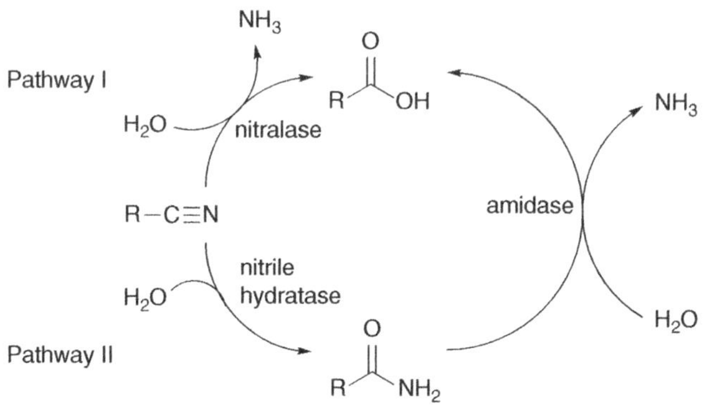

Scheme 4.21 Enzymatic hydrolysis pathways for nitrile to form carboxylic acid.

enzymatic transformation with microbial cells has been applied to many organic syntheses either through the kinetic resolution for a variety of chiral amides and chiral functionalized nitriles, amino- and hydroxynitriles, and nitriles containing small rings or by the desymmetrization for prochiral and meso dinitriles and diamides [78, 80].

Moreover, it has been found that the constituent of incubation medium for microbial cell will affect the nitrile-degrading activity toward different substrates. For example, when the bacterium strain *Nocardia globerula* NHB-2 cultured in a medium containing 1% glucose and 0.1% yeast extract, the cells displayed nitrile hydratase/amidase activity specific for saturated aliphatic nitriles or amides. However, when adding 0.4% *v/v* acetonitrile to the medium, the cells exhibited nitrile hydratase/amidase activity specific for not only saturated aliphatic nitriles-amides but also for aromatic amide hydrolysis. If 0.4% *v/v* propionitrile was added to the culture medium, the cells showed nitrilase activity that was able to hydrolyze aromatic nitrile and unsaturated aliphatic nitrile [79]. Similarly, an isolated bacterium strain *Bacillus subtilis* ZJB-063 exhibited nitrilase activity without adding ε-caprolactam and was highly specific for arylacetonitriles. While the cells yielded the nitrile hydratase/amidase activity as ε-caprolactam was added to the culture media that not only hydrolyzed arylacetonitriles but also some aliphatic nitriles and heterocyclic nitriles [81].

One of the potential industrial applications for nitriles converted into corresponding carboxylic acids through the two-step reaction pathway is the production of nicotinic acid from the starting material 3-cyanopyridine via the intermediate nicotinamide. Nicotinic acid, also known as niacin, belongs to the water-soluble vitamin B-complex which is an important dietary nutrient for our body. Nicotinic acid has been approved by the FDA for the treatment of pellagra which is caused by its dietary deficiency. It is also used for the treatment of hypercholesterolemia, schizophrenia, diabetes, auto-immune diseases, and osteoarthritis. In addition, nicotinic acid has been thought to reduce the risk of heart disease. The worldwide year production of niacin and nicotinamide is tremendous and the demand is still rising. The microorganism *Microbacterium imperial* CBS 498-74 has been found to contain both nitrile hydratase and amidase that are necessary for the transformation of nitrile to its corresponding acid following the two-step pathway. Thus, resting cells of *M. imperial* CBS 498-74 were employed as the catalyst for converting nicotinamide into nicotinic acid. Total conversion was achieved by using a batch reactor. While in the controlled continuous stirred membrane bioreactor (CSMR), the conversion could be as high as 88% with an optimized residence time. Amidase was stable up to 50 °C but partially substrate inhibition was observed as nicotinamide concentrations ≥300 mM [82].

4.3 Hydrolysis of Phosphate Esters

The catalytic enzyme exploited for the hydrolysis of a phosphate ester bond is phosphatase which is a subcategory of hydrolases because water is used to cleave the phosphate monoester linkage during the hydrolysis (Scheme 4.22). In the reaction, the –OH group of water is attached to the product phosphate ion and the H^+ is associated to the hydroxyl group of the other hydrolysis product. The dephosphorylation with phosphatases and the phosphorylation with protein kinases together serve diverse roles for cellular regulation and signaling in molecular biology systems [83].

With the ability of hydrolyzing a variety of organic phosphate esters into inorganic orthophosphate and the ester alcohol, phosphatases have been key-enzymes in modern biotechnology. Substrates that have been used for research studies involving alkaline phosphatase (ALP) include phenyl phosphate, *p*-nitrophenyl phosphate, *p*-aminophenyl phosphate, hydroquinone diphosphate, 1-naphthyl phosphate, 2-phospho-L-ascorbic acid, 3-indoyl phosphate, *O*-phosphorylethanolamine, 2-naphthyl phosphate, *p*-cresyl phosphate, *p*-chlorophenyl phosphate, *p*-*t*-butylphenyl phosphate, and *O*-methoxy-*p*-methylphenyl phosphate [84–86]. However, seldom reports about the applications of alkaline phosphatases were in the organic synthesis; most of the applications involved with alkaline phosphatases were in the area of biotechnology, diagnostic industry, and biosensors [84, 87]. For examples, the elasticity of PE-PAAm hydrogels can be changed by ALP that was demonstrated by the change of the morphology of human mesenchymal stem cells (hMSCs) into an elongated shape without cell damage when hMSCs were cultured with medium containing the hydrogels and ALP [88]; a collagen/calcium phosphate composite film was fabricated by the ALP-catalyzed hydrolysis of disodium glycerophosphate in a collagen solution containing calcium ions. This film was found to support the osteoblastic differentiation when the preosteoblast cell line, MC3T3-E1 cells were cultured on the film [89]. Similarly, multilayer sheets consisted of alternatively cumulated collagen and calcium phosphate layers were prepared. The inorganic layer was mineralized by means of an ALP-catalyzed hydrolysis of water-soluble phosphate esters such as glycerol 1-phosphate in the presence of calcium ions. The sheets with the calcium phosphate layer on the top were used as a scaffold to support the attachment and growth of fibroblast [90]; ALP was utilized as a marker by its activity on the hydrolysis of phosphate esters for normal skeletal mineralization

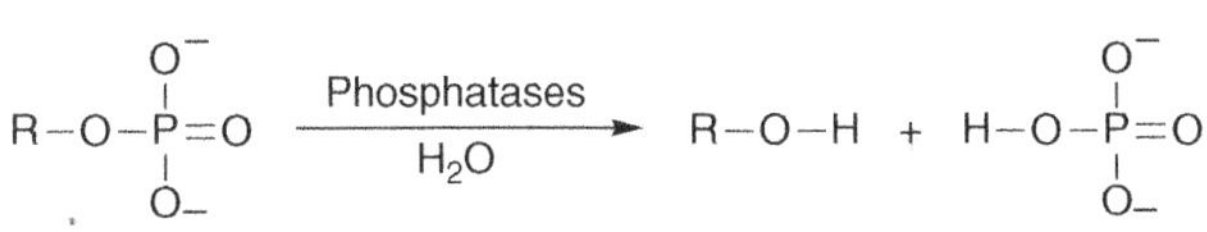

Scheme 4.22 Phosphatases-catalyzed hydrolysis of a phosphate ester bond.

which was demonstrated by investigating histochemically its temporal and local appearance during the early events in bone healing and neogenesis in a guided bone regeneration model system [91]; the chemical warfare nerve agent paraoxon, an organophosphate compound, has been efficiently degraded into inorganic phosphoric acid by sequentially applying three enzymes (an organophosphate hydrolase, phosphodiesterase, and ALP) [92, 93].

Acid phosphatase (AP) has been coupled to L-rhamnulose-1-phosphate aldolase (RhaD) for a straightforward and highly efficient synthesis of L-fructose from dihydroxyacetone phosphate (DHAP) and racemic glyceraldehyde (Scheme 4.23). The exclusive production of L-fructose is based on the stereoselectivity of RhaD on the L-glyceraldehyde [94]. Following this protocol, two rare sugars (D-sorbose and D-psicose) can be simultaneously produced from DHAP and D-glyceraldehyde with a D-sorbose/D-psicose ratio about 2/3 [95, 96]. Secretory AP from *Candida mycoderma* has been exploited for increasing the orthophosphate content in soil by hydrolyzing the organic phosphorus compounds present in soil.

The results show that the yeast strain *C. mycoderma* is potential to enhance the phosphorus utilization in those plants grown in rice soil [97]. A purple acid phosphatase with phytase activity was purified from tobacco (*Nicotiana tabacum*) root exudates which catalyzes the hydrolysis of phytate, the major organic phosphorus in soil, to liberate inorganic phosphates [98]. Thus, it releases the phosphorus starvation of tobacco. The purple acid phosphatase in tobacco root exudates also has high affinity for *p*-nitrophenyl phosphate. In addition to the plant purple acid phosphatase, mammalian purple acid phosphatase found from porcine uteri (uteroferrin) catalyzes the hydrolysis of various phosphate esters including phenyl phosphate (and the *p*-nitro derivative), pyrophosphate, tripolyphosphate, adenosine 5′-triphosphate (ATP), and the prototype $[H_2PO_4]^-$ [99].

There are also many other specific phosphatases that have important biological functions in organisms to lead to potential applications in the pharmaceutical, medicinal, and food industries. Trehalose-6-phosphate phosphatase (T6PP) catalyzes the

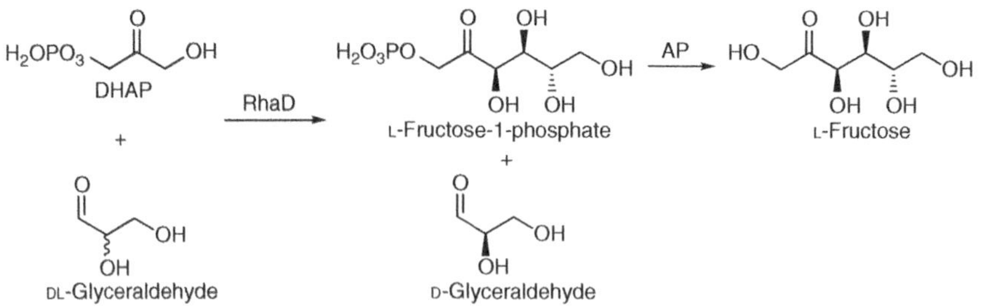

Scheme 4.23 Coupled enzymes synthesis of L-fructose from *rac*-glyceraldehyde and DHAP.

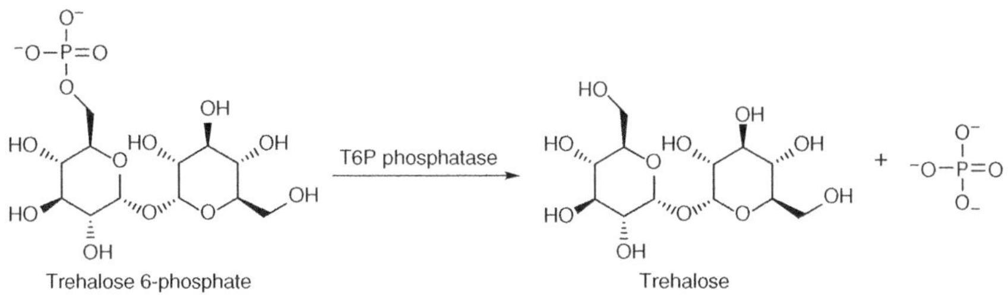

Scheme 4.24 T6PP catalyzed hydrolysis of T6P for trehalose biosynthesis.

hydrolysis of trehalose-6-phosphate (T6P) to produce trehalose and inorganic phosphate (Scheme 4.24). The D-glucose disaccharide, α,α-trehalose, exists in various kinds of bacteria, fungi, plants, and invertebrates but not in mammals. Trehalose is essential to support fungal, bacterial, and nematodal pathogenic cell survival by serving as a fuel, metabolic regulator, antidesiccant, or outer membrane component. T6P is thought of as a toxic material to the host organisms by acting as "anti-metabolites" to lead inhibition of metabolic enzymes, in particular phosphotransferase [100]. The synthesis of bacterial cell wall glycan, for example, peptidoglycan, liposaccharide, teichoic acid, and capsular polysaccharide, starts in cytosol.

This cell wall glycan is constructed from a glycan lipid carrier, undecaprenyl phosphate ($C_{55}P$), which is essential for cell growth and survival. It has been found that an undecaprenol kinase (UK) in the Gram-positive bacterium *Streptococcus mutans* was not only an undecaprenol kinase but also an Mg-ADP-dependent undecaprenyl phosphate phosphatase (UpP) which catalyzes the hydrolysis of $C_{55}P$ to undecaprenol ($C_{55}OH$) and a free phosphate. The lipid derivative $C_{55}OH$ exists predominant in many Gram-positive bacteria but not in Gram-negative bacteria. Therefore, the reciprocal conversion of $C_{55}P$ by the unique bifunctional membrane enzyme UK/UpP and the $C_{55}OH$ pool can efficiently regulate cell wall synthesis of the Gram-positive bacteria [101]. Fructose 2,6-bisphosphate has been found present in most eukaryotic cells and discovered a regulator of carbohydrate metabolism. Fructose-2,6-bisphosphatase purified from *S. cerevisiae* is responsible for the highly specific degradation of fructose 2,6-bisphosphate to convert it to fructose 6-phosphate and inorganic phosphate without showing product inhibition [102].

4.4 Hydrolysis of Epoxides

Epoxide hydrolases (EHs) (EC 3.3.2.3) are enzymes that catalyze the enantio- and regio-selective addition of water to the oxirane ring of either racemic or *meso*-epoxides, resulting in the formation of corresponding enantiopure vicinal diols and the less reactive enantiopure epoxides. These enzymes are ubiquitously found in all types of living organisms, including yeasts, fungi, bacteria, plants, mammals, and insects. Both the microsomal epoxide hydrolase (mEH) and soluble epoxide hydrolase (sEH) belong to α/β-fold hydrolase family of enzymes and are co-factor independent enzymes. Generally, epoxides hydrolases originated from higher organisms are known as detoxification enzymes (animals) and defense enzymes (plants). However, the physiological role of EHs from microbial sources is still not understood [103–107].

Enantiopure epoxides and vicinal diols play a very important role in synthetic organic chemistry because they represent highly versatile chiral synthons and intermediates for preparing biologically active molecules in the pharmaceutical and agrochemical industries. For the epoxide hydrolases catalyzed hydrolysis of epoxides, the proposed reaction mechanism involves a two-step mechanism. In the first step, an S_N2-type nucleophilic attack of an aspartate carboxylic anion in the active site toward the oxirane ring of an epoxide produces a covalent linked β-hydroxylalkyl ester intermediate. Then, in the following second step, an activated water molecule attacks the esters intermediate to release the *vicinal*-diol product. According to this mechanism, one will expect an inversion stereochemistry of the epoxide molecules that *cis*-epoxides will be converted to *threo*-diols and *trans*-epoxides will be transformed into *erythro*-diols [105–107].

For epoxides, the strain of the three-membered oxirane and the polarization of the C–O bond ring make the epoxide molecules quite reactive. The enantiopure form of oxiranes and its hydrolysis product vicinal diols have been used for the synthesis of natural products and pharmacologically active compounds such as protease inhibitors, receptor agonists and antagonists, antiobesity drugs, substituted tetrahydrofurans, nematocides, and anticancer agents. Generally, enantiopure epoxides and vicinal diols are obtained from inexpensive racemic epoxides using EHs through kinetic resolution [107–110]. However, many research studies have demonstrated that the enantioselectivity and regio-selectivity of kinetic hydrolysis of epoxides with EHs are related with the type of substrate and the type or source of EH [108].

Since aryl glycidyl ethers and their related compounds are potentially useful intermediates for synthesizing optically pure amino alcohols and β-blockers, in a study the whole cell yeast *Trichosporon loubierii* ECU1040 catalyzed kinetic resolution of three 3-(nitrophenoxy)propylene oxide isomers (**a-c**) showed a preferential hydrolysis for (*S*)-enantiomers of the epoxides (**a-c**) that yields (*S*)-diols and (*R*)-epoxides with an increased activity for nitro group in the phenyl ring from the 4′-position to the 2′-position (Scheme 4.25). When racemic 3-(2′-nitrophenoxy) propylene oxide was performed at Gram-scale, the product yield of the enantiopure (*R*)-**a** was 41% and the *e.e.* value was 97% [111, 112].

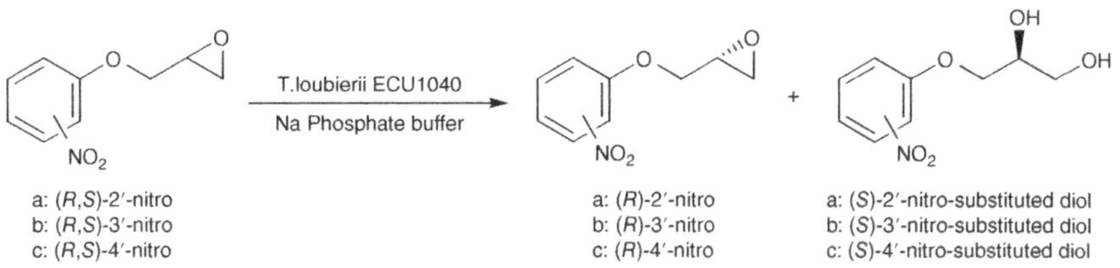

Scheme 4.25 Kinetic resolution of racemic nitro-substituted phenoxypropylene oxides catalyzed by lyophilized cells of *T. loubierii* ECU1040.

In another study, the enantiopreferences of EHs in two different yeasts isolated from soil for the hydrolytic kinetic resolution of 2,2-disubstituted epoxide, (±)-2-methyl-2-pentyl oxirane (**1a**) and (±)-2-methyl-2-phenyl-1,2-epoxypropane (**2a**), were found to be opposite. When yeast *Cryptococcus laurentii* was used for the reaction, the (*R*)-epoxides of both 2,2-disubstituted epoxides were hydrolyzed to their corresponding optically active diols; while both (*S*)-epoxides of both 2,2-disubstituted epoxides were hydrolyzed by yeast *Cryptococcus podzolicus* to their corresponding diols (Scheme 4.26). In addition, some EHs originating from C. *podzolicus* displayed high enough selectivity to be considered for practical use [113]. The effect of the type of enzyme on the enantioselectivity of EHs for epoxides hydrolysis was also demonstrated by the kinetic resolution of (±)-2-methylglycidyl benzyl ether. Epoxide hydrolases obtained from microbes and plants show opposite enantiopreference and stereo-complementary mode of action that led to hetero- and homo-chiral product mixtures. By optimizing the reaction conditions for *Rhodococcus* sp. R312, the deracemization of (±)-2-methylglycidyl benzyl ether can proceed with retention of configuration and enhanced enantioselectivity ($E > 200$) which furnish (*R*)-1-benzyloxy-2-methylpropane-2,3-diol in >97 *e.e.* and 78% yield from racemate by subsequent inverting acid-catalyzed hydrolysis of (*R*)-2-methylglycidyl benzyl ether obtained after the biohydrolysis [114]. The EH activity in marine bacteria isolated from the oil-spilled foreshore of South Korea showed different but complementary enantioselectivity toward the kinetic resolution of chlorostyrene oxides is another good example [115].

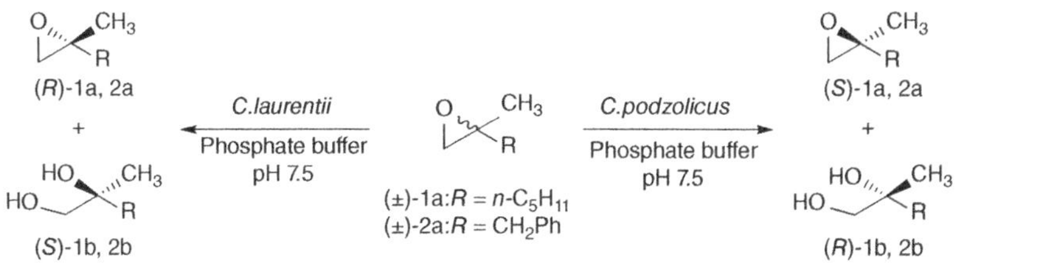

Scheme 4.26 The enantioselectivity of epoxide hydrolases from two different yeasts for the hydrolytic kinetic hydrolysis of 2,2-disubstituted epoxides.

Since the EHs catalyzed kinetic resolution of epoxides in single aqueous solution may be subjected to the problems of spontaneous nonenzymatic hydrolysis to vicinal diols and a very low solubility of the epoxides in aqueous media, a lot of researchers perform the hydrolytic kinetic resolution of epoxides in organic/aqueous biphasic medium to increase the yield and enantioselectivity [116–118]. For instance, racemic 2-, 3-, and 4-chlorostyrene oxides have been enantioselectively hydrolyzed by epoxide hydrolase from *Sphingomonas* sp. HXN-200 in *n*-hexane/aqueous buffer two-phase solution. The result of the hydrolytic kinetic resolution of 3-chlorostyrene epoxide to (*S*)-3-chlorostyrene oxide and (*R*)-3-(3-chlorophenyl)-1,2-propanediol was the best (Scheme 4.27) that gave a 41% yield and 99.1% *e.e.* of (*S*)-3-chlorostyrene oxide, an intermediate for IGF-1R kinase inhibitor BMS-536924, with a preparative biotransformation of 100 mM substrate [116].

In order to increase the efficiency of the biocatalytic resolution of epoxides with EHs, recombinant EH has been widely employed recently and has been used not only in aqueous or organic/aqueous biphasic media but also used in pure organic medium [119–125]. Particularly, recombinant EHs showed broad substrate specificity toward epoxides [120, 121]. For the synthesis of enantiopure epoxide (*R*)-epichlorohydrin, a key intermediate widely used for the synthesis of β-blockers, L-carotine, and ferroelectric liquid crystals, racemic epichlorohydrin (ECH) can be kinetic hydrolyzed by the recombinant EH obtained from *Rhodosporidium toruloides* in aqueous buffer solution [119]. However, it has been found that under the optimized conditions, the kinetic resolution of (*R,S*)-ECH to (*R*)-ECH in isooctane/buffer biphasic medium by whole cells from a recombinant *E. coli* BL21 (DE3) expressing EH produced a yield of 37.5 and 99.3% *e.e.* using an 1-L stirred-tank reactor (Scheme 4.28). When the biphasic reaction is compared to the monophasic aqueous reaction, the substrate concentration and the average productivity

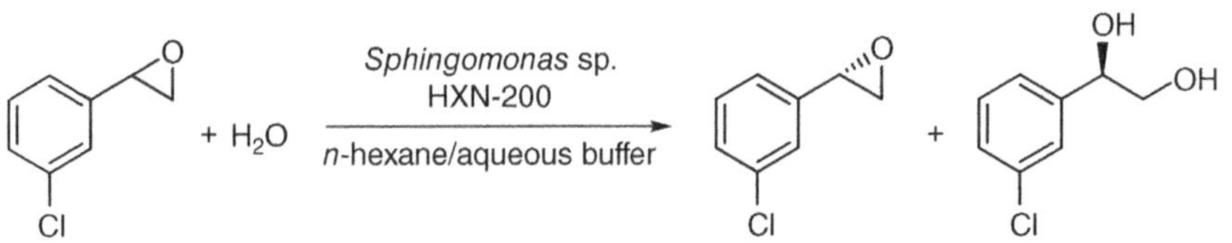

Scheme 4.27 Enantioselective hydrolysis of racemic 3-chlorostyrene oxide to corresponding (*R*)-diols and (*S*)-epoxide by epoxide hydrolase.

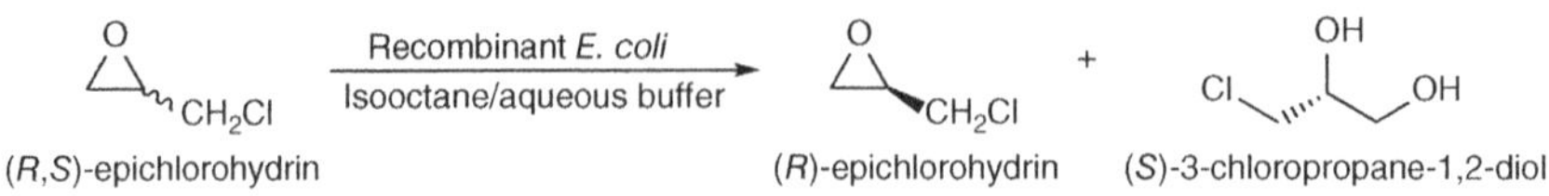

Scheme 4.28 Kinetic resolution of racemic epichlorohydrin for producing (*R*)-epichlorohydrin using recombinant *E. coli* whole cells in biphasic medium.

of (*R*)-ECH are all improved [122]. While in another study, the recombinant EH from *Agromyces mediolanus* ZJB120203 showed completely different enantioselectivity toward the hydrolytic kinetic resolution of (*R*,*S*)-ECH to produce enantiopure (*S*)-ECH, a useful intermediate for the synthesis of atorvastatin, in aqueous buffer solution with an enantiomeric excess (*e.e.*) >99% and a yield 21.5% [120]. The discovery and development of "green" solvents ionic liquids (ILs) have been extensively studied and used for biocatalytic transformations with both whole cells and isolated enzymes [126].

The advantages of ILs in biotransformations include better enzyme stability, substrate and/or product selectivity, and suppression of side reactions. Therefore, hydrophobic and hydrophilic ILs have been employed for the asymmetric hydrolysis styrene oxide to (*R*)-1-phenyl-1,2-ethanediol in either IL/buffer or IL/*n*-hexane/buffer biphasic system by providing better solubility of the substrate and reducing the nonenzymatic hydrolysis to improve the hydrolytic product yield and e.e. value [127, 128].

In addition to the enantioselectivity, regioselectivity is also an important concern for the preparation of chiral epoxides and vicinal diols [129]. In one study, four racemic substituted epoxides, including (±)-hexyl-oxirane, (±)-phenyl-oxirane, (±)-*trans*-1-phenyl-2-methyl-oxirane, and (±)-*cis*-1-phenyl-2-methyl-oxirane, have been used for investigating the regioselectivity of kinetic resolution by whole cells of fungus *Beauveria bassiana* ATCC 7159 and with EH "A" in cell debris and the soluble EH "B." Results indicate that significant differences in regioselectivity of EH "A" and EH "B" were for hexyl-oxirane, phenyl-oxirane, and *cis*-1-phenyl-2-methyl-oxirane but not for *trans*-1-phenyl-2-methyl-oxirane [130]. Thus, the combination of the enantioselectivity and regioselectivity of EHs provides a method to overcome the inherent limitation of a 50% maximum theoretical yield by the kinetic resolution hydrolysis of epoxides using EH that leads to an enantioconvergent process for a 100% theoretical yield of enantiopure diol [108]. Depending on the epoxide structure, there are in general three kinds of strategy can be employed for the enantioconvergent process: the mono-enzymatic approach [131], the bi-enzymatic approach [132, 133], and the chemo-enzymatic approach [134]. Since the regioselectivity of EHs can be different, the attack of EHs to the carbon atoms on the oxirane ring can be different from one enantiomer to the other and may lead to a partially enantioconvergent reaction [135, 136]. In order to achieve an ideally enantioconvergent process of a theoretical 100% yield for either product diols, careful selection of EHs with different enantioselectivities and regioselectivities can be exploited. Therefore, an enantioconvergent production of (*R*)-diol with a 70% *e.e.* and a 100% yield for the enzymatic resolution of racemic 3-pyridyloxirane by EH from *Solanum tuberosum* can be obtained (Scheme 4.29). The enantiopure 2-, 3-, and 4-pyridyloxiranes are key building blocks for the synthesis of β-adrenergic receptor agonists or antiobesity drugs [137]. The cress sEH

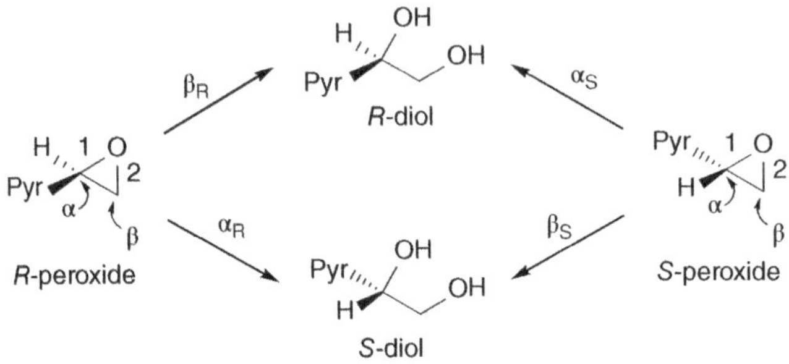

Scheme 4.29 Enantioconvergent processes for 2-, 3-, or 4-pyridyloxirane by the selection of suitable epoxide hydrolase.

catalyzed hydrolysis of *trans*-β-methylstyrene oxide in ILs was also found to be a stereoconvergent process to form the corresponding optically active (1*S*,2*R*)-*erythro*-1-phenylpropane-1,2-diol [138].

Epoxide hydrolases from various sources have been immobilized on a variety of materials such as nano-/microscale UiO-66-NH_2 metal-organic framework (MOF), cross-linked enzyme aggregates (CLEAs), nonporous DEAE-cellulose, Accurel EP-100, gelatin gel, and silica gel, in order to enhance their hydrolytic resolution efficiency and reusability [139–143]. Therefore, the immobilized soybean epoxide hydrolase on nano-/microscale UiO-66-NH_2 MOF has been efficiently used for the asymmetric hydrolysis of 1,2-epoxyoctane to (*R*)-1,2-octanediol in a novel deep eutectic solvent with 81.2% *e.e.* and a 41.4% yield [139]. Epoxide hydrolases from mung bean immobilized on CLEAs have been used for styrene oxide hydrolysis to (*R*)-1-phenyl-1,2-ethanediol in *n*-hexane/buffer biphasic system and a 250-mL preparative scale to give a 93.5% *e.e.* and a 46% yield. The CLEAs can be reused for eight batches and maintained 53% of their activity in biphasic system [140]. When *Aspergillus niger* EH immobilized on epoxide-activated silica gel was employed to perform the kinetic resolution of racemic *p*-nitrostyrene oxide, both the enantioselectivity for unreacted (*S*)-epoxide and formed (*R*)-diol had very high e.e. values of 99 and 92%, respectively [143]. The epoxide hydrolase in crude extract mung bean meal was immobilized in gelatin gel and employed for the stereoselective epoxide ring opening of ethyl *trans*-(±)-3-phenyl glycidate in diisopropyl ether. The results showed an improved enantioselectivity to produce ethyl (2*S*,3*R*)-phenyl glycidate with an e.e. >99% and a yield 45% and corresponding ethyl *anti*-(2*R*,3*R*)-2,3-dihydroxy-3-phenyl propanoate with a diastereomeric excess (d.e.) of 78%, an *e.e.* 94%, and a yield 40% (Scheme 4.30). However, about 5% minor product of ethyl *syn*-(±)-2,3-dihydroxy-3-phenyl propanoate was produced by the spontaneous hydrolysis. Note that both the (2*S*,3*R*)- and

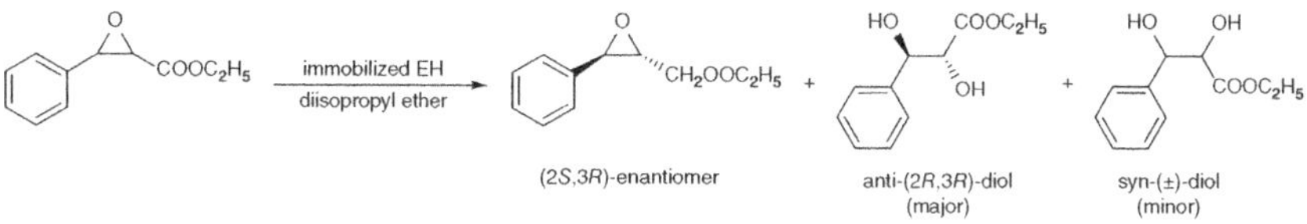

Scheme 4.30 Kinetic resolution of glycidate ester with immobilized crude extract mung bean epoxide hydrolase in diisopropyl ether. *Source:* Based on Clemente-Jiménez et al. [152].

(2*R*,3*S*)-enantiomers of ethyl phenyl glycidate can be used for the Taxol side chain synthesis [142].

Since limonene epoxide hydrolase (LEH) isolated from *R. erythropolis* DCL 14 has unusual secondary structure that catalyzes the hydrolysis of epoxide by a one-step mechanism. In nature, LEH catalyzes only the natural substrate limonene epoxide. However, the engineered LEH mutants show broad substrate scope with high stereoselectivity and except for cyclopentene-oxide prove to be excellent catalysts for the desymmetrization of other meso-epoxides and for the kinetic resolution of racemic substrates [144]. An EH from a variant of *Bacillus megaterium* can be used for the preparative catalytic kinetic resolution of racemic α-naphthyl glycidyl ether in biphasic system (isopropyl ether/isooctane/aqueous) to its enantiopure (*S*)-epoxide and (*R*)-3-(1′-naphthyloxy)-propane-1,2-diol products with a high turnover number to give both an enantioselectivity >99% *e.e.* and high yields of 45.3 and 42.4%, respectively. The (*S*)-epoxide and (*R*)-1,2-diol were subsequently used for the chemical synthesis of (*R*)- and (*S*)-propranolol, a typical β-blocker, with overall yields of 31.4 and 44.8%, respectively. Thus, this chemoenzymatic method using EH provides a new and eco-friendly route for the simultaneous synthesis of biologically active propranolol [145].

4.5 Hydrolysis of Hydantoins

Hydantoinases (dihydropyrimidinase EC 3.5.2.2) are commonly used enzymes in organic synthesis and belong to the cyclic amidohydrolases superfamily that take part in either pyrimidines or purines catabolism in microorganisms, plants, and animals. They are essentially used for the biocatalytic conversion of hydantoins to optically pure amino acids both natural and nonnatural which has been recognized in recent decades as important intermediates for the production of semisynthetic antibiotics, peptide, hormones or enzyme receptors, pyrethroids, pesticides, developmental drugs, as well as food ingredients such as sweeteners and vitamins [146]. Animal and plant hydantoinases have long been partially purified and used for catalyzing the hydrolysis of hydantoin to hydantoic acid as early as 1949. The same enzyme has also been prepared from calf liver and termed hydropyrimidine hydrase later. Based on the study of the metabolism of racemic *N*-substituted 5-phenylhydantoins, the hydrolysis of hydantoins with hydantoinases gave *R*-form of phenyl hydantoic acid and led the finding that dihydropyrimidinase could produce several *N*-carbamoyl-D-amino acids from the corresponding DL-5-monosubstituted hydantoins [147]. In 1978, researchers showed that microbial cells such as bacteria are good catalysts to hydrolyze *N*-carbamoyl-D-amino acids to D-amino acids; thus, enzymatic production of D-amino acids from DL-5-monosubstituted hydantoins were established [148].

In organic synthesis, optically pure amino acids can be produced *via* chemoenzymatic routes that hydantoinase is first used to catalyze the reversible stereospecific hydrolysis of chemically synthesized racemic cyclic hydantoins that yields optically pure carbamoyl amino acids. Then, the corresponding optically pure amino acids can be efficiently obtained by chemical process using acid/base hydrolysis [149, 150]. However, due to the many advantages such as environmentally friendly, energy-efficient, easy product separation and recovery, and, particularly, with almost 100% optical purity, enzymatic synthesis of optically pure amino acids involving hydantoinases has become increasing popular over the past couple of decades. In general, two groups of enzymes, that is, hydantoinases and carbamoylases, with both providing broad spectrum of substrate specificity, are grouped together for α-amino acids synthesis because they form a natural pair. The hydantoinases break the ring of amino acid hydantoins to form the *N*-carbamoyl derivative of amino acids which are further hydrolyzed to produce the free amino acids with a strictly enantioselective step by *N*-carbamoylase [146, 151]. In order to obtain 100% conversion of optically pure α-amino acids from DL-5-monosubstituted hydantoins, a third enzyme called as hydantoin racemase is needed to racemize the remaining nonhydrolyzed 5-monosubstituted hydantoin to the corresponding enantiomeric form which can be hydrolyzed by the hydantoinase. This kind of multienzymatic pathway appeared in 1978 is named as "Hydantoinase Process" and exhibited in many microorganisms (Scheme 4.31) [152].

Since enantiomerically pure D-α-amino acids are important building blocks for a variety of biologically active pharmaceuticals, microbial preparation of D-α-amino acids has been interested in and extensively used for the pharmaceutical industries *via* the enantioselective ring cleaving of hydantoins by D-hydantionases. Two new D-hydantoinases from thermophilic microorganisms have been employed for the production of 14 D-α-amino acids from the corresponding chemically synthesized 5-monosubstituted DL-hydantoins. A quantitative conversion of the racemic starting hydantoins has been obtained by the spontaneous racemization of the hydantoins under a reaction condition of pH 8.5 and 50 °C [153]. Whole cells of recombinant *E. coli* co-expressed three genes encoding D-specific hydantoinase, *N*-carbamoyl-D-amino acid amidohydrolase, and

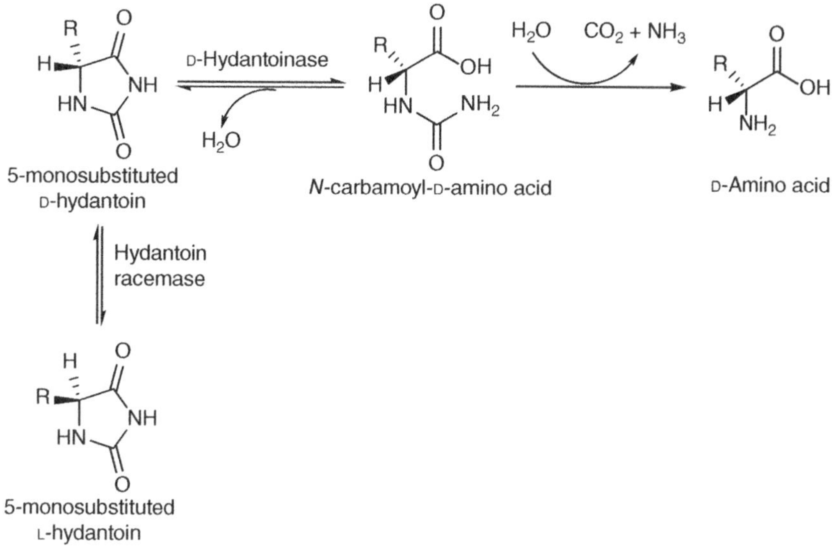

Scheme 4.31 Hydrolysis of DL-5-monosubstituted hydantoins for optically pure D-amino acids production with "Hydantoinase Process."

hydantoin racemase were used for the efficient production of eight D-amino acids including D-phenylalanine, D-tyrosine, D-tryptophan, *O*-benzyl-D-serine, D-valine, D-norvaline, D-leucine, and D-norleucine through the "Hydantoinase Process" from the corresponding DL-5-monosubstituted hydantoins [154]. The substrate specificity of commercial D-hydantoinase immobilized, recombinant, cloned, and expressed in *E. coli* from *Vigna angularis* was explored toward a series of 5-benzylhydantoin derivatives with halogen and methyl substituents in the phenyl ring (Scheme 4.32). The results show that the activity and stereoselectivity toward 5-monosubstituted hydantoins varied significantly with the type of substrate [155]. A series of 17 ring-mono and disubstituted D-phenylglycine derivatives in high optical pure were prepared by enantioselective hydrolysis of corresponding DL-hydantoins using D-hydantoinase followed by *N*-carbamoylase

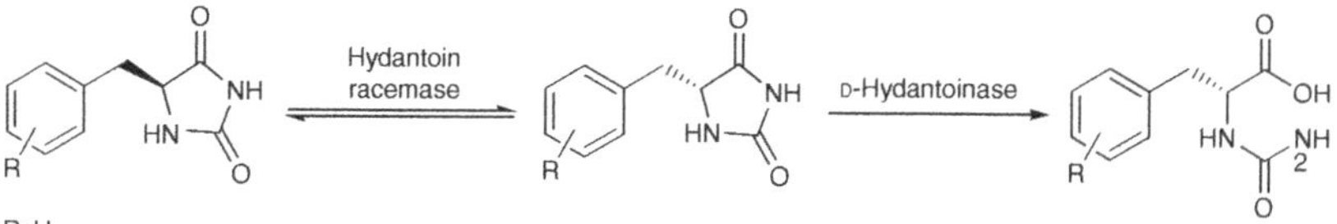

R: H
R: 2-CH_3, 3-CH_3, 4-CH_3, 2,4-CH_3
R: 2-F, 3-F, 4-F
R: 2-Br, 3-Br, 4-Br

Scheme 4.32 Production of *N*-carbamoyl-D-phe derivatives using D-hydantoinase.

of microorganisms or purified enzymes [156]. Free and immobilized D-hydantoinase were used for the stereoselective hydrolysis of DL-benzylhydantoin to produce *N*-carbamoyl-D-phenylalanine in order to survey the activity and stability of the immobilized D-hydantoin. D-Hydantoinase was covalently immobilized on to polystyrene anion-exchange resin *via* glutaraldehyde and used in a circulated packed-bed reactor. The reactions catalyzed by both free and immobilized D-hydantoinases were accelerated under microwave irradiation [157].

Among the D-amino acids, the production of D-*p*-hydroxyphenylglycine using microbial enzymes is of particularly interested and important in industrial application because D-*p*-hydroxyphenylglycine and its derivatives are important side-chain precursors for semisynthetic penicillins and cephalosporins. The industrial large-scale production of D-*p*-hydroxyphenylglycine involves the chemical synthesis of hydantoin substrates, the enzymatic stereospecific hydrolysis of hydantoins, and the enzymatic decarbamoylation. One advantage of this process is the spontaneous racemization of DL-5-*p*-hydroxyphenylhydantoin substrate through base catalysis instead of using hydantoin racemase as described in Scheme 4.30. This process was originally designed for the synthesis of L-amino acid, but in the course of screening, only D-isomers were found to occur during the hydrolysis of various hydantoins. This serendipitous outcome resulted in the commercial production of D-*p*-hydroxyphenylglycine [158, 159]. The "Hydantoinase Process" involving three enzymes (a racemase, an hydantoinase, and a D-*N*-carbamoylase) to perform the dynamic kinetic resolution has also been a prominent industrial application for the production of D-*p*-hydroxyphenylglycine with a 100% yield. This process can be carried out with wild-type strains such as *Agrobacterium radiobacter* or *P. putida*, or recombinant *E. coli*. Cloning of enzymes from thermophiles and intrinsic enzyme modifications by either directed or random mutagenesis has led to improved results [160–162]. The method of directed evolution is the iterative modification of the amino acid sequence of an enzyme until the enzyme functions in the desired manner. An enzyme modified by directed evolution opens the opportunities to reach a variety of required properties such as activity, substrate specificity, thermal and oxidative stability, enantioselectivity, pH range, and tolerance to solvents. Further, the directed evolution of the genomes of whole organisms provides the potential to evolve whole-cell biocatalysts for a sequential process [163]. In a study of the commercial production of D-hydroxyphenylglycine *via* hydantoin-hydrolysis of DL-*p*-hydroxyphenyl hydantoin by wild-type *Agrobacterium tumefaciens* RU-OR cells, the level of hydantoinase and *N*-carbamoylase activity was enhanced by manipulating the growth medium or overexpressing the global nitrogen regulatory factors, NtrBC. The study also shows that the strain used encodes two distinct D-selective *N*-carbamoylases and one of them is a new class of D-*N*-carbamoylases with potentially novel biocatalytic properties [164].

Resting cells of *Sinorhizobium morelense* S-5 have been used for the biotransformation of DL-5-*p*-hydroxyphenylhydantoin to D-*p*-hydroxyphenylglycine with an enhanced activity by the addition of DL-5-(2-indolymethyl)hydantoin. Hydantoinase and carbamoylase involved in this biotransformation showed remarkable thermostability and was strictly D-stereospecific [165]. However, in order to increase the production performance for D-*p*-hydroxyphenylglycine, immobilized cells or immobilized enzymes have been exploited to conduct the reaction. For instances, recombinant *E. coli* encoding D-hydantoinase from *P. putida* was immobilized within calcium alginate beads to produce *N*-carbamoyl-D-hydroxyphenylglycine from DL-*p*-hydroxyphenyl hydantoin with 90% conversion in 24 hours and with increased thermostability and reusability [166]; D-Hydantoinase immobilized on a granulated organic polymer particle was used for the conversion of poorly soluble substrate *p*-hydroxyphenyl hydantoin into soluble *N*-carbamoyl-*p*-D-hydroxyphenylglycine with an increased reaction rate and higher conversion when performed in a stirred-tank pressure-swing reactor [167, 168]; a multienzyme extract from *A. radiobacter* containing D-hydantoinase and *N*-carbamoyl-D-amino acid amidohydrolase was immobilized on chitin for the synthesis of D-*p*-hydroxyphenylglycine by a multistep biotransformation. The reaction with the immobilized enzymes showed greater pH-stability and higher activity at low temperature as compared to the soluble multienzymatic extract [169]. Immobilization of D-hydantoinase from *Vigna angularis* on amminopropyl glass beads using glutaraldehyde as the spacer agent was also used for the enantiospecific hydrolysis of fluorinated *rac*-5-arylhydantoins in a preparative scale reaction to synthesize *N*-carbamoyl-D-*p*-fluorophenylglycine and *N*-carbamoyl-D-*p*-trifluoromethylphenylglycine with a 100% conversion and over 98% enantiomeric excess (Scheme 4.33). These fluorine containing compounds have unique physical and pharmacological characteristics that have potential utility in biological and medicinal application [170].

Although a variety of microorganisms have hydantoinase activity to produce *N*-carbamoyl-D-amino acids and/or D-amino acids, there are a group of microorganisms that are able to convert hydantoins to L-amino acids. The use of resting cells of a novel *P. putida* strain converts 5-alkylhydantoins to corresponding

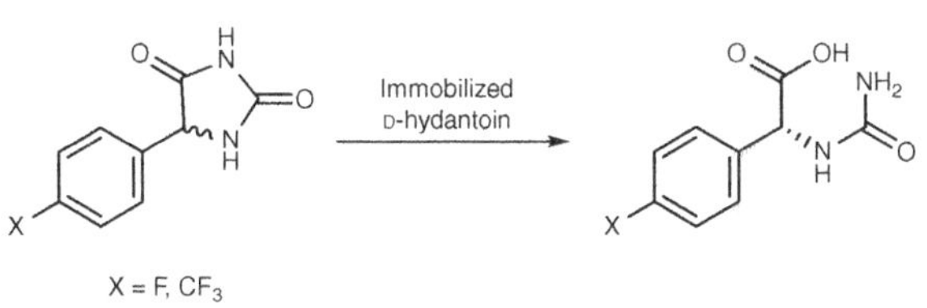

Scheme 4.33 Enantiospecific hydrolysis of fluorinated *rac*-5-arylhydantoins with immobilized D-hydantoinase. *Source:* Lo et al. [175].

L-amino acids such as alanine, valine, and norleucine with yields between 60 and 100%. The hydantoinase is highly active but nonstereoselective, and does not show substrate inhibition or product inhibition by NH_3 while the *N*-carbamoyl amino acid amidohydrolase which converts the *N*-carbamyl amino acids to corresponding amino acids exhibits L-selectivity [171]. The study of the mechanism of stereospecific conversion of DL-5-substituted hydantoins to the corresponding L-amino acids by *Pseudomonas* sp. strain NS671 showed that with the help of hydantoin racemase hydantoinase converts exclusively DL-5-substituted hydantoins to the L-forms of the corresponding *N*-carbamyl-amino acids. Subsequently, *N*-carbamyl-L-amino acid amidohydrolase converts the *N*-carbamyl-L-amino acids to corresponding L-amino acids [172]. The method of directed evolution has also been employed for improving the hydantoinase process to produce L-methionine in *E. coli*. The D-selective hydantoinase from *Arthrobacter* sp. DSM 9771 was converted into an L-selective enzyme by using random mutagenesis, saturation mutagenesis, and screening together with an increase of its total activity fivefold. The clone of the evolved L-hydantoinase, an L-*N*-carbamoylase, and a hydantoin racemase together into *E. coli* led to a strain capable of 100% conversion of 5-(2-methylthioethyl)hydantoin to L-methionine [173, 174]. (*S*)-2-amino-4-phenylbutanoic acid is a useful precursor for the synthesis of many antihypertensive drugs such as enalapril, lisinopril, delapril, and temocapril that act as angiotensin-converting enzyme inhibitors. The "Hydantoinase Process" has also been used for the enantioselective synthesis of (*S*)-2-amino-4-phenylbutanoic acid (L-homophenylalanine) by the combination of recombinant nonenantioselective hydantoinase and L-*N*-carbamoylase to give quantitative yield (Scheme 4.34) [175]. The four enzymes, DL-hydantoinase (preferably hydrolyzing D-hydantoins), hydantoin racemase (racemization of the L-hydantoins to D-hydantoins), carbamoyl racemase (racemization of *N*-carbamoyl-D-amino acids to *N*-carbamoyl-L-amino acids), and L-carbamoylase, were immobilized separately to perform the "double-racemase hydantoinase process" for the

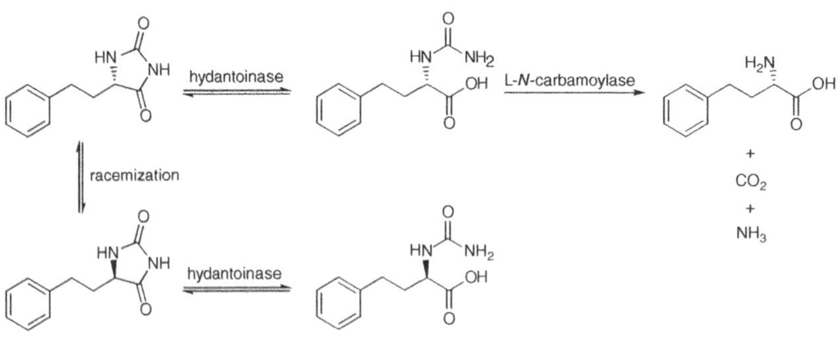

Scheme 4.34 Enantioselective synthesis of L-homophenylalanine by hydantoinase process.

production of L-amino acids from corresponding racemic mixture of hydantoins. The immobilized system was compared with the free enzymes system and results indicated that the reaction velocity increased for L-norvaline, L-norleucine, L-aminobutyric acid, and L-homophenylalanine, while the reaction velocity decreased for L-valine and remained unchanged for L-methionine [176].

Enantiomerically pure β-amino acids are of increasing importance, since they are valuable building blocks for pharmaceuticals and fine chemicals. Some β-amino acids like β-alanine exist as free form in nature. More often, β-amino acids are found as essential components of bioactive compounds or constituents of biologically active secondary metabolites such as taxol, jaspamide, theopalauamide, and dolastatins. β-Amino acids are also used as precursors for synthesizing β-peptides which are able to form very stable and predictable secondary structures and are promising in applications as peptidomimetics. Furthermore, there are cyclized β-amino acids like β-lactam type antibiotics which are important pharmaceutical drugs [177, 178].

It has been proved that the enantioselective hydrolysis of 6-substituted dihydrouracils can be carried out by hydantoinase to their corresponding *N*-carbamoyl derivatives as a route to enantiomerically pure β-amino acids. For example, a range of racemic 6-substituted dihydrouracils have been enantioselective hydrolyzed by the hydantoinase from *V. angularis* to yield the corresponding *N*-carbamoyl-(*S*)-β-amino acids and unreacted (*R*)-dihydrouracils. Free β-amino acids were then obtained by the subsequent treatment of the *N*-carbamoyl derivatives with nitrous acid (Scheme 4.35). High enantioselectivity ($E > 100$) was achieved as the C-6 substituent was an aryl group [179]. A modified hydantoinase process using three recombinant D-hydantoinases and several bacterial strains was used to test the hydrolysis of two aryl-substituted dihydropyrimidines

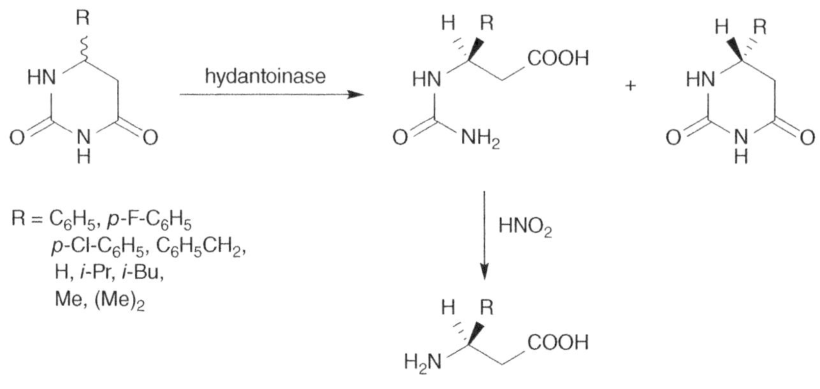

Scheme 4.35 The route of hydantoinase catalyzed hydrolysis of racemic 6-substituted dihydrouracils to β-amino acids. *Source:* Lee et al. [224].

DL-6-phenyl-5,6-dihydrouracil (PheDU) and *para*-chloro-DL-6-phenyl-5,6-dihydrouracil (*p*ClPheDU) to corresponding *N*-carbamoyl-β-amino acids. PheDU can be converted to *N*-carbamoyl-β-phenylalanine (*N*CβPhe) by all applied recombinant D-hydantoinases and eight bacterial strains and some biocatalysts showed an enantioselectivity for either the D- or L-PheDU enantiomers.

The substrate *p*ClPheDU can be hydrolyzed by all three recombinant D-hydantoinases and six of the bacterial strains [178]. Later, whole cell biotransformation with recombinant *E. coli* expressing the hydantoinase from *Arthrobacter crystallopoietes* DSM20117 was used for the enantioselective hydrolysis of chemically synthesized 6-(4-nitrophenyl)dihydropyrimidine-2,4(1*H*,3*H*)-dione (*p*NO_2PheDU). The enzymatic hydrolysis showed (*S*)-enantioselectivity toward *p*NO_2PheDU [177].

In addition to the proteinogenic amino acids, nonproteinogenic amino acids are also useful building blocks in peptide chemistry. They may alter the parent structure of natural proteins and produce interesting modified properties. Silicon-containing amino acids are one of the interesting nonnatural amino acids which show several advantages by their large hydrophobic groups when incorporated into peptides such as the prevention of hydrophobic pocket collapse, higher lipophilicity, and enhanced stability toward proteolytic enzymatic degradation. Therefore, the hydantoinase process with two hydantoinases of different origins and stereoselectivities and one L-*N*-carbamoylase was successfully employed for the highly stereoselective bioconversion of racemic 5-(dimethyl)phenylsilyl-, 5-(1-methyl-1-silacyclopentyl)-, and 5-(trimethylsilyl)-substituted hydantoins with good yields [180]. Sulfur-containing amino acids like D-2-thienylglycine and its derivatives are useful intermediates for the synthesis of β-lactam type antibiotics. D-2-Thienylglycine can be conveniently synthesized by the enantioselective hydrolysis of DL-5-(2-thienyl)hydantoin using D-hydantoinase in washed cells of *Pseudomonas striata* IFO 12996 to form *N*-carbamoyl-D-2-thienylglycine and followed by the chemo-decarbamylation of *N*-carbamoyl-D-2-thienylglycine with nitrite in acidic conditions [181].

4.6 Hydrolysis of Glycosidic Bonds and Natural Polysaccharide

The enzymatic hydrolysis of glycosidic bonds in complex sugars is catalyzed by glycoside hydrolases (GHs) (also called glycosidases or glycosyl hydrolases, EC 3.2.1.-) that leads to the formation of sugar hemiacetal or hemiketal and the corresponding free aglycone. Glycoside hydrolases can catalyze the hydrolysis of not only *O*-linked glycosides but also *N*- and *S*-linked glycosides. According to the site to cleave the glycosidic linkage of the substrate chain, GHs can be classified as either exo- or endo-acting that cleave the nonreducing end of the substrate chain at the end or in the middle, respectively [182–184]. Furthermore, GHs can also be classified as either retaining or inverting enzymes according to the stereoselective hydrolysis mechanism of a glycoside with net retention or inversion of anomeric configuration, respectively [185, 186]. Glycoside hydrolases are a widespread group of enzymes found in nature (microorganisms, plants, and animals) with biological functions including the degradation of biomass such as starch, cellulose, and hemicellulose, the antibacterial defense strategies, the pathogenesis mechanisms, and the normal cellular function. Glycoside hydrolases are found potential use in a variety of industrial applications such as bioethanol production, fine chemical production, food processing, animal feed, paper and pulp processing, laundry and detergents, textile, medical, and pharmaceutical [187–191].

4.6.1 Hydrolysis of Starch and Applications

Starch is a major storage product of many crops such as rice, wheat, maize, potato, tapioca, and yam which is an important nutrition carbohydrate source for human being stored in leaves, glands, seeds, and roots of plants. Starch is the second plentiful heterogeneous polysaccharide produced by plants after cellulose in the form of water insoluble granules. The molecules of starch are constituted by the link of hundreds and thousands of glucose with glycosidic bond that makes the constituent C, H, and O atoms in the ratio of 6:10:5 $(C_6H_{10}O_5)n$. Natural starches are made of about 20–25% amylose and 75–80% amylopectin. Amylose is a linear chain polymer of glucose units linked by α-D-1,4-glycosidic bonds with a molecular weight around 10^6 Dalton (Figure 4.1a) and amylopectin is a branched chain polymer of glucose units linked by 5% α-D-1,6-glycosidic bonds in addition to α-D-1,4-glycosidic bonds and has a molecular weight about 10^8 Dalton (Figure 4.1b). Starch is a major industrial raw material and can be chemically and/or

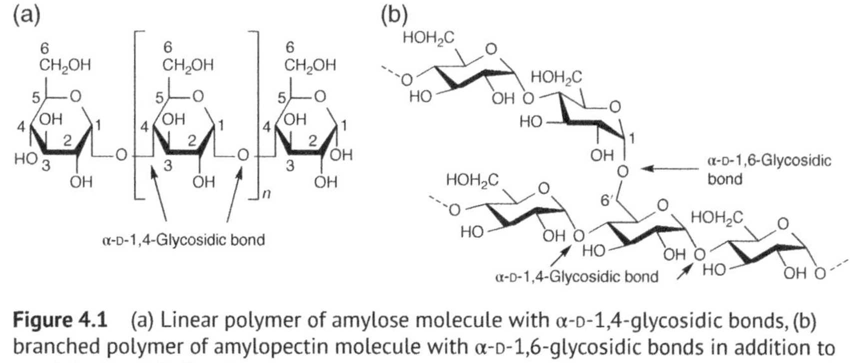

Figure 4.1 (a) Linear polymer of amylose molecule with α-D-1,4-glycosidic bonds, (b) branched polymer of amylopectin molecule with α-D-1,6-glycosidic bonds in addition to α-D-1,4-glycosidic bonds.

enzymatically processed to produce various products for use in a variety of industries such as brewing, biofuel, food and beverage, washing and detergent, paper, and medicinal [182, 188, 189].

Basically, there are four types of starch-converting enzymes: (i) endoamylases; (ii) exoamylases; (iii) debranching enzymes; and (iv) transferases. The enzymes commonly used for catalyzing the hydrolysis of glycosidic bonds of starch into sugars are amylases. Within the category of amylases, α-amylases (EC 3.2.1.1), also known as 1,4-α-D-glucan glucanohydrolase or glycogenase, are calcium metalloenzymes that randomly cleave α-1,4-glycosidic bonds in the inner part (endo-) of amylose or amylopectin chain to produce linear and branched oligosaccharides such as maltotriose, maltose, and α-limit dextrins and glucose. These enzymes are important industrial enzyme obtained from bacteria and fungi [192]. These enzymes are maltogenic and their main products are maltose and other oligomers; so they are not used for liquefaction but are used for producing high maltose syrups in the food industry. β-Amylases (EC 3.2.1.2), also known as 1,4-α-D-glucan maltohydrolase, glycogenase, and saccharogen amylase, are another form of amylase and are found in bacteria, fungi, and plants. These enzymes catalyze the hydrolysis of starch chain to break externally (exo-) the second α-1,4-glycosidic bonds exclusively from the nonreducing end of amylose or amylopectin and therefore produce maltose (two glucose units) and β-limit dextrins. The third type amylase is γ-amylases (EC 3.2.1.3) with alternative names such as amyloglucosidase and glucoamylase. These enzymes are exo-acting amylases that cleave both α-1,4- and α-1,6-glycosidic bonds on the external glucose residues of amylose or amylopectin at the nonreducing end thus yield only glucose. Note that α-glucosidases (EC 3.2.1.20) are another exoamylases which act like γ-amylases [182, 188, 189].

The major market for α-amylases is in the starch industry which is used for the hydrolysis of starch in the starch liquefaction process to produce fructose and glucose syrups. Because of the high sweetening property of syrup, they are used in huge quantities in the food and beverage industry, particularly as sweeteners for soft drinks. The process requires the use of highly thermostable α-amylase for starch liquefaction, which can act at temperatures around 70–100 °C [189, 192]. Maltodextrins, with a general formula of $[(C_6H_{10}O_5)nH_2O, 2 < n < 20]$, are hydrolysis products of starches with low dextrose equivalent (DE) lower than 20 (DE 3-20). Since maltodextrins are digested by human pancreatic α-amylase faster than starch, they are not suitable diet food for diabetic patient owing to their ability to increase blood glucose. However, maltodextrins with low DE (<5) can be used as a fat replacer that have been sold commercially such as from corn starch, potato starch, whole rice, tapioca, and waxy corn. Recently, banana starch has been used as raw material to prepare low DE maltodextrin. Banana starch was extracted from unripe banana fruit var. uli (*Musa* sp.) after steeping in NaOH. Then, the gelatinized banana starch was hydrolyzed by α-amylase from *B. subtilis* using a batch reactor to produce maltodextrin with DE ≈ 3. This banana maltodextrin had a lower digestibility than that of the commercial maltodextrin from waxy corn and potato starch [193]. High amylose starch has been chemically modified with epichlorohydrin to form cross-linked amylose (CLA) which has been used as a polymeric matrix in the oral drug controlled release system. Drug tablets are prepared by direct compression of the dry mixture of CLA, α-amylase, and drug powders. The α-amylase within the tablets is able to hydrolyze α-1,4-glycosidic bonds of the CLA substrate to release drugs. The drug release rate can thus be modulated by the amount of α-amylase which seems to be controlled by two sequential mechanisms: (i) hydration and swelling of CLA tablets followed by (ii) internal enzymatic hydrolysis of the hydrated gel phase [194].

4.6.2 Hydrolysis of Cellulose and Applications

Lignocellulosic materials refer to the plant dry matter and are the most abundantly renewable raw material on Earth for the production of fuels (mainly bioethanol), chemicals, and biochemicals. The main composition of lignocellulose contains two kinds of carbohydrate polymers (cellulose and hemicellulose) and one aromatic polymer (lignin). The two carbohydrate polymers have different sugar monomers (hexose and pentose) which are tightly bound to lignin. Cellulose, a carbohydrate with formula $(C_6H_{10}O_5)n$, is a linear chain water insoluble homopolysaccharide of several hundreds to many thousands β-D-glucose units interlinked by β-1,4-glycosidic bonds (Figure 4.2a). Cellulose is the major structural component of cell-wall of green plants. The hydrogen bonds formed by the multiple hydroxyl groups on glucose from one chain to a neighbor chain side by side make the formation of microfibrils with high tensile strength which are further meshed into a polysaccharide matrix of macrofibril or fiber (Figure 4.2b). In its native state, cellulose consists of alternative crystalline and amorphous regions. Thus, the crystalline nature of cellulose makes it resistant to degradation [190, 195–198].

The degradation of cellulose into small polysaccharides or glucose units means a cheap and nontoxic raw material to produce numerous useful products. However, the recalcitrant structure of cellulose presents a principal challenge for the hydrolysis of the very stable β-1,4-glycosidic bonds between glucose units and the intra- and intermolecular hydrogen bonds between cellulose chains. There are two methods for converting cellulose to glucose viz. chemical and enzymatic hydrolysis. The hydrolysates from chemical hydrolysis are complex including not only fermentable sugars but also other toxic to microbes products such as furfural due to the use of inorganic acids under harsh conditions for the hydrolysis. Application of environmentally friendly enzymes to hydrolyze cellulose can be

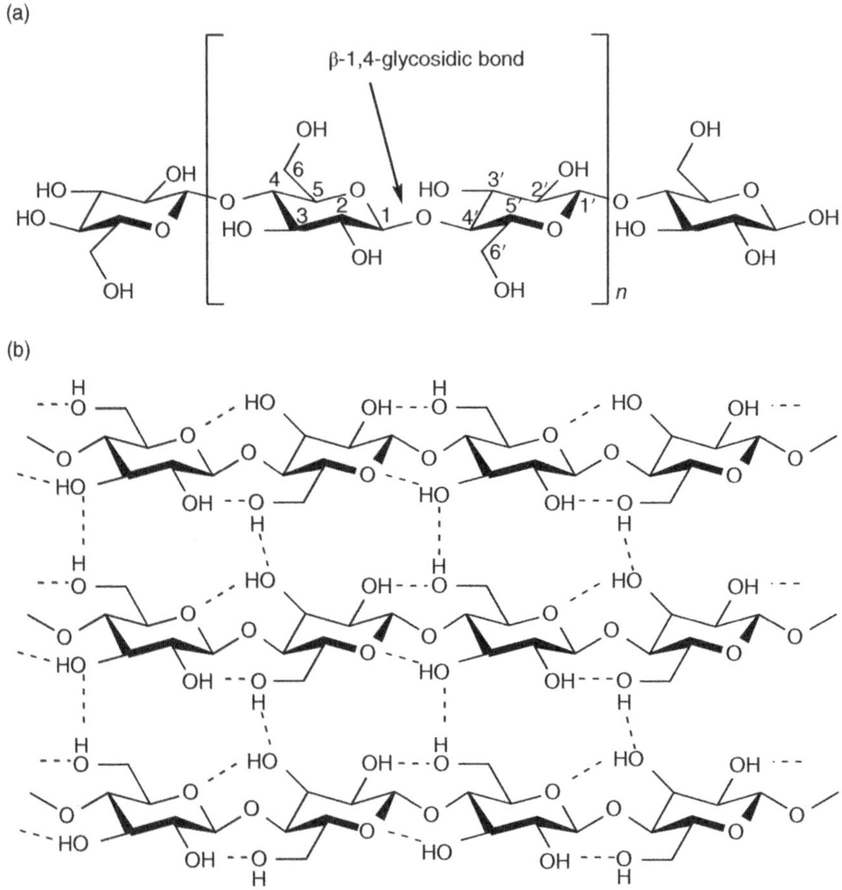

Figure 4.2 (a) Molecular structure of cellulose with β-1,4-glycosidic linkage of D-glucose repeating units, (b) cellulose microfibrils formed by the hydrogen bonds between adjacent linear polymer chains.

performed at mild conditions without producing byproducts that makes the enzymatic hydrolysis more attractive for large scale production [190, 195].

Cellulases, one kind of glycoside hydrolases, are the enzyme used to serve for the purpose of cellulose hydrolysis and are produced chiefly by fungi, bacteria, and protozoans. Conventionally, cellulases are an enzyme complex of three principal enzymes acting synergistically for the complete hydrolysis of cellulose: endoglucanases (EC 3.2.1.4), exoglucanases or cellobiohydrolases (EC 3.2.1.91), and β-glucosidases (EC 3.2.1.21). Endoglucanases attack randomly the internal glycosidic bonds at the amorphous regions of the cellulose that generates new long chain oligomers of nonreducing end. Exoglucanases cleave the β-1,4-glycosidic bonds from the ends of the long chain oligosaccharides produced by endoglucanases to form cellobiose. Exoglucanases are further classified into two types: one that act unidirectionally from the reducing end of the cellulose chain and one that act progressively from the nonreducing end. The liberated cellobioses by synergistic action of endoglucanases and exoglucanases are further hydrolyzed to form glucoses by β-glucosidases [190, 195, 198]. The enzymatic degradation of crystalline cellulose using cellulase is advantageous but there still presents a great challenge by employing it for the effective and economical industrial production of glucose from lignocellulosic materials. In order to obtain an effective enzymatic hydrolysis of cellulose, lignocellulosic materials must be pretreated to improve cellulose accessibility to cellulase for hydrolysis.

At present, the processes used for pretreating lignocellulosic materials including steam explosion, acid or alkaline hydrolysis, ammonia fiber explosion, and organosolv pretreatment. Of these pretreatment processes, steam explosion or hydrothermolysis and acid or alkaline hydrolysis were generally used in early days in order to afford lignocellulosic materials easily degradable by cellulases from fungus *Trichoderma* strains. Steam explosion seems the most used process for the bioconversion of hardwood and agricultural residues to ethanol. However, the high lignin content of steam explosion pretreated substrates has an adverse effect on both the enzymatic cellulose hydrolysis and enzyme recycling. On the other hand, the organosolv pretreatment for lignocellulosic materials produces not only a high hydrolysable cellulose substrate and low lignin content but also the production of potential value-added byproducts with the precipitated lignin from this process [199, 200].

Most of the applications of organosolv pretreatment for delignification of lignocellulosic biomasses such as agricultural and forest residues, grasses, and wood from angiosperms and gymnosperms species were to convert them into a form that can be efficiently enzymatically hydrolyzed to release glucose which then can be readily fermented by yeast *Saccharomyces cerevisiae* to produce ethanol in high yield. Ethanol produced in this way is recognized as bioethanol and is called the second-generation fuel ethanol because they are produced alternative to the sucrose or starch-to-ethanol process. The primitive organosolv pretreatment uses ethanol-water solutions of various ethanol volume percentages (25–70%) that were heated at high temperature (160–200 °C) for different times (15–120 min). The lignocellulosic materials pretreated by organosolv process for effective saccharification and fermentation include *Pinus radiata*, *Acacia dealbata*, *Eucalyptus globulus*, wheat straw, and bagasse [201–205]. In order to strengthen the efficacy of ethanol organosolv pretreatment for lignocellulosic, a catalyst such as sulfuric acid was added to the ethanol-water solution to refine the process. The addition of a catalyst can reduce the use of ethanol and run the pretreatment at lower temperature (160–180°) for just an hour long. Pine trees such as Lodgepole pine (*Pinus contorta*) and Loblolly pine (*Pinus taeda*), sweetgum, hybrid poplar, and sugarcane bagasse have been pretreated with the refined process efficiently to produce fuel ethanol [206–211]. Instead of sulfuric acid various other substances including HCl, NaOH, and $MgCl_2$ have been used for serving as the catalyst of ethanol organosolv process to pretreat sugarcane bagasse, pitch pine (*Pinus rigida*), willow wood, and wheat straw and their effect on the digestibility of biomass, the glucose yield, and the ethanol yield was found to depend upon the kind of biomass, the concentration of catalyst, and the operational conditions [212–214]. Recently, green liquor (GL) produced from pulping process has been used as a catalyst for the ethanol organosolv pretreatment to improve the enzymatic digestibility of biomass and retain maximum levels of polysaccharide in the substrate for enzymatic hydrolysis. GL produced from soda pulping process is a mixture of sodium carbonate and sodium hydroxide which when combined with ethanol and adding anthraquinone as auxiliary agent can efficiently pretreat furfural residues (FRs) to improve delignification and reduce cellulose decomposition thus increase the glucose yield for cellulose hydrolysis with cellulase and β-glucosidase [215]. GL combined ethanol organosolv pretreatment method has been applied for sugarcane bagasse pretreatment that showed an improve of the glucose yield as well [216].

The ethanol organosolv pretreatment for lignocellulosic materials has been modified to improve the glucose yield of enzymatic hydrolysis and the ethanol production. A two-step procedure involving a dilute-acid presoaking step and the ethanol organosolv treatment has been applied for the pretreatment of *Miscanthus x giganteus*. This modified process allowed an efficient fractionation of the raw material into three fractions: a cellulose-rich residue, an ethanol organosolv lignin fraction, and a water-soluble fraction mainly containing hemicellulose sugars [217]. Horticultural waste (HW) used as a raw material for ethanol production has been pretreated using ethanol organosolv process with or without acid catalysts (H_2SO_4, HCl, H_3PO_4) followed by H_2O_2 posttreatment. The results showed that the addition of acid catalysts in the pretreatment process was not critical and H_2O_2 posttreatment was essential for the enhancement of HW digestibility and subsequent fuel ethanol production [218]. Ethanol organosolv process with H_2SO_4 catalyst was modified by the addition of lignin for the pretreatment of loblolly pine. The enhancement of the enzymatic hydrolysis of ethanol organosolv lignin pretreated loblolly pine was influenced by pH [196]. Glycerol or acetone has been substituted for ethanol in the ethanol organosolv pretreatment to form glycerol organosolv pretreatment or acetone organosolv pretreatment for lignocellulosic materials. The results of glycerol organosolv pretreatment for sugarcane bagasse showed good component selectivity that removes approximately 70% lignin and hemicellulose to yield a near-intact preservation (94%) of the overall cellulose and retains extraordinarily few fermentation inhibitors. The remarkable hydrolyzability and fermentability of pretreated substrate from glycerol organosolv pretreatment indicates the glycerol organosolv pretreatment a promising candidate toward industrially relevant enzyme-based lignocelluosic biorefineries [219]. Acidified aqueous glycerol also has been applied for the pretreatment of coconut coir fibers for the bioethanol production *via* simultaneous saccharification and fermentation (SSF) [220]. The cost-effective mild acetone organosolv process including sulfuric acid catalyst has been employed for the pretreatment of herbaceous biomass (wheat straw and corn stover), hardwood (beech, poplar, and birch), and softwood (spruce and pine) to give an effective fractionation of the three main components of lignocellulose constituents for subsequent enzymatic cellulose hydrolysis [221]. Recently, organic electrolyte solutions (OES) formed by room temperature ionic liquid 1-ally-3-methylimidazolium chloride [AMIM]-Cl and polar solvent DMSO was used for the pretreatment of eastern white pine (*Pinus strobes* L., EWP). The glucose yields of the pretreated EWP by OES after enzymatic hydrolyses were remarkably improved by 460–500% and no negative effect on downstream ethanol fermentation was observed [222]. In addition to ethanol organosolv pretreatment, eucalyptus chips were hydrotropic pretreated by using 30% (w/w) sodium xylene sulfonate and 2.0% (w/w) formic acid at 160 °C for 90 minutes to give a more effective removing of lignin and xylose thus a better enzymatic saccharification [204].

Another approach to improve the performance of cellulose hydrolysis with cellulase is to immobilize the cellulase in or on solid matrices which makes them reusable and hence reduces the enzyme-related operating costs. A study showed that the immobilization of cellulase on vinyltrimethoxysilane functionalized FDU-12 mesoporous silica appeared to be a promising approach in terms of high immobilization efficiency, maintained cellulase activity, and provided temporal enzyme stability [223]. Mesoporous silica nanoparticles (MSNs) have been synthesized as well and used for the immobilization of cellulase and magnetic nanoparticles (Fe_3O_4). The synthesized biocatalyst, that is, cellulase-Fe_3O_4@MSN, was applied for the cellulose hydrolysis to produce glucose in a novel integrated, cascading enzyme-/chemocatalytic cellulose conversion into 5-hydroxymethylfurfural (HMF) with a high yield (Scheme 4.36) [224].

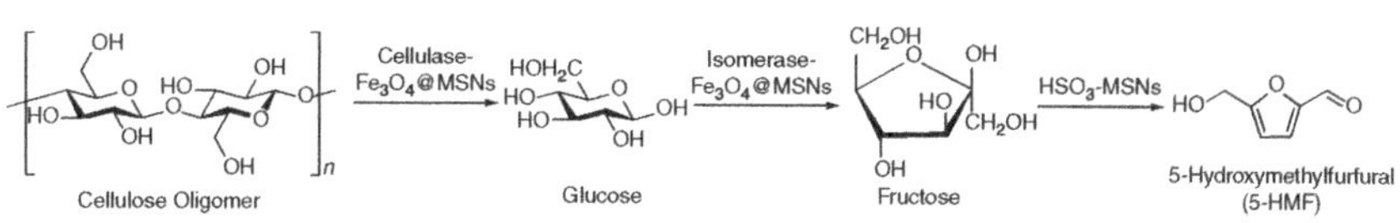

Scheme 4.36 Integrated enzyme cascade-chemocatalytic cellulose conversion into HMF using catalysts based on mesoporous silica nanoparticles. *Source:* Amiri et al. [226].

The production of acetone, butanol, and ethanol (ABE) through fermentation using corn starch and molasses as raw materials has been the second largest biotechnological process in the world after the production of ethanol. However, during the 1960s, the ABE fermentation process was no longer economically viable due to the development of cheaper petrochemical-based process. In recent years, there has been a revival of interest in the ABE fermentation by the increasing concerns about unstable crude oil supplies and dramatic environmental impact of fossil fuels and the use of cost-effective renewable resources from forest and agricultural wastes as alternative substrates instead of oil-derived feedstocks. From this respect, domestic organic waste (DOW) has been used as substrate for ABE production. Two different samples of DOW (fresh and dried DOW) were treated by extrusion in order to expand the polymer fibers present and to obtain a homogeneous mixture. Then the extruded fresh DOW was simultaneously hydrolyzed by a combination of commercial cellulases and β-glucosidases and fermented by solventogenic strain *Clostridium acetobutylicum* ATCC 824 for ABE production [225]. Enzymatic hydrolysis of the rice straw after organosolv pretreatment by 75% (v/v) aqueous ethanol containing 1% w/w sulfuric acid at 180 °C for 30 minutes resulted in a 46.2% glucose yield, which was subsequently fermented by *C. acetobutylicum* to produce 80.3 g butanol, 21.1 g acetone, and 22.5 g ethanol (Scheme 4.37) [226]. Organosolv pretreatment of reed using 70% (v/v) ethanol without catalyst was performed at 40 °C for 6 h. Enzymatic hydrolysis of the pretreated reed using crude cellulase at 50 °C and pH 5.5 for 72 hours resulted in a hydrolysate containing glucose and xylose. Fermentation of the hydrolysate by *C. acetobutylicum* gave a yield of 0.21 and 0.33 $g\ g^{-1}$ for biobutanol and ABE, respectively [227].

In a study, one-step production of lactate from regenerated amorphous cellulose (RAC) can be achieved by using the recombinant cellulolytic *B. subtilis* via anaerobic fermentation. *B. subtilis* XZ7(pBscel5-MT2C) slowly hydrolyzed RAC to soluble sugars and then utilized soluble sugars for cell growth, cellulase synthesis, and lactate production. About 3.1 $g\ L^{-1}$ lactate with a yield of 60% of the theoretical maximum was obtained at the end of fermentation and the lactate yield was enhanced to 63% of the theoretical maximum when 0.1% (w/v) yeast extract was added to the medium [228]. In another study, high enantiomeric excess (>99%) and high yields (0.90 and 0.84 $g\ g^{-1}$) of L-(+)-lactic acid (LA) were obtained by the enzymatic hydrolysis of organosolv pretreated hemp hurds using the enzymes blend CTec2 and further fermentation with thermophilic bacteria *Bacillus coagulans* strain XZL4 and DSM1, respectively. The organosolv pretreatment of hemp hurds was with 65% (v/v) methanolic solution containing 0–3% (w/w) concentrate sulfuric acid. Overall, 42.0 g L-LA/100 g of raw hemp hurds was obtained from the sugar streams generated by organosolv pretreatment, enzymatic hydrolysis of cellulose (C6) and hemicellulose (C5), and fermentation using *B. coagulans* which is

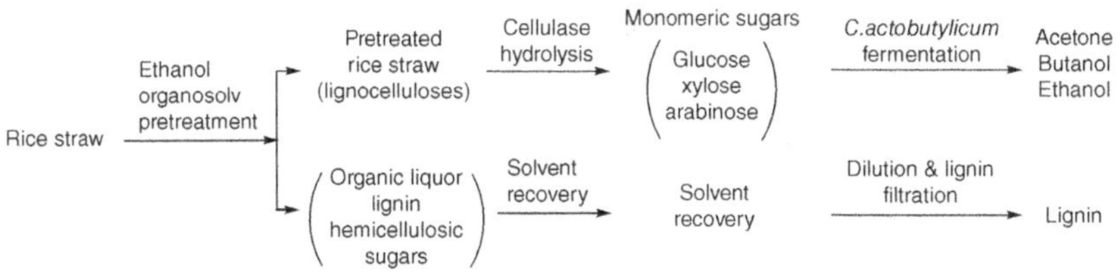

Scheme 4.37 Acetone-butanol-ethanol (ABE) produced from rice straw via ethanol organosolv pretreatment, enzymatic hydrolysis, and fermentation.

promising for lignocellulosic feedstock valorization toward the production of polymer-grade LA [229].

4.6.3 Hydrolysis of Hemicellulose and Applications

Hemicellulose is one of the major components of lignocellulose. Hemicellulose is the second most abundant polysaccharides after cellulose in nature and consists of shorter sugar unit chains than cellulose. Hemicelluloses including xylan, glucuronoxylan, arabinoxylan, glucomannan, and xyloglucan are branched heteropolymers which except small amounts of acetic and uronic acids (D-glucuronic, D-4-*O*-methyl-glucuronic, and D-galacturonic) consist of many different pentose (D-xylose and L-arabinose) and hexose (D-glucose, D-mannose, and D-galactose) sugar monomers with xylose being the most abundant. Xylans, hemicelluloses hydrolyzed to yield xylose, present primarily in the secondary plant cell wall. Xylans are the main hemicelluloses in hardwoods from angiosperms (15–30%) but are less abundant in softwoods from gymnosperms (7–12%). Xylan, a complex highly branched heteropolysaccharide, varies in structure between different plant species with a linear homopolymeric backbone chain of β-D-xylopyranosyl units linked by 1,4-glycosidic bonds that can be substituted to varying degrees with 4-*O*-methyl-D-glucuronopyranosyl units, acetyl group, α-L-arabinofuranosyl units, etc. (Figure 4.3) [191, 230].

Owing to the complex heterogeneous structure of hemicellulose, the complete degradation of hemicellulose involves several hemicellulase enzymes synergistically acting on it. The members of this enzyme system include endo-xylanase (endo-1,4-β-xylanase, EC 3.2.1.8), β-xylosidase (xylan-1,4-β-xylosidase, EC 3.2.1.37), α-glucuronidase (α-glucosiduronase, EC 3.2.1.39), α-arabinofuranosidase (α-L-arabinofuranosidase, EC 3.2.1.55), and acetylxylan esterase (EC 3.1.1.72) and endo-xylanase and β-xylosidase, collectively as xylanases, are the key enzymes responsible for the hydrolysis of xylan (the major constituent of hemicellulose). Endo-xylanase cleaves the 1,4-glycosidic bonds of the homopolymeric xylan backbone resulting in the production of xylooligomers, while β-xylosidase further acts on these degraded xylooligomers thus releasing xylose.

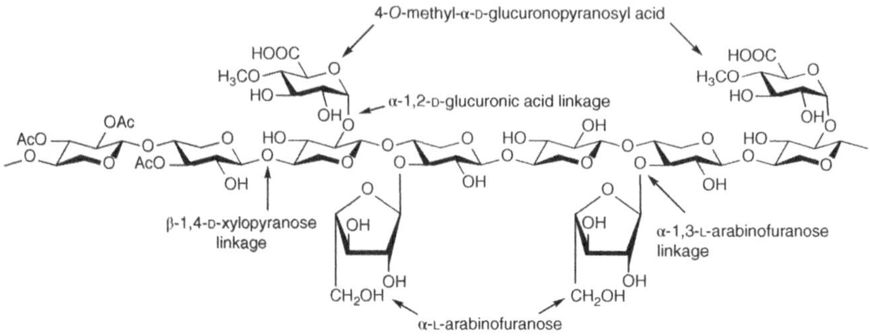

Figure 4.3 Structure of xylan with different side chains attached. *Source:* Based on Juturu and Wu [191]; Polizeli et al. [230].

In nature, hemicellulases are produced by fungi, bacteria, yeast, marine algae, protozoans, snails, crustaceans, insect, seeds, etc.; however, xylanases from filamentous fungi and extremophilic bacteria are industrially interested by their commercial applications as a supplement in animal feed, for the manufacture of bread, food and drinks, textiles, bleaching of cellulose pulp, ethanol, and xylitol production [191, 230].

Novel xylanase and α-glucuronidase from thermophilic bacteria *Alicyclobacillus acidocaldarius* and α-glucuronidase from *Aldicellulosiruptor saccharolyticus* were found remarkable stability at high temperatures (best at 65 °C) and good specificity on beechwood xylan. These enzymes efficiently catalyzed the hydrolysis of (methyl)-glucurono xylans of beechwood and together with a thermophilic β-xylosidase, they catalyzed the smart cooking pretreated *Arudo donax* biomass to produce xylose for second-generation biorefineries [231]. Endo-1,4-β-xylanase from *Trichoderma reesei* expressed in *Pichia pastoris* was exploited for the hemicellulose hydrolysis of the alkaline-acid pretreated kenaf (*Hibiscus cannabinus*). The use of a single enzyme with a batch feeding strategy was proven to yield 59% conversion of pretreated hemicellulosic kenaf after 48 hours incubation at 50 °C in pH 4.0 buffer [232]. Mechanical grinding of the pulp powder was first applied to refine the bamboo prehydrolysis kraft pulp in order to improve the efficiency of subsequent enzymatic hemicellulose removal using xylanase. Results revealed that the xylan content of the bamboo- prehydrolysis kraft pulp (after first stage oxygen delignification) could be decreased to 2.72% (w/w) due to the increasing of accessibility of xylan by the fibrillation of mechanical refining [233].

In order to maintain the enzyme activity, improve the enzyme recovery, and reduce the cost of enzyme reaction, a new strategy was applied to immobilize xylanase within a hydrophilic and nonswelling polyethylene glycol (PEG) net-cloth grafted on a polypropylene nonwoven fabric membrane by a visible light-induced surface graft cross-linking polymerization. The xylanase was *in situ* entrapped within the PEG net-cloth which can effectively maintain the xylanase without linkage in long-term operation. This immobilized xylanase was applied for the hydrolysis of alkali-extracted corncob hemicelluloses. The experimental results showed that the immobilized xylanase retained 80% of its original activity after 25 cycles reuse and 60% after 50 cycles reuse based on the liberated reducing sugars which is far better than other entrapment immobilization methods [234, 235]. The *in vitro* xylanase isolated from fungus *Trichoderma viride* was immobilized on the zeolite matrix activated with HCl as well as to study the hydrolysis of hemicellulose into xylo-oligosaccharide and xylose. The immobilized xylanase activity determined by the reducing sugar (xylose) produced was found greater than the purified xylanase. The immobilized xylanase was stable and reusable for as much as 6 times which afforded the activity 21.3 $\mu g\ g^{-1}\ min^{-1}$ and the conversion efficiency of 56.8% [236].

References

1 Méndez-Sánchez, D., López-Iglesias, M., and Gotor-Fernández, V. (**2016**). *Curr. Org. Chem. 20*: 1186–1203.

2 Vaquero, M.E., Barriuso, J., Martínez, M.J., and Prieto, A. (**2016**). *Appl. Microbiol. Biotechnol. 100*: 2047–2061.

3 Bornscheuer, U.T. (**2002**). *FEMS Microbiol. Rev. 26*: 73–81.

4 Torres, S., Martínez, M.A., Pandey, A., and Castro, G.R. (**2009**). *Bioresour. Technol. 100*: 896–902.

5 Saminathan, M., Muthukumaresan, K.T., Deepar, A.P.J. et al. (**2016**). *Int. J. Adv. Res. Biol. Sci. 3*: 27–37.

6 Romano, A., Gandolfi, R., Molinari, F. et al. (**2005**). *Enzyme Microb. Technol. 36*: 432–438.

7 Da Costa, B.Z., Rodrigues, V.D., de Oliveira, V.M. et al. (**2016**). *Braz. J. Microbiol. 47*: 846–852.

8 Yi, L., Cao, L., Liu, L., and Xi, Z. (**2008**). *Tetrahedron 64*: 8947–8951.

9 Pawlak, J.L. and Berchtold, G.A. (**1987**). *J. Org. Chem. 52*: 1765–1771.

10 Jackson, M.A., Labeda, D.P., and Becker, L.A. (**1995**). *Enzyme Microb. Technol. 17*: 175–179.

11 Vretik, L. and Ritter, H. (**2006**). *Polymer 47*: 1886–1891.

12 Sukumaran, J., van Gool, J., and Hanefeld, U. (**2005**). *Enzyme Microb. Technol. 37*: 254–260.

13 Cozzi, P., Hilgraf, R., and Zimmermann, N. (**2007**). *Eur. J. Org. Chem.* 5969–5994.

14 Herter, S., Nguyen, G.-S., Thomson, M.L. et al. (**2011**). *Appl. Microbiol. Biotechnol. 90*: 929–939.

15 Fillat, A., Romea, P., Urpí, F. et al. (**2014**). *Appl. Microbiol. Biotechnol. 98*: 4479–4490.
16 Jurezak, J., Pikul, S., and Bauer, T. (**1986**). *Tetrahedron 42*: 447–488.
17 Godinho, L.F., Reis, C.R., Tepper, P.G. et al. (**2011**). *Appl. Environ. Microbiol.* 6094–6099.
18 Herdan, J.-M., Balulescu, M., and Cira, O.J. (**1996**). *Mol. Catal. A-Chem. 107*: 409–414.
19 Bousquet, O.R., Braun, J., and Goffic, F.L. (**1995**). *Tetrahedron Lett. 36*: 8195–8196.
20 Jaeger, K.E., Dijkstra, B.W., and Reetz, M.T. (**1999**). *Annu. Rev. Microbiol. 53*: 315–351.
21 Kumar, S., Kikon, K., Upadhyay, A. et al. (**2005**). *Protein Expr. Purif. 41*: 38–44.
22 Sikora, A., Siódmiak, T., and Marszall, M.P. (**2014**). *Chirality 26*: 663–669.
23 Barriuso, J., Vaquero, M.E., Prieto, A., and Martínez, M.J. (**2016**). *Biotechnol. Adv. 34*: 874–885.
24 Tan, C.H., Show, P.L., Ooi, C.W. et al. (**2015**). *Biotechnol. J. 10*: 31–44.
25 Qayed, W.S., Aboraia, A.S., Abdel-Rahman, H.M., and Youssef, A.F. (**2015**). *J. Chem. Pharm. Res. 7*: 311–322.
26 Abdelmoez, W. and Mustafa, A. (**2014**). *J. Oleo Sci. 63*: 545–554.
27 Romano, D., Bonomi, F., de Mattos, M.C. et al. (**2015**). *Biotechnol. Adv. 33*: 547–565.
28 Sharma, R., Chisti, Y., and Banerjee, U.C. (**2001**). *Biotechnol. Adv. 19*: 627–662.
29 Benito-Gallo, P., Franceschetto, A., Wong, J.C.M. et al. (**2015**). *Eur. J. Pharm. Biopharm. 93*: 353–362.
30 Martinelle, M., Holmquist, M., and Hult, K. (**1995**). *Biochim. Biophys. Acta 1258*: 272–276.
31 Hun, C.J., Abd. Rahman, R.N.Z., Salleh, A.B., and Basri, M. (**2003**). *Biochem. Eng. J. 15*: 147–151.
32 Brockman, H. (**2013**). *Lipases*, 729–732. Waltham: Academic Press.
33 Gupta, R., Kumari, A., Syal, P., and Singh, Y. (**2015**). *Prog. Lipid Res. 57*: 40–54.
34 Yousefi, M., Mohammadi, Z.M., and Habibi, Z.J. (**2014**). *Mol. Catal. B-Enzym. 104*: 87–94.
35 Iriuchijima, S. and Kojima, N. (**1982**). *Agric. Biol. Chem. 46*: 1153–1157.
36 Kourist, R. and Bornscheuer, U.T. (**2011**). *Appl. Microbiol. Biotechnol. 91*: 505–517.
37 Eum, H., Kazlauskas, R.J., and Ha, H.-J. (**2014**). *Adv. Syn. Catal. 356*: 3585–3599.
38 Wang, S.-Z., Wu, J.-P., Xu, G., and Yang, L.-R. (**2012**). *Biochem. Eng. J. 65*: 57–62.
39 You, P., Qiu, J., Su, E., and Wei, D. (**2013**). *Eur. J. Org. Chem.* 557–565.
40 Forró, E. and Fülöp, F. (**2012**). *Curr. Med. Chem. 19*: 6178–6187.
41 Ozturk, T.K. and Kilinc, A.J. (**2010**). *Mol. Catal. B-Enzym. 67*: 214–218.
42 Akita, H., Umezawa, I., and Matsukura, H. (**1997**). *Chem. Pharm. Bull. 45*: 272–278.
43 Itoh, T. (**2017**). *Chem. Rev. 117*: 10567–10607.

44 Sookkheo, B., Sinchaikul, S., Phutrakul, S., and Chen, S.-T. (**2000**). *Protein Expres. Purif. 20*: 142–151.

45 Beg, O.K., Sahai, V., and Gupta, R. (**2003**). *Proc. Biochem. 39*: 203–209.

46 Tejaswini, A.V.N., Ranjan, S., and Sridevi, V. (**2014**). *Int. J. Adv. Res. 2*: 456–462.

47 Stauffer, C.E. and Zeffren, E. (**1970**). *J. Biol. Chem. 245*: 3282–3284.

48 Barel, A.O. and Glazer, A.N. (**1968**). *J. Biol. Chem. 243*: 1344–1348.

49 Farmer, D.A. and Hageman, J.H. (**1975**). *J. Biol. Chem. 250*: 7366–7371.

50 Bender, M.L. and Kemp, K.C.J. (**1957**). *Amer. Chem. Soc. 79*: 111–116.

51 Sethi, M.K., Bhandya, S.R., Shukla, R. et al. (**2014**). *Mol. Catal. B-Enzym. 108*: 77–81.

52 Bordusa, F. (**2002**). *Chem. Rev. 102*: 4817–4867.

53 Miyazawa, T., Imagawa, K., Minowa, H. et al. (**2005**). *Tetrahedron 61*: 10254–10261.

54 Miyazawa, T., Imagawa, K., Minowa, H., and Yamada, T.J. (**2006**). *Mol. Catal. B-Enzym. 38*: 73–75.

55 Nishino, N., Arai, T., Ueno, Y., and Ohba, M. (**1996**). *Chem. Pharm. Bull. 44*: 212–214.

56 Rivard, M., Maloň, P., and Čeřovský, V. (**1998**). *Amino Acids 15*: 389–392.

57 Koszelewski, D., Paprocki, D., Brodzka, A., and Ostaszewski, R. (**2017**). *Tetrahedron-Asymmetry 28*: 809–818.

58 Bauer, C.-A., Thomson, R.C., and Blout, E.R. (**1976**). *Biochemistry 15*: 1291–1295.

59 Berezin, I.V., Kazanskaya, N.F., Klyosov, A.A., and Švedas, V.K. (**1973**). *Eur. J. Biochem. 38*: 529–536.

60 Case, A. and Stein, R.L. (**2003**). *Biochemistry 42*: 3335–3348.

61 Pacaud, M. (**1978**). *Eur. J. Biochem. 82*: 439–451.

62 Mϕss, J. and Bundgaard, H. (**1990**). *Pharm. Res. 7*: 751–755.

63 Kudryashova, E.V., Mozhaev, V.V., and Balny, C. (**1998**). *Biochim. Biophys. Acta 1386*: 199–210.

64 Kanelli, M., Vasilakos, S., Ladas, S. et al. (**2017**). *Process Biochem. 59*: 97–103.

65 Wu, Z.-M., Zheng, R.-C., and Zheng, Y.-G. (**2016**). *Enzyme Microb. Technol. 86*: 93–102.

66 Chebrou, H., Bigey, F., Arnaud, A., and Galzy, P. (**1996**). *Biochim. Biophys. Acta 1298*: 285–293.

67 Sharma, M., Sharma, N.N., and Bhalla, T.C. (**2009**). *Rev. Environ. Bio/Technol. 8*: 343–366.

68 Shen, W., Chen, H., Jia, K. et al. (**2012**). *Appl. Microbiol. Biotechnol. 94*: 1007–1018.

69 Novo, C., Farnaud, S., Tata, R. et al. (**2002**). *Biochem. J. 365*: 731–738.

70 Thuku, R.N., Brady, D., Benedik, M.J., and Sewell, B.T. (**2009**). *J. Appl. Microbiol. 103*: 703–727.

71 Nel, A.J.M., Tuffin, I.M., Sewell, B.T., and Cowan, D.A. (**2011**). *Appl. Environ. Microbiol. 77*: 3696–3702.

72 Effenberger, F., Graef, B.W., and Oβwald, S. (**1997**). *Tetrahedron-Asymmetry 8*: 2749–2755.

73 Chen, P., Gao, M., Wang, D.-X. et al. (**2012**). *J. Org. Chem. 77*: 4063–4072.

74 Zheng, R.-C., Jin, J.-Q., Wu, Z.-M. et al. (**2018**). *Bioorg. Chem. 76*: 81–87.

75 Ciskanik, L.M., Wilczek, J.M., and fallon, R.D. Appl. (**1995**). *Environ. Microbiol. 61*: 998–1003.

76 Palermo, G., Campomanes, P., Cavalli, A. et al. (**2015**). *J Phys. Chem. B 119*: 789–801.

77 Dato, F.M., Maaßen, A., Goldfuß, B., and Pietsch, M. (**2018**). *Anal. Biochem. 546*: 50–57.

78 Sugai, T., Yamazaki, T., Yokoyama, M., and Ohta, H. (**1997**). *Biosci. Biotech. Biochem. 61*: 1419–1427.

79 Bhalla, T.C. and Kumar, H. (**2005**). *Can. J. Microbiol. 51*: 705–708.

80 Wang, M.-X. (**2015**). *Acc. Chem. Res. 48*: 602–611.

81 Zheng, Y.-G., Chen, J., Liu, Z.-Q. et al. (**2008**). *Appl. Microbiol. Biotechnol. 77*: 985–993.

82 Cantarella, M., Cantarella, L., Gallifuoco, A. et al. (**2008**). *Enzyme Microb. Technol. 42*: 222–229.

83 Liberti, S., Sacco, F., Calderone, A. et al. (**2013**). *FEBS J. 280*: 379–387.

84 Preechaworapun, A., Dai, Z., Xiang, Y. et al. (**2008**). *Talanta 76*: 424–431.

85 Levine, D., Reid, T.W., and Wilson, I.B. (**1969**). *Biochemistry 8*: 2374–2380.

86 Barrett, H., Butler, R., and Wilson, I.B. (**1969**). *Biochemistry 8*: 1042–1047.

87 Sharifian, S., Homaei, A., Kim, S.-K., and Satari, M. (**2018**). *Proc. Biochem.* 64: 103–115.

88 Toda, H., Yamamoto, M., Uyama, H., and Tabata, Y. (**2016**). *Acta Biomater. 29*: 215–227.

89 Tomomatsu, O., Tachibana, A., Yamauchi, K., and Tanabe, T. (**2008**). *J. Ceram. Soc. Jpn. 116*: 10–13.

90 Yamauchi, K., Goda, T., Takeuchi, N. et al. (**2004**). *Biomaterials 25*: 5481–5489.

91 Stucki, U., Schimid, J., Hämmerle, C.F., and Lang, N.P. (**2001**). *Clin. Oral Impl. Res. 12*: 121–127.

92 Kim, K., Tsay, O.G., Atwood, D.A., and Churchill, D.C. (**2011**). *Chem. Rev. 111*: 5345–5403.

93 De la Peña Mattozzi, M., Tehara, S.K., Hong, T., and Keasling, J.D. (**2006**). *Appl. Environ. Microbiol. 72*: 6699–6706.

94 Franke, D., Machajewski, T., Hsu, C.-C., and Wong, C.-H. (**2003**). *J. Org. Chem. 68*: 6828–6831.

95 Li, a., Cai, L., Chen, Z. et al. (**2017**). *Carboh. Res 452*: 108–115.

96 Li, Z., Cai, L., Qi, Q., and Wang, P.G. (**2011**). *Bioorg. Med. Chem. Lett. 21*: 7081–7084.

97 Yan, M., Yu, L., Zhang, L. et al. (**2014**). *J. Environ. Sci. 26*: 2315–2321.

98 Lung, S.-C., Leung, A., Kuang, R. et al. (**2008**). *Phytochemistry 69*: 365–373.

99 Lim, J.-S., Aquino, M.A.S., and Sykes, A.G. (**1996**). *Inorg. Chem. 35*: 614–618.

100 Liu, C., Dunaway-Mariano, D., and Mariano, P.S. (**2017**). *Tetrahedron 73*: 11324–11330.

101 Huang, L.-Y., Wang, S.-C., Cheng, T.-J.R., and Wong, C.-H. (**2017**). *Biochemistry 56*: 5417–5427.

102 François, J., van Schaftingen, E., and Hers, H.-G. (**1988**). *Eur. J. Biochem. 171*: 599–608.

103 Mischitz, M., Kroutil, W., Wandel, U., and Faber, K. (**1995**). *Tetrahedron-Asymmetry 6*: 1261–1272.

104 Finney, N.S. (**1998**). *Chem. Biol. 5*: R73–R79.

105 Williamson, K.C., Morisseau, C., Maxwell, J.E., and Hammock, B.D. (**2000**). *Tetrahedron-Asymmetry 11*: 4451–4462.

106 Weijers, C.A.G.M. and de Bont, J.A.M.J. (**1999**). *Mol. Catal. B-Enzym. 6*: 199–214.

107 Lee, E.Y. and Shuler, M.L. (**2007**). *Biotechnol. Bioeng. 98*: 318–327.

108 Kotik, M., Archelas, A., and Wohlgemuth, R. (**2012**). *Curr. Org. Chem. 16*: 451–482.

109 Kamble, M.P. and Yadav, G.D. (**2018**). *Catal. Today 309*: 236–241.

110 Zocher, F., Enzelberger, M.M., Bronscheuer, U.T. et al. (**2000**). *J. Biotechnol. 77*: 287–292.

111 Gurjar, M.K., Sasalapure, K., Adhikari, S. et al. (**1998**). *Heterocycles 48*: 1471–1476.

112 Xu, Y., Xu, J.-H., Pan, J. et al. (**2004**). *Mol. Catal. B-Enzym. 27*: 155–159.

113 Botes, A.L., Lotter, J., Rhode, O.H.J., and Botha, A. (**2005**). *Syst. Appl. Microbiol. 28*: 27–33.

114 Simeó, Y. and Faber, K. (**2006**). *Tetrahedron-Asymmetry 17*: 402–409.

115 Woo, J.-H., Kang, K.-M., Kwon, T.-H. et al. (**2015**). *J. Ind. Eng. Chem. 28*: 225–228.

116 Jia, X., Wang, Z., and Li, Z. (**2008**). *Tetrahedron-Asymmetry 19*: 407–415.

117 Liu, Z., Michel, J., Wang, Z. et al. (**2006**). *Tetrahedron-Asymmetry 17*: 47–52.

118 Gong, P.-F. and Xu, J.-H. (**2005**). *Enzyme Microb. Technol. 36*: 252–257.

119 Woo, J.-H., Kang, J.-H., Hwang, Y.-O. et al. (**2010**). *J. Biosci. Bioeng.* 109: 539–544.

120 Xue, F., Liu, Z.-Q., Zou, S.-P. et al. (**2014**). *Process Biochem. 49*: 409–417.

121 Liu, Z.-Q., Zhang, L.-P., Cheng, F. et al. (**2011**). *Catal. Commun.* 16: 133–139.

122 Zou, S., Yan, H., Hu, Z., and Zheng, Y. (**2013**). *Chin. J. Catal. 34*: 1339–1347.

123 Hu, D., Wang, R., Shi, X.-L. et al. (**2016**). *J. Biotechnol. 236*: 152–158.

124 Beloti, L.L., Costa, B.Z., Toledo, M.A.S. et al. (**2013**). *Protein Expr. Purif. 91*: 175–183.

125 Karboune, S., Archelas, A., and Baratti, J. (**2006**). *Enzyme Microb. Technol. 39*: 318–324.

126 van Rantwijk, F. and Sheldon, R.A. (**2007**). *Chem. Rev. 107*: 2757–2785.

127 Chen, W.-J., Lou, W.-Y., and Zong, M.-H. (**2012**). *Bioresour. Technol. 115*: 58–62.

128 Chen, W.-J., Lou, W.-Y., Yu, C.-Y. et al. (**2012**). *J. Biotechnol.* 162: 183–190.

129 Li, C., Hu, D., Zong, X.-C. et al. (**2017**). *Catal. Commun. 102*: 57–61.

130 Moussou, P., Archelas, A., Furstoss, R., and Baratii, J.C. (**2000**). *Enzyme Microb. Technol. 26*: 414–420.

131 Wu, Y.-W., Kong, X.-D., Zhu, Q.-Q. et al. (**2015**). *Catal. Commun. 58*: 16–20.

132 Manoj, K.M., Archelas, A., Baratti, J., and Furstoss, R. (**2001**). *Tetrahedron 57*: 695–701.

133 Min, J.Y. and Lee, E.Y. (**2012**). *J. Ind. Eng. Chem. 18*: 160–164.

134 Orru, R.V.A., Kroutil, W., and Faber, K. (**1997**). *Tetrahedron Lett. 38*: 1753–1754.

135 Shimizu, K.-I., Sakamoto, M., Hamada, M. et al. (**2010**). *Tetrahedron-Asymmetry 21*: 2043–2049.

136 Kotik, M., Štěpánek, V., Marešová, H. et al. (**2009**). *Mol. Catal. B-Enzym. 56*: 288–293.

137 Genzel, Y., Archelas, A., Broxterman, Q.B. et al. (**2002**). *Mol. Catal. B-Enzym. 16*: 217–222.

138 Chiappe, C., Leandri, E., Lucchesi, S. et al. (**2004**). *Mol. Catal. B-Enzym. 27*: 243–248.

139 Cao, S.L., Yue, D.-M., Li, X.-H. et al. (**2016**). *ACS Sustain. Chem. Eng. 4*: 3586–3595.

140 Yu, C.-Y., Li, X.-F., Lou, W.-Y., and Zong, M.-H. (**2013**). *J. Biotechnol. 166*: 12–19.

141 Karboune, S., Archelas, A., and Baratti, J.C. (**2010**). *Process Biochem. 45*: 210–216.

142 Devi, A.V., Lahari, C., Swarnalatha, L., and Fadnavis, N.W. (**2008**). *Tetrahedron-Asymmetry 19*: 1139–1144.

143 Petri, A., Marconcini, P., and Savadori, P.J. (**2005**). *Mol. Catal. B-Enzym. 32*: 219–224.

144 Zhang, H. and Reetz, M.T. (**2010**). *J. Am. Chem. Soc. 132*: 15744–15751.

145 Kong, X.-D., Yu, H.-L., Yang, S. et al. (**2015**). *Mol. Catal. B-Enzym. 122*: 275–281.

146 Engineer, A.S., Dhakephalkar, A.P., Gaikaiwari, R.P., and Dhakephalkar, P.K. (**2013**). *J. Ind. Microbiol. Biotechnol. 40*: 1467–1372.

147 Cecere, F., Galli, G., and Morisi, F. (**1975**). *FEBS Lett. 57*: 192–194.

148 Ogawa, J. and Shimizu, S.J. (**1997**). *Mol. Catal. B-Enzym. 2*: 163–176.

149 Gokhale, D.V., Bastawde, K.B., Patil, S.G. et al. (**1996**). *Enzym. Microb. Technol. 18*: 353–357.

150 Ohishi, T., Nanba, H., Sugawara, M. et al. (**2007**). *Tetrahedron Lett. 48*: 3437–3440.

151 Martínez-Rodríguez, S., Martínez-Gómez, A.I., Rodríguez-Vico, F. et al. (**2010**). *Appl. Microbiol. Biotechnol. 85*: 441–458.

152 Clemente-Jiménez, J.M., Martínez-Rodríguez, S., Rodríguez-Vico, F., and Heras-Vázquez, F.J.L. (**2008**). *Recent Pat. Biotechnol. 2*: 35–46.

153 Keil, O., Schneider, M.P., and Rasor, J.P. (**1995**). *Tetrahedron-Asymmetry 6*: 1257–1260.

154 Nozaki, H., Takenaka, Y., Kira, I. et al. (**2005**). *J. Mol. Catal. B-Enzym. 32*: 213–218.

155 Latacz, G. and Kieć-Kononowicz, K. (**2015**). *Appl. Biochem. Biotechnol. 175*: 698–704.

156 Garcia, M.J. and Azerad, R. (**1997**). *Tetrahedron-Asymmetry 8*: 85–92.

157 Jia, H.-H., Ni, F., Chen, M.-J. et al. (**2006**). *Mol. Catal. B-Enzym. 43*: 74–79.

158 Ogawa, J. and Shimizu, S. (**1999**). *Trends Biotechnol. 17*: 13–20.

159 Ogawa, J. and Shimizu, S. (**2002**). *Curr. Opin. Biotechnol. 13*: 367–375.

160 Schule, B. and Wubbolts, M.G. (**1999**). *Curr. Opin. Biotechnol. 10*: 609–615.

161 Hartley, C.J., Kirchmann, S., Burton, S.G., and Dorrington, R.A. (**1998**). *Biotechnol. Lett. 20*: 707–711.

162 Jiwaji, M. and Dorringto, R.A. (**2009**). *Appl. Microb. Cell Physiol. 84*: 1169–1179.

163 Hibbert, E.G., Baganz, F., Hailes, H.C. et al. (**2005**). *Biomol. Eng. 22*: 11–19.

164 Jiwaji, M., Hartley, C.J., Clark, S.-A. et al. (**2009**). *Enzym. Microb. Technol. 44*: 203–209.

165 Wu, S., Yang, L., Liu, Y. et al. (**2005**). *Enzym. Microb. Technol. 36*: 520–526.

166 Chen, Y.-C., Yin, B.-D., Lin, S.-C., and Hsu, W.-H. (**1999**). *Process Biochem. 35*: 285–290.

167 Lee, C.-K. and Lin, K.-C. (**1996**). *Enzym. Microb. Technol. 19*: 623–627.

168 Lee, C.-K., Fan, C.-H., and Yang, P.-F. (**2001**). *Biochem. Eng. J. 7*: 233–239.

169 Aranaz, I., Ramos, V., De La Escalera, S., and Heras, A. (**2003**). *Biocatal. Biotransform. 21*: 349–356.

170 Arcuri, M.B., Sabino, S.J., Antunes, O.A.C., and Oestreicher, E.G.J. (**2003**). *Flor. Chem. 121*: 55–56.

171 Buchanan, K., Burton, S.G., Dorrington, R.A. et al. (**2001**). *Mol. Catal. B-Enzym. 11*: 397–406.

172 Ishikawa, T., Watabe, K., Mukohara, Y., and Nakamura, H. (**1997**). *Biosci. Biotech. Biochem. 61*: 185–187.

173 May, O., Nguyen, P.T., and Arnold, F.H. (**2000**). *Nat. Biotechnol. 18*: 317–320.

174 Razor, J.P. and Voss, E. (**2001**). *Appl. Catal. A-Gen. 221*: 145–158.

175 Lo, H.-H., Kao, C.-H., Lee, D.-S. et al. (**2003**). *Chirality 15*: 699–702.

176 Rodríguez-Alonso, M.J., Rodríguez-Vico, F., Heras-Vázquez, F.J.L., and Clemente-Jiménez, J.M. (**2017**). *Catalysis 7*: 192–202.

177 Slomka, C., Zhong, S., Fellinger, A. et al. (**2015**). *AMB Expr. 5*: 85.

178 Engel, U., Syldatk, C., and Rudat, J. (**2012**). *Appl. Microbiol. Biotechnol. 94*: 1221–1231.

179 O'Neill, M., Hauer, B., Schneider, N., and Turner, N.J. (**2011**). *ACS Catal. 1*: 1014–1016.

180 Smith, R.J., Pietzsch, M., Waniek, T. et al. (**2001**). *Tetrahedron-Asymmetry 12*: 157–165.

181 Shimizu, S., Shimada, H., Takahashi, S. et al. (**1980**). *Agric. Biol. Chem. 44*: 2233–2234.

182 Hii, S.L., Tan, J.S., Ling, T.C., and Ariff, A.B. (**2012**). *Enzym. Res.* 921362.

183 Franková, L. and Fry, S.C. (**2013**). *J. Exp. Bot. 64*: 3519–3550.

184 Kanda, T., Brewer, C.F., Okada, G., and Hehre, E.J. (**1986**). *Biochemistry 25*: 1159–1165.

185 Sinnott, M.L. (**1990**). *Chem. Rev. 90*: 1171–1202.

186 Davies, G. and Henrissat, B. (**1995**). *Structure 3*: 853–859.

187 Cobucci-Ponzano, B., Perugino, G., Rossi, M., and Moracci, M. (**2011**). *Protein Eng. Des. Sel. 24*: 21–26.

188 Esmaeili, S. and Noorolahi, z. (**2017**). *Carpath. J. Food Sci. Technol. 9*: 114–127.

189 Van der Maarel, M.J.E.C., van der Veen, B., Uitdehaag, J.C.M. et al. (**2002**). *J. Biotechnol. 94*: 137–155.

190 Juturu, V. and Wu, J.C. (**2014**). *Renew. Sust. Energ. Rev. 33*: 188–203.

191 Juturu, V. and Wu, J.C. (**2012**). *Biotechnol. Adv. 30*: 1219–1227.

192 Gupta, R., Gigras, P., Mohapatra, H. et al. (**2003**). *Process Biochem. 38*: 1599–1616.

193 Yusraini, E., Hairiyadi, P., and Kusnandar, F. (**2013**). *Starch-Starke 65*: 312–321.

194 Dumoulin, Y., Cartilier, L.H., and Mateescu, M.A. (**1999**). *J. Control. Release 60*: 161–167.

195 Bégiun, P. and Aubert, J.-P. (**1994**). *FEMS Microbiol. Rev. 13*: 25–58.

196 Lai, C., Tu, M., Yong, Q., and Yu, S. (**2018**). *RSC Adv. 8*: 13835–13841.

197 Zhang, K.-D., Li, W., Wang, Y.-F. et al. (**2018**). *Biomacromolecules 19*: 1686–1696.

198 Niu, H., Shah, N., and Kontoravdi, C. (**2016**). *Biochem. Eng. J. 105*: 455–472.

199 Tu, M., Pan, X., and Saddler, J.N.I. (**2009**). *Agic. Food Chem. 57*: 7771–7778.

200 Bonn, G., Hörmeyer, H.F., and Bobleter, O. (**1987**). *Wood Sci. Technol. 21*: 179–185.

201 Muñoz, C., Mendonça, R., Baeza, J. et al. (**2007**). *J. Chem. Technol. Biotechnol. 82*: 767–774.

202 Muñoz, C., Baeza, J., Freer, J., and Mendonça, T.R. (**2011**). *J. Ind. Microbiol. Biotechnol. 38*: 1861–1866.

203 Huijgen, W.J.J., Smit, A.T., deWild, P.J., and den Uil, W.H. (**2012**). *Bioresour. Technol.* 114: 389–398.

204 Mou, H. and Wu, S. (**2016**). *Bioresour. Technol. 220*: 637–640.

205 Wei, W., Wu, S., and Xu, S. (**2017**). *J. Chem. Technol. Biotechnol. 92*: 580–587.

206 Pan, X., Gilkes, N., Kadla, J. et al. (**2006**). *Biotechnol. Bioeng. 94*: 851–861.

207 Pan, X., Xie, D., Yu, R.W. et al. (**2007**). *Ind. Eng. Chem. Res. 46*: 2609–2617.

208 Pan, X., Xie, D., Yu, R.W., and Saddler, J.N. (**2008**). *Biotechnol. Bioeng. 101*: 39–48.

209 Mesa, L., Gonzáler, E., Cara, C. et al. (**2010**). *J. Chem. Technol. Biotechnol. 85*: 1092–1098.

210 Sannigrahi, P., Miller, S.J., and Ragauskas, A.J. (**2010**). *Carbohyd. Res. 345*: 965–970.

211 Lai, C., Tu, M., Li, M., and Yu, S. (**2014**). *Bioresour. Technol. 156*: 92–99.

212 Mesa, L., Gonzáler, E., Ruiz, E. et al. (**2010**). *Appl. Energy 87*: 109–114.

213 Park, N., Kim, H.-Y., Koo, B.-W. et al. (**2010**). *Bioresour. Technol. 101*: 7046–7053.

214 Huijgen, W.J.J., Smit, A.T., Reith, J.H., and den Uil, H. (**2011**). *J. Chem. Technol. Biotechnol. 86*: 1428–1438.

215 Yu, H., Xing, Y., Lei, F. et al. (**2014**). *Bioresour. Technol. 167*: 46–52.

216 Yu, H., You, Y., Lei, F. et al. (**2015**). *Bioresour. Technol. 187*: 161–166.

217 Brosse, N., Sannigrahi, P., and Ragauskas, A. (**2009**). *Ind. Eng. Chem. Res. 48*: 8328–8334.

218 Geng, A., Xin, F., and Ip, J.-Y. (**2012**). *Bioresour. Technol. 104*: 715–721.

219 Sun, F.F., Zhao, X., Hong, J. et al. (**2016**). *Biotechnol. Biofuels 9*: 1–13.

220 Ebrahimi, M., Caparanga, A.R., Ordono, E.E., and Villaflores, O.B. (**2017**). *Renew. Energy 109*: 41–48.

221 Smit, A. and Hijgen, W. (**2017**). *Green Chem. 19*: 5505–5514.

222 Tian, X., Rehmann, L., Xu, C.C., and Fang, Z. (**2016**). *ACS Sustainable Chem. Eng. 4*: 2822–2829.

223 Nartono, S.B., Qia, S.Z., Liu, J. et al. (**2010**). *J. Phys. Chem. C 114*: 8353–8362.

224 Lee, Y.-C., Dutta, S., and Wu, K. (**2014**). *C-W. Chem. Sus. Chem. 7*: 3241–3246.

225 López-Contreras, A.M., Claassen, P.A.M., Mooibroek, H., and De Vos, W.M. (**2000**). *Appl. Microbiol. Biotechnol. 54*: 162–167.

226 Amiri, H., Karimi, K., and Zilouei, H. (**2014**). *Bioresour. Technol. 152*: 450–455.

227 Zhu, Y., Xin, F., Chang, Y. et al. (**2015**). *Biomass Bioenerg. 76*: 24–30.

228 Zhang, X.-Z., Sathitsuksanoh, N., Zhu, Z., and Zhang, Y.-H.P. (**2011**). *Metab. Eng. 13*: 364–372.

229 Gandolfi, S., Pistone, L., Ottolina, G. et al. (**2015**). *Bioresour. Technol. 191*: 59–65.

230 Polizeli, M.L.T.M., Rizzatti, A.C.S., Monti, R. et al. (**2005**). *Appl. Microbiol. Biotechnol. 67*: 577–591.

231 Cobucci-Ponzano, B., Strazzulli, A., Iacono, R. et al. (**2015**). *Enzym. Microb. Technol. 78*: 63–73.

232 Azelee, N.I.W., Jahim, J.M., Ismail, A.F. et al. (**2016**). *Am. J. Agric. Biol. Sci. 11*: 54–66.

233 Zhao, L., Yuan, Z., Kapu, N.S. et al. (**2017**). *Bioresour. Technol. 223*: 40–46.

234 Graebin, N.G.; da N. Schöffer, J.; de Andrades, D.; Hertz, P.F.; Ayub, M.A.Z.; Rodrigues, R.C (**2016**). *Molecules 21*, 1074-1111.

235 Zhang, L., Ma, Y., Zhao, C. et al. (**2016**). *Ind. Eng. Chem. Res. 55*: 6354–6364.

236 Na'imah, J., Prasetyawan, S., and Srihardyastutie, A. (**2017**). *J. Pure Appl. Chem. Res. 6*: 181–188.

5

Organic Synthesis with Lyases

Lyases are enzymes that catalyze the bond generation and the bond breaking with the addition or removal of groups from their substrates. Lyases catalyze the bond breaking by means other than hydrolysis or redox reactions, often forming new double bonds or ring structures. Lyases are classified as EC 4 according to the EC number classification of enzymes, and they can be further classified into seven subclasses, depending upon the bond type involved. EC 4.1.x.x are lyases with carbon–carbon bonds, and they are the most important and abundant subclass. EC 4.2.x.x are lyases with carbon–oxygen bonds. EC 4.3.x.x are lyases with carbon–nitrogen bonds. All these three subclasses have four subcategories. EC 4.4.x.x, EC 4.5.x.x, EC 4.6.x.x, and EC 4.99.x.x are lyases involved with carbon–sulfur bonds, carbon–halide bonds, phosphorus–oxygen bonds, and other bonds, respectively [1–3].

5.1 Lyases with Carbon–Carbon Bonds

5.1.1 Aldolases in Organic Syntheses

Enzyme-catalyzed stereoselective reactions forming new carbon–carbon bonds are of utmost important in synthetic organic chemistry [1, 4–6]. However, the enzymatic aldol reaction is recognized as one of the most useful tools for the construction of C–C bonds to generate up to two new stereogenic centers that allows the design of a wide range of both natural and novel poly-hydroxylated compounds in the synthesis of building blocks for natural product and drug development [1, 4, 7]. Therefore, for the preparation of those biologically organic compounds, such as carbohydrates and amino acids, with structurally elaborated and water-soluble and with several functional groups, the enzymatic aldol reaction is particularly attractive to avoid the requirement of extensive use of

Enzyme-Based Organic Synthesis, First Edition. Cheanyeh Cheng.

protective groups and the demanding of stereochemical control by conventional synthetic approaches.

Aldolases, ubiquitous enzymes in nature, are rooted to the sugar metabolic cycle and are found in the biosynthetic pathways of carbohydrates, keto acids, and amino acids and in a variety of catabolic pathways of carbohydrates. Among the known aldolase, most of them catalyze the stereoselective aldol reaction with the addition of a ketone to an aldehyde. The aldolases are divided into two classes on a mechanistic basis. The class I aldolases activate the donor substrates by forming a Schiff base as an intermediate with a strictly conserved ε-NH_2 group of a lysine residue in the active site. The enamine tautomer formed in the active site then attacks with high stereoselectivity to the bound acceptor aldehyde. The enzyme-bound imine is then hydrolyzed to release the product. The class II aldolases contain a Zn^{2+} cofactor coordinated to three histidine nitrogen atoms in the active site which acts as a Lewis acid to activate the bound donor substrate [1, 4]. According to the structure of donor substrate, aldolases are divided into four groups: dihydroxyacetone phosphate- (DHAP), pyruvate-, acetaldehyde-, and glycine-dependent aldolases [1, 7, 8].

The class I fructose 1,6-biphosphate (FBP) aldolase isolated from rabbit muscle is the most employed enzyme for reversibly catalyzing the addition of dihydroxyacetone phosphate (DHAP) to D-glyceraldehyde 3-phosphate (D-G3P) to produce D-fructose 1,6-biphosphate (FBP) (Scheme 5.1) [1, 4, 7–9]. The broad range of aldehyde acceptors for this DHAP-dependent enzyme has made the reaction a versatile tool in the synthesis of monosaccharides and sugar analogs [10], alkaloids [11], and other natural products. For example, the FBP aldolase-mediated reaction of DHAP and an aldehyde is a key step for the total synthesis of microbial elicitor (−)-syringolide 2 [1, 12]. The other class I tagatose 1,6-diphosphate (TDP) aldolase also uses D-G3P as the acceptor when catalyzes DHAP aldol reaction to give D-tagatose 1,6-diphosphate (TDP) with complementary (3*S*, 4*S*)-*erythro* configuration to the (3S, 4*R*)-*threo* configuration of FBP (Scheme 5.1) [1, 8, 9]. Both L-fuculose 1-phosphate (Fuc 1-P) aldolase and L-rhamnulose 1-phosphate (Rha 1-P) aldolase are class II DHAP-dependent enzymes that use (*S*)-lactaldehyde as the natural acceptor. The configuration of the vicinal diols produced in the DHAP aldol reaction of Fuc 1-P aldolase with (*S*)-lactaldehyde generates the (3*R*, 4*R*)-diol, while Rha 1-P aldolase creates the corresponding (3*R*, 4*S*)-isomer (Scheme 5.1) [1, 8, 9].

The DHAP-dependent aldolases have several drawbacks: (i) the strict specificity toward the donor substrate, (ii) expensive and unstable donor substrate, and (iii) necessary removal of the phosphate group of the product. The discovery of a novel class I D-fructose 6-phosphate aldolase (FSA) isolated from *Escherichia coli* genome by directed evolution approaches overcomes these disadvantages [13]. The FSA aldolase accepts the readily available unphosphorylated dihydroxyacetone (DHA) as donor substrate instead of the expensive DHAP and the readily obtained products have the unbiased stereoselectivity (3*S*, 4*R*)-configuration and without subsequent

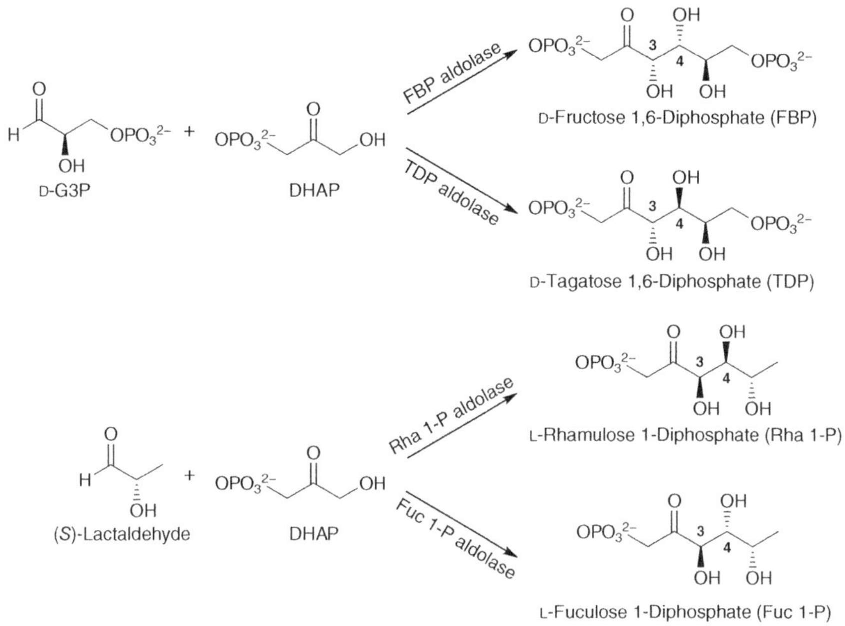

Scheme 5.1 Stereochemical complementary sets of DHAP-dependent aldolases. *Source:* Based on Brovetto et al. [1]; Dean et al. [4]; Fischer and Pietruszka [7]; Clapés et al. [8]; Schimidt et al. [9].

dephosphorylation [1, 8, 9]. Besides DHA, FSA also weakly takes hydroxyacetone (HA), hydroxybutanone (HB), or glycoaldehyde (GA) as donor substrate [1, 9, 14, 15]. The versatility of the wild-type FSA has been demonstrated by its use in the multiple synthesis yielding carbohydrates and analogues, such as catalyzing the direct stereoselective self-aldol addition of glycoaldehyde to yield D-(−)-threose (Scheme 5.2), that glycoaldehyde acts as both the donor and acceptor substrate for FSA [1, 14, 16].

The group of pyruvate-dependent aldolases is usually class I aldolases that reversibly catalyze the aldol reaction between the donor substrate pyruvate to various polyhydroxylated aldehyde acceptors producing α-oxo acids.

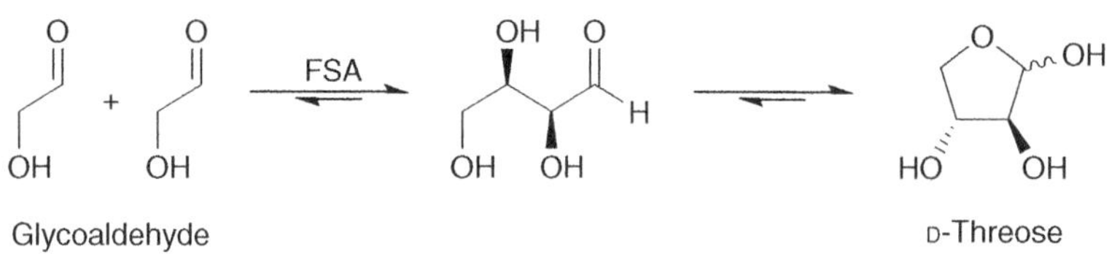

Scheme 5.2 FSA-catalyzed self-aldol addition using glycoaldehyde as the donor substrate. *Source:* Based on Brovetto et al. [1]; Garrabou et al. [14]; Concia et al. [16].

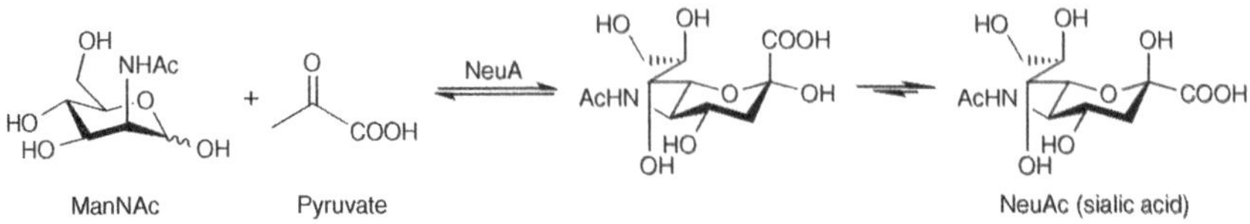

Scheme 5.3 *N*-Acetylneuraminic acid (NeuAc) aldolase-catalyzed aldol reaction. *Source:* Based on Brovetto et al. [1]; Clapés et al. [8].

N-Acetylneuraminic acid aldolase (NeuA), also named sialic acid aldolase, and its mutants catalyze the reversible aldol addition of *N*-acetyl-D-mannosamine (ManNAc) to pyruvate yielding *N*-acetylneuraminic acid (NeuAc or sialic acid) (Scheme 5.3) [1, 8]. NeuAc aldolase isolated from bacteria and mammals accepts only pyruvate as donor substrate that forms a Schiff-base/enamine intermediate with pyruvate to promote a *si*-face attack to the aldehyde resulting in the formation of a 4*S* stereocenter for the natural substrate. However, the stereospecificity of NeuAc aldolase is not absolute and depends on the structure of the acceptor aldehyde [17, 18]. Directed evolution has been applied to *E. coli* D-2-keto-3-deoxy-6-phosphogluconate (KDPG) aldolase in order to solve the narrow substrate acceptance and the need for phosphorylated aldehydes with aldol reactions.

By screening for mutants of KDPG aldolase, the variants showed not only the ability to accept nonphosphorylated substrate but also an increase in relative activity for L-glyceraldehyde and pyruvate in addition to the naturally substrate D-glyceraldehyde, and thus were used to synthesize the L-sugar 3-deoxy-L-*threo*-2-hexulosonic acid (Scheme 5.4) [4, 15, 19]. The structurally guided mutagenesis has been applied to the NeuAc aldolase to obtain a mutant aldolase that is capable of accepting substituted amides of the natural D-ManNAc as the acceptor to synthesize sialic acid mimetics [4, 20]. The improvement in pyruvate-dependent NeuAc aldolase by directed evolution thus disclosed a new route to the core structure of important pharmaceuticals such as zanamivir (Relenza®) [7].

Nowadays, the only known acetaldehyde-dependent aldolases having been applied in organic synthesis is 2-deoxy-D-ribose 5-phosphate aldolase (DERA). DERA isolated from animal tissues and microorganisms is an acceptably stable class I aldolase and is commercially available. DERA, which catalyzes the reversible aldol reaction of acetaldehyde and D-G3P to give 2-deoxyribose-5-phosphate, has been rationally designed toward nonphosphorylated D-glyceraldehyde with 2.5-fold higher activity (Scheme 5.5) [1, 4, 7, 8, 15]. Another attractive feather of DERA compared to other aldolases proved to be the unique ability to catalyze the sequential aldol reactions with acetaldehyde as the only substrate to produce (3*R*,5*R*)-2,4,6-trideoxyhexose (Scheme 5.6) [1, 8].

The DERA mutant S238D showed the ability to accept 3-azido-propionaldehyde as a substrate for a sequential one-pot aldol condensation for the synthesis of a

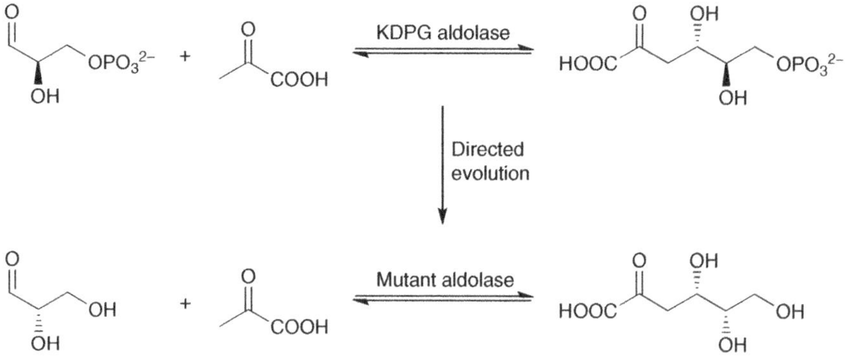

Scheme 5.4 KDPG mutant aldolase accepting nonphosphorylated and altered substrate specificity by directed evolution. *Source:* Based on Dean et al. [4]; Stroheier et al. [15]; Fong et al. [19].

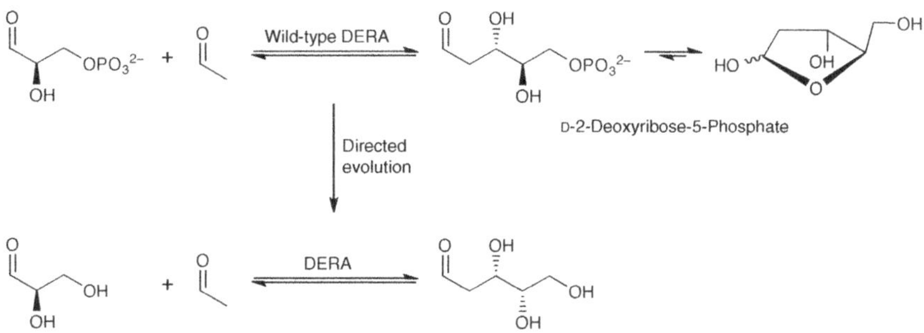

Scheme 5.5 Aldol reactions catalyzed by acetaldehyde-dependent DERA. In comparison to other aldolases, DERA has a rather broad substrate range that made DERA-catalyzed aldol reactions great effect in the preparation of key intermediates for the synthesis of potent antitumor agent epothilones A [1, 4, 7, 21]. *Source:* Based on Brovetto et al. [1]; Fischer and Pietruszka [7]; Clapés et al. [8]; Stroheier et al. [15].

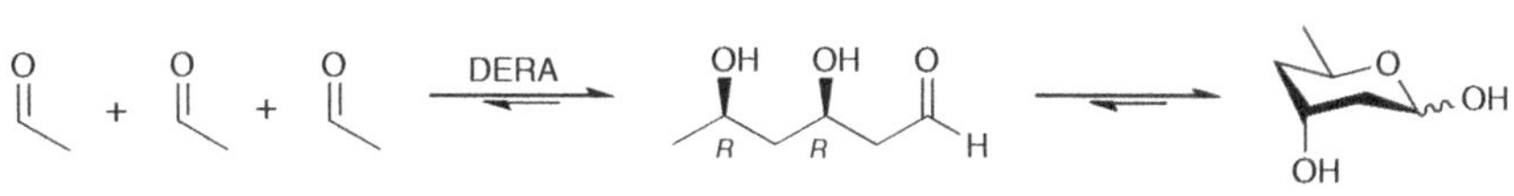

Scheme 5.6 Sequential aldol reaction catalyzed by DERA with acetaldehyde as the only substrate. *Source:* Based on Brovetto et al. [1]; Clapés et al. [8].

key fragment of the cholesterol-lowering drug atorvastatin (Scheme 5.7) [4, 7, 22]. DERA has also been evolved for an industrial application in the synthesis of statin drug reported by DSM [4, 7, 8, 15, 23]. This enzyme has been exploited for the tandem aldol addition of 2 equiv. of acetaldehyde to 1 equiv. of 3-chloropropanal

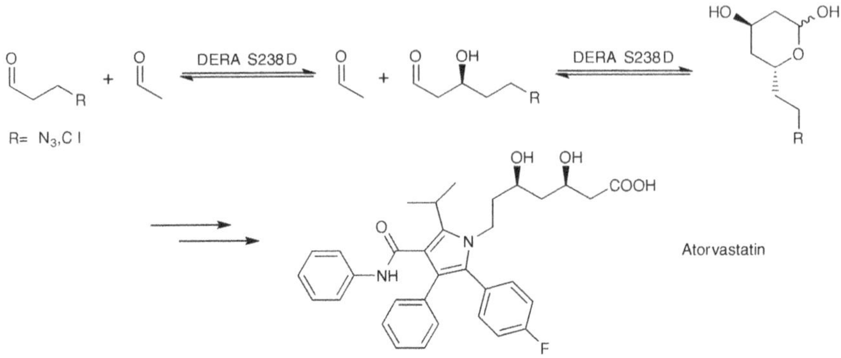

Scheme 5.7 Application DERA mutant S238D catalyzed sequential on-pot reactions in the synthesis of atorvastatin. *Source:* Based on Dean et al. [4]; Fischer and Pietruszka [7]; Liu et al. [22].

in the preparation of (3*R*,5*S*)-6-chloro-2,4,6-deoxyhexpyranoside, a key statin intermediate, for the production of atorvastatin (Scheme 5.7) [1, 4, 7, 8, 15].

Threonine aldolases (ThrAs) are the glycine-dependent aldolases that by working with pyridaoxal-5′-phosphate (PLP) catalyze the reversible aldol addition of glycine to aldehydes to form β-hydroxy-α-amino acids. ThrAs display complete stereospecificity for the four stereo-complementary β-hydroxy-α-amino acids using formally a single aldehyde and either L-threonine/D-threonine or corresponding *allo*-threonine selective aldolases (Scheme 5.8) [1, 8].

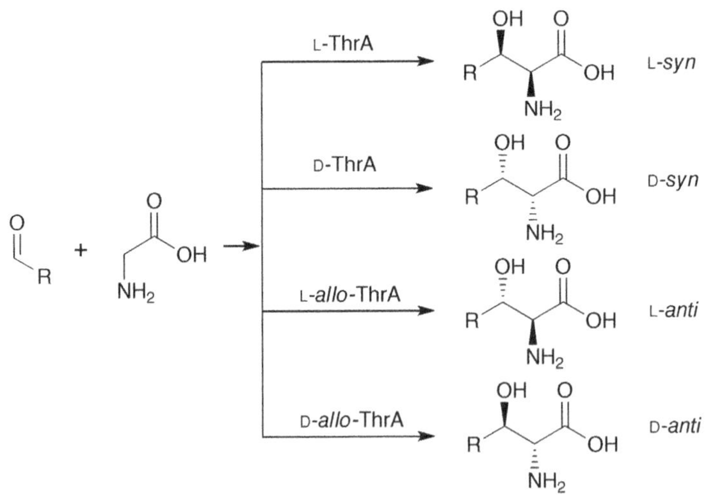

Scheme 5.8 Stereocomplementary aldol reactions with glycine-dependent ThrAs. *Source:* Based on Brovetto et al. [1]; Clapés et al. [8].

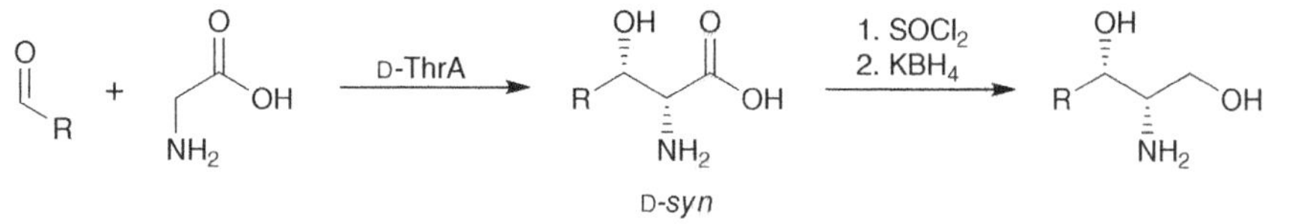

Scheme 5.9 Synthetic application of glycine-dependent ThrA. *Source:* Based on Clapés et al. [8]; Nishiyama et al. [24]; Toshihiro et al. [25].

When ThrAs are used in synthetic applications, it is necessary to work with an excess of glycine to compensate the unfavorable equilibrium constant. L-Threonine aldolases have employed for the introduction of both multifunctionality in chemoenzymatic and chirality in one single step such as in the preparation of the antifungal precursor of thymine deoxypolyoxin C (Scheme 5.9) and imino-deoxydigitoxose [8, 24, 25]. ThrAs have also been used to synthesize a number of γ-halogenated and long-chain β-hydroxy-α-amino acids (log chain L-3-alkylserine and D-3-alkylserine derivatives). For D-threonine aldolase the *syn*-selectivity was observed exclusively that further chemically yielded the 2-amino-1,3-diols as potential precursors for short-chain sphingosine derivatives synthesis [7, 8, 26].

5.1.2 Enzymatic Acyloin Condensation

Bifunctional acyloins (α-hydroxyketones) are compounds of ubiquitous structure with a range of applications, particularly their enantiopure forms are versatile intermediates in natural product synthesis for fine chemicals as well as pharmaceuticals production. They can be used for the synthesis of antidepressants, antifungal agents, antitumor antibiotics, or inhibitors of farnesyl transferase or amyloid-β-protein (for Alzheimer's disease) [7, 9, 27]. In addition, they also can be used for the preparation of diols and aminoalcohols. In nature, the enzymes involved for efficient formation of α-hydroxyketones are a group of thiamine diphosphate (ThDP)-dependent lyases that catalyze the C–C bond formation related to the acyloin condensation between an aldehyde and an adequate donor, usually from inexpensive aldehydes. The ThDP-dependent lyases include pyruvate decarboxylase (PDC), benzoylformate decarboxylase (BFD), phenylpyruvate decarboxylase (PhDC), indole-3-pyruvate decarboxylase (InPDC), and benzaldehyde lyase (BAL) [1, 7, 9]. The key step of the catalytic mechanism for acyloin condensation involves the reaction of the "donor" aldehyde with the thiazolium ring of the ThDP cofactor forming a highly reactive zwitterion, followed by the attack of this active zwitterion to the acceptor aldehyde, releasing the cofactor to generate the (*R*)-α-hydroxyketone [1, 9].

5.1.2.1 Acyloin Condensation Catalyzed by Pyruvate Decarboxylase

Pyruvate decarboxylases (PDC, EC 4.1.1.1) are commonly found in plants, yeasts, and fungi, but absent in mammals [27]. PDC is the first enzyme of the branched glycolytic pathway, catalyzing the nonoxidative decarboxylation of pyruvic acid into acetaldehyde and CO_2 under anaerobic conditions, thus is an important enzyme in glycolysis and ethanol fermentation [28]. In addition to the decarboxylase activity, the catalytic activity of PDC for C–C bond formation by being able to catalyze an acyloin condensation-like carboligation of aldehydes was noticed in 1921. Since then, PDC has been widely used as biocatalyst in carboligation reactions that leads to an industrial fermentation process for the preparation of (*R*)-phenylacetylcarbinol (PAC), a key intermediate for ephedrine and pseudoephedrine production (Scheme 5.10) [7, 29, 30].

PDC catalyzes a number of decarboxylations of α-keto acids with different aldehydes to form chiral α-hydroxyketones using either whole cell systems or isolated enzymes [28, 31, 32]. A broad range of substrates have been employed for the decarboxylation reaction of PDC such as α-keto acids (α-keto butyric and α-keto valeric acids) and branched substrates (α-ketoisocaproic acid and α-ketoisovaleric acid). It has been found that PDC from bacteria showed lower affinities for longer-chain-aliphatic α-keto acids than those of yeast. In the PDC-catalyzed acyloin condensations, α-hydroxyketones with a *sec*-alcohols moiety (*sec*-α-hydroxyketones) will be formed with the acceptor aldehyde. A variety of *sec*-α-hydroxyketones can be obtained depending on the donor/

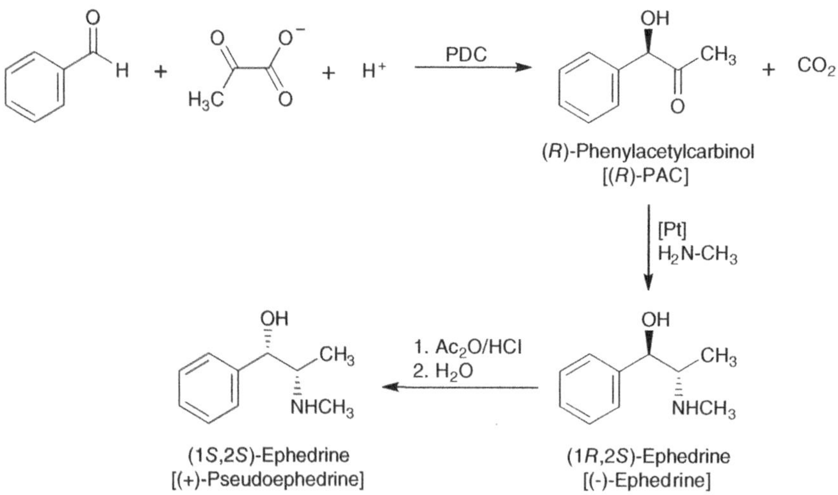

Scheme 5.10 PDC-catalyzed synthesis of (*R*)-PAC and its use for the production of ephedrine and pseudoephedrine. *Source:* Based on Fischer and Pietruszka [7]; Ward and Singh [29]; Demir et al. [30].

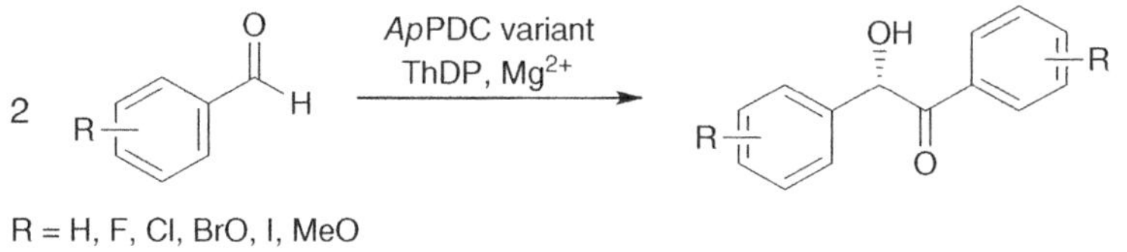

Scheme 5.11 Direct asymmetric synthesis of meta- and *para*-substituted (*S*)-benzoins catalyzed by *Ap*PDC variant. *Source:* Based on Schimidt et al. [9]; Westphal et al. [35].

acceptor and the enzymes used such as phenylacetylcarbinol (PAC) derivatives, 2-hydroxypropiophenone analogues (2-HPPs), δ-hydroxy-γ-keto acids, benzoins, or aliphatic acyloins [9]. For example, benzoin is a precursor used for the synthesis of benzyl, a common photoinitiator in polymerization. The direct preparation of (*R*)-benzoin and analogues through enzymatic cross-coupling reaction of aldehydes have been developed in the past using either BAL or BFD [33, 34]. The direct asymmetric synthesis of (*S*)-benzoins with excellent enantiomeric excess (>99%) and very good conversion have been identified by a rational designed (*S*)-selective variant of pyruvate decarboxylase from *Acetobacter pasteurianus* (*Ap*PDC) via the direct enzymatic homocoupling of commercially available benzaldehyde derivatives (Scheme 5.11) [9, 35].

5.1.2.2 Acyloin Condensation Catalyzed by Benzoylformate Decarboxylase

Benzoylformate decarboxylases (BFD, EC 4.1.1.7) have been found in bacteria catalyzing the nonoxidative decarboxylation of benzoylformate to benzaldehyde with a reaction mechanism similar to that of pyruvate decarboxylases and playing an important role in the mandelate pathway [36–38]. BFD also exhibits the enantioselective C–C formation activity that accepts aldehydes as donors to catalyze the acyloin and benzoin condensations to give corresponding (*S*)-acyloins and (*R*)-benzoins within a broad range of pH (5–8) and temperature (20–40°C) (Scheme 5.12) [1, 29, 30, 39, 40]. The addition of organic solvent dimethylsulfoxide (DMSO) increased the conversion rate but had no effect on the enantioselectivity of the product. The carboligase activity of BFD from *Pseudomonas putida* was first reported in 1992 and the enantiomeric excess of (*S*)-2-hydroxypropiophenone ((*S*)-2-HPP) from benzaldehyde and acetaldehyde by BFD from *P. putida* was found as high as 91–92% [41]. The range of substrates is very broad including *m*-substituted aromatic aldehydes, heteroaromatic aldehydes, acyclic aldehydes, and α,β-unsaturated aldehydes but preferring aromatic aldehydes [32, 34, 42–45].

The carboligase activity of BFD from *P. putida* has been improved by a combination of directed evolution and site-directed mutagenesis [46]. The substrate

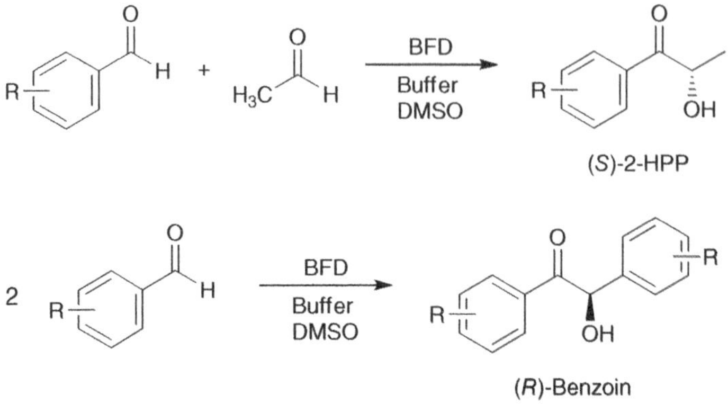

Scheme 5.12 Benzoylformate decarboxylase-mediated asymmetric acyloin condensations. Sources: Based on Brovetto et al. [1]; Ward and Singh [29]; Demir et al. [30]; Wilcocks and Ward [39]; Demir et al. [40].

specificity was also broadened by using genetic engineering techniques, thus the limitation of the wild type BFD not accepting *o*-substituted benzaldehyde derivatives, except 2-fluorobenzaldehydes, has been circumvented by the directed evolution to alter the substrate specificity of BFD [47]. Even BFD mutants accept long-chain aliphatic 2-ketoacids with improved decarboxylation activity by site-directed mutagenesis [48, 49].

5.1.2.3 Acyloin Condensation Catalyzed by Benzaldehyde Lyase

Benzaldehyde lyase (BAL. EC4.1.2.38) from *Pseudomonas fluorescens* Biovar I reported first time in 1989 was another extensively investigated ThDP-dependent enzymes that also uses divalent cations as cofactors. The first purified enzyme catalyzed the cleavage of the acyloin linkage of benzoin to produce two molecules of benzaldehyde [50]. The enantioselective catalytic activity of BAL in C–C bond forming reactions using generally acetaldehyde as acceptor that afforded (*R*)-benzoin and (*R*)-2-hydroxypropiophenone derivatives was first described in 2001 [1, 30, 51]. The carboligation activity of this enzyme has been extensively studied since then with a broad range of substrates. Research findings also demonstrated that BAL and BFD are homologous ThDP-dependent enzymes catalyzing the formation of chiral 2-hydroxyketones from aldehydes. However, the reverse reaction has only been observed with BAL. In addition, the stereospecificity of BAL is a strictly specific for *R* configurations (Scheme 5.13) [1, 52].

The catalytic mechanism of BAL has been widely studied and found in accordance with other ThDP-dependent enzymes [51, 53–57]. The initial step of the catalytic cycle is the attack of the ylide form of ThDP to the carbonyl carbon of the (*R*)-benzoin to produce the corresponding adduct. Then, the electrophilic

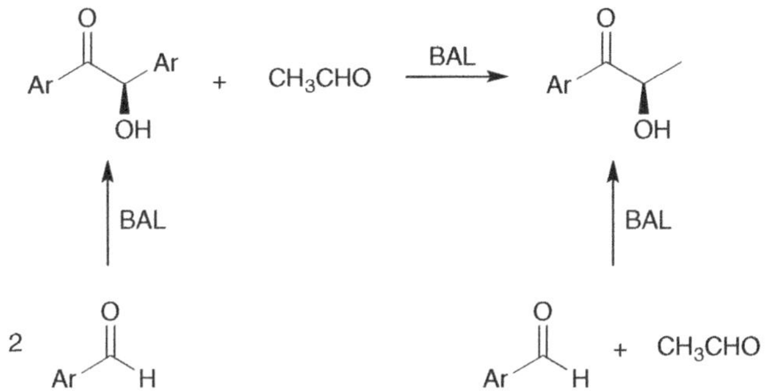

Scheme 5.13 General reaction scheme for BAL-catalyzed carboligations. *Source:* Based on Brovetto et al. [1]; Knoll et al. [52].

nitrogen atom of the ThDP moiety promotes a rearrangement to form the enamine intermediate and a free benzaldehyde. The enamine intermediate reacts with the acceptor aldehyde and undergoes the carboligation reaction, which produces the acyloin and regenerates the ylide to complete the catalytic cycle. In the absence of an acceptor aldehyde, the protonation of enamine intermediate leads to the release of the bound benzaldehyde and regenerates the ylide. The formation of the enamine-carbanion intermediate can also be formed by the attack of the ylide form of ThDP on the carbonyl group of the released benzaldehyde [1, 30].

Both BAL and BFD catalyze a wide range of substrates [32–34, 42, 52, 58–66]. However, BAL-catalyzed carboligations not only show complementary stereoselectivity to BFD but also exhibit a broader substrate specificity, for examples, aromatic and *o*-substituted aromatic aldehydes, heteroaromatic aldehydes, simple aliphatic aldehydes such as propionaldehyde, mono- and dimethoxyaldehyde, and also glyceraldehyde and butyraldehyde derivatives are accepted [33, 34, 40, 51, 59, 60, 62, 67]. For aliphatic aldehydes of not highly sterically hindered are suitable as both donors and acceptors [1, 42]. In addition to the enantioselectivity of BAL-catalyzed carboligations, BAL affords highly diastereoselective α-hydroxyketones by simultaneously performing ligation and kinetic resolution of one enantiomer of a racemic α-chiral aldehyde. In this case, two chiral centers in one reaction can be set up and perfect enantioselectivity can be obtained by using benzaldehyde as the donor (Scheme 5.14) [9, 68].

The wide ranges of substrate specificity and (*R*)-enantioselective catalytic activity in C–C bond forming reactions have made BAL many industrial synthetic applications. For instance, 2,3-dioxygenated aryl propanones are very important chiral building blocks for the synthesis of various biologically active compounds. BAL-catalyzed acyloin condensation using aromatic aldehydes

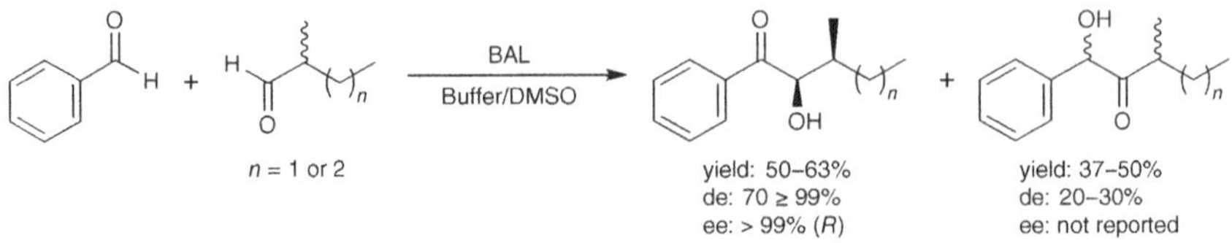

Scheme 5.14 Diastereoselective condensation of benzaldehyde and racemic (±)-2-methyl-alkyl-aldehydes using BAL for producing enantiopure 2-HPP derivatives. *Source:* Based on Schimidt et al. [9]; Muller et al. [68].

donor and methoxy or dimethoxy acetaldehyde acceptor furnishes either (*R*)-2-hydroxy-3-methoxy-1-arylpropan-1-one or (*R*)-2-hydroxy-3,3-dimethoxy-1-aryl propan-1-one derivatives in high yields and with 89 to >98% *e.e.* (Scheme 5.15a) [30, 59]. These products were used as synthons for the synthesis of 1,4-benzodioxane lignans (5′-methoxyhydnocarpin), an antihepatotoxic with insecticidal activity, or cytoxazone, a novel cytokine modulator, or for the production of aryl isoserines which are precursors to the side-chain of paclitaxel (Taxol®) (Scheme 5.15b) [7, 30, 59].

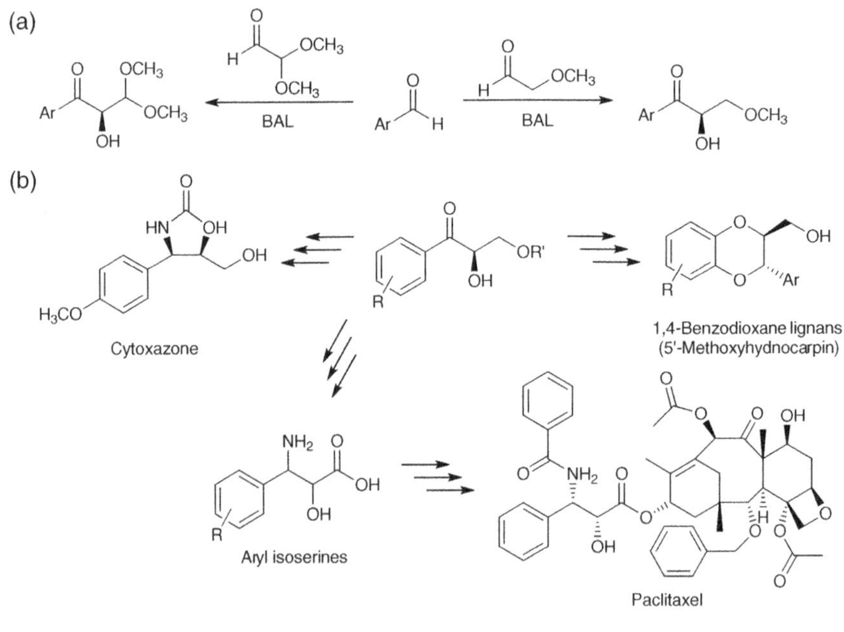

Scheme 5.15 (a) BAL-catalyzed 2,3-dioxygenated aryl propanones. *Source:* Based on Demir et al. [30]; Demir and Şeşenoglu [59]. (b) applications of 2,3-dioxygenated aryl propanones in natural product synthesis. *Source:* Based on Fischer and Pietruszka [7]; Demir et al. [30]; Demir and Şeşenoglu [59].

Many biocatalytic practical techniques have been developed to improve BAL as a biocatalyst in industrial applications. For instances, an aqueous-organic two-phase system has been found the best reaction system for the preparative asymmetric synthesis of (*R*)-3,3′-dimethoxybenzoin and (*R*)-3-methoxy-2′-chlorobenzoin [69]. It has been found BAL tolerates a variety of organic co-solvents such as DMSO, MTBE, butanol, 2-MeTHF, or deep-eutectic solvents for carboligation under aqueous conditions [70, 71]; the development of a continuous bioreactor in combination with membrane technology enables high space time yields of the product (*R*)-2-HPP as well as high total turnover numbers of the BAL [72]; a further development of gas reactors was used for the continuous production of propioin from two propanal molecules or benzoin from benzaldehyde [73, 74]; the bioprocess parameters on the enzymatic activity and stability of BAL-catalyzed carboligation were optimized by means of reaction engineering [62, 72, 75, 76]; the immobilization of BAL either by entrapping in polyvinyl alcohol (PVA) to give a better productivity of novel benzoins for a gel-stabilized two-phase reaction system or by means of metal ion affinity binding to a nickel(II)-nitrilotriacetic acid derivatized carrier to perform repetitive batch reactions and a continuously operated plug-flow reactor [30, 67, 77, 78]; and the direct use of *E. coli* whole cells overexpressing BAL in a biphasic system eliminates the external addition of expensive cofactor ThDP and affords high productivity and enantioselectivity [27, 63].

In addition, BAL has been embedded in a one-pot two-step cascade reaction to transform bio-based aliphatic alcohols, which is *in situ* oxidized by an alcohol oxidase and a catalase to corresponding aldehydes, and benzaldehyde to optically pure 2-hydroxypropiophenone and analogues [9, 79].

5.1.2.4 Acyloin Condensation Catalyzed by Phenylpyruvate Decarboxylase

Phenylpyruvate decarboxylase (PhDC, EC 4.1.1.43) catalyzing the acyloin condensation of phenylpyruvic acid and acetaldehyde is another ThDP-dependent α-ketoacid decarboxylase [30, 80, 81]. This enzyme produced from *Thauera*, *Achromobacter*, *Pseudomonas*, and *Aromatoleum* spp. by the induction with L-phenylalanine, tryptophan, or mandelate participates in the anaerobic metabolism of aromatic compounds via benzoyl-CoA [82–84]. The crystal structure of this enzyme from *Azospirillum brasilense* has been reported first time at 1.5 Å resolution and also found involving in the biosynthesis of the plant hormone indole-3-acetic acid and the antimicrobial compound phenylacetic acid [85].

The stereoselective acyloin condensation of phenylpyruvic acid and acetaldehyde-catalyzed by *Achromobacter eurydice* and *P. aromatic* grown on L-phenylalanine-containing medium provided the acyloin product 3-(*R*)-hydroxy-1-phenyl-2-butanone with only 1–7% yield but 95 and 84% *e.e.*, respectively (Scheme 5.16). However, the yield and *e.e.* of acyloin product obtained with a

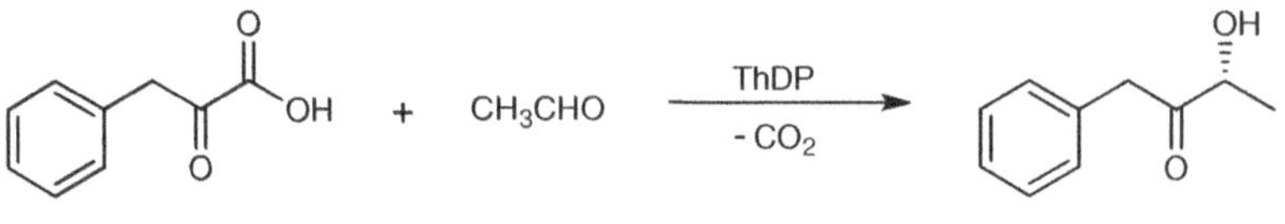

Scheme 5.16 Acyloin condensation of phenylacetic acid with PhDC.

partially purified PhDC prepared from the cell-free extract of *A. eurydice* were 45 and 91%, respectively [80]. With phenylacetic acid acyl donor, a variety of straight chain aliphatic aldehydes have been found as suitable acyl acceptor but no aromatic aldehydes for the asymmetric acyloin condensation using purified PhDC from *A. Eurydice* to give optically pure acyloins $PhCH_2COCH(OH)R$. However, the yield decreased with increasing chain length for straight chain aliphatic aldehydes from 76% for acetaldehyde to 24% for valeraldehyde and the *e.e.* values of the products were in the range of 87–98%. For chloroaldehyde and glycoaldehyde, the yields of the acyloin products were low as 13 and 16%, respectively. Indole-3-pyruvate can be used as an acyl donor for the purified enzyme to produce the acyloin condensation product 3-hydroxy-1-(3-indolyl)-2-butanone with aldehyde in a yield of 19%.

The proposed catalytic mechanism of PhDC is analogous to PDC and BFD, with only minor changes occur in the interactions with cofactors, thiamine diphosphate, and Mg^{2+} [1, 30, 85]. That is, it involves the decarboxylation of phenylacetic acid to the thiamine–enzyme bound complex in the first step, followed by the reaction of the acceptor aldehyde with the thiamine–enzyme bound complex to provide the acyloin product.

5.1.3 Cyanohydrin Formation with Hydroxynitrile Lyases

Hydroxynitrile lyases (HNLs, EC 4.1.2.x), also known as oxynitrilases, are a versatile of enzymes that catalyze the reversible enantioselective cleavage of α-cyanohydrins into hydrocyanic acid (HCN) and the corresponding aldehyde or ketone. HNLs are typically found in plants, but recent investigations also have reported nonplant sources organisms like bacteria, fungi, lichen, millipedes, arthropods, and insects [1, 7, 86, 89–95]. In plants, the release of HCN from the HNL-catalyzed cleavage of cyanohydrins is related to a defense mechanism of these plants against several herbivores, commonly known as cyanogenesis, which also functions a nitrogen source.

The classification of HNLs falls into two major groups based on the presence or absence of flavin adenine dinucleotide (FAD) cofactor. The FAD-containing (*R*)-selective HNLs such as *Prunus amygdalus* (*Pa*-HNL) as well as (*R*)-(+)-mandelonitrile lyases (MDL) have been isolated exclusively from plants belonging to *Prunoideae* and *Maloideae* subfamilies of *Rosaceae*. However, FAD is not involved in the mechanism of HNL-catalyzed reaction, but their removal will cause the inactivation of HNL [96, 97]. The FAD-independent HNL found in *Sorghum bicolor* (*Sb*-HNL), *Manhot esculenta* (*Me*-HNL), *Phlebodium aureum* (*Pha*-HNL), and *Linum usitatissimum* (*Lu*-HNL) is (*R*)- and (*S*)-selective enzymes and is less prevalent protein [1, 89]. The catalytic mechanism of HNL studied by structure, spectroscopy, molecular modeling, and density functional theoretical calculations reveals general acid–base catalysis via three elementary steps: (i) deprotonation of the OH-Ser80 by His235 and concomitant abstraction of aproton from the substrate hydroxyl by Ser80; (ii) the C–C bond cleavage of the cyanohydrin compound; and (iii) protonation of the cleaved cyanide by His235. The rate-limiting step is the C–C bond cleavage [1, 89, 97, 98].

The *in vitro* reversibility of the cyanohydrin cleavage enables the enantioselective condensation of HCN with aldehydes or ketones to produce either (*R*)- or (*S*)-α-cyanohydrins (α-hydroxy nitriles) (Scheme 5.17). α-Cyanohydrins formed by the HNL-catalyzed addition of HCN to aldehydes have been studied for a long time. Although a wide spectrum of aliphatic and aromatic aldehydes has been converted into their corresponding cyanohydrins with excellent enantiopurities, still lower yields and enantioselectivities were found in some cases. These limitations have been overcome by using a biphasic reaction system [87, 91, 92, 94, 99–101]. Recently, the incorporation of the biphasic systems with the immobilization of purified HNL or whole cell as crossed-linked enzyme aggregates (CLEAs) has greatly improved the cyanohydrins synthesis [86, 102–104]. The substitution of one bulky group for hydrogen atom in the aldehyde molecule causes a higher degree of steric hindrance of the ketones that has proven the asymmetric condensation of HCN with ketones to be more difficult.

With the help of biphasic system and immobilized enzyme, a variety of enantiopure cyanohydrins have been synthesized from aliphatic methyl ketones and aromatic ketones [100, 105–110]. The enantiomeric selectivity of cyanohydrin formation for methyl ketones was very good when the other group on the ketones is larger than ethyl group [107, 109, 110]. For 2-butanone, the size of the two substituents is similar that causes only minor difference in the spatial orientation selectivity of the substrate and thus only 54% *e.e.* was obtained for the

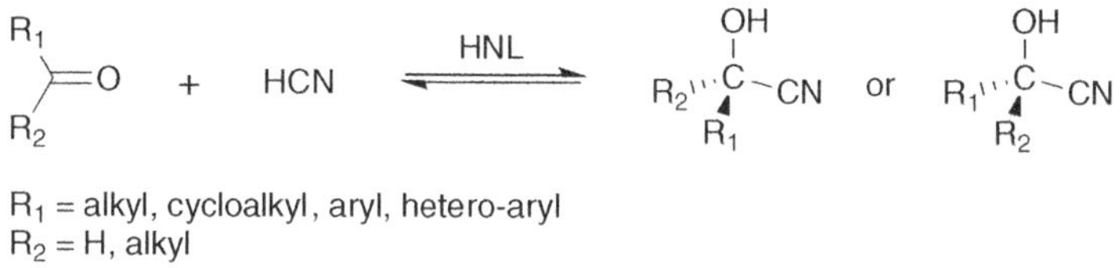

Scheme 5.17 HNL-catalyzed enantioselective cyanohydrin formation.

corresponding (*S*)-cyanohydrin when using *Hb*-HNL [107, 109]. Nevertheless, a better enantiomeric selectivity could be acquired for the corresponding (*R*)-cyanohydrin when using *Pa*-HNL or *Lu*-HNL (Scheme 5.18) [109, 110]. The asymmetric synthesis of (*R*)- and (*S*)-cyanohydrins from a range of aromatic methyl and ethyl ketones such as acetophenone, phenylacetone, benzylacetone, and propiophenone, with different substituted variants on the phenyl ring, has been recently studied using *Lu*-HNL, *Me*-HNL, or *Pa*-HNL as catalyst (Scheme 5.19).

The results show that both reaction yields and stereoselectivities were influenced by the carbon chain length between the ketone and phenyl functional groups, and the type of aromatic substitution of the substrates. Substrates converted with the greatest degree of productivity and selectivity were phenylacetones containing large, electron withdrawing meta-substituents [1, 108]. HNL-catalyzed addition of HCN to 2-, 3-, and 4-substituted cyclohexanones to give corresponding cyanohydrins has been performed for further synthesis of pharmaceuticals and plant protective agents [86, 111–113].

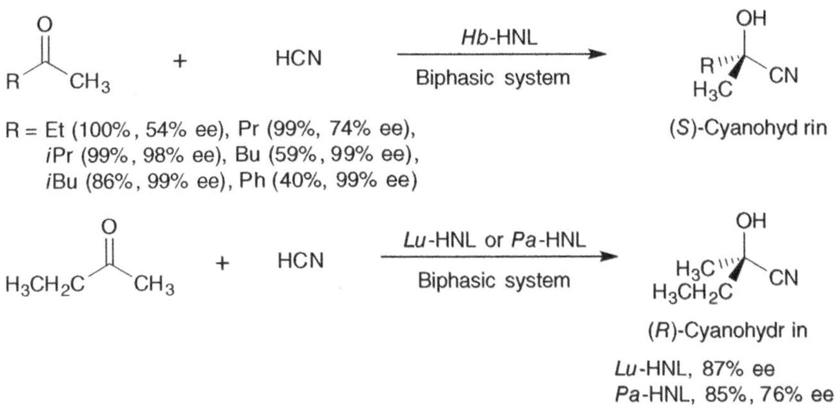

Scheme 5.18 HNL-catalyzed enantioselective synthesis of cyanohydrins from aliphatic ketones. *Source:* Based on Fechter et al. [109]; Cabirol et al. [110].

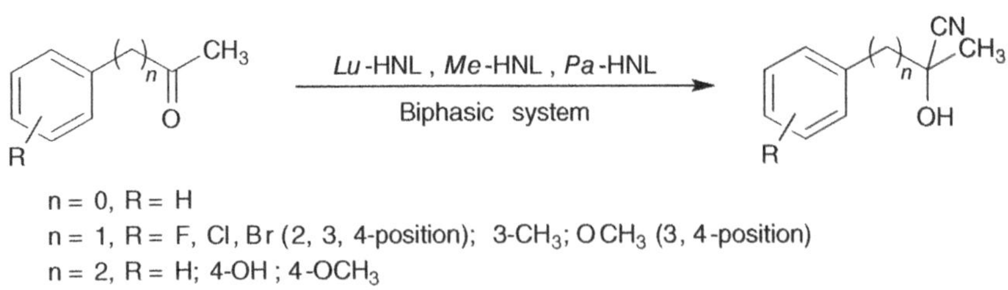

Scheme 5.19 HNL-catalyzed enantioselective synthesis of cyanohydrins from aromatic ketones.

The cyanohydrins obtained from the addition reactions contain two asymmetric centers thus result four diastereomers (Scheme 5.20). With *Pa*-HNL from bitter almonds, the addition to 2-alkyl cyclohexanones is highly (*R*)-selective, whereas the 2-methyl cyclohexanone shows (*S*)-selective. With *Me*-HNL from cassava, all 2-alkyl cyclohexanones are (*S*)-selective. The results also show that the increasing size of the substituents causes a decrease of the catalytic activity of both enzymes. However, the diastereoselectivity of HCN addition to 3-substituted cyclohexanones is only moderate for both enzymes [111]. The addition of HCN to 4-substituted cyclohexanones is strongly catalyzed by hydroxynitrile lyase from both *Pa*-HNL and *Me*-HNL. However, *trans*-addition occurs almost exclusively to yield *trans*-addition product with *Pa*-HNL from bitter almond; using *Me*-HNL from cassava, *cis*-addition is preferred to give nearly quantitative *cis*-addition product, especially for cyclohexanones with larger 4-substituents [112].

The carbon–carbon bond formation ability of HNLs makes it one of the classes of enzymes extensively used for synthetic and industrial processes [1, 86–88]. Particularly, cyanohydrins are a compound having a cyano and a hydroxyl group attached to the same carbon atom. With their dual functional groups, they can be readily converted to yield α-hydroxy aldehydes or ketones, β-amino alcohols, β-hydroxyamines, α-aminonitriles, α-hydroxycarboxylic acids, α-sulfonyloxynitriles,

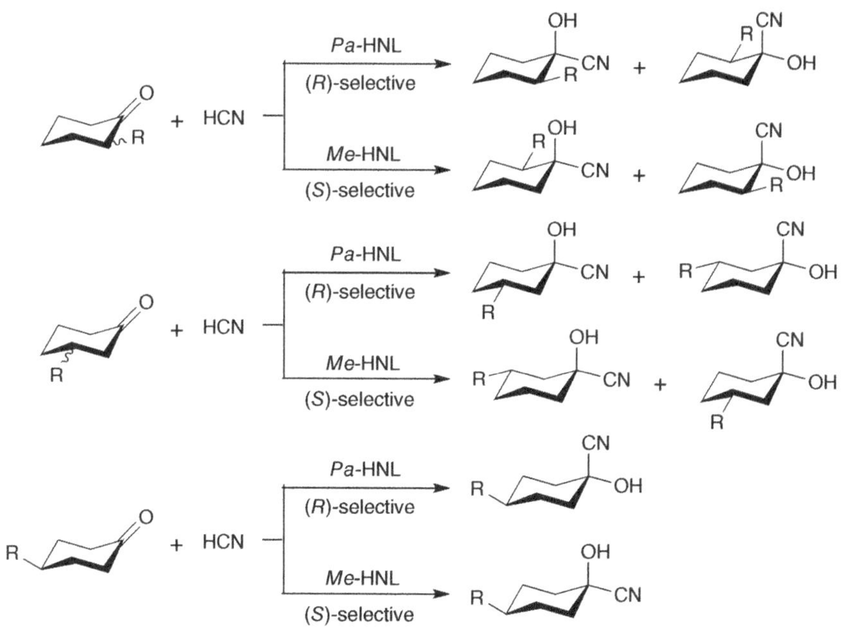

Scheme 5.20 HNL-catalyzed cyanohydrin formation from 2-, 3-, 4-monosubstituted cyclohexanones.

α-fluorocyanides, α-hydroxyesters, aziridines, α-azidonitrile, and many other compounds [1, 86]. Thus, enantiopure cyanohydrins are valuable and versatile building blocks in asymmetric organic synthesis for the production of fine and bulk chemicals. As a result, they are central importance in academic and industrial research [114].

For examples, aliphatic unsaturated aldehydes can be transformed into chiral unsaturated cyanohydrin intermediates by HNLs [115–119], from which a myriad of products such as L- and D-sphingosines (Scheme 5.21) [116], pentoses [115], and nucleosides [117] can be further accessed via various chemoenzymatic protocols; since (*R*)-pantolactone is the most important precursor in the synthesis of vitamin B_5, β-substituted pivaldehydes have been employed to convert to the corresponding (*R*)-cyanohydrins using highly purified and immobilized (*R*)-HNL from wild-type *Pa*-HNL and the acid-catalyzed hydrolytic cyclization of the cyanohydrins to (*R*)-pantolactone [101]. However, the highest chemical and enantiomeric excess yields obtained for (*R*)-pantolactone were 84 and 89%, respectively. However, a highly stable and efficient (*R*)-selective *Pa*-HNL5 was developed by site-saturation mutagenesis, which allowed performing the quantitative conversion of hydroxypivaldehyde to the corresponding (*R*)-cyanohydrins at pH 2.5 and in an aqueous mono-phasic system with an enantiomeric excess of 97% (Scheme 5.22) [120, 121].

Currently, this reaction system has been rated as the most efficient biocatalytic route for the industrial production of vitamin B_5; another (*R*)-selective HNL from kernels of *Badamu* (*Prunus communis* L. var. dulcis Borkh, almond from Xinjiang, China) was used for catalyzing the hydrocyanation reaction of 4-methylsulfanyl-benzaldehyde in a biphasic system at room temperature to prepare the enantiopure 4-methylsulfanyl-mandelonitrile, a key intermediate of the total synthesis toward antibiotics thiamphenicol and florphenicol (Scheme 5.23).

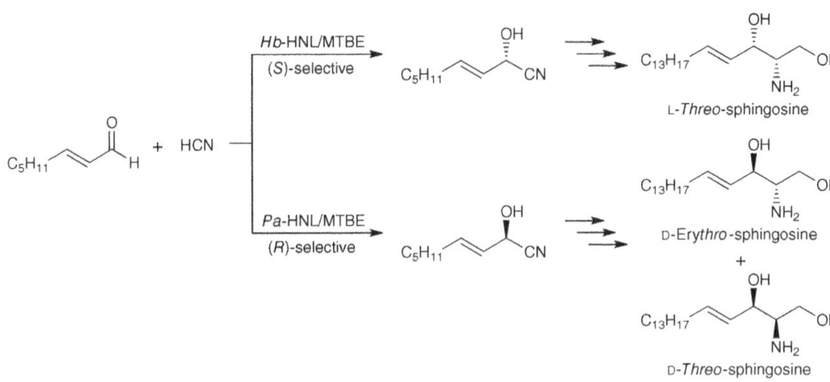

Scheme 5.21 HNL-catalyzed production of chiral unsaturated cyanohydrins as intermediate for L- and D-sphingosines synthesis. *Source:* Based on Johnson et al. [116].

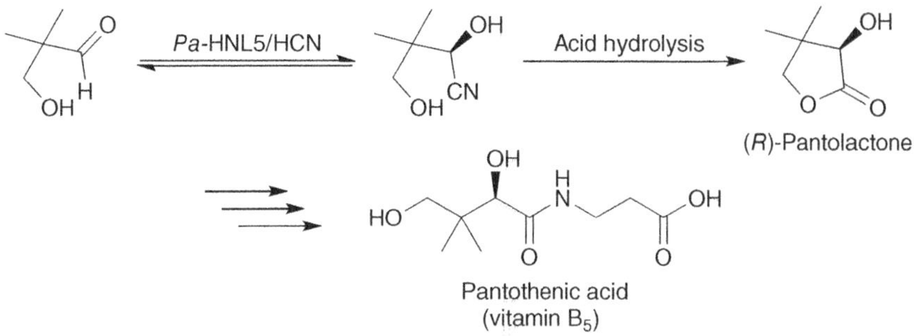

Scheme 5.22 Chemoenzymatic synthesis of (*R*)-pantolactone for vitamin B_5 production. *Source:* Based on Pscheidt et al. [120]; Scheidt et al. [121].

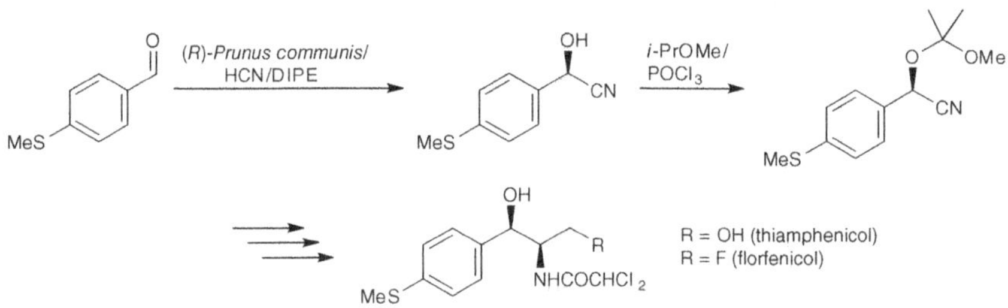

Scheme 5.23 (*R*)-HNL-catalyzed synthesis of (*R*)-4-methylsulfanyl-mandelonitrile for total synthesis of thiamphenicol and florfenicol.

This reaction is highly efficient to give the cyanohydrin with 96% *e.e.* and 98% yield after an extensive screening of the biocatalyst [122].

Hydroxynitrile lyases are very versatile and green catalysts in organic synthesis that make them potential in industrial applications [123, 124]. However, the implementation of hydroxynitrile lyases-catalyzed reactions in large-scale production still suffers from diverse limitations related to process parameters or HNL itself, especially when performing reactions with nonnatural substrates and under nonphysiological conditions [125].

Many efforts from the aspects of reaction engineering, enzyme engineering, or substrate engineering have been employed in order to overcome these limitations. In the following are just a few cases used for demonstration: for the synthesis of 2-hydroxy-(4′-oxocyclohexy)acetonitrile from either 4-methoxycyclohex-3-ene carbaldehyde or 4-trimethylsilyoxycyclohex-3-ene carbaldehyde using *Hb*-HNL as catalyst, substrate and enzyme engineering were coupled to increase the selectivity of *Hb*-HNL by masking the carbonyl functionality as an enol ether to give an enantiomeric excess 90% and a conversion yield 86% [126]. Industrial large-scale production of the key pharmaceutical intermediate (*R*)-2-chloromandelic acid based on the HNL-catalyzed 2-chlorobenzaldehyde to (*R*)-2-chloromandelonitrile was hampered by low turnover rate and moderate enantiomeric excess selectivity [100, 124]. These problems have been overcome by the rationally designed mutation and the mutational anticipation of posttranslational modification of an (*R*)-*Pa*-HNL5-L1Q-A111G that tremendously raised the productivity and stereoselectivity for the synthesis of (*R*)-2-chloro-mandelonitrile [127, 128]. Structure-guided design was employed to generate recombinant almond (*R*)-*Pa*-HNL variants suitable for large-scale synthesis of aromatic cyanohydrins in water-based systems under low pH. Among the variants, the variant *Pa*-HNL5-V360I showed extraordinary catalytic properties that convert 3-phenylpropionaldehyde and 3-phenylpropenal (*trans*-cinnamaldehyde) to corresponding cyanohydrins (*R*)-2-hydroxy-4-phenylbutyronitrile and (*R*)-2-hydroxy-4-phenyl-3-butene nitrile, respectively. The enantiomeric excess for the conversion of 3-phenylpropionaldehyde to (*R*)-2-hydroxy-4-phenylbutyronitrile and thus (*R*)-2-hydroxy-4-phenylbutyric acid at very low amount of enzyme/substrate ratio was improved to >96% with a 98% conversion rate. Similarly, very low amount of variant *Pa*-HNL5-V360I was needed to convert *trans*-cinnamaldehyde to its corresponding cyanohydrin with an improved optical purity about 98% *e.e.* and 97% conversion rate within a greatly reduced time 3 hours [129]. *Pa*-HNL5-V360I is therefore suggested for industrial large-scale production of pharmaceutically active "prils."

5.2 Lyases with Carbon–Oxygen Bonds

Enzymes that catalyze the selective addition of water to carbon–carbon double bonds, and thereby generate primary, secondary, or tertiary alcohols from prochiral substrates, or the reverse water elimination are called hydratases or hydro-lyases (EC 4.2.1.x) [130, 131]. These enzymes are divided into two groups depending on their substrates. One group of enzymes catalyzes the water addition to electron-rich, isolated C=C bonds and is an electrophilic addition following the rule of Markovnikov. The other group of enzymes catalyzes the water addition to an electro-deficient, activated (conjugated) C=C bond with α,β-unsaturated carbonyl group and is performed via nucleophilic Michael addition (Scheme 5.24) [131]. Depending on the enzyme, the activation of water is performed with the help of a metal ion cofactor located in the active site, or the reaction is catalyzed without the need of any cofactors, that of course leads to different reaction mechanisms [132]. In general, the addition and elimination of water can occur either in *syn*- or in *anti*-fashion for both groups [131, 132].

Because hydratases allow for hydration of alkenes with up to 100% atom efficiency, they are highly interest to synthetic chemists for the synthesis of chiral building blocks and products chemical industrial applications. In addition, they

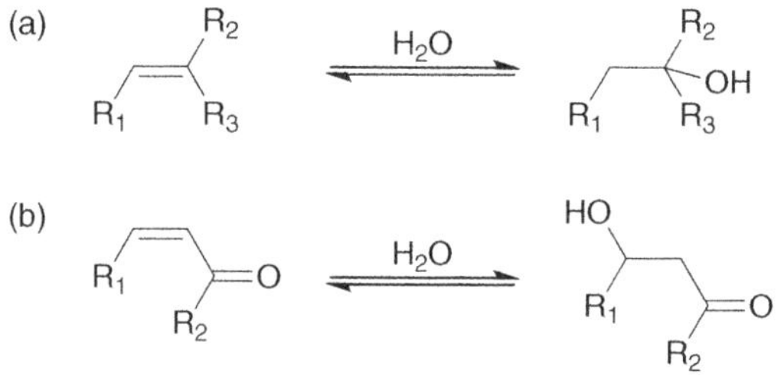

Scheme 5.24 (a) Electrophilic water addition to isolated C=C bonds following Markovnikov's rule, (b) Nucleophilic Michael addition of water to polarized C=C bonds in α,β-unsaturated carbonyl compounds. *Source:* Based on Chen et al. [131].

can be well expressed in commonly used recombinant hosts showing good activity for physiological substrates and performing reactions without requiring cofactor recycling that lead them viable alternatives to other enzyme systems for biocatalytic hydroxylation [130].

5.2.1 Enzyme-Catalyzed Water Addition to Electron-Rich C=C Bonds

A variety of hydratases such as oleate hydratase (EC 4.2.1.53), carotenoid hydratase (EC 4.2.1.131), kievitone hydratase (EC 4.2.1.95), phaseollidin hydratase (EC 4.2.1.97), limonene hydratase, linalool dehydratase-isomerase, and acetylene hydratase (EC 4.2.1.112) have been found to catalyze the addition of water to electron-rich carbon–carbon double bonds. Research on this kind of hydratases has been remarkably over the past years; however, only those hydratases employed either in industrial processes or in the preparative scale will be discussed herein.

5.2.1.1 Oleate Hydratase

Oleate hydratases (OAHs) can be found in many different organisms. The first research reported about oleate hydratase for its hydration activity on oleic acid (OA) can be dated back in 1962 [133]. However, it was not until 2009 detailed information about this enzyme was known when OAH isolated from *Elizabethkingia meningoseptica*, formerly known as *Pseudomonas* sp. 3266, catalyzing the hydration of OA to 10-hydroxystearic acid (10-HSA) was described [134]. OAH contains a flavin adenine dinucleotide (FAD) cofactor thus is FAD-dependent enzyme, but the most probable role of FAD is the correct assembly of amino acids in the active site of enzyme and the additional beneficial effect of reduction to its two-electron reduced state [130, 135–139]. Oleate hydratases either in microbes or isolated enzymes catalyze the hydration of *cis*-double bond of OA at the C9 and C10 positions and stereoselectively produce the (*R*)-10HSA (Scheme 5.25) [130, 132, 140, 141]. However, the microbial hydration of OA to10-HAS by *Nocardia aurantia* ATCC 12647 gave enantiomeric mixtures of (*R*)- and (*S*)-10-HAS [142]. When linoleic acid Δ9 hydratase (conjugated linoleic acid hydratase, CLA-HY) was used to catalyze the

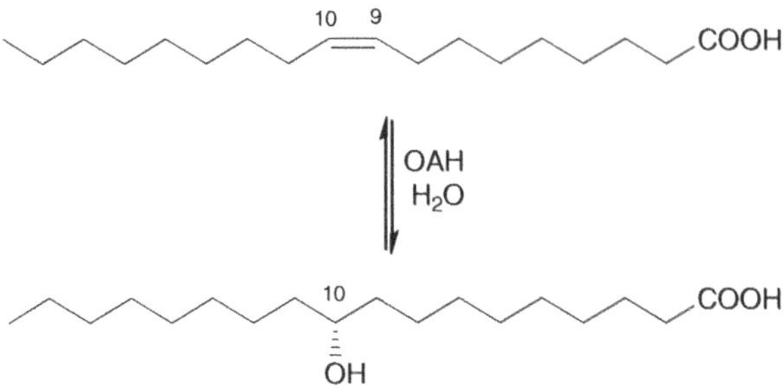

Scheme 5.25 Hydration of oleic acid by OAH yielding (*R*)-hydroxystearic acid. *Source:* Based on Engleder and Pichler [130]; Resch and Hanefeld [132]; Schmid et al. [140]; Kim and Oh [141].

hydration of *cis*-double bond on linoleic acid (LA), α-, and γ-linolenic acid (LnA), corresponding (*S*)-10-hydroxy fatty acids were obtained [143]. The regioselectivity for hydration of different free fatty acids by OAHs from various sources has been reviewed, sorted, and discussed. A strict regioselectivity for water addition to the *cis*-9 double bond of unsaturated fatty acids was observed for most enzymes with just a few exceptions that catalyzed the hydration on the *cis*-12 double bond of LA, α-LnA, and γ-LnA to form a 13-hydroxy fatty acid [130, 135, 138, 144–146]. In addition, the functionalized FA-HY1 from *Lactobacillus acidophilus* NTV001 after heterologous expression in *E. coli* showed a unique exemption from the apparently strict regioselectivity which hydrated *cis*-9, *cis*-11, *cis*-12, *cis*-13, and *cis*-14 double bonds in a total different mono-, di-, and poly-unsaturated fatty acids with chain lengths between C16 and C22 [130, 138].

Because OAH is easily expressed in *E. coli*, reasonably stable and active, without the need to supply FAD, and with excellent regio- and stereoselective lipid modification, OAH offers a highly interesting biotechnological route to functionalized fatty acids. A successful application of OAHs in large-scale biotransformations was the production of optically pure (*R*)-γ-dodecalactone, which is a flavor compound of malt whisky. In this process, a two-step conversion starts from the OAH-catalyzed hydration of oleic acid to produce (*R*)-10-hydroxysteric acid, which is followed by the baker's yeast-mediated fermentation to further convert it to (*R*)-γ-dodecalactone with an optical purity of 87.6% *e.e.* (Scheme 5.26) [132, 147, 148]. In order to improve the yield or productivity for the large-scale production of 10-hydroxystearic acid from oleic acid, native or expressed OAHs from different bacteria have been employed including *Stenotrophomonas nitritireducens* and *Stenotrophomonas maltiphilia* [149–151]. In another application, OAHs have also been employed in a novel multistep conversion process as an enzyme embedded in cascade reactions for the synthesis of plastic monomers, α,ω-dicarboxylic acids for polyesters and ω-aminocarboxylic acids for polyamides,

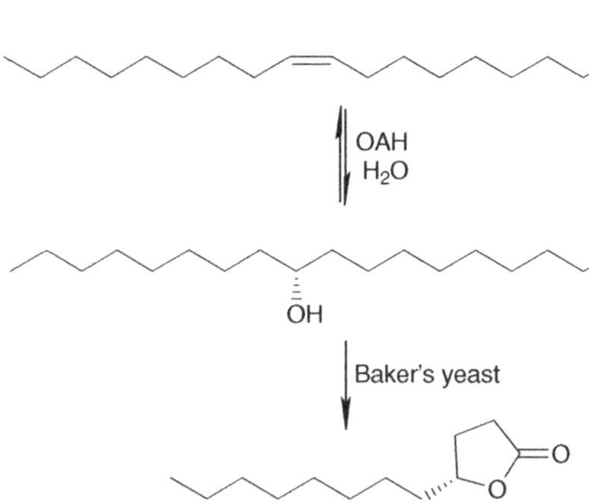

Scheme 5.26 A two-step process with the combination of oleate hydratase from bacterium and baker's yeast for the production of γ-dodecalactone from oleic acid. *Source:* Based on Resch and Hanefeld [132]; Gocho et al. [147]; Wanikawa et al. [148].

using renewable fatty acids and plant oils (Scheme 5.27) [152, 153]. Furthermore, the bio-based isobutene production by the dehydration of short chain isobutanol using partial purified OAH from recombinant *E. coli* as catalyst has been described [154].

The combination of innovative reaction engineering strategies with the newly discovered hydratases or a more detailed characterization of known enzymes allows the OAH as catalyst for the hydration of nonnatural short chain unsaturated fatty acid (*Z*)-undec-9-enoic acid (C11) to (*S*)-10-hydroxyundecanoic acid with >95% purity upon short chain saturated fatty acids was added as reaction additives [155]. The expressed OAH through full-length gene encoding was purified and used for the hydration of *cis*-9 polyunsaturated fatty acids (PUFAs) to 10-hydroxy fatty acids with higher activity than other available OAHs. The expressed OAH has been used effectively to convert PUFAs such as OA, LA, α-LnA, and/or γ-LnA in plant oils including soybean oil and perilla seed oil hydrolysates into 10-hydroxy fatty acids [156]. In order to make the enzymatic hydration of oleic acid to corresponding 10-HSA commercially practical, recombinant oleate hydratase from *E. meningoseptica* expressed in *E. coli* was purified and immobilized by different immobilization strategies to enhance its stability and reusability. Among the tested immobilization methods, the immobilization by covalent bonding onto chitosan magnetic composites showed the best results. The biomagnetic OAH demonstrated higher stability and activity over a larger pH and temperature range compared to the native enzyme which preserved 75% of the initial activity after five reuses [157].

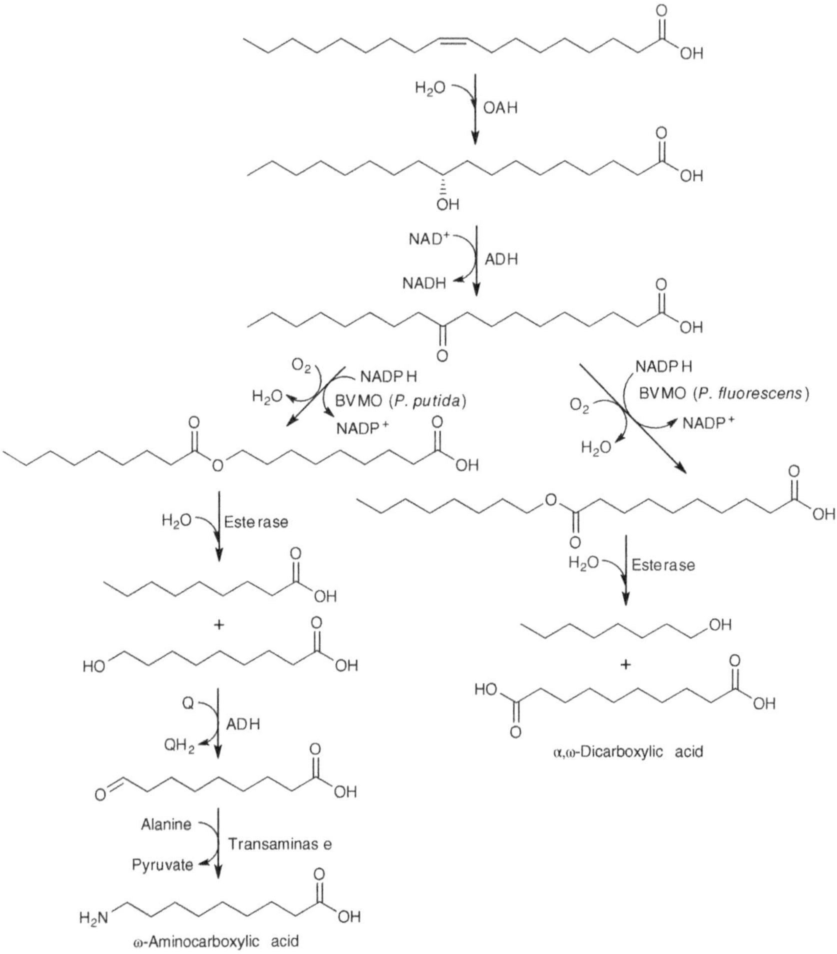

Scheme 5.27 Multistep enzymatic cascade synthesis of α,ω-dicarboxylic acid, ω-hydroxycarboxylic acid, and ω-aminocarboxylic acid starting with oleate hydratase and unsaturated fatty acid. *Source:* Based on Song et al. [152]; Song et al. [153].

5.2.1.2 Linalool Dehydratase-Isomerase

A unique bifunctional enzyme, linalool dehydratase-isomerase (LDI, EC 4.2.1.127), catalyzes the reversible dehydration and isomerization of (*S*)-(+)-linalool to produce β-myrcene and geraniol, respectively (Scheme 5.28) [158, 159]. The crystal structure of LDI from *Castellaniella defragrans* was independently solved by two groups and both showed a cyclic homopentamer consisting a central hole, with each monomer showing an $(\alpha,\alpha)_6$ barrel fold. The active site of the enzyme is located at the interface of two subunits. According to the amino

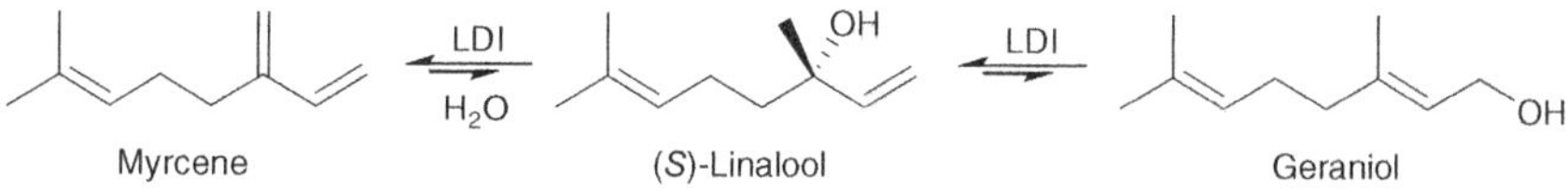

Scheme 5.28 Linalool dehydratase-isomerase catalyzed transformation of (*S*)-linalool to myrcene or geraniol. *Source:* Based on Demming et al. [158]; Brodkorb et al. [159].

sequence of the active site, two hypothetic reaction mechanisms were proposed by two groups, respectively, one via a covalent intermediate, and one acid–base mechanism via a carbocation intermediate. For both mechanisms, amino acid residue C171, Y45, and D39 at the active site act as general acid and base for the protonation of the hydroxyl leaving group of the substrate (*S*)-(+)-linalool and the dehydration at the chiral carbon atom. Water is activated by H129 or C180 and added to the covalent or carbocation intermediate [158, 160, 161].

The study of the transformation with purified enzyme showed that the dehydration of myrcene is thermodynamically favored; however, in *C. defragrans* as geraniol is further oxidized shifting the reaction equilibrium [159]. Further investigation of the LDI-catalyzed hydration, β-myrcene can be enantioselectively transformed to (*S*)-(+)-linalool with an *e.e.* at least 95.4% and virtually without (*R*)-(−)-linalool detected [162]. The pleasant odor of (*S*)-(+)-linalool displays an incentive for its use in the cosmetics and fragrance industries. In addition, (*S*)-(+)-linalool is commercially hardly available that suggests the hydration of β-myrcene using LDI would be an intriguing route to produce (*S*)-(+)-linalool and could be potential in industrial application [158, 162]. The detailed study on the substrate specificity with different linalool analogues and derivatives revealed that the α-methylally alcohol signature motif was essential for dehydration. Enzymatic kinetic resolutions of truncated and elongated aromatic and aliphatic tertiary alcohols (C5–C15) that contain a specific signature motif were dehydrated with selectivity factors ranging from 5 to > 200 [130, 158, 161].

5.2.1.3 Limonene Hydratase

Limonene hydratase (LIH) catalyzes the very specific regio- and stereoselective hydration of the 8,9-double bond of (*R*)-(+)-limonene to form (*R*)-(+)-α-terpineol (Scheme 5.29) [130, 132, 163, 164]. Sources of LIH have been reported in a variety of bacteria, fungi (particularly those growing on rotting citrus peel), and yeasts [165]. In nature, the conversion of limonene to α-terpineol is part of the degradation pathway of the monoterpene. (*R*)-(+)-Limonene is the most

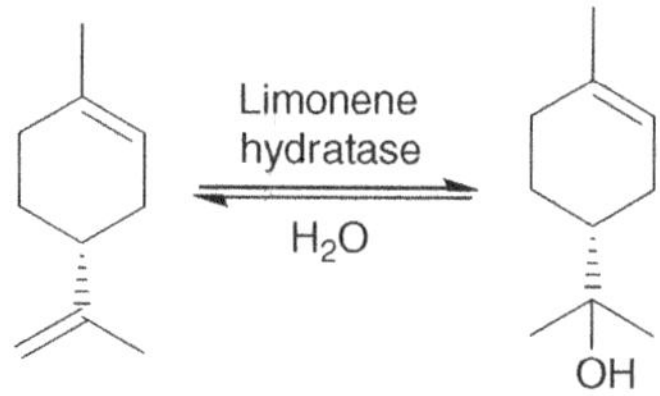

Scheme 5.29 Limonene hydratase catalyzed addition of water to (*R*)-(+)-limonene to form (*R*)-(+)-α-terpineol. *Source:* Based on Engleder and Pichler [130]; Marmulla and Harder [163]; Kaspera et al. [164].

abundant monocyclic monoterpene in nature, accounting for more than 90% of orange peel oil, and is a major by-product during processing of citrus oil or wood. Therefore, it represents an attractive inexpensive precursor for biosynthesis of high added value flavor and fragrance compounds such as (*R*)-(+)-α-terpineol [130, 166–169]. The strong lilac-like smell of (*R*)-(+)-α-terpineol makes it a common fragrance in the perfume industry and is an essential raw material for the food and cosmetics industries [163, 167, 170]. Almost exclusively, the biotransformation of limonene to α-terpineol was performed by whole cell system with the co-production a variety of other limonene derivatives [164, 165, 170–173]. In order to make the whole-cell biotransformation of (*R*)-(+)-limonene to (*R*)-(+)-α-terpineol efficient for industrial applications, a variety of strategies and efforts have been used including the screening of microorganisms [172], a thermostable limonene hydratase gene expressed *E. coli* cells [174], optimization of process conditions using response surface methodology (RSM) [175], submerged culture production [176], immobilized fungal mycelia cells for repeated-batch and continuously fed systems [171], and the use of a closed gas loop bioreactor [177]. Besides the production of (*R*)-(+)-α-terpineol using limonene hydratase, research has shown that the use of concentrated resting cells of *Sphingobium* sp. can convert (*R*)-(+)- and (*S*)-(−)-limonene to corresponding (*R*)-(+)- and (*S*)-(−)-α-terpineol, respectively, in a biphasic medium [168]. The use of isolated enzyme for the stereoselective hydration of (*R*)-(+)-limonene to (*R*)-(+)-α-terpineol was reported in an earlier work that the enzyme isolated from *Pseudomonas gladioli* was named as α-terpineol dehydratase [178].

5.2.2 Enzyme-Catalyzed Water Addition to Electron-Deficient C=C Bonds

5.2.2.1 Fumarase

Fumarase (EC 4.2.1.2) has been used industrially for catalyzing the reversible, stereospecific hydration of fumarate to form (*S*)-malate or L-malate (Scheme 5.30). It exists in all organisms and participates in part of the mitochondria matrix tricarboxylic acid (TCA) cycle [132, 179]. Three distinct types of fumarase (fumarase A, B, and C) were purified from *E. coli* and categorized in two different classes [180, 181]. Fumarase A and B belong to class I fumarases which are Fe^{2+}-dependent and sensitive to heat. Fumarase C is class II fumarases which are independent of Fe^{2+} and are not sensitive to elevated temperatures [182–186].

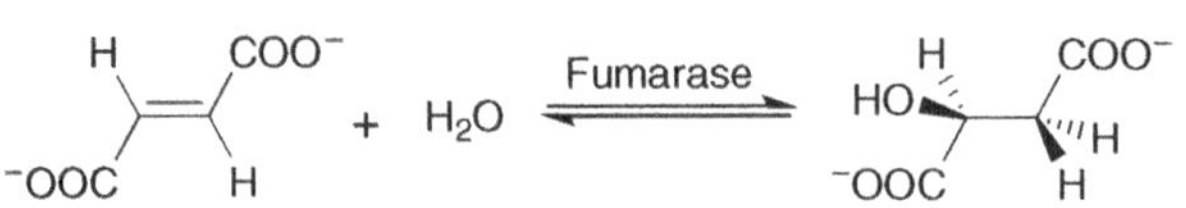

Scheme 5.30 Fumarase-catalyzed hydration of fumarate to form (*S*)-malate.

Only class II fumarases have been used for industrial applications by their high stability. The studies on the structure of fumarase C allowed detailed insight of the catalytic mechanism of fumarase-catalyzed water addition of fumarate to form (*S*)-malate [131, 132, 187–191].

L-Malic acid is major used as an acidulant in the food industry holding approximately 10% of the market, thus a potential alternative to citric acid. It is also used as an ingredient in pharmaceutical and cosmetic products. Furthermore, it is a potential monomer for biodegradable polymers [132]. Since the chemical hydration of fumaric acid produces a racemic mixture of D,L-malic acid and often requires harsh conditions, the employment of fumarase for the production of L-malic acid is favored industrially. To improve the productivity of L-malic acid from fumaric acid, the immobilized cells coupled to the design of continuous reaction system has been developed and investigated [192–195]. It has been found that an 80% conversion rate can be obtained when *Brevibacterium flavum* cells were immobilized on κ-carrageenan gel [196]. Instead of using immobilized whole-cell biocatalyst, the glutaraldehyde-treated *E. coli* having *Thermus thermophilus* fumarase (TtFTA) has been applied to a continuous reactor equipped with a cell-separation membrane filter to perform the enzymatic hydration of fumarate to malate which could be operated for more than 600 minutes with a molar conversion yield of 60% or higher [197]. For isolated enzyme approaches, an enzyme membrane reactor was developed as early in 1984 by the former Degussa company that combines enzyme recycling with homogeneous enzyme catalysis to produce the L-malic acid from fumaric acid with fumarase [132, 198]. Studies showed that the catalytic activity of purified fumarase for the hydration of fumarate in nonconventional one-phase media (water/polar organic solvent) exhibits a decreasing effect by the organic solvent [199, 200]. However, the inhibition effect of organic co-solvent on the catalytic activity of purified fumarase can be improved by immobilization of fumarase either covalently bonded on polymer surface or physical entrapment into cross-linked polymer gels. In addition, the immobilized fumarase with physical-entrapped enzyme demonstrates higher activity and stability than the covalently bonded enzyme [201].

The substrate specificity of fumarase isolated from pig heart has been studied since 1968, which revealed the activity toward fluorofumarate, fumarate, chlorofumarate, bromofumarate, acetylenedicarboxylate, iodofumarate, and mesaconate. Nearly all tested substrates showed a *trans* water addition product with the exception of fluorofumarate which gives α-fluorofumarate and spontaneously decomposes to oxaloacetate [202]. Later studies showed that 2,3-difluorofumarate can also be accepted as substrate [203, 204]. With chlorofumarate, the hydration product was L-*threo*-chlorofumarate, which can be further chemically converted to 2-deoxy-D-ribose and *trans*-D-*erythro*-sphingosine (Scheme 5.31) [204]. Besides fumarate and its derivatives, fumarase isolated from swine heart muscle catalyzed

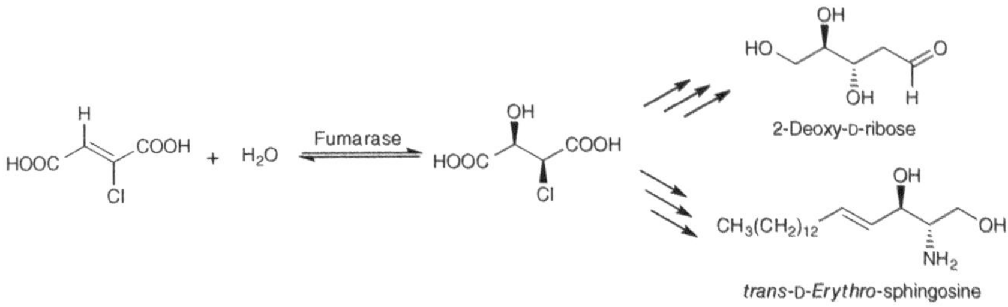

Scheme 5.31 Fumarase-catalyzed synthesis of L-*threo*-chloromaleic acid and its conversion to 2-deoxy-D-ribose and *trans*-D-*erythro*-sphingosine. *Source:* Based on Findeis and Whitesides [204].

the stereospecific hydration of L-*trans*-2,3-epoxysuccinate to yield meso tartrate [205, 206].

5.2.2.2 Malease

Malease (maleic acid hydratase, EC 4.2.1.31) catalyzes the stereospecific reversible water addition to maleic acid (*Z*-isomer) to produce (*R*)-malate (D-malate) (Scheme 5.32) which can be dated back to 1951 [131, 132, 207]. Just as fumarase, malease is a cofactor-independent enzyme and shows good stability. Since malease is not involved in the primary metabolism of organisms, it is in contrast to fumarase less abundant in nature and can be found only in bacteria or mammals [208–213]. The stereochemical study of the malease-catalyzed hydration indicates the addition of water to maleic acid is via an antiaddition to form the (*R*)-malate product [209, 214–216].

The substrate specificity of malease was described, as the case of fumarase, rather narrow that strictly performs the reversible water addition of maleic acid [209, 217]. However, later surveys showed the hydration activity was also allowed for citraconate, chloromaleate, and bromomaleate, but no activity toward dimethyl maleate, acetylenedicarboxylate, fumarate, mesaconate, *Z*-epoxysuccinate, or *E*-epoxysuccinate. The results also reveal the water addition catalyzed by malease is exclusively to *Z*-double bonds in maleic acid and its derivatives [131, 211–213,218, 219].

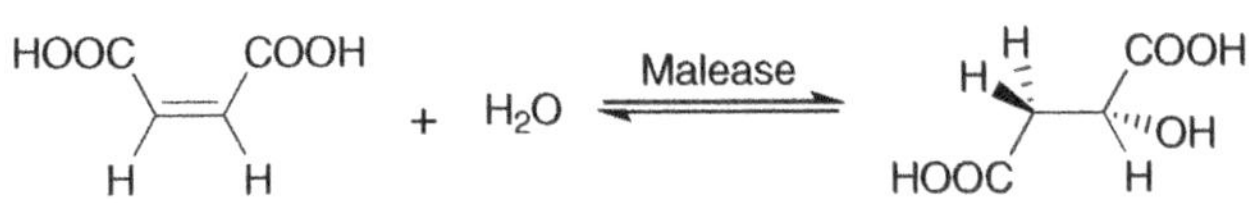

Scheme 5.32 Malease-catalyzed water addition of maleic acid to form (*R*)-malic acid. *Source:* Based on Engleder and Pichler [130]; Resch and Hanefeld [132]; Marmulla and Harder [163]; Kaspera et al. [164].

For industrial application, malease from *Pseudomonas pseudoalcaligenes* is employed for the large-scale production of D-malate. The maleic acid obtained from the spontaneous hydrolysis of maleic anhydride under aqueous conditions allows a cost-efficient process [132]. A continuous stirred tank reactor (CSTR) has also been developed for the more efficient production of solid D-malate from a Ca-maleate suspension in a liquid-solid-solid three phase system using permeabilized *P. pseudoalcaligenes* as catalyst [220–224].

5.2.2.3 Enoyl-CoA Hydratase

The β-oxidation pathway of fatty acid degradation in living organisms consists of a set of four enzymatic reactions. Enoyl-CoA hydratase (ECH, EC 4.2.1.17) catalyzes the second step for the reversible addition of water molecule to the double bond of *trans*-2-enoyl-CoA thioesters to the corresponding (*S*)-3-hydroxyacyl-CoA thioesters. Subsequent steps complete the oxidation of β-carbon and release acetyl-CoA with a shortened acyl-CoA [225–227].

In fact, the investigations of this hydratase have demonstrated that there are two different types of enoyl-CoA hydratases either (*S*)-selective (ECH1, EC 4.2.1.17) or (*R*)-selective (ECH2, EC 4.2.1.119) (Scheme 5.33) [225, 226, 228]. In addition, the substrate specificity for these two types of hydratases is also different. The ECH1 (also known as crotonase) catalyzes the water addition to form α,β-hydroxyacyl-CoA thioesters with an alkyl chain length varying between 4 and 20 carbon atoms via a concerted or a more likely sequential mechanism by a flexible loop separating the active site pocket from the surroundings, allowing space for larger substrates [131, 229, 230]. The studies of stereoselectivity of ECH1 indicate that the rate of 3(*R*)-hydroxybutyryl-CoA formation is 400,000-fold slower than the normal hydration reaction to form 3(*S*)-hydroxybutyryl-CoA and ECH1 is capable of catalyzing the epimerization between 3(*R*)-hydroxybutyryl-CoA and 3(*S*)-hydroxybutyryl-CoA [231]. In another study using hexadienyl-CoA as the

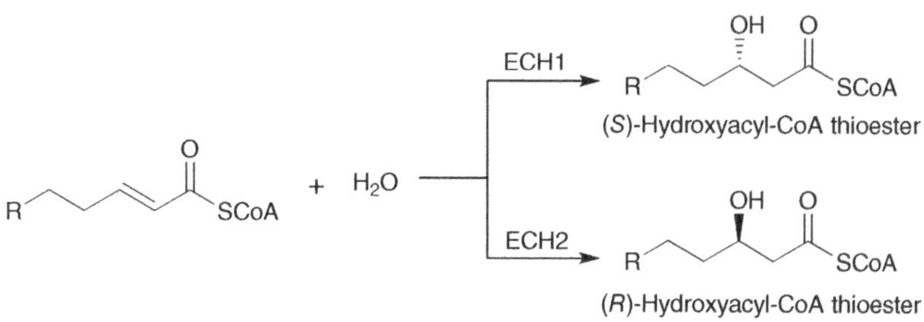

Scheme 5.33 ECH1- and ECH2-catalyzed water addition to *trans*-2-enoyl-CoA thioester to form corresponding alcohols with different enantioselectivity. *Source:* Based on Hiltunen and Qin [225]; Agnihotri and Liu [226]; Bhaumik et al. [228].

substrate, results reveal that both the *s-cis* and *s-trans* conformers of the substrate are bound to the enzyme, but that only the *s-cis* conformers can be activated to proceed the (*S*)-selective enzyme-catalyzed hydration [232]. Furthermore, an isomerization-hydration reaction follows this kind of mechanism from the crotonase family [233]. Due to a hot-dog fold of the active site of ECH2, the ECH2 shows a preference substrate of straight medium-length carbon atoms via a concerted mechanism, while depending on the enzyme sources, it also catalyzes the hydration of long-chain substrates [131, 234–236]. From the structural information of these two hydratases, the difference in enantioselectivity between ECH1 and ECH2 can be easily understood by the geometry of the active sites which behave like mirror image of each other [236, 237].

The exploration of the catalytic mechanism of enoyl-CoA hydratases showed that both types of hydratase catalyze *syn*-addition/elimination of water molecule to the double bond of a *trans*-enoyl-CoA thioester without requiring any cofactors or metal ion [226, 231, 238, 239]. This quite simple catalysis system using enoyl-CoA hydratases has long been applied to industry for the production of (*R*)-3-hydroxybutyric acid and (*R*)-3-hydroxyisobutyric acid from butyric acid and isobutyric acid, respectively. The processes run by Kanegafuchi Japan were a three-step whole-cell system using *Candida rugose* IFO 0750M. (*R*)-3-hydroxybutyric acid and (*R*)-3-hydroxyisobutyric acid are important building block for the synthesis of a carbapenem intermediate and a precursor for captopril, respectively [131, 240, 241].

Recently, because of energy and sustainability concerns, there is increasing interest in the synthesis of higher chain volatile alcohols by engineered microorganisms. An engineered *E. coli* strain was used to synthesize 1-hexanol from glucose by extending the coenzyme A (CoA)-dependent 1-butanol synthesis reaction sequence. In the modified reaction sequence, the butyryl-CoA first synthesized via acetyl-CoA acetyltransferase (AtoB), 3-hydroxybutyryl-CoA dehydrogenase (Hbd), crotonase (Crt), and *trans*-enoyl-CoA reductase (Ter) was further extended to hexanoyl-CoA via β-ketothiolase (BktB), Hbd, Crt, and Ter, and finally was reduced to yield 1-hexanol by aldehyde/alcohol dehydrogenase (AdhE2) and 1-hexanol was secreted to the fermentation medium under anaerobic conditions [242]. Since hexanoic acid is a precursor for fine chemicals, it has diverse industrial applications. A yeast strain *Kluyceromyces marxianus* was engineered to simultaneously integrate five to seven genes into the chromosomes to construct a synthetic pathway for the production of hexanoic acid from either glucose or galactose. In this pathway, hexanoic acid synthesis starts from acetyl-CoA and is catalyzed by AtoB from *E. coli*, BktB from *Ralstonia eutropha*, Hbd and Crt from *Clostridium acetobutylicum*, Ter from *Treponema denticola*, malonyl CoA-acyl carrier protein transacylase (MCT1) from *Saccharomyces cerevisiae*, and acetyl-CoA thioesterase (TES1) from *K. marxianus* [243].

The recombinant bacterium *Pediococcus acidilactici* BD16 (*fcs*$^+$/*ech*$^+$) encoded with synthetic genes of feruloyl-CoA synthetase (*fcs*$^+$) and enoyl-CoA hydratase (*ech*$^+$) has also been implemented for wine production from a variety of substrates (orange, rose, and black grapes) to enhance the flavor characterization of wine during secondary malolactic fermentation (MLF). The expressed synthetic *fcs*$^+$ and *ech*$^+$ genes of the recombinant *P. acidilactici* BD16 are responsible for the bioconversion of ferulic acid to vanillin during the MLF [244]. Industrial crude glycerol by-product from palm oil biodiesel production, an inexpensive, abundant, and promising carbon source, was used to decrease the high production cost of bio-base plastic poly(3-hydroxybutyrate) (P(3HB)) using recombinant *E. coli* encoded with the polyhydroxyalkanoates (PHAs) synthetic genes of β-ketothiolase ($PhaA_{Re}$) and acetoacetyl-CoA reductase ($PhaB_{Re}$) from *R. eutropha* and PHA synthase ($PhaC_{Re}$) from *Aeromonas hydrophila*. Then, P(3HB-co-3-hydroxyhexanoate) (P(3HB-co-3HHx)) copolymer was achieved using *E. coli*-$ABC_{Ah}J_{Ah}$ obtained by further encoding an extra (*R*)-specific enoyl-CoA hydratase ($PhaJ_{Ah}$) gene from *A. hydrophila* [245]. The production of P(3HB) from nonchiral substrate β-butyrolactone via (*R*)-3HB-CoA with recycling of CoA can be achieved by using the immobilized multiple enzyme system. Three enzymes including an (*R*)-specific $PhaC_{Re}$ from *R. eutropha*, an (*S*)-specific dehydrogenase (FadB) from *P. putida*, and an (*R*)-specific enoyl-CoA hydratase ($PhaJ4_{Pa}$) from *Pseudomonas aeruginosa* were immobilized by affinity tag-assisted binding to self-assembled antiparallel type β-sheets with a coiled fiber structure formed from a decapeptide [246].

5.3 Lyases with Carbon–Nitrogen Bonds

Ammonia lyases (ALs, EC 4.3.1.x) are a class of enzymes capable of catalyzing the reversible cleavage of C–N bonds, typically α-amino acids, to release ammonia and corresponding α,β-unsaturated or cyclic derivatives without employing hydrolysis or oxidation mechanism. The structure and function of this class of enzymes are quite different that they are further divided to 31 subclasses and grouped into seven groups based on their specific analogies in their catalyzed reactions and the types of accepted substrate [247, 248]. The use of ALs to enantiomerically pure amino acids provides an increasingly attractive route for industrial needs. In addition, the improvement in functional properties of ALs by the advancement of discovery and engineering technologies leads to their applications in treating phenylketonuria (PKU) and cancer [249, 250]. In the following context, those ammonia lyases of the most popularly studied and already applied in synthetic chemistry will be discussed in detail.

5.3.1 Aspartate Ammonia Lyase

Aspartate ammonia lyases (AAL, EC 4.3.1.1), also referred to as aspartases, catalyze the reversible deamination of L-aspartic acid to fumaric acid and ammonia (Scheme 5.34) [251]. The ubiquitous aspartases play an important role in bacterial nitrogen metabolism and belong to a broad aspartase/fumarase superfamily with a characteristic tertiary and quaternary structure, as well as a similar active site architecture [251, 252]. The structural similarities among the superfamily members suggest that all of them follow a mechanism in which a general base catalyzes proton abstraction and formation of an enediolate intermediate [249, 252].

Aspartase is one of the most specific enzymes known that only L-aspartic acid can be used to perform the deamination reaction. This high selectivity of AAL toward the natural substrate has narrowed its practical application. Nonetheless, the reverse reaction, the amination of fumaric acid to enantiopure L-aspartic acid, catalyzed by aspartase from *E. coli* (AspA) has been exploited as a starting material in the industrial preparation of artificial sweetener aspartame (*N*-L-α-aspartyl-L-phenylalanine 1-methyl ester) (Scheme 5.35) [251]. In this reaction, in order to drive the equilibrium from fumaric acid to L-aspartic acid, an excess of ammonia is added to the reaction system.

L-Aspartic acid and its derivatives are highly valuable tools for biological research and as chiral precursor for many pharmaceuticals [251]. Many methods have been employed to improve the catalytic efficiency of aspartase for L-aspartic acid production.

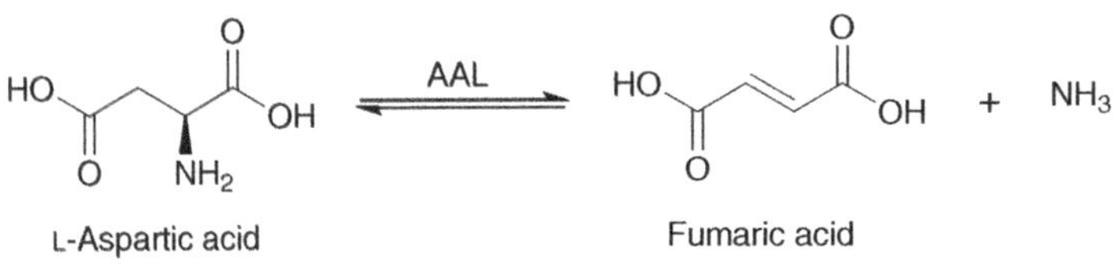

Scheme 5.34 Aspartate ammonia lyase catalyzed reversible deamination reaction. *Source:* De Villier et al. [251].

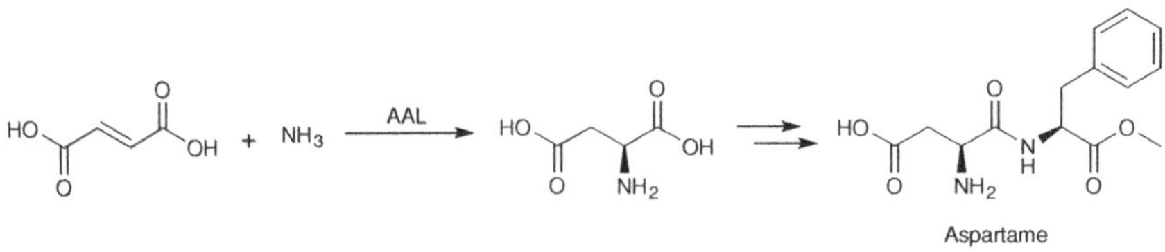

Scheme 5.35 Industrial synthesis of artificial sweetener aspartame with the starting material L-aspartic acid formed by aspartase-catalyzed amination of fumaric acid. *Source:* Based on De Villier et al. [251].

One strategy was based on the alginate-immobilized PSCat (AspA) constructed by expressing *E. coli* aspartase in psychrophilic *Shewanella livingstonensis* Ac10 which has led to the improved yield of L-aspartic acid to 99.3% and the catalyst could be re-used nine times [247, 253]. There were also patented systems using chemical recycling of the waste stream from the aspartase-catalyzed reaction. In this recycled system, maleic anhydride was added to product effluent to lower the pH of the L-aspartate solution followed by precipitating the L-aspartic acid, then the effluent containing only maleic acid and low concentration ammonia was chemically converted to fumaric acid and pH adjustment and recycled back to the reaction system. A modified but more elegant system, maleic anhydride was used as the only substrate and reacted by the coupled maleate isomerase and aspartase enzyme system to make the recycled system more efficient for producing much higher purity L-aspartic acid [247].

Although the high substrate specificity of aspartase limits the substrate range, it avoids cross-reactions with other enzymatic systems. Therefore, aspartase is suitable for use in multienzymatic cascade processes to synthesize value-added chemicals from inexpensive materials. One example is the coupled enzyme synthesis of L-phenylalanine by the co-immobilization of aspartase (AspB) with a microbial aspartase transaminase (TA) from phenylpyruvate via the *in situ* formation of amino donor L-aspartate from fumarate (Scheme 5.36) [254]. The yield of L-phenylalanine in the continuous stirred-tank reaction using the co-immobilized enzymes on an amino-epoxy support at 0.1 M phenylpyruvate was 95%.

The multienzymatic cascade reaction strategy was applied for the synthesis of important naturally occurring nonessential amino acid β-alanine as well in favor of pharmaceuticals, food and feed additives, polymeric materials, and electroplating industry. Nitrilase BjNIT3397 from *Bradyrhizobium japonicum* strain USDA110 was used for hydrolyzing 3-aminopropionitrile to produce β-alanine. However, the substrate specificity is not ideal, up to 33% of the by-product 3-aminopropanamide was formed when using 3 M 3-aminopropionitrile.

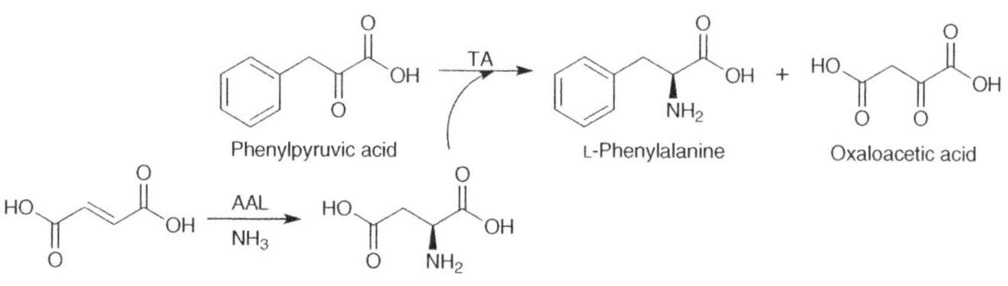

Scheme 5.36 Multienzymatic cascade synthesis of L-phenylalanine from phenylpyruvic acid using L-aspartic acid as amino acid donor via *in situ* formation from fumaric acid. *Source:* Based on Cárdenas-Fernández et al. [254].

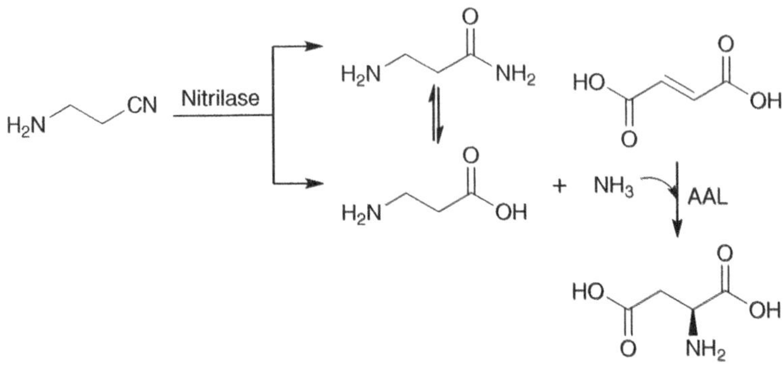

Scheme 5.37 Multienzymatic cascade reaction for the production of β-alanine and L-aspartic acid using nitrilase and aspartate ammonia lyase.

To solve this high amide by-product problem, AAL and fumaric acid were added to the reaction system to consume the normal by-product NH_3 that will shift the equilibrium between the amide, the acid, and the ammonia toward the formation of β-alanine (Scheme 5.37). The multienzymatic cascade reaction system by coupling AAL and nitrilase resulted in an effective reduction of 3-aminopropanamide from 33 to 3% without influencing the activity of nitrilase. In addition, this reaction system also demonstrates the ability of one-pot synthesis β-alanine and L-aspartic acid [247, 255].

Despite the high specificity of AAL for its natural reaction, numerous attempts have been put to expand its application to the synthesis of substituted aspartic acid derivatives. The earliest example should be the synthesis of *N*-hydroxyaspartic acid by the addition of hydroxylamine with fumaric acid using partially purified AAL (AspA) from *Bacillus cadaveris* [256]. However, the product was too unstable to be isolated and was only identified by its physicochemical properties. It is not until the year 2008, another similar experiment was performed to test a variety of alternative nucleophiles for the addition reaction catalyzed by AAL (AspB). The results showed that AspB accepts not only hydroxylamine but also methoxylamine, hydrazine, and methylamine in addition to fumaric acid to yield L-*N*-hydroxyaspartic acid, L-*N*-methoxyaspartic acid, L-2-hydrazinosuccinic acid, and L-*N*-methylaspartic acid in a manner of complete conversion (Scheme 5.38). Whereas L-*N*-hydrxyaspartic acid could not be isolated due to its instability [257]. In order to expand the substrate spectrum and the consideration of β-carboxyl group of aspartic acid but not the α-carboxyl group involved in the intermediate of AAL-catalyzed ammonia addition to fumaric acid, the combination of structure-guided engineering and a cluster screening method affords researchers to synthesize enantiopure (*R*)-3-aminobutyric acid from crotonic acid and ammonia using mutant enzyme BSASP-C6 (Scheme 5.39) [247, 249, 258].

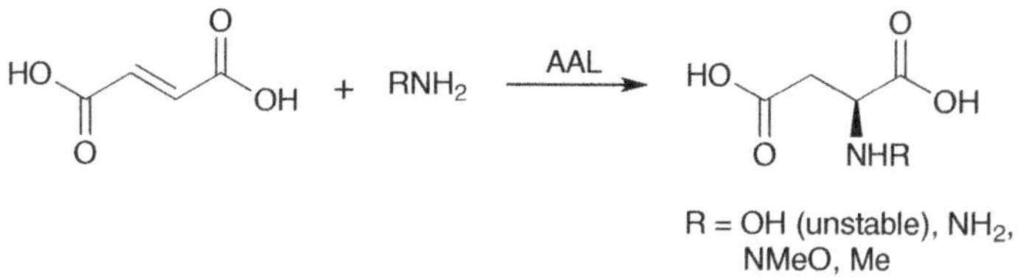

Scheme 5.38 Synthesis of *N*-substituted aspartic acid derivatives with aspartase.

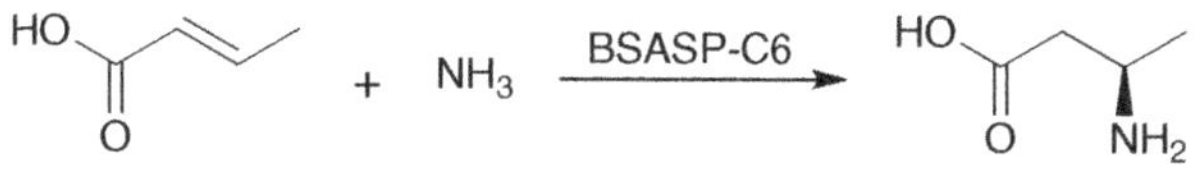

Scheme 5.39 Synthesis of enantiopure (*R*)-3-aminobutyric acid with mutant aspartase. *Source:* Based on Hiltunen and Qin [225]; Agnihotri and Liu [226]; Bhaumik et al. [228].

5.3.2 Methylaspartate Ammonia Lyase

3-Methylaspartate ammonia lyase (MAL, EC 4.3.1.2), also called 3-methylaspartase, catalyzes the reversible α,β-elimination of ammonia from L-*threo*-3-methylaspartic acid and L-*erythro*-3-methylaspartic acid to yield 2-methylfumaric acid (mesaconic acid) without specific cofactor, except that Mg^{2+} and K^+ are required for activity (Scheme 5.40) [247, 251]. MAL was first identified in the anaerobic bacterium *Clostridium tetanomorphum* H1 with the activity of deamination in L-glutamate catabolic pathway to form L-*threo*-3-methylaspartate and finally to acetyl-coenzyme A. However, MAL has been found in a number of facultative anaerobes later. Although MAL shows similar transformation as AAL, the protein structure of MAL is quite different from AAL and shares mechanistic and structural features with the enolase superfamily [251, 259, 260]. In contrast to AAL, MAL has been proved to be much wider substrate spectrum and further expanded to the synthesis of many substituted aspartic acid derivatives by structure-based protein engineering [247].

The ability of MAL accepting nonnatural substrates is interested in synthetic applications. The preparation of L-*threo*-3-methylaspartic acid from mesaconic

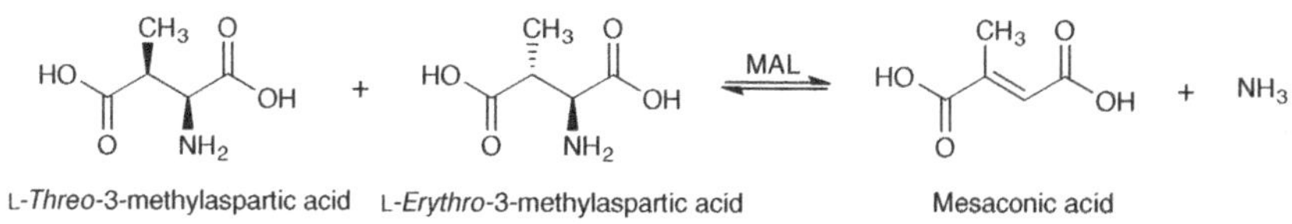

Scheme 5.40 3-Methylasparte ammonia lyase catalyzes the reversible deamination of L-*threo*-3-methylaspartic acid and L-*erythro*-3-methylaspartic acid to yield mesaconic acid and ammonia. *Source:* Based on Bhaumik et al. [228]; Akhtar et al. [266]; Gulzar et al. [267].

acid with a yield of 76% and D-*threo*-3-methylaspartic acid from racemic *threo*-diastereomer by the removal of L-enantiomer with a yield of 36% using CtMAL (MAL from *C. tetanomorphum*) in early studies is obviously very efficient [261, 262]. Later report showed that a variety of β-substituted aspartic acids, such as ethyl, *n*-propyl, and isopropyl groups, were also accepted by CtMAL with progressively low rates [263]. Then, only about two decades later came the first asymmetric addition of ammonia to 2-halofumaric acids using CtMAL. Results showed that only with 2-chlorofumaric acid the product could be isolated in high purity and yield [264, 265]. The same group also demonstrated the synthesis of a number of 3-alkylaspartic acids from various 2-alkylfumaric acids and the synthesis of *N*-substituted aspartic acids with nitrogen nucleophiles other than simple ammonia, such as hydrazine, methylamine, hydroxylamine, and methoxyamine, with CtMAL (Scheme 5.41) [247, 266, 267]. Although a much wider substrate range was obtained than aspartases by these works, the addition reaction was still limited to small nucleophiles. Analogously, the thermostable ChMAL (MAL from the thermophilic bacterium *Carboxydothermus hydrogenoformans*) shows similar substrate scope by accepting various substituted fumaric acids and amines with high regio- and stereoselectivity that makes ChMAl also is a promising enzyme for biocatalytic applications [268].

Recent advances in protein engineering have led to the alteration of the activity and stereoselectivity and the broadening of the substrate scope of CtMAL. The diastereoselectivity of CtMAL has been altered by site-directed mutagenesis to allow the exclusive production of L-*threo*-3-methylaspartic acid [269]. The structure-based engineering has been used for redesigning the CtMAL and led to the development of a single active site mutant (Leu384Ala) that has a wide electrophile scope at the C2 position of the fumaric acid (Scheme 5.42a) [247, 270]. These investigators used the same technique to engineer the nucleophile (amine) binding site of CtMAL to form a single active site mutant (Gln73Ala) that showed a remarkably broad nucleophile tolerance, accepting amines from simple alkylamines to

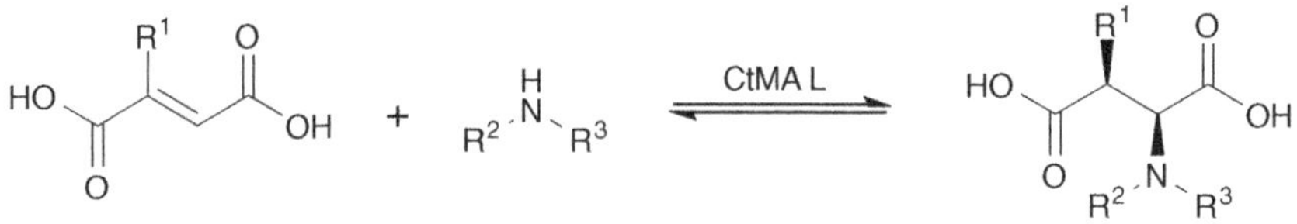

R^1 = H, Me, Et, *n*-Pr, *i*-Pr
R^2 = H, NH_2, Me, Et, OH, OMe
R^3 = H, Me,

Scheme 5.41 Synthesis of aspartic acid derivatives with wild-type CtMAL. *Source:* Based on Parmeggiani et al. [247]; Raj et al. [270].

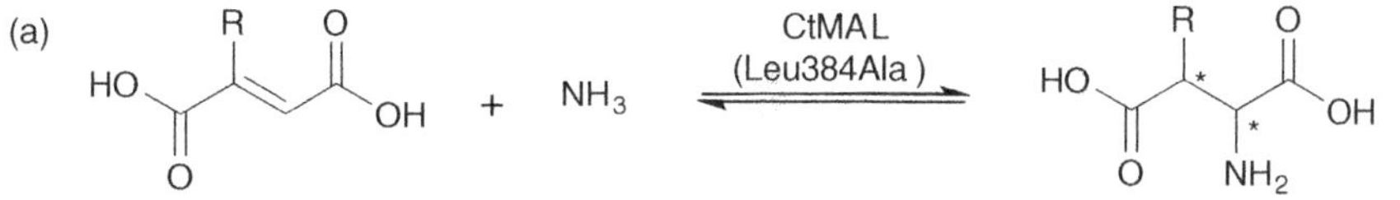

R = Et, *n*-Pr, *n*-Bu, *n*-Pent, *n*-Hex, Bn, EtO, PhO, BnO, EtS, PhS, BnS

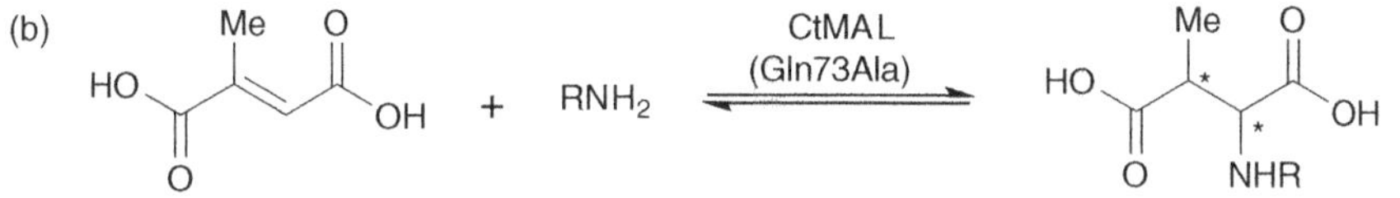

R = Me, Et, *n*-Pr, *i*-Pr, *n*-Bu, *n*-Pent, *n*-Hex, cyclopropyl, cyclobutyl, cyclopentyl, cyclohexyl, (cyclopropyl)methyl, EtO, $(CH_2)_2OH$, $(CH_2)_3OH$, $(CH_2)_2OMe$, $(CH_2)_2NH_2$, $(CH_2)_2NHMe$, $(CH_2)_3NH_2$

Scheme 5.42 Engineered MAL variants catalyzed amination with broad nucleophile/electrophile tolerance to aspartic acid derivatives. (a) For substituted fumaric acid substrates. (b) Broad amine nucleophiles. *Source:* Based on Parmeggiani et al. [247]; Heberling et al. [249]; Turner [250]; Cooke et al. [272]; MacDonald and D'Cunha [273].

cyclohexylamine and benzylamine for L-*threo*-3-methylaspartic acid (Scheme 5.42b) [247, 270]. These results indicate that CtMAL could be the most potential enzyme for employing in the asymmetric synthesis of optical pure L-aspartic acid derivatives in industrial applications. Furthermore, accompany the growth of synthetic biology MALs has also been exploited in the metabolic engineering production of mesaconic acid from glucose. In this research, recombinant expression of glutamate mutase (for the conversion of glutamic acid to 3-methylaspartic acid) and MAL from *C. tetanomorphum* in *E. coli* has been employed to perform a cascade multienzyme reaction system which efficiently converted glucose to mesaconic acid with a productivity of 6.96 g L^{-1} [271].

5.3.3 Aromatic Amino Acid Ammonia Lyases

Among the aromatic amino acid ammonia-lyases (or arylalanine ammonia-lyases), phenylalanine ammonia-lyase (PAL, EC 4.3.1.5), histidine ammonia-lyase (HAL, EC 4.3.1.3), and tyrosine ammonia-lyase (TAL, EC 4.3.1.23) catalyze the reversible nonoxidative deamination of L-phenylalanine, L-histidine, and L-tyrosine toward corresponding *trans*-form of α,β-unsaturated carboxylic acids, cinnamic acid, urocanic acid, and *p*-coumaric acid, respectively (Scheme 5.43) [247, 249, 250, 272, 273]. There was a sole available literature concerning the conversion of L-tryptophan to 3-indoleacrylic acid and ammonia using tryptophan ammonia-lyase (WAL, EC 4.3.1.31) from *Rubrivivax benzoatilyticus* [274]. Of these arylalanine

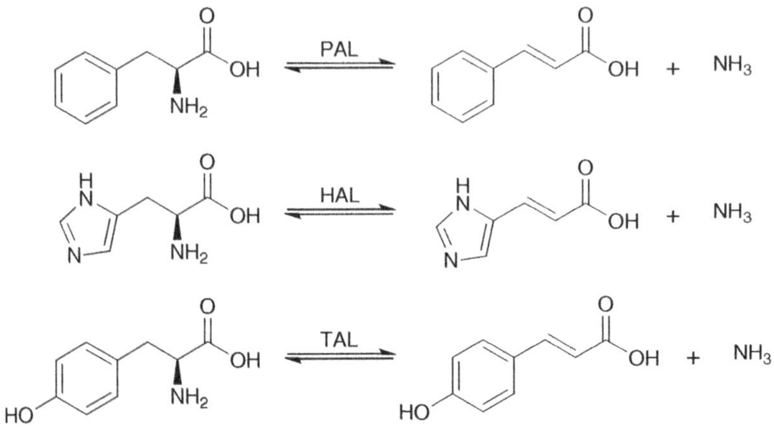

Scheme 5.43 Aromatic amino acid ammonia-lyases catalyzed reversible deamination of L-phenylalanine, L-histidine, and L-tyrosine toward corresponding cinnamic acid, urocanic acid, and *p*-coumaric acid, respectively. *Source:* Based on Renard et al. [282].

ammonia-lyases, PAL and TAL are the most popular enzymes already applied in organic synthesis. These arylalanine ammonia-lyases contain a highly electrophilic heterocycle 4-methylideneimidazol-5-one (MIO) group for substrate activation [275–277]. MIO is formed by posttranslational cyclization and dehydration of a conserved alanine–serine–glycine tripeptide in the active site, which is responsible for the amination/deamination reactions [272, 276, 278]. Two kinds of reaction mechanisms were proposed for MIO-dependent ammonia lyases: (i) an E1cb mechanism with an electrophilic attack of the MIO group on the amine substrate; (ii) a Friedel–Crafts-type attack of the MIO group on the aromatic ring [279]. A later study based on docking and molecular dynamic simulations on the PAL from *Petroselinum crispum* suggests that the elimination of an ammonia group from the aromatic amino acids can be performed by both mechanisms [280].

5.3.3.1 L-Arylalanines Synthesis

Form the synthetic chemistry point of views, arylalanine ammonia-lyases catalyzed reversible reactions are more useful in the direction toward amination to produce higher value L-arylalanine from the inexpensive arylacrylic acid substrate. A wide range of L-phenylalanine derivatives and nonnatural L-arylalanines have been synthesized by the addition of ammonia to their corresponding unsaturated precursors using PAL. With its high specificity, PAL from *Rhodotorula glutinis* (RgPAL) was first employed by Yamada et al. to rapidly convert *trans*-cinnamic acid to L-phenylalanine and the conversion yield was ~70% [281]. A decade later, RgPAL was applied to the amination of some substituted cinnamic acid derivatives to synthesize substituted L-phenylalanine analogues and thiophene substituted amino acids (Scheme 5.44) [282]. The variation of conversion yield is correlated to the steric effect of the substituent and the active site residues, increasing further the size of the aromatic ring, say 1-naphthyl, may dramatically reduce the conversion yield or results no conversion at all, for example, indol-3-yl and 3,4-methylenedioxyphenyl. A number of nonnatural halophenylalanine derivatives were synthesized in preparative scale under optimized conditions by PAL using ammonium carbamate as the ammonia source instead of the conventional ammonia or other ammonium salts. In each case, a superior full conversion was obtained with a space-time yields from 5 g L^{-1} d^{-1} up to 237.6 g L^{-1} d^{-1} which is due to the release of two equivalents of ammonia and keeping a lower ionic strength [247, 283]. The improved conversion yield and much simpler downstream processing show the industrial availability of PALs in the production of enantiopure halophenylalanines from corresponding acrylic acids.

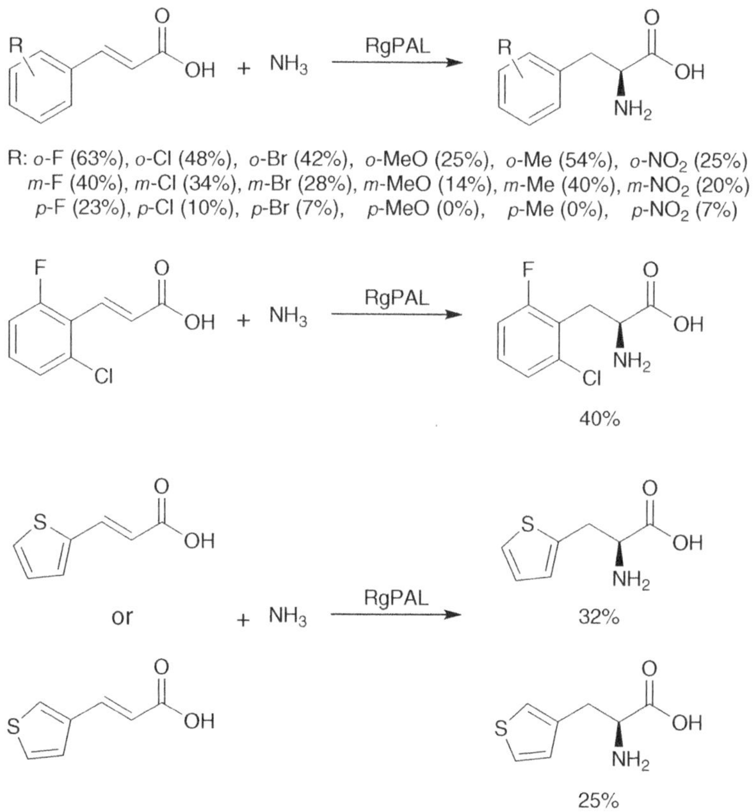

Scheme 5.44 RgPAL-catalyzed amination for the synthesis of substituted phenylalanine analogues. *Source:* Based on D'cunha et al. [284].

PAL from *R. glutinis* (RgPAL) was also utilized for the direct, one-step enzymatic amination of methyl *trans*-cinnamate to L-phenylalanine methyl ester using ammonium sulfate as the ammonia source. Since the solubility of the substrate in aqueous solution is poor, an organic-aqueous biphasic solution was employed to overcome this problem. After solvent screening, *n*-heptane was selected as the cosolvent for the organic-aqueous biphasic system which gives a product yield of about 70% under the optimized conditions (Scheme 5.45) [284]. This reaction favors the application for aspartame synthesis. On the other hand, synthesis of phenylalanine analogues with nonaromatic cyclic systems using PAL and corresponding substrates was also explored by several groups. It has been demonstrated that L-(2,5-dihydrophenyl)alanine can be deaminated by RgPAL to form corresponding acrylic acid with an 85% yield (Scheme 5.46a), while its regio-isomer L-(1,4-dihydrophenyl)alanine is not and is a moderately good competitive inhibitor of the reaction [285, 286]. The reaction mechanism was explained by the Friedel–Crafts mechanism [287, 288]. In addition to RgPAL, the mechanistic investigations on PAL from *P. crispum* (PcPAL) were implemented by the deamination of two aromatic pyrimidine substrate analogues, L-(pyrimidin-2-yl)alanine and L-(pyrimidin-5-yl)alanine, indicating L-(pyrimidin-5-yl) alanine was a moderately good substrate of PcPA to give a 57% yield of the acrylic acid (Scheme 5.46b) and L-(pyrimidin-2-yl)alanine was determined as an inhibitor [288]. The results also support the Friedel–Crafts reaction mechanism. Further exploration

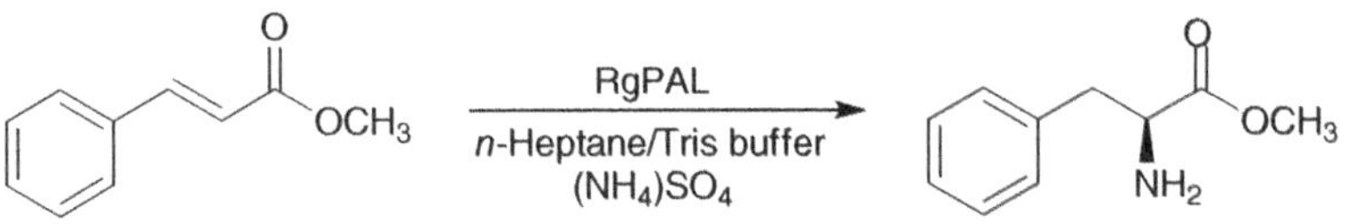

Scheme 5.45 RgPAL-catalyzed biphasic amination for L-phenylalanine methyl ester. *Source:* Based on Hanson et al. [285]; Skolaut and Rétey [286]; Based on Gloge et al. [288].

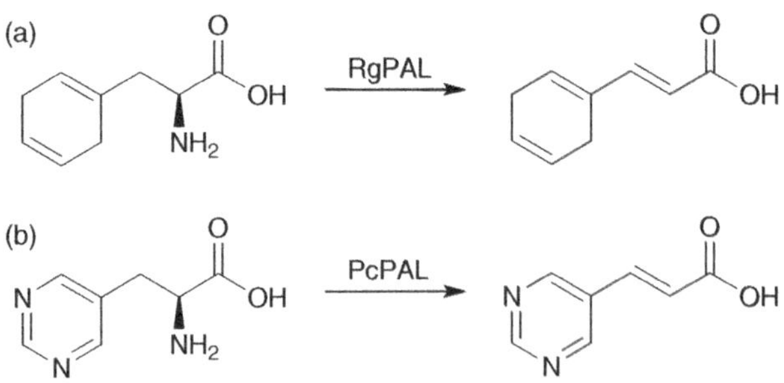

Scheme 5.46 PAL-catalyzed deamination of nonaromatic cyclic phenylalanine analogues to corresponding acrylic acids: (a) L-(2,5-dihydrophenyl)alanine, (b) L-(pyrimidin-5-yl) alanine. *Source:* Based on Weiser et al. [297].

of the substrate scope of PcPAL demonstrates its broad substrate tolerance that several polyfluoro- and monochlorophenylalanine analogues [288], all three regioisomers of pyridylacrylic acid [289], furan as well as thiophene substituted systems including larger fused aromatic rings [290], and 5-arylfuran-2-yl derivatives [291] can be afforded to their corresponding products with moderate to good yields from 38 to 89% and high optical purity using purified enzyme. Besides, nitro-substituted phenylalanines have been synthesized in very good yields [292, 293]. All the results described above highlight potential sustainable routes and ability using PcPAL to produce a broad spectrum of nonnatural amino acids for the pharmaceutical and agrochemical industries. The expansion of substrate scope of PAL for the amination to manufacture nonnatural amino acids was further implemented by cyanobacterial AvPAL from *Anabaena variabilis*. The catalytic activity of AvPAL toward several nonnatural substrates proved an even better substrate scope as compared with yeasts RgPAL and plants PcPAL that can be exemplified by the preparative synthesis of L-(4-trifluoromethyl)phenylalanine, a substrate not accepted by either RgPAL or PcPAL, with a moderate yield of 42% [294].

The covalent immobilization of PcPAL on carboxylated single-walled carbon nanotubes ($SwCNT_{COOH}$) has been employed for the ammonia addition of (*E*)-3-(thiophen-2-yl)acrylic acid in a continuous-flow microreactor system to yield enantiopure (*S*)-2-amino-3-(thiophen-2-yl)propanoic acid with a conversion about 65% at 60 °C without significant loss of activity over a period of 72 hours. $SwCNT_{COOH}$-PAL was stable which maintains >90% the original activity after three cycles of reaction [295]. Later, an even higher stability and durability of the immobilized PcPAL was demonstrated that uses aminated SwCNTs as the immobilization carrier instead of the carboxylated SwCNTs [296]. In addition to the use of SwCNTs as carrier for the immobilization of PcPAL, PcPAL was immobilized on magnetic nanoparticles (MNPs) as well. Then, the MNPs-PAL was fixed in a microfluidic reactor and successfully utilized for catalyzing the ammonia addition to a novel (*E*)-pent-2-ene-4-yonate producing enantiopure L-propargylglycine with a 14% isolated yield (Scheme 5.47) [297]. This application not only broadens the substrate scope but also opens a new perspective for PAL toward the ammonia addition to acrylic acid. The activity of isolated and lyophilized PcPAL in organic-aqueous biphasic media was explored in order to increase the product yields for previous one-step synthesis of L-phenylalanine and its methyl ester derivative using yeast whole cells. The isolated PcPAL in organic-aqueous

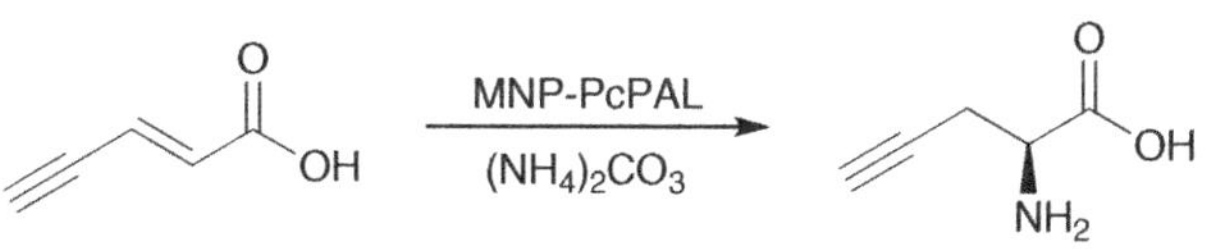

Scheme 5.47 Synthesis of L-propargylglycine by PcPAL immobilized on MNPs in a microfluidic reactor system. *Source:* Based on Zhang [301].

biphasic system was about fourfold higher activity relative to the whole-cell system and methanol-treated, ethanol-treated, and heptane-treated enzymes retained 89–99% activity for both the amination and deamination reactions [298].

The strategy of immobilization of isolated enzymes was also applied to RgPAL that cross-linked enzyme aggregates (CLEAs) of crude RgPAL were prepared to improve the synthesis of L-phenylalanine. When compared to the free enzyme, PAL-CLEAs exhibited higher storage and operational stability. In addition, PAL-CLEAs could be recycled at least 12 consecutive batch reactions without significant loss of activity [299]. The activity of RgPAL in ionic liquids (ILs) for the synthesis of L-phenylalanine from amination of *trans*-cinnamic acid was also investigated by comparing four commonly used ILs. The results showed that 1-butyl-3-methylimidazolium hexafluorophosphate ([BMIM][PF_6]) was the best IL under the optimized conditions for this application giving about 59% substrate conversion [300].

5.3.3.2 D-Arylalanines Synthesis

When lyases are applied for the asymmetric synthesis of nonproteinogenic D-amino acid derivatives (AADs), the kinetic resolution with the stereoselective removal of the L-enantiomer and deracemization with a two-step conversion of the L-enantiomer to the D-enantiomer or the racemic mixture through a nonchiral intermediate were generally adopted. PAL in the cell free homogenate of the yeast *Rhodotorula graminis* served as an enantioselective scavenger (ES) was employed to remove the L-enantiomer in a racemic mixture of AADs, thus producing D-enantiomer with high enantiomeric excess (*e.e.*). Thirteen D-arylalanines have been synthesized using this ES homogenate to give >99% *e.e.* and yields from 36 to 50% (Scheme 5.48) [301]. The results show an efficient and convenient kinetic resolution method for

preparing D-AADs. The same strategy was used for the simultaneous production of D-histidine and urocanic acid from racemic histidine in high yields by exploiting HAL from *Achromobacter liquidum* and a shake-culture. In this biotransformation, the surfactant triethanolamine lauryl sulfate (TEALS) was also added to shorten the conversion time due to an increase of the permeability of the cell membrane (Scheme 5.49) [302]. In addition, a variety of enantiomerically

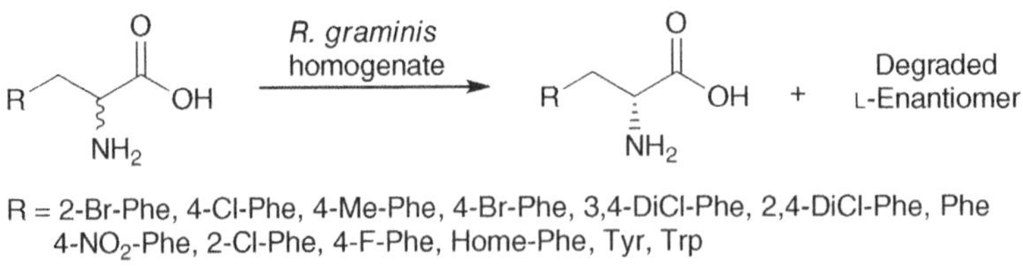

Scheme 5.48 Preparation of D-AADs by kinetic resolution of racemic AADs. *Source:* Based on Shibatani et al. [302].

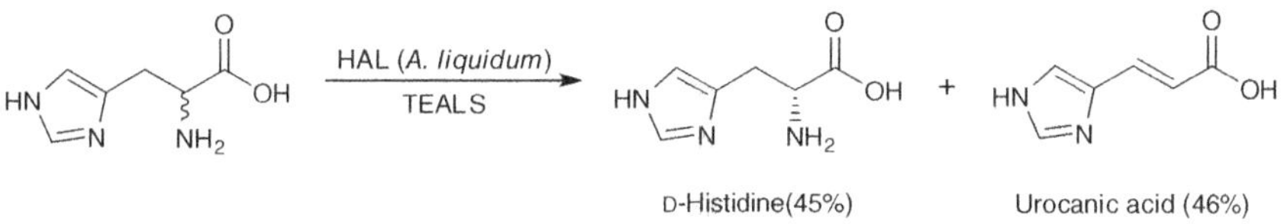

Scheme 5.49 Kinetic resolution of *rac*-histidine using HAL for simultaneous synthesis of D-histidine and urocanic acid. *Source:* Based on Parmeggiani et al. [304].

pure D-heteroarylalanines have been synthesized by similar method that isolated PcPAL was employed to remove the L-enantiomer from racemic mixtures [290, 291, 293, 295, 297].

Moreover, the kinetic resolution process was combined with the enzyme immobilization technique for the production of enantiopure D-phenylalanine from racemic phenylalanine in a large-scale recirculating packed-bed reactor (RPBR). Phenylalanine ammonia lyase from *Rhodotorula glutinis* JN-1 (RgPAL) was covalently immobilized on a modified mesoporous silica support and demonstrated higher resolution efficiency in RPBR than in the stirred-tank reactor to give a productivity of D-phenylalanine reaching 7.2 g L^{-1} h^{-1} and an optical purity (ee_D) >99% [303]. Since the inherent theoretical yield for D-enantiomer by kinetic resolution is limited to 50%, a chemoenzymatic amination-deracemization cascade strategy was developed to overcome this limitation. A number of nonnatural substituted L-phenylalanines were first formed by PAL-catalyzed amination with inexpensive cinnamic acids, then the substituted L-phenylalanine products were selectively oxidized by the L-amino acid deaminase (LAAD) from *Proteus mirabilis* to their corresponding amino acids, which were chemically nonselectively reduced to a racemic substituted phenylalanine mixture. Repeating the L-selective enzymatic oxidation and nonselective chemical reduction several times, highly optical pure nonnatural substituted D-phenylalanines can be obtained (Scheme 5.50) [304].

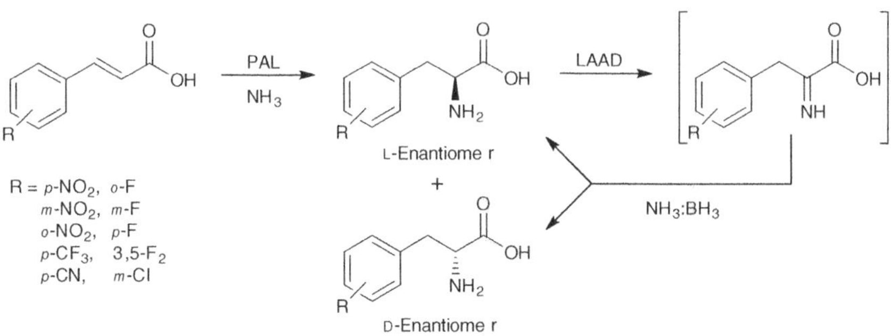

Scheme 5.50 Synthesis of substituted D-phenylalanines from corresponding cinnamic acids with chemoenzymatic cascade strategy. *Source:* Modified from Weise et al. [307].

5.3.3.3 β-Arylalanines Synthesis

Although unnatural aromatic β-amino acids are not abundant in nature, they are found in pharmacologically important natural products. In addition, they are moieties in a lot of pharmaceuticals such as the antiglycemic sitagliptin and the antimitotic paclitaxel/Abraxane [305]. As isomers of proteinogenic α-amino acids, β-amino acids are particularly useful as building blocks for synthetic β-peptide oligomers that are used as biologically active antibiotics with the advantages of bioavailability and *in vivo* stability for therapeutic purposes [306]. Thus, they exhibit an important class of compound in the area of medicinal chemistry.

Lyases used for the direct β-amination of nonchiral arylacrylates to produce corresponding chiral β-arylalanines derivatives is convenient and attractive, but the low regioselectivity of the wild-type enzyme or the structure-based rational designed variants expressed in *E. coli* for some substrates due to the inherent strength of electron-withdrawing/-donating groups on the ring of each substrate in the reaction mechanism limits the applications at small scale and a concentration of 1 mM [307]. Except aminomutases the enzyme exploited for the direct β-amination of arylacrylates was EncP which has been displayed to be a PAL in the thermotolerant bacterium *Streptomyces maritimus* [308, 309]. The β-amination for a range of arylacrylates by whole cell biocatalyst EncP results a mixture of (*S*)-α- and (*S*)-β-arylalanines that could be tuned to achieve a selectivity between 99:1 1nd 1:99 for β:α-product ratios (Scheme 5.51) [307]. Generally, the industrially interested catalyst used for the direct preparation of β-arylalanines via amination with arylacrylic acids is phenylalanine aminomutase from *Taxus wallichiana* (TwPAM) [247].

In addition, there are the other two strategies that can be conducted to synthesize optical pure β-arylalanines, one is the isomerization of L-phenylalanine or its derivatives using either an (*R*)-selective PAM from *Taxus canadensis* (TcPAM) or (*S*)-selective PAM (specifically AdmH by Japanese company Mitsui Chemicals Ltd), and the other one is the kinetic resolution of a racemic mixture of the starting compound using a tandem biocatalytic process [247]. An example reported for the

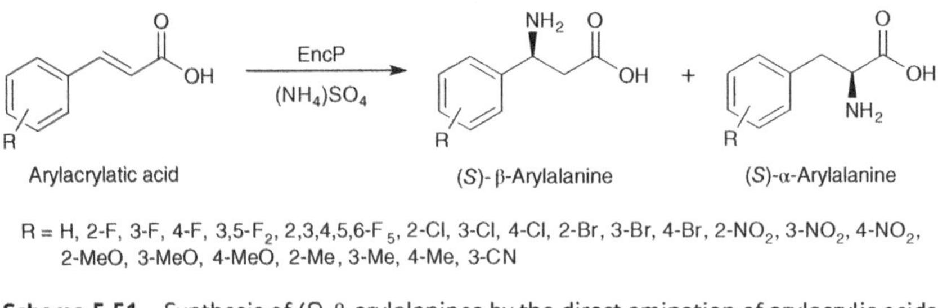

Scheme 5.51 Synthesis of (*S*)-β-arylalanines by the direct amination of arylacrylic acids using whole cell catalyst EncP. *Source:* Based on Turner [250]; Wu et al. [310].

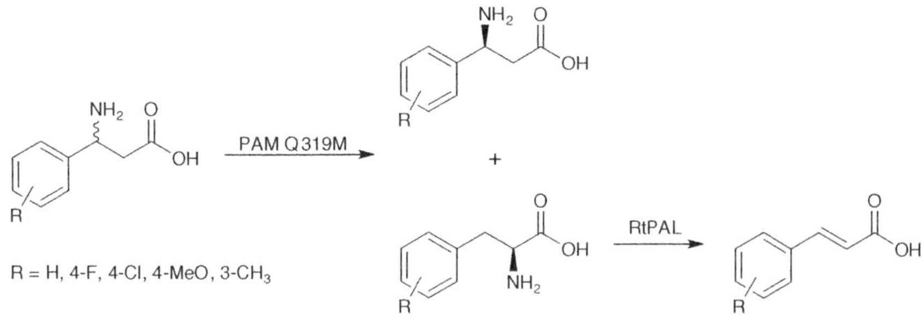

Scheme 5.52 Kinetic resolution of racemic β-phenylalanines using coupled PAM and PAL for synthesizing enantiopure (*S*)-β-phenylalanines.

kinetic resolution is the coupled PAM and PAL reaction system, which the stereoselective isomerization of (*R*)-β-phenylalanines to (*S*)-α-phenylalanines was initially performed by PAM Q319M and the later were immediately deaminated *in situ* to the corresponding cinnamic acids by PAL from *Rhodosporidium toruloides* (RtPAL) to leave (*S*)-β-phenylalanines in solution (Scheme 5.52) [250, 310]. The *in situ* generated cinnamic acids can be easily removed by acidification and ion exchange.

Preparative scale conversions of a range of β-amino acids were performed using PAM Q319M to acquire an enhanced catalytic activity with excellent *e.e.* > 99% and in *ca.* 50% isolated yield [310, 311]. Another example of the kinetic resolution using combined PAM and PAL enzymes was for the production of enantiopure (*R*)-β-arylalanines. In this process, the racemic mixture of β-arylalanines was chemically synthesized using achiral aldehydes as the starting material. Then, (*S*)-β-arylalanine in the racemic mixture was isomerized by the (*S*)-PAM activity of EncP from *S. maritimus* to the (*S*)-α-arylalanine, which was immediately deaminated with a strict α-lyase activity of PAL from *A. variabilis* to β-arylacrylic acid, leaving (*R*)-β-arylalanine in the reaction mixture (Scheme 5.53). This kinetic

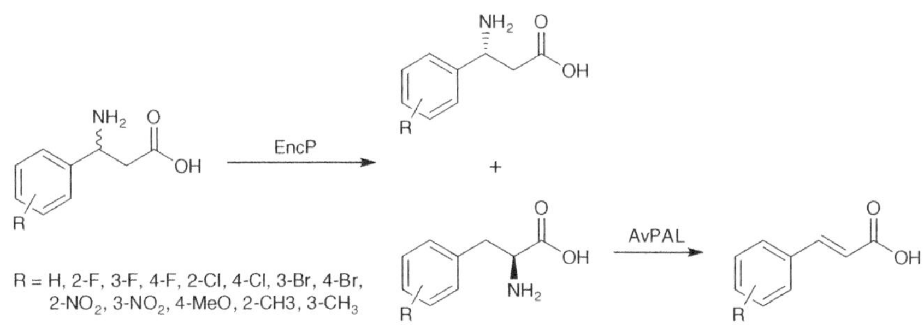

Scheme 5.53 Kinetic resolution of racemic β-phenylalanines using coupled EncP and PAL for synthesizing enantiopure (*R*)-β-phenylalanines. *Source:* Based on Paizs et al. [313].

resolution strategy allowed five out of 13 tested racemic substrates to be resolved in a preparative-scale with (*R*)-enantiomeric excess (*e.e.*) >98% [312].

5.3.3.4 Applications for Arylalanines Synthesis

There are many enantiopure arylalanines or arylalanine derivatives that have been synthesized by chemoenzymatic cascade strategy which combines the robustness of chemical catalysis with the elegant enantioselectivity of biocatalysis to provide new, efficient, and telescopic routes for optical pure products from inexpensive starting materials.

The first demonstration of the chemoenzymatic synthesis is the one-pot production of a number of halophenylalanines and arylalanines from the corresponding aromatic aldehydes. In this route, the aldehyde was first reacted with (triphenylphosphoranylidene)ethyl acetate in water using a Wittig-type reaction to form the corresponding ethyl cinnamate. The ethyl cinnamate ester was then hydrolyzed with pig liver esterase (PLE) to afford the cinnamic acid. Finally, the cinnamic acid was aminated with ammonia by a purified phenylalanine ammonia lyase from parsley (PcPAL) to give corresponding enantiopure L-arylalanines (Scheme 5.54) [313]. The pure L-arylalanines obtained by this synthetic strategy were phenyl-, 4-chlorophenyl-, 3-fluorophenyl-, and thiophen-2-yl-alanines with a yield of 88, 78, 72, and 91%, respectively, and >98% *e.e.* over a period of 48–96 hours. Later, an alternative two-step process was proposed using a Knoevenagel–Doebner condensation reaction for the synthesis of the cinnamic acid from the corresponding aldehyde starting material, followed by the organic-aqueous biphasic PAL-mediated amination to produce the corresponding optically pure L-arylalanines (Scheme 5-55) [314].

The two consecutive reactions can also be run in one-pot preparative synthesis for five L-dihalophenylalanines with 71–84% yield and 98–99% *e.e.* The same one-pot chemoenzymatic telescopic strategy was further applied for the synthesis of 12 L-pyridylalanine analogues and five other L-heteroarylalanines with heterocycle moieties including isoxazole, thiophene, and quinoline from corresponding aldehydes using AvPAL as the biocatalyst to give isolated yields of 32–60% and

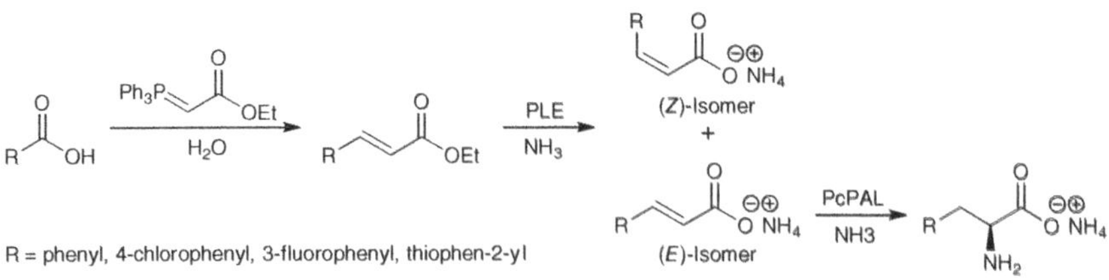

Scheme 5.54 Three-step one-pot chemoenzymatic synthesis of enantiopure aryalanines from corresponding aldehydes. *Source:* Based on Parmeggiani et al. [314].

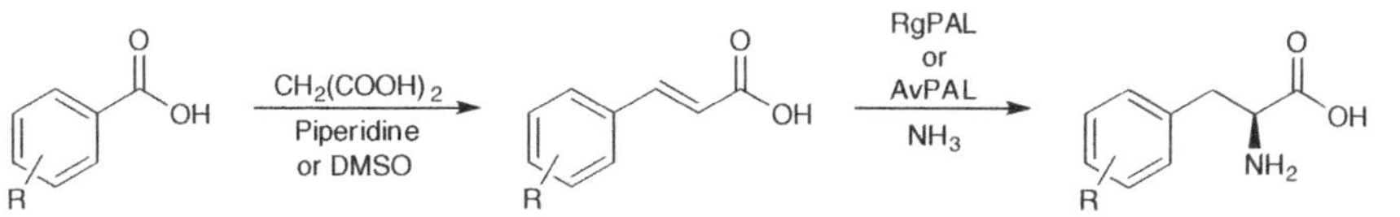

R = H, 2-F, 3-F, 4-F, 2-Cl, 3-Cl, 4-Cl, 2-Br, 4-Br, 3,4-F_2, 2,4-Cl_2, 3,4-Cl_2, 3,5-Cl_2, 2-Cl-4-F, 3-Cl-4-F, 4-Cl-3-F

Scheme 5.55 Two-step one-pot chemoenzymatic synthesis of L-arylalanines from benzaldehydes. *Source:* Based on Parmeggiani et al. [247]; Heberling et al. [249]; Cooke et al. [272]; MacDonald and D'Cunha [273].

excellent enantiopurity of >99% *e.e.* It should be mentioned that the perfect enantioselectivity was obtained by an additional one-pot deracemization cascade process using D-amino acid oxidase to selectively oxidize the D-enantiomer to the amino acid followed by a nonselective reduction with ammonia-borne [315].

Generally, in the chemoenzymatic cascade process, the chemical steps usually involve the use of metallic or organometallic catalyst, organic solvent, and high temperature. In order to fulfill green chemistry, it is necessary to eliminate completely the use of organic solvents and run under moderate temperatures to make the one-pot chemoenzymatic process suitable for industrial fine chemicals production. One typical example demonstrated by the chemical company DSM is the synthesis of enantiopure (*S*)-2-indolinecarboxylic acid, a key chiral intermediate for angiotensin-converting enzyme inhibitors such as perindopril and indolapril (Scheme 5.56) [247, 248, 316, 317]. Traditionally, the synthesis of chiral indoline carboxylic acid was prepared by the formation of racemic precursor through a Fischer indole synthesis followed by a classical or enzymatic resolution; however, the yields never exceeded 50%. This problem was overcome by an alternative approach, which the enantioselective amination of ammonia to the *ortho*-halocinnamic acids was first catalyzed by RgPAL

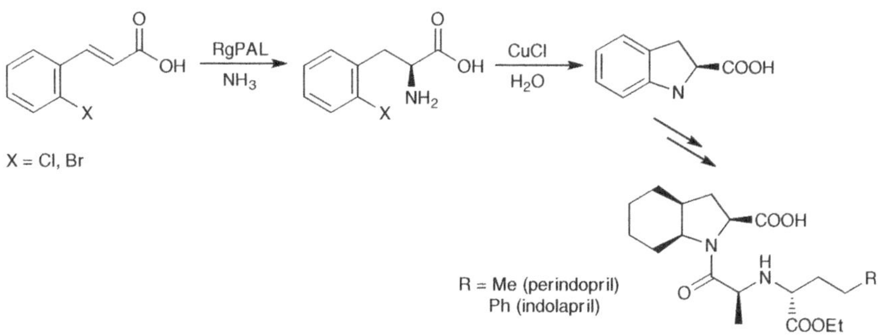

Scheme 5.56 RgPAL-mediated chemoenzymatic cascade synthesis of (*S*)-2-indolinecarboxylic acid. *Source:* Based on Busto et al. [320].

DSMPAL01, followed by an aqueous compatible copper-catalyzed cyclization for the corresponding optical pure L-*o*-halophenylalanines. Under an excess of aqueous NH_3, a CuCl catalyst, and optimized conditions, the overall production yield achieved for (*S*)-2-indolinecarboxylic acid from 2-chlorocinnamic acid was 60% with an enantiomeric excess of 99%. Using a similar approach, (*S*)-1,2,3,4-tetrahydroisoquinoline-3-carboxylic acid was chemoenzymatically prepared by an AvPAL-mediated reaction followed by a Pictet–Spengler reaction with acidic formaldehyde solution to give a 53% isolated yield and enantiopure tetrahydroisoquinoline scaffold [247]. A chemoenzymatic strategy was implemented for the synthesis of a range of *N*-Boc-protected nonnatural L-biarylalanine derivatives, active pharmaceutical ingredient (API) precursors, by combining the AvPAL-mediated amination for the starting *p*-bromocinnamic acid with the subsequent palladium-catalyzed Suzuki–Miyaura coupling of a panel of arylboronic acids under mild aqueous conditions [247, 318].

Although the applications in using PALs were concentrated on conducting the reverse (nonnatural) amination of ammonia to cinnamic acids for synthesizing huge amount of valuable amino acid enantiomers, the PALs-mediated natural reaction has not been overlooked for its synthetic importance, particularly with the assistance of synthetic biology of the multienzymatic cascade process. The formation of cinnamic acid analogues by the deamination of corresponding phenylalanine with PALs can afford a variety of commercially relevant compounds via additional chemical and/or enzymatic processes. As an example, the coexpression of PAL, *trans*-cinnamic acid decarboxylase, and styrene monooxygenase in *E. coli* can be employed for the *in vivo* synthesis of (*S*)-styrene oxide, a key fine chemical used for producing functionalized chiral diols, from glucose via the primary metabolite phenylalanine [319]. Furthermore, the synthesis of a number of *para*-vinylphenol derivatives from readily available simple substituted phenols and pyruvate was performed through a three-step enzymatic cascade process involving recombinant tyrosine phenol lyase M379V from *Citobacter freundii* (CfTPL), tyrosine ammonia lyase from *Rhodobacter sphaeroides* (RsTAL) expressed in *E. coli*, and ferulic acid decarboxylase from *Enterobacter* sp. (EsFAD) expressed in *E. coli*, in which CfTPL was responsible for the initial C–C formation between phenol and pyruvate in the presence of ammonia leading to the corresponding L-tyrosine derivatives, followed by RsTAL-catalyzed deamination and EsFAD-catalyzed decarboxylation to form corresponding coumaric acid derivatives and substituted vinylphenols, respectively, with >99% conversion and isolated yields of 65–83% (Scheme 5.57) [320]. The multienzymatic cascade strategy was even applied for the sustainable asymmetric synthesis of high-value nonnatural chiral chemicals from biobased L-phenylalanine. The designed and engineered set of unique nonnatural enzymatic cascades was an eight-step process using ten enzymes including PAL, phenylacrylic acid decarboxylase (PAD), styrene

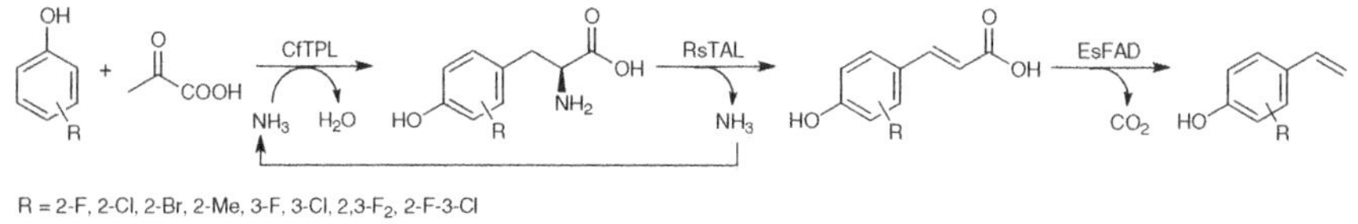

Scheme 5.57 Multienzymatic cascade synthesis for substituted *para*-vinylphenols.

monooxygenase (SMO), epoxide hydrolase (EH), alcohol dehydrogenase (ADH), aldehyde dehydrogenase (ALDH), hydroxymandelate oxidase (HMO), transaminase (TA), glutamate dehydrogenase (GluDH), and catalase (CT) to produce a variety of useful fine chemical intermediates such as styrene oxide, diols, α-hydroxyacids, and phenylglycine derivatives. This biocatalytic cascade leads to the synthesis of enantiopure (*S*)-phenylglycine with 85% conversion and >99% *e.e.* It has also been proposed to combine this nonnatural enzyme cascade with the natural metabolic pathway of the host strain in order to achieve the chiral fine chemicals from glucose fermentation [321].

The advances and combination of synthetic biology and metabolic engineering allowed the microbial *in vivo* synthesis of bioactive compounds such as resveratrol and pinosylvin from tyrosine and phenylalanine, respectively, with the potential of industrial-scale production [322]. In this application, the PAL/TAL-catalyzed deamination is the key step to the artificial stilbene biosynthetic pathway, involving steps of phenylalanine/tyrosine ammonia-lyase, 4-coumarate:CoA ligase, and stilbene synthase that were co-expressed in *E. coli* (Scheme 5.58). The results show 20 mg L^{-1} pinosylvin and 37 mg L^{-1} resveratrol were obtained from 490 mg L^{-1} phenylalanine and 540 mg L^{-1} tyrosine, respectively [323]. Microbial production of resveratrol was improved by engineering the yeast *S. cerevisiae* to produce it directly from glucose or ethanol via tyrosine intermediate. The stilbene biosynthetic pathway was first introduced in a similar manner into *S. cerevisiae*, followed by the control of metabolic fluxes through overexpression of the genes for the enzymes involved in the feedback-inhibition-resistant versions of 3-deoxy-D-arabino-heptulosonate-7-phosphate synthase and chorismate mutase and in malonyl-CoA synthesis. The final fed-batch fermentation using the final strain with glucose or ethanol as carbon source resulted in the production of 415.65 and 531.41 mg L^{-1} resveratrol, respectively [324]. The conversion of resveratrol was further improved by applying a pull-push-block strain engineering strategy that the yeast *S. cerevisiae* was engineered by overexpression of the resveratrol biosynthetic pathway, optimization of the electron transfer to the cytochrome P450 monooxygenase, increase the precursors supply, and decrease of the pathway intermediates degradation. The fed-batch fermentation using the final strain produced 800 mg L^{-1} resveratrol from glucose [325]. Similarly, a flavonoid biosynthetic pathway was engineered by expressing flavone synthase I from *P. crispum* into recombinant

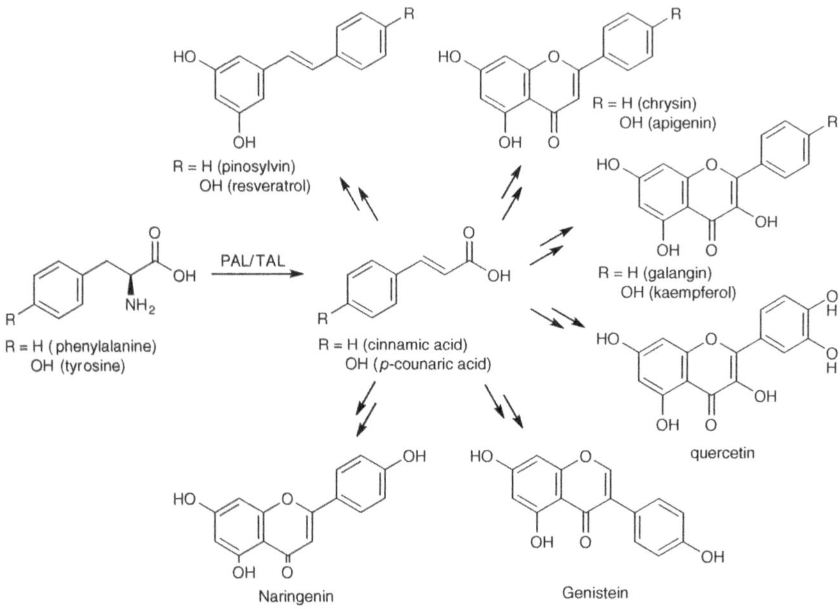

Scheme 5.58 PAL/TAL-mediated *in vivo* biosynthesis of various secondary metabolites via metabolic engineering and synthetic biology. *Source:* Based on Kiddle et al. [328]; Yamagata et al. [330]; Steegborn et al. [331].

E. coli cells containing PAL, cinnamate/coumarate: CoA ligase, chalcone synthase, chalcone isomerase and employed for the fermentative *in vivo* synthesis of flavones (9.4 mg L^{-1} chrysin from phenylalanine and 13 mg L^{-1} apigenin from tyrosine); when flavanone 3β-hydroxylase and flavonol synthase from the plant *Citrus* species were simultaneously expressed into the recombinant *E. coli* cells, flavonols of 15.1 mg L^{-1} kaempferol from tyrosine and 1.1 mg L^{-1} galangin from phenylalanine were obtained (Scheme 5.27) [326]. In addition, the biosynthesis of stilbenoid resveratrol, flavanone naringenin, isoflavone genistein, and flavonols kaempferol and quercetin from phenylalanine has been conducted by different metabolically engineering approach for four yeast strains (Scheme 5.27) [327].

5.4 Lyases with Carbon–Sulfur Bonds

Enzymes that catalyze the cleavage of a carbon–sulfur bond by means other than hydrolysis or oxidation are carbon–sulfur lyases (C–S lyases, EC 4.4.x.x). Specifically, carbon–sulfur bond lyases are pyridoxal-5′-phosphate (PLP)-dependent enzyme, which catalyze the elimination reactions of C–S bond in various sulfur-containing amino acids to yield corresponding sulfur-containing molecules, α-keto acid, and ammonia. The C–S lyases are ubiquitous enzymes existing in microorganisms, plants, and animals (including human). They are involved in the biosynthesis of key primary and secondary metabolites in plants bacteria, fungi, and mammals, which are important in reducing oxidative stress, protein biosynthesis, antimicrobial activity, odor and flavor characteristic, medical disease, and signaling agent [328–336]. Generally, the reaction mechanism of PLP-dependent lyases proceeds via three key chemical transformations between the covalently bound PLP-amino acid intermediates, which involves the transamination of the amino group between the substrate and the enzyme-bound cofactor PLP, followed by the α-proton abstraction of the substrate to yield a carbanionic intermediate stabilized as the pyridoxal-ketamine *p*-quinonoid intermediate, and the subsequent release of the resulting thiol molecule by protonating the sulfur atom to generate the PLP-aminoacrylate. Finally, reverse transamination of the PLP-aminoacrylate takes place by attacking the C4' atom with enzyme-bound lysine to convert the aminoacrylate into an iminopropionate and regenerate the enzyme-bound PLP. The hydrolysis of the released iminopropionate outside the active site produces pyruvate and ammonia [334, 337–339].

Cysteine lyases (EC 4.4.1.x) cleave cysteine containing compounds, mainly including cystathionases (cystathionine γ-lyase, EC 4.4.1.1; cystathionine β-lyase, EC 4.4.1.8), alliinases (EC 4.4.1.4), and several other subfamilies [329, 340]. Cystathionine γ-lyase (CGL, also known as γ-cystathionase, cysteine desulfhydrase, or cystalysin), which is often found in animals and fungi, catalyzes the cleavage of the L-cystathionine Cγ–S bond to produce L-cysteine, α-ketoglutarate, and ammonia (Scheme 5.59) and is essential for the conversion of dietary L-methionine to L-cysteine via a metabolic transsulfuration [328, 330, 331]. CGL is a multifunctional enzyme able to catalyze a range of substrates including cystathionine, L-cystine, L-cysteine, and *S*-alkyl-L-cysteine conjugate, *S*-sulfo-L-cysteine, and L-cysteine sulfoxide [328, 341, 342].

In humans, this enzyme is related with the metabolic disorder diseases such as cystathioninuria, cystinosis, and homocystinuria, which potentially result in mental and physical impairments [331]. Cystalysin isolated from human oral pathogen *T. denticola* was identified as a virulence factor by its hemeolytic and

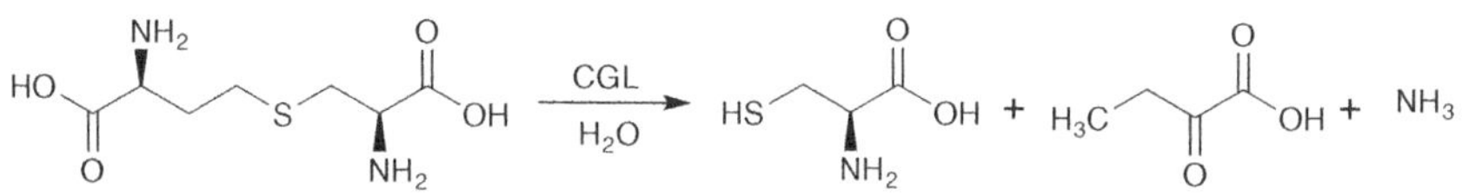

Scheme 5.59 γ-Cleavage of L-cystathionine catalyzed by cystathionine γ-lyase. *Source:* Based on Kiddle et al. [328]; Ramírez and Whitaker [347]; Hamamoto and Mazelis [348]; Based on Kiddle et al. [328]; Yoshida et al. [346].

heneoxidative activity, which leads to both tooth and gum damage and causes the etiology of periodontal diseases. *In vitro* studies revealed that cystalysin is a homodimeric PLP-dependent Cβ–S cleavage lyase, which catalyzes the α,β-elimination of L-cysteine to produce pyruvate, ammonia, and H_2S. Since H_2S is highly toxic for mammalian cells, the formation of H_2S by cystalysin catalysis in the septic subgingival sulcus might contribute to the initiation and progression of periodontal diseases [328, 343–345]. The enzyme has a broad substrate specificity and accepts several sulfur- and nonsulfur-containing amino acids as well as disulfidic amino acids as substrate [344].

In plants and bacteria, cystathionine β-lyase (CBL) predominantly catalyzes the cleavage of the L-cystathionine Cβ–S bond to produce L-homocysteine, pyruvate, and ammonia (Scheme 5.60a) [328, 346]. In plants, L-homocysteine is involved in methionine biosynthesis which is further converted to L-methionine [328, 330]. In higher plants, L-cystine is also degraded by CBL through β-elimination to yield L-thiocysteine, pyruvate, and ammonia (Scheme 5.60b) [328, 329. 347, 348] and L-thiocysteine can be converted to L-cysteine and H_2S [344]. For cysteine, the products formed by the β-elimination of the CBL are pyruvate, ammonia, and H_2S. This enzyme is responsible for the initial reaction that produces characteristic flavors and aromas in vegetables. In addition to L-cystathionine, L-cysteine, and L-cysteine, this enzyme catalyzes a broad spectrum of L-cysteine derived substrates including *S*-aryl-, *S*-aralkyl-, and *S*-alkyl-L-cysteine to corresponding thiols, pyruvate, and NH_3 [349–351]. Leukotriene E4 (LTE4) is a cysteinyl leukotriene involved in inflammation. Increased production and excretion of LTE4 have been linked to several respiratory diseases, such as severe asthma attacks. Thus, when LTE4 was incubated with CBL isolated from the intestinal bacterium *Eubacterium limosum*, the product 5-hydroxy-6-mercapto-7,9,11,14-eicosatetra

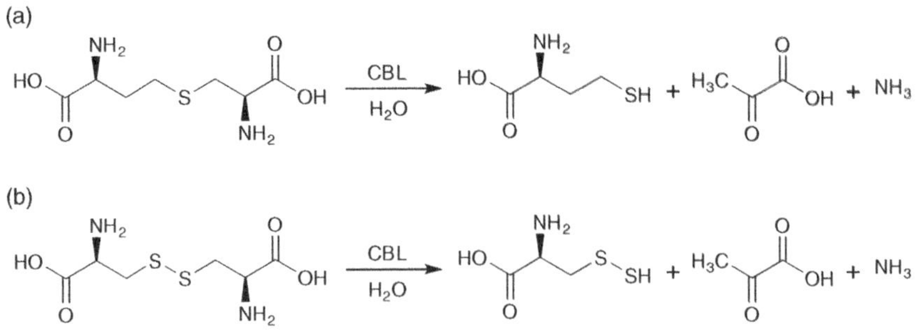

Scheme 5.60 β-elimination reaction with cystathionine β-lyase to produce corresponding thiol, pyruvate, and ammonia: (a) L-cystathionine, (b) L-cystine. *Source:* Based on Kiddle et al. [328]; Nock and Mazelis [332]; Ramírez and Whitaker [347].

enoic acid, together with pyruvate and NH_3, was formed by the β-cleavage of the cysteine C–S bond in LTE4. This reaction was also occurred by rat cecal contents [352].

Alliin lyases (EC 4.4.1.4), also known as alliinases, are another well-known class of C–S lyases, which catalyze the β-elimination of *S*-alkyl-L-cysteine sulfoxides to corresponding *S*-alkyl-sulfonic acids, pyruvate, and ammonia (Scheme 5.61) [328, 332, 347]. The source of this class of lyases is usually from *Allium* species, such as *A. cepa* (onion), *A. sativum* (garlic), *A. porrum* (leek), and *A. ursinum* (wild garlic) [328]. The endogenous substrate specificities of alliin lyases for garlic and onion are different. The *A. sativum* alliin lyase preferentially cleaves *S*-ally-L-cystine sulfoxide, whereas the *A. cepa* alliin lyase prefers *S*-propenyl-L-cystine [332]. Earlier studies indicate that the principal ingredient of fruiting bodies of shiitake mushroom, *Lentinus edodes*, flavor lenthionine arises from enzymatic action on the nonvolatile precursor (2-(γ-glutamylamino)-4,6,8,9,10-pentaoxo-4,6,8,10-tetrathioundecanoic acid (lentinic acid), which is related with two different enzymes: a γ-glutamyltransferase and a kind of alliin lyase catalyzing β-elimination of desglutamyl-lentinic acid [353, 354].

In enzymology, L-methionine γ-lyase (MGL, EC 4.4.1.11) is a PLP-dependent enzyme that cleaves L-methionine at the Cγ–S bond to form methylthiol, α-ketobutyrate, and ammonia using a five-step mechanism (Scheme 5.62) [355]. Evidences demonstrate that MGL shows deaminase activity in the L-methionine degradation reaction as well [355, 356]. *Porphyromonas gingivalis* is a Gram-negative anaerobic oral pathogenic bacterium that causes halitosis and periodontitis in humans. Halitosis is due to the formation of volatile methylthiol by the catabolism of methionine with MGL and the generated methylthiol can lead to

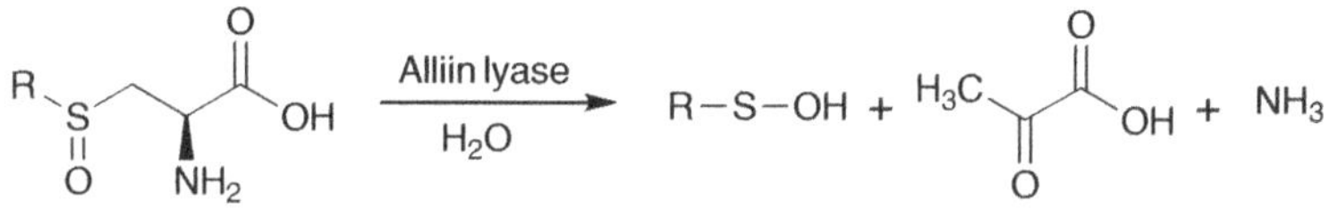

Scheme 5.61 β-elimination of *S*-alkyl-L-cysteine sulfoxides with alliin lyases to form corresponding sulfonic acids, pyruvate, and ammonia. *Source:* Based on Morcos et al. [355].

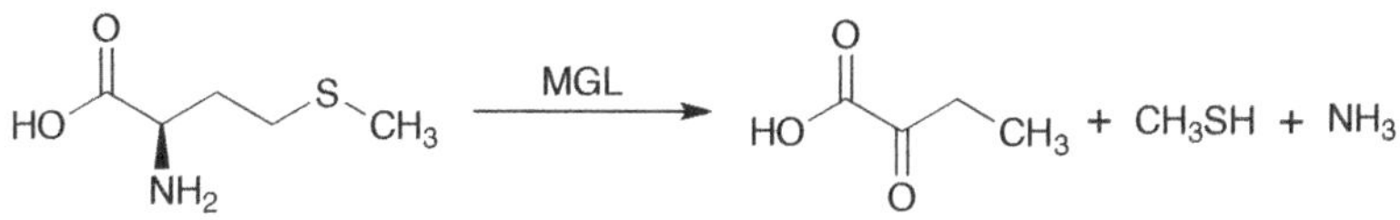

Scheme 5.62 L-Methionine degradation catalyzed by methione γ-lyase to yield α-ketobutyrate, methylthio, and ammonia.

inflammation by oxidative damage to the oral mucosa and cause periodontitis. It was suggested that the MGL inhibitor DL-propargylglycine could be useful for controlling halitosis and periodontitis [355].

Ergothioneine and ovothiols are histidine derivatives with a sulfur substitution at the ε- and δ-carbons of the histidine imidazole side chain, respectively. Ergothioneine isolated from ergot functions as a cellular antioxidant thus plays beneficial roles in human health. Humans obtain ergothioneine from the diet through a specific ergothioneine-specific transporter [357, 358]. Ovothiol was first isolated from the eggs and ovaries of sea urchins. There are three different forms of ovothiol according to the degree of methylation at the amino group, which ovothiol A, B, and C are unmethylate, monomethylated, and dimethylated amino group, respectively. Ovothiols function as a potent radical and peroxide scavenger and are believed to protect themselves from oxidative stress. In addition, ovothiol A has also been demonstrated to induce autophagy in human liver carcinoma cell lines [357, 359].

The mycobacterial biosynthesis of ergothioneine involves a five-step pathway (EgtA-E) characterized with two novel reactions: the oxidative C–S bond formation using a mononuclear nonheme iron enzyme (EgtB) and a PLP-dependent reductive C–S lyase (EgtE) reaction (Scheme 5.63). However, fungus *Neurospora crassa* uses a more simpler pathway employing a nonheme iron enzyme Egt1 and cysteine instead of γ-Glu-Cys as the substrate [357, 360].

The biosynthesis of ovothiol involves an initial nonheme iron enzyme OvoA catalyzing the C–S bond formation oxidative coupling between cysteine and histidine to yield a sulfoxide coupling product and a minor product cysteine sulfinic acid. The C–S bond of the sulfoxide coupling product in the first step is cleaved by the C–S lyase OvoB to produce a mercaptohistidine intermediate, which is

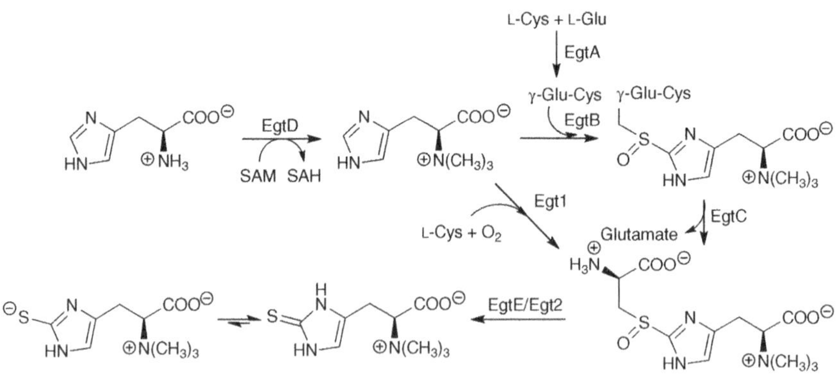

Scheme 5.63 Two aerobic biosynthetic pathways for ergothioneine. *Source:* Based on Naowarojna et al. [357]; Chen et al. [358]; Naowarojna et al. [359].

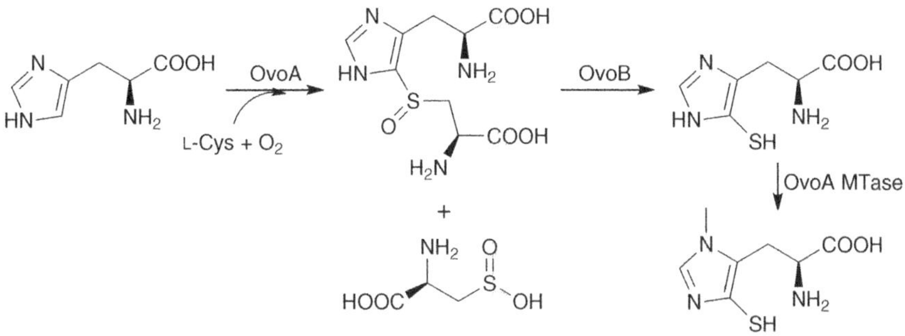

Scheme 5.64 Proposed biosynthetic pathway for ovothiol A. *Source:* Based on You et al. [362]; Schallmey and Schallmey [363].

followed by a π-*N*-methylation at the imidazole group by OvoA methyltransferase (OvoA MTase) to produce ovothiol A (Scheme 5.64) [357–359].

5.5 Lyases with Carbon–Halide Bonds

Halogenated organic compounds (HOCs) represent a large group of substance combining one or more halogens (fluorine, chlorine, bromine, or iodine) with carbon atom which occur either naturally or synthetically. Industrially, they have been widely used for the manufacture of insecticides, fungicides, herbicides, and intermediates for pharmaceuticals [361, 362]. Since HOCs are toxic, they are harmful to human health. Therefore, the release of HOCs into environments either by organisms or chemical industries causes environmental pollution problem. HOCs are stable chemicals and do not decompose spontaneously. Dehalogenases existing in many microorganisms are enzymes catalyzing the cleavage of carbon–halogen bonds, thus are important in the biodegradation of HOCs. They are roughly categorized into haloalkane dehalogenases, haloacid dehalogenases, and halohydrin dehalogenases which degrade halogenated alkanes, halogenated alkanoic acids, and halogenated alcohols, respectively [362].

Halohydrin dehalogenase (HHDH, EC 4.5.1.x), also called haloalcohol dehalogenase, halohydrin exposidase, or halohydrin hydrogen-halide-lyase, catalyzes the reversible dehalogenation of halohydrins to their corresponding epoxides through intramolecular nucleophilic displacement of the halogen substituent by the neighboring hydroxyl group without the need of any cofactors (Scheme 5.65) [362, 363]. In addition, HHDHs can also catalyze the reverse irreversible epoxide ring-opening reaction through other negatively charged nucleophiles such as cyanide, azide, and nitrite to form novel C–C, C–N, C–O, or C–S

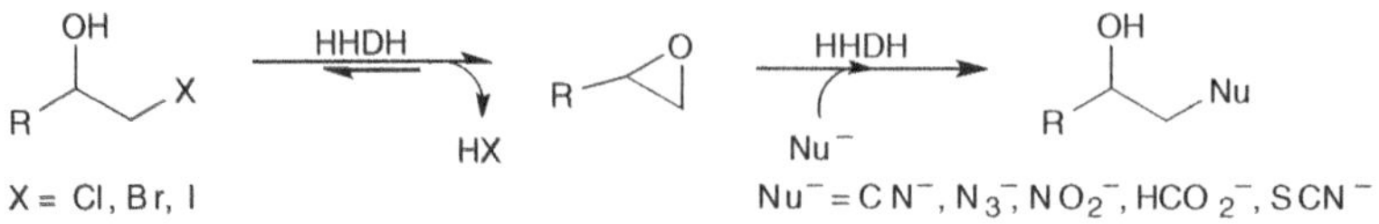

Scheme 5.65 The dehalogenation of a vicinal halohydrin by halohydrin dehalogenase. *Source:* Based on You et al. [362]; Kabat et al. [369]; Barbachyn and Ford [370]; Assis et al. [371].

bonds (Scheme 5.65) [364]. This remarkable catalytic ability not only allows microorganisms to survive in a harsh environment but also affords the formation of various epoxides and β-substituted alcohols for synthetic chemistry. The earliest report of the HHDH was about the degradation of 2,3-dibromo-1-propanol to form corresponding epoxide epibromohydrin by removing the bromo-substituent at the C2-position. In the presence of chloride ion, the enzyme can also rapidly convert epibromohydrin into epichlorohydrin [365]. With significant sequence similarity, HHDHs belong to the short-chain dehydrogenase/reductase (SDR) superfamily and share mechanistic features. HHDHs are homotetramers that can be regarded as a dimer of dimers, and the Ser-Tyr-Arg catalytic triad is similar to the Ser-Tyr-Lys catalytic triad of many SDR enzymes. However, HHDHs contain a spacious nucleophile binding pocket instead of the NAD(P)H-binding site found in SDR enzymes [366, 367].

The catalytic process with HHDHs can be divided into ring closure and ring opening. During the ring closure by dehalogenation, the haloalcohol substrates are fixed by the formation of a hydrogen bond with the Ser residue within the active site, while the Tyr residue acts as a catalytic base abstracting a proton from the hydroxyl group adjacent to the halide atom and in the meantime the substrate oxygen proceeds a nucleophilic attack on the halogen-bearing carbon to cause a ring closure and release a halide ion. Note that the Arg residue does not interact with the substrate itself but only lowers the pK_a of the tyrosine-OH via hydrogen bonding in order to activate it for proton abstraction. In the reverse epoxide ring-opening reaction, the nucleophile attacks the epoxide ring to result in ring opening, while the Tyr residue donates a proton to the oxygen anion after the nucleophilic attack to form the cyanohydrin rapidly [362, 363].

5.5.1 Synthesis of Chiral Epoxides

Due to their high regio- and enantioselectivity toward aliphatic and aromatic substrates, HHDHs are useful tools for the synthesis and industrial production of enantiopure epoxides and their ring-opening compounds, in particular, the β-substituted alcohols as building blocks of pharmaceuticals [248, 368]. Chiral epoxides have been extensively used as building blocks for synthesizing enantiomerically pure pharmaceuticals and agrochemicals. HHDHs-catalyzed ring-closure reaction of halohydrins or kinetic resolution of racemic epoxides is one of the potential synthetic strategies to produce chiral epoxides. Chiral (*R*)-epichlorohydrin (ECH), which has been employed for the synthesis of optically active pharmaceuticals such as (*R*)-carnitine and Linezolid, was directly preparaed from 1,3-dichloro-2-propanol (1,3-DCP) by the action of the haloalcohol dehalogenase (DehA) from *Arthrobacter erithii* H10a with 10.7% yield and > 95% *e.e.* in an excess of KBr (Scheme 5.66a) [362, 369–371]. Using the same enzyme but with epibromohydrin and KCl as the starting substrates, (*S*)-ECH can also be formed by the transhalogenation [371]. The kinetic resolution of racemic ECH in the presence of NO_2^- by the halohydrin dehalogenase HheC from *Agrobacterium radiobacter* AD1 and expessed in *E. coli* is another approach for the production of (*R*)-ECH. Uner optimal conditions, (*R*)-ECH obtained was in 99% *e.e.* optical purity and 41% yield for an 18 minutes reaction (5-65B) [372]. The limitation of a 50% maximum theoretical yield for the traditional kinetic resolution can be overcome by the dynamic kinetic resolution (DKR) process to give a 100% theoretical yield that the unreacted starting chiral β-haloalcohols is concurrently racemized by the added metal complex catalyst iridacycle prepared by cycloiridation of *N*-methylbenzylamine with an iridium precursor during the reaction [373].

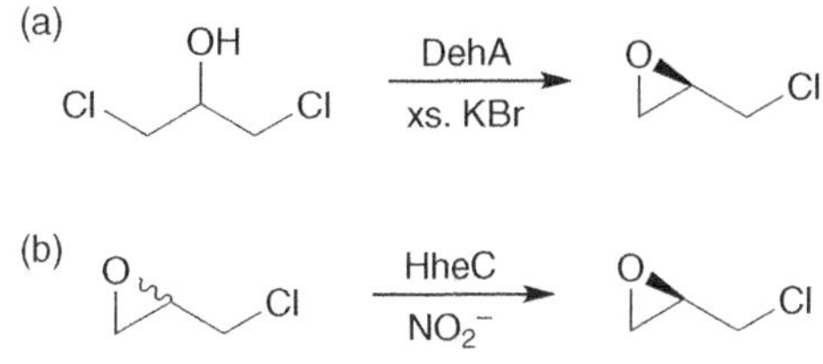

Scheme 5.66 Synthesis of chiral epoxides by HHDH: (a) DehA-catalyzed reaction for (*R*)-ECH from 1,3-DCP, (b) Kinetic resolution of racemic ECH to produce (*R*)-ECH. *Source:* Based on Haak et al. [373]; Jerphagnon et al. [374].

When the DKR process was applied for the kinetic resolution of aromatic chlorohydrins to enantiopure aromatic substituted epoxides using a mutant of HheC and the iridacycle as racemization catalyst, the maximum yield can be as high as 90% and an *e.e.* of 98% (Scheme 5.67) [373, 374]. In contrst to the highly *R*-enantioselectivity of the enzyme HheC from *A. radiobacter* AD1 in the resolution of halohydrins and epoxides, a novel halohydrin dehalogenase of HheA10 from *Tsukamurella* sp. 1534 has been heterologously expressed, purified, characterized and demonstrated to be a highly *S*-enantioselective enzyme toward

Scheme 5.67 DKR process of aromatic chlorohydrins catalyzed by HHDH for producing enantiopure epoxides. *Source:* Based on Wan et al. [375].

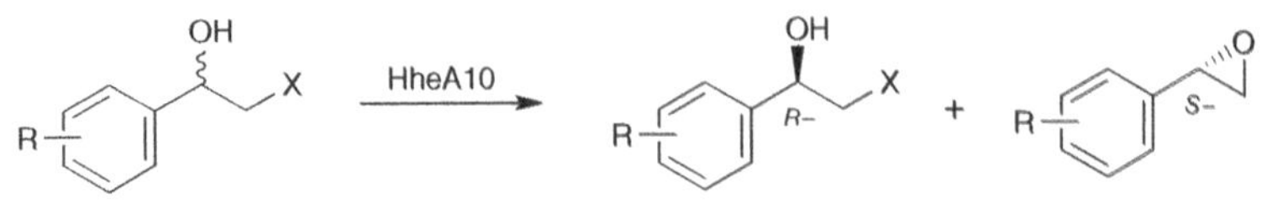

1. X = Cl, R = H; 2. X = Cl, R = 3-Br; 3. X = Cl, R = 4-F
4. X = Br, R = H; 5. X = Br, R = 2-CH_3; 6. X = Br, R = 2-Br
7. X = Br, R = 3-Br; 8. X = Br, R = 4-F; 9. X = Br, R = 4-Cl

Scheme 5.68 HheA10-catalyzed kinetic resolution of racemic aromatic halohydrins for producing the corresponding epoxides.

2-chloro-phenylethanol derivatives. The exploitation of HheA10 for the kinetic resolution of racemic aromatic halohydrins produced the corresponding epoxides with the *e.e.* and *E* values up to >99% and >200, respectively (Scheme 5.68) [375]. Thus, this enzyme provides a complementary enantioselectivity for enzyme HheC.

5.5.2 Synthesis of Enantiopure β-Substituted Alcohols

The stereoselective ring opening of epoxides catalyzed by HHDHs with various small anionic nucleophiles has been widely used for the asymmetric synthesis of enantiopure β-substituted alcohols, which are important building blocks of pharmaceuticals and biologically active compounds [364, 376].

A series of compounds such as β-halide alcohols, β-hydroxy nitriles, β-nitro alcohols, and β-zaido alcohol were synthesized based on different nucleophiles (Cl^-, Br^-, I^-, CN^-, NO_2^-, N_3^-, OCN^-, SCN^-, and $HCOO^-$) with HheC from *A. radiobacter*. When using 1,2-epoxybutane as the substrate, the preparative nucleophilic ring-opening reactions demonstrated a predominant attack at the less-substituted carbon atom and an enantioselectivity for the (*R*)-epoxide (Scheme 5.69). Note that the ring opening of 1,2-epoxybutane with cyanate ion yielded (*R*)-5-ethyl-oxazolidin-2-one as the sole product. In addition, the ring-opening reaction for more sterically demanding aromatic substrates and disubstituted epoxides were successfully performed with the same enzyme using different nucleophiles and resulting in high enantioselectivity as well as high yields (Scheme 5.70) [364].

Among the various anionic nucleophiles, cyanide is the most interesting one due to the facile transformation of nitrile group in the β-hydroxy nitriles into amino, amide, or carboxy group by nitrile-amide converting enzymes [377]. The ring opening of ECH with cyanide catalyzed by HHDH expressed in *E. coli* from *Corynebacterium* sp. strain N-1074 was employed for the enantioselective synthesis of (*R*)-γ-chloro-β-hydroxybutyronitrile in (*R*)-*e.e.* > 95% [378]. The enantioselective ring opening of epoxides using cyanide to produce nonracemic β-hydroxy nitriles catalyzed by three groups of HHDH (HheA, HheB, and HheC) was studied and found HheC was the most selective one [376]. One of the most important biotechnological applications

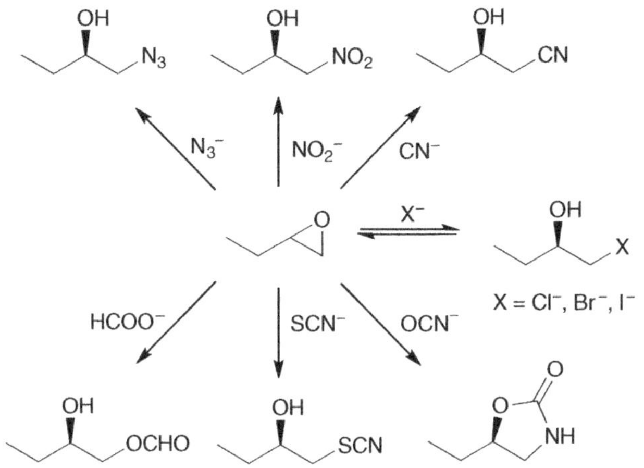

Scheme 5.69 HHDH-catalyzed 1,2-epoxybutane ring opening for yielding various β-substituted alcohols with different nucleophiles. *Source:* Hasnaoui-Dijoux et al. [364].

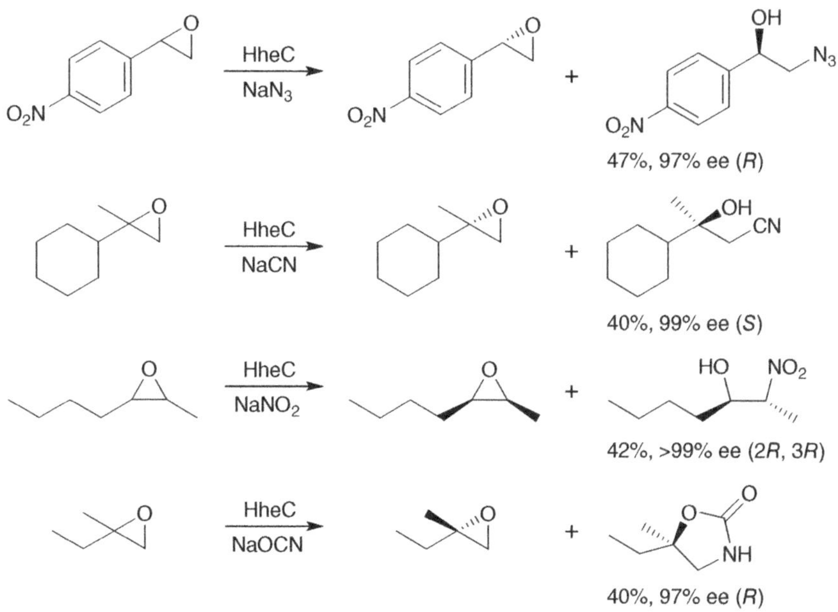

Scheme 5.70 Enantioselective epoxide ring-opening reactions catalyzed by HheC. *Source:* Based on Wan et al. [382].

involving the cyanide-mediated epoxides ring opening catalyzed by HHDH is the commercial enantioselecive synthesis of the cholesterol-lowering drug atorvastatin calcium (Lipitor) precursor ethyl (*R*)-4-cyano-3-hydroxybutyrate ((*R*)-HN) from the ethyl (*S*)-4-chloro-3-hydroxybutyrae ((*S*)-CHBE) via the corresponding intermediate (*S*)-epoxide (Scheme 5.70) [362, 363, 379]. The HHDH used for this enzymatic conversion was a heavily engineered HheC mutant generated by ProSAR-based protein engineering in order to meet the high cyanolysis activity or increased enzyme stability requirements [380]. This cyano-introducing reaction was performed at neutral pH and ambient temperature for five hours with 92% recovered yield and >99.5% *e.e.* In fact, this is a two-step reaction starting from ethyl 4-chloroacetoacetate, which was first asymmetrically converted to (*S*)-CHBE using a ketoreductase in combination with a NADP-dependent glucose dehydrogenase for cofactor regeneration [380]. In a later process, (*R*)-HN was successfully produced in a preparative scale of 200 g L^{-1} (*S*)-CHBE using a high catalyst loading whole-cell biocatalytic system of HheA3 from *Parvibaculum lavamentivorans* DS-1, which resulted in 95% conversion and 85% yield within 14 hours [381]. Alternatively, (*S*)-CHBE can be synthesized via (*S*)-4-chloro-3-hydroxybutyronitrile ((*S*)-CHBN) from cheap 1,3-dichloropropan-2-ol (1,3-DCP) using HheC. This reaction route involves the HheC-catalyzed dehalogenation and ring closure of prochiral 1,3-DCP to epichlorohydrin followed by the enantioselective ring opening of the formed (*S*)-epichlorohydrin (ECH) using cyanide as the nucleophile (Scheme 5.71) [382]. The produced (*S*)-CHBN is easily esterified with ethanol in acidified solution to give (*S*)-CHBE. The enantioselectivity of this route toward (*R*)-HN can be >99% by applying the more selective mutant HheC W249F. When scaling up the reaction to 200 mL, the yield of (*S*)-CHBN from 20 mM 1,3-DCP within 1 hour was 86% with only a slight decrease of product *e.e.* to 97.5% [363].

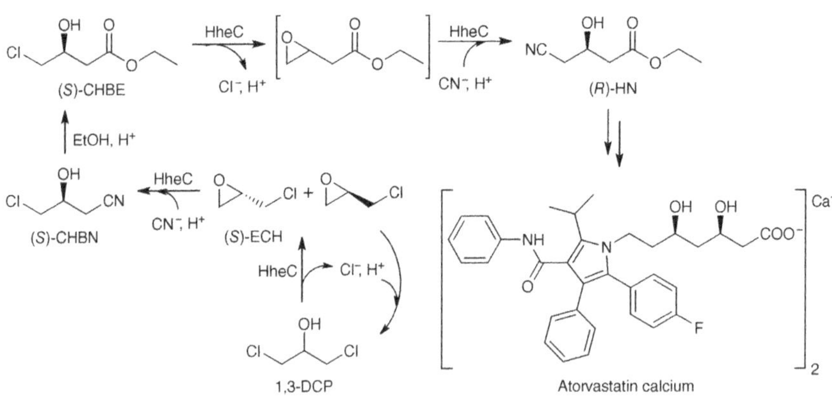

Scheme 5.71 HHDH-catalyzed synthesis of atorvastatin side-chain precursor ethyl (*R*)-4-cyano-3-hydroxybutyrate. *Source:* Based on Campbell-Verduyn et al. [385].

The second most popular nucleophiles in the enantioselective epoxide ring-opening reaction are azide and nitrite. The recombinant HHDH from *A. radiobacter* AD1 was used for catalyzing the highly enantioselective (E > 200) and β-regioselective azidolysis of (*para*-substituted) styrene oxides. The kinetic resolution performed in the scale of 7.8 g L^{-1} *p*-nitrostyrene resulted in a yield up to 47% of (*R*)-2-azidoalcohol and 97% *e.e.* and 46% of (*S*)-*para*-nitrostyrene oxide as well as 98% *e.e.* [383] Later, an enzymatic dynamic kinetic resolution (DKS) by combining the enantioselective irreversible ring opening of an epihalohydrin by N_3^- and a racemization caused by a reversible ring opening by a halide was performed in order to yield optically active (*S*)-1-azido-3-halo-2-propanol. However, when ECH was used as the substrate, the reaction resulted in a mixed kinetic resolution and DKS due to a higher ring-opening rate by N_3^- than the racemization rate. While with epibromohydrin as the substrate, the reaction showed an efficient DKS because the racemization was higher than the rate of ring opening. Therefore, under the optimal conditions, the product (*S*)-1-azido-3-bromo-2-propanol could be obtained in 84% yield and 94% *e.e.* The enantioselectivity of (*S*)-1-azido-3-bromo-2-propanol can be increased to >99% with 77% yield by the selective ring closure of (*R*)-enantiomer of bromoalcohol [384]. When HheC-catalyzed azide-induced ring opening of aromatic epoxides was combined with a one-pot "click" reaction to ligate the 2-azido alcohols and alkynes in the presence of Cu(I), optically pure hydroxyl triazoles were produced with excellent enantiomeric excess (80–99%) (Scheme 5.72) [385]. Furthermore, a one-pot four-step multienzymatic cell cascade with alcohol dehydrogenase and HHDHs in designer single cells was combined with the click chemistry to synthesize (*R*)- and (*S*)-enantiomers of different β-hydroxytriazoles from α-halo ketones via the corresponding azide intermediates in high enantiomeric excess [386]. β-Hydroxytriazoles are useful as β-adrenergic receptor blockers and thus important pharmacophores.

The enzymatic resolution of ten 2-alkyl-2-aryl-disubstituted epoxides using the Codex HHDH P2E2 and sodium azide was applied for the synthesis of novel regio- and enantioselective 1-azido-2-arylpropan-2-ols with the resulting (*R*)-azido alcohols in 19–45% yield and 88–99% *e.e.* from test substrates [387].

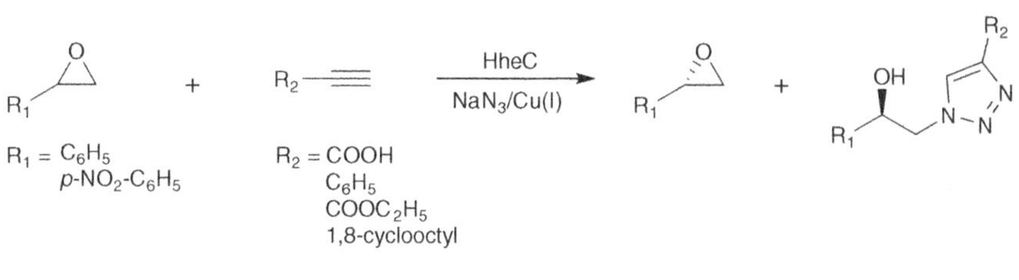

Scheme 5.72 HHDH-catalyzed one-pot tandem enantioselective epoxide ring opening and [3+2] azide alkyne cycloaddition. *Source:* Elenkov et al. [389].

Generally, the HHDH from *A. radiobacter* AD1 (HheC) catalyzes the kinetic ring opening of broad 2,2-disubstituted epoxides with cyanide and azide, which exhibits excellent enantioselectivity (*E* up to >200) and gives access to various enantiopure epoxides and β-substituted tertiary alcohols (*e.e.* up to 99%) [388]. It has been found that HHDH from *Arthrobacter* sp. (HheA) catalyzed highly regioselective azidolysis of spiroepoxides containing 5-, 6-, and 7-membered cycloalkane rings. In addition, the HHDH from *A. radiobacter* (HheC) not only displayed high regioselectivity but also showed moderate-to-high enantioselectivity (*E* up to >200) in the kinetic resolution of chiral spiroepoxides (Scheme 5.73) [389]. Nucleophile nitrite was used for the ring opening of styrene oxide derivatives by HHDH obtained from *A. radiobacter* AD1 (HheC).

When *para*-nitrostyrene oxide (*p*-NSO) was attacked by nitrite, two reaction routes were found for this nitrite-mediated ring opening (Scheme 5.74) [390]. One route is mainly the conversion of (*R*)-*p*-NSO to 2-hydroxy-2-(*para*-nitrophenyl)-ethylnitrite ester through the attack of the oxygen atom of nitrite on the β-position of the epoxide ring. The ring-opening product of this attack is unstable which is spontaneously hydrolyzed to corresponding 2-(*para*-nitrophenyl)-1,2-ethanediol and the nitrite is recovered. The other route is via the attack of nitrite on the β-position of *p*-NSO through the nitrogen atom to form the product 2-hydroxy-2-(*para*-nitrophenyl)-nitroethane but with low amounts. This conversion is exhibited with high regioselectivity on the β-position and high enantioselectivity of the

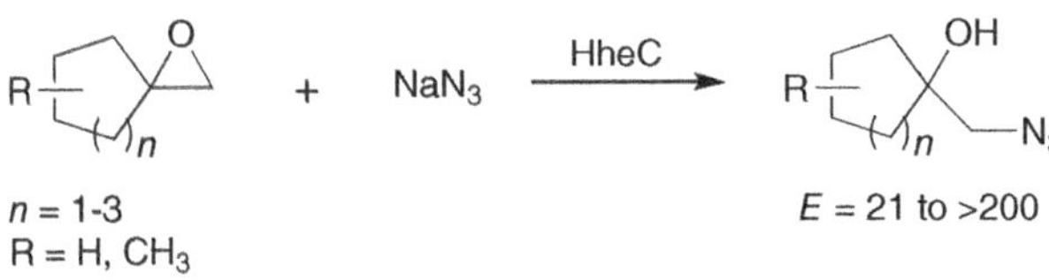

Scheme 5.73 HHDH-catalyzed regio- and enantioselective azidolysis of spiroepoxides. *Source:* Based on Hasnaoui et al. [390].

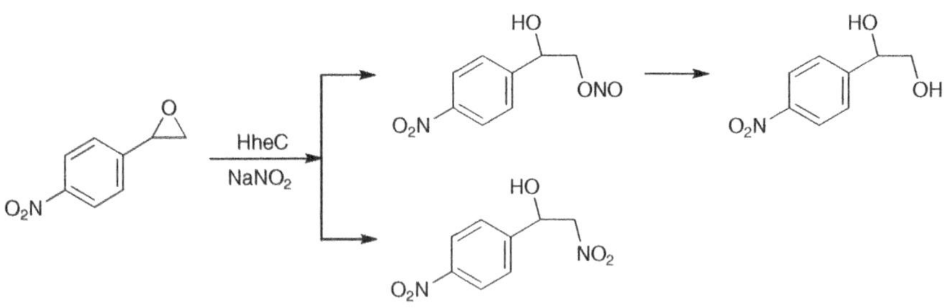

Scheme 5.74 HheC-catalyzed nitrite-mediated ring opening of *p*-nitrostyrene oxide. *Source:* Based on Elenkov et al. [391].

(*R*)-*p*-NSO that leaves the unreacted (*S*)-*p*-NSO in an *e.e.* of 99% and a calculated yield of 48% as well as the calculated total yield for the diol and nitroalcohol products of 40 and 10%, respectively. Nitrite-mediated epoxide ring opening using HheC from *A. radiobacter* AD1 was also applied for the synthesis of chiral epichlorohydrin and the ring-opening product 1-chloro-3-nitro-2-propanol by kinetic resolution of racemic epichlorohydrin which was proven to be more efficient than the nucleophile cyanide or azide [372].

Besides those inorganic ions mentioned above, cyanate ion (OCN^-) is also exploited as nucleophiles in the ring opening of epoxides catalyzed by halohydrin dehalogenase. One report showed that HHDH from *A. radiobacter* (HheC) was explored as the biocatalyst for the enantioselective kinetic resolution ring opening of different terminal aliphatic epoxides with cyanate to form highly enantio-enriched 5-substituted 2-oxazolidinones in high yields. The ring opening of epoxides can proceed via nitrogen or oxygen attack which leads to two isomeric products, β-hydroxy isocyanate and β-hydroxycyanate, respectively. The organic cyanates are unstable at room temperature that undergoes isomerization to form isocyanate. In the meantime, β-hydroxy isocyanates are spontaneously and rapidly cyclized to form 2-oxazolidinones as the final product. (Scheme 5.75a) [391]. This type of kinetic resolution was further employed to

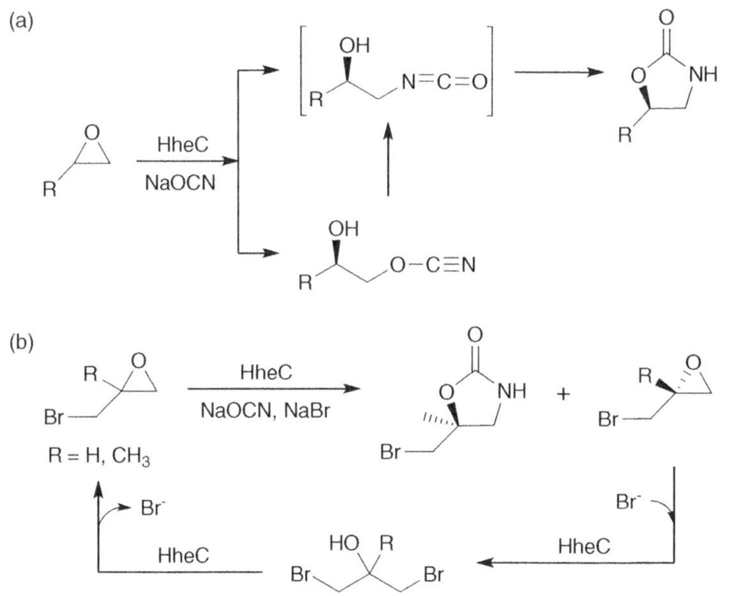

Scheme 5.75 HHDH-catalyzed epoxide ring opening with cyanate for chiral oxazolidinones by (a) kinetic resolution and (b) dynamic kinetic resolution. *Source:* Schrittwieser et al. [395].

an efficient DKR by including an *in situ* racemization of the slower reacting epoxide enantiomer through reversible epoxide ring opening using bromide and also catalyzed by HheC (Scheme 5.75b). Thus, substrates epibromohydrin and 2-bromomethyl-2-methyloxirane were afforded corresponding (*S*)-5-substituted 2-oxazolidinones in high yields (97 and 87%) and high enantiopurity (89 and >99% *e.e.*) [392]. Note that (*S*)-oxazolidinones are important synthetic building blocks of oxazolidinone antibiotics and important chiral auxiliaries and ligands in chemical synthesis [370, 393].

5.5.3 Application of Cascade Reactions

The one-pot cascade synthesis of chemicals involving multienzymatic reaction steps has merits such as the omission of separation and purification for intermediates, the shifting of reaction equilibria to lead high production yield, and cost-effective. An early application of the cascade reactions was the kinetic resolution of halohydrins for the synthesis of optically pure (*S*)-2,3-dichloro-1-propanol ($E > 100$) and (*S*)-2-chloro-1-phenylethanol ($E = 73$), as well as corresponding enantiopure epoxides, with an enantiomeric excess larger than 99% by the combination of HheC and a recombinant epoxide hydrolase [394]. Later, a one-pot three-step biocatalytic cascade using two enzymes was employed for the enantioselective synthesis of optically active β-azidolalcohols and β-hydroxynitriles. The first step involves the asymmetric bioreduction of prochiral α-chloroketones via hydrogen transfer catalyzed by an alcohol dehydrogenase (ADH) to form the stereogenic center, which gives an enantiopure chlorohydrin intermediates. The second step is the ring closure to corresponding optically active epoxides, which are followed by HheC catalyzed nucleophilic ring opening with either azide ion, N_3^-, or cyanide ion, CN^- to produce the final enantiopure β-azidolalcohols and β-hydroxynitriles, respectively (Scheme 5.76) [395]. Similar biocatalytic cascade strategy was adopted for a one-pot four-step preparation of enantiopure β-hydroxytriazoles from prochiral α-haloketones with high enantiomeric excess

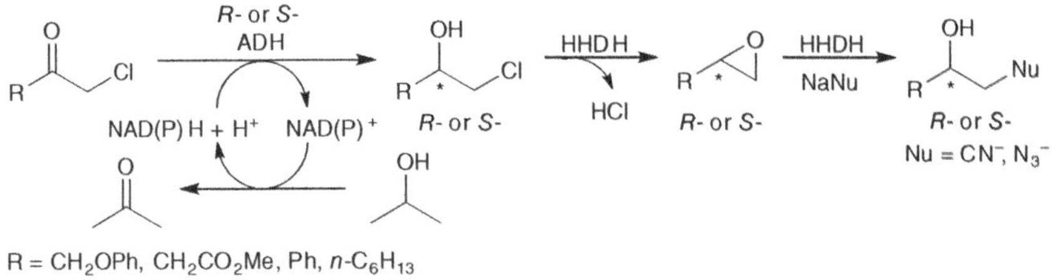

Scheme 5.76 Multienzymatic cascade synthesis of β-substituted alcohols by the combination of a stereoselective ADH and an HHDH. *Source:* Based on Yao et al. [397].

using two kinds of designer cells that overexpress either the *R*-selective ADH and HHDH or *S*-selective ADH and HHDH [386].

Following the same concept, a multienzymatic biosynthesis was designed for the synthesis of both (*R*)- and (*S*)-β-hydroxy nitriles through the co-expression of all necessary enzymes (ADH with opposite stereoselectivities, HHDH, and cofactor-regenerating enzyme) in one *E. coli* host. After the preparation of two engineered *E. coli* strains, they were separately used for the one-pot cascade production of chiral β-hydroxy nitriles from parochial α-haloketones with different co-enzyme regeneration systems to obtain high enantiomeric excess and substrate conversion >99% [396]. A very efficient one-pot bienzymatic synthesis of ethyl (*R*)-3-hydroxyglutarate (EHG), a key intermediate for the synthesis of rosuvastatin, from ethyl (*S*)-4-chloro-3-hydroxybutyrate (ECHB) achieved by using recombinant *E. coli* whole cells expressing separately or co-expressing a mutant HHDH gene from *A. radiobacter* AD1 (HheC) and a nitrilase gene from *Arabidopsis thaliana* (AtNIT2). The bienzymatic cascade involving three reactions which was started with the dehalogenation ring closure of (*S*)-ECHB using HheC and the subsequent ring opening of the epoxide intermediate by cyanide was also catalyzed by HheC, while the final hydrolysis of the introduced nitrile group was catalyzed by nitrilase AtNIT2 (Scheme 5.77) [397]. Since nitrilase exhibits with substrate inhibition by high concentration of ECBH and cyanide, a fed-batch of ECBH and NaCN to the one-pot one-step was performed for the bienzymatic cascade reaction. The production of EHG by a one-pot two-step process from ECHB can be implemented at a substrate concentration of 1.2 M with full conversion. Furthermore, the immobilization of recombinant *E. coli* cells harboring both HheC and epoxide hydrolase (EH) was combined with a two-step bienzymatic cascade reaction and was used for preparing (*R*)-epichlorohydrin from 1,3-dichloro-2-propanol.

The immobilization carrier used for cells was perlite and the immobilization method was by physical adsorption. In addition, a two-phase solution reaction

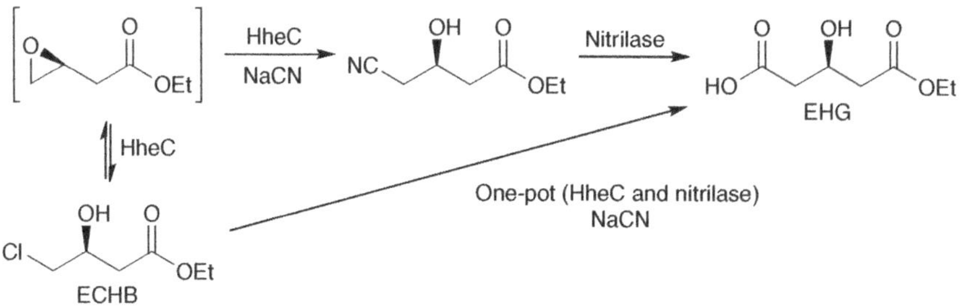

Scheme 5.77 One-pot bienzymatic cascade synthesis of ethyl (*R*)-hydroxyglutarate catalyzed by halohydrin dehalogenase and nitrilase.

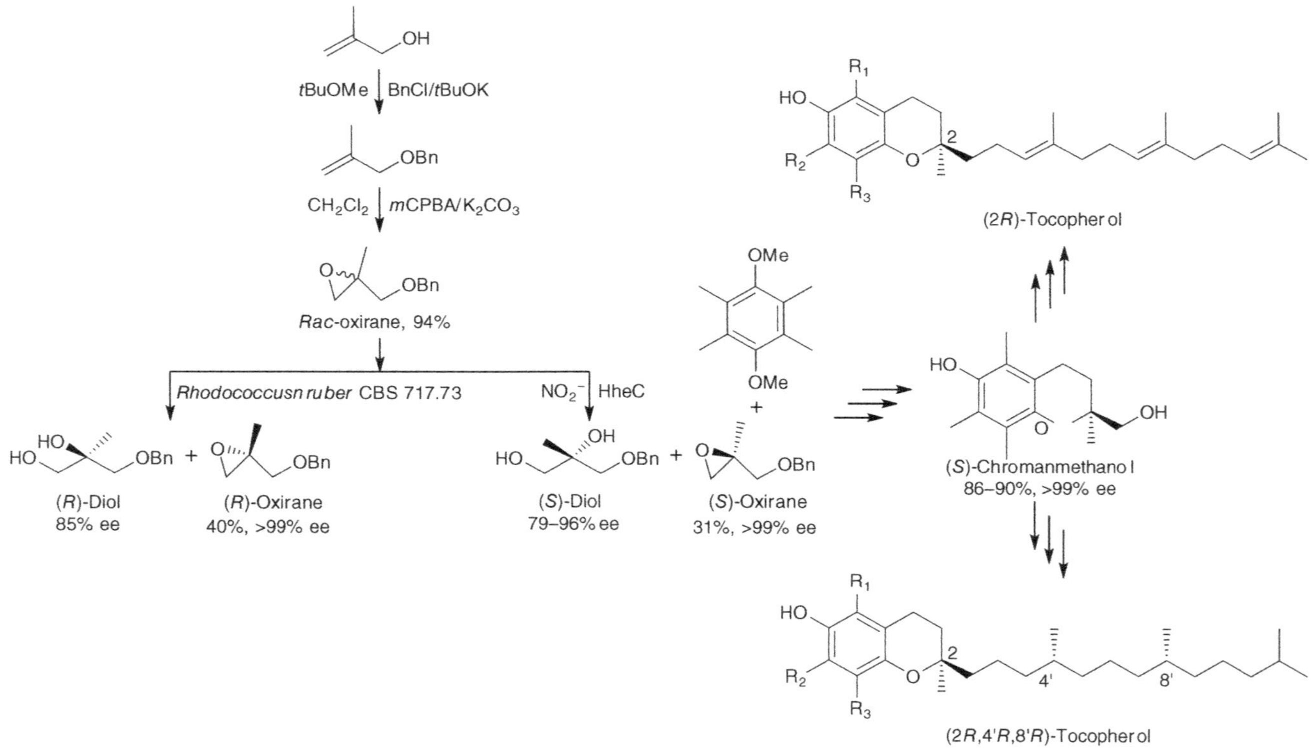

Scheme 5.78 Chemoenzymatic cascade synthesis of (S)-chromanmethanol toward optically pure α-tocopherols.

system, coupled with a specially designed reactor, was used for the bienzymatic cascade reaction in order to improve the yield of (*R*)-ECH by reducing the spontaneous chemical hydrolysis and the reverse reaction of the formed ECH catalyzed by HheC. Under optimal conditions, the yield of racemic ECH formed in the first step was 73% and the yield of (*R*)-ECH was 25.1% with enantiomeric excess >99% in the second step. Moreover, the yield of (*R*)-ECH can be further increased to 26.4% by optimizing the flow rate of air and the amount of immobilized cells [398].

The chemoenzymatic cascade reaction is another powerful tool for the synthesis of bioactive compounds. The enantiopure (*R*)- and (*S*)-chromanemethanol are the key building blocks for the synthesis of optical pure isomers of α-tocopherol, which can be obtained from the starting material 2-methylprop-2-enol by the enantiocomplementary chemoenzymatic route via the biocatalytic resolution of racemic 2,2-disubstituted oxirane using either a halohydrin dehalogenase or epoxide hydrolase (Scheme 5.78). The introduction of chirality at an early stage of the synthesis by the opposite enantioslectivity of EH and HHDH ensured a high efficiency, which leads to total overall yield of 16 and 26% for (*R*)- and (*S*)-chromanemethanol, respectively [399].

References

1 Brovetto, M., Gamenara, D., Méndez, P.S., and Seoane, G.A. (**2011**). *Chem. Rev. 111*: 4346–4430.
2 Nomenclature Committee of the International Union of Biochemistry and Molecular Biology (http://www.sbcs.qmul.ac.uk/iubmb/enzyme/EC4).
3 Wikipedia (https://en.wikipedia.org/wiki/Lyase).
4 Dean, S.M., Greenberg, W.A., and Wong, C.-H. (**2007**). *Adv. Synth. Catal. 349*: 1308–1320.
5 Li, C.-J. (**2005**). *Chem. Rev. 105*: 3095–3165.
6 Alcaide, B. and Almendros, P. (**2003**). *Angew. Chem. Int. Ed. 42*: 858–860.
7 Fischer, T. and Pietruszka, J. (**2010**). *Top. Curr. Chem. 297*: 1–43.
8 Clapés, P., Fessner, W.-D., Sprenger, G.A., and Samland, A.K. (**2010**). *Curr. Opin. Chem. Biol. 14*: 154–167.
9 Schimidt, N.G., Eger, E., and Kroutil, W. (**2016**). *ACS Catal. 6*: 4286–4311.
10 Fessner, W.-D. and Helaine, V. (**2001**). *Curr. Opin. Biotechnol. 12*: 574–586.
11 Romero, A. and Wong, C.-H. (**2000**). *J. Org. Chem. 65*: 8264–8268.
12 Chênevert, R. and Dasser, M. (**2000**). *J. Org. Chem. 65*: 4529–4531.
13 Schürmann, M. and Sprenger, G.A. (**2001**). *J. Biol. Chem. 276*: 11055–11061.
14 Garrabou, X., Castillo, J.A., Guérard-Hélaine, C. et al. (**2009**). *Angew. Chem. Int. Ed. 48*: 5521–5525.

15 Stroheier, G.A., Pichler, H., May, O., and Gruber-Khadjawi, M. (**2011**). *Chem. Rev. 111*: 4141–4164.
16 Concia, A.L., Lozano, C., Castillo, J.A. et al. (**2009**). *Chem. Eur. J. 15*: 3808–3816.
17 Machajewski, T.D. and Wong, C.-H. (**2000**). *Angew. Chem. Int. Ed. 39*: 1352–1374.
18 Smith, B.J., Lawrence, M.C., and Barbosa, J.A.R.G. (**1999**). *J. Org. Chem. 64*: 945–949.
19 Fong, S., Machajjewski, T.D., Mak, C.C., and Wong, C.-H. (**2000**). *Chem. Biol. 11*: 873–883.
20 Woodhall, T., Williams, G., Berry, A., and Nelson, A. (**2005**). *Angew. Chem. Int. Ed. 44*: 2109–2112.
21 Liu, J. and Wong, C.-H. (**2002**). *Angew. Chem. Int. Ed. 41*: 1404–1407.
22 Liu, J., Hsu, C.-C., and Wong, C.-H. (**2004**). *Tetrahedron Lett. 45*: 2439–2441.
23 Jennewein, S., Schürmann, M., Wolberg, M. et al. (**2006**). *Biotechnol. J. 1*: 537–548.
24 Nishiyama, T., Kajimoto, T., Mohile, S.S. et al. (**2009**). *Tetrahedron-Asymm. 20*: 230–234.
25 Toshihiro, N., Surendra, S., Tetsuya, K., and Manabu, N. (**2007**). *Heterocycles 71*: 1397–1405.
26 Steinreiber, J., Fesko, K., Mayer, C. et al. (**2007**). *Tetrahedron 63*: 8088–8093.
27 Hoyos, P., Sinisterra, J.-V., Molinari, F. et al. (**2010**). *Accounts Chem. Res. 43*: 288–299.
28 Gocke, D., Graf, T., Brosi, H. et al. (**2009**). *Mol. Catal. B-Enzym. 61*: 30–35.
29 Ward, O.P. and Singh, A. (**2000**). *Curr. Opin. Biotechnol. 11*: 520–526.
30 Demir, A.S., Ayhan, P., and Sopaci, S.B. (**2007**). *Clean 35*: 406–412.
31 Rosche, B., Sandford, V., Breuer, M. et al. (**2002**). *Mol. Catal. B-Enzym. 19-20*: 109–115.
32 Cosp, A., Dresen, C., Pohl, M. et al. (**2008**). *Adv. Synth. Catal. 350*: 759–771.
33 Demir, A.S., Şeşenoglu, Ö., Eren, E. et al. (**2002**). *Adv. Synth. Catal. 344*: 96–103.
34 Dünkelmann, P., Kolter-Jung, D., Nitsche, A. et al. (**2002**). *J. Am. Chem. Soc. 124*: 12084–12085.
35 Westphal, R., Vogel, C., Schmitz, C. et al. (**2014**). *Angew. Chem. Int. Ed. 53*: 9376–9379.
36 Weiss, P.M., Garcia, G.A., Kenyon, G.L. et al. (**1988**). *Biochemistry 27*: 2197–2205.
37 Reynolds, L.J., Garcia, G.A., Kozarich, J.W., and Kenyon, G.L. (**1988**). *Biochemistry 27*: 5530–5538.
38 Tsou, A.Y., Ransom, S.C., and Gerlt, J.A. (**1990**). *Biochemistry 29*: 9856–9862.
39 Wilcocks, R. and Ward, O.P. (**1992**). *Biotechnol. Bioeng. 39*: 1058–1063.
40 Demir, A.S., Dünnwald, T., Iding, H. et al. (**1999**). *Tetrahedron-Asymm. 10*: 4769–4774.
41 Wilcocks, R., Ward, O.P., Collins, S. et al. (**1992**). *Appl. Environ. Microbiol. 58*: 1699–1704.

42 De María, P.D., Pohl, M., Gocke, D. et al. (**2007**). *Eur. J. Org. Chem.* 2940–2944.

43 Dünnwald, T. and Müller, M. (**2000**). *J. Org. Chem. 65*: 8608–8621.

44 Dünnwald, T., Demir, A.S., Siegert, P. et al. (**2000**). *Eur. J. Org. Chem.* 2161–2170.

45 Iding, H., Dünnwald, T., Greiner, L. et al. (**2000**). *Chem. Eur. J. 6*: 1483–1495.

46 Lingen, B., Grötzinger, J., Kolter, D. et al. (**2002**). *Protein Eng. 15*: 585–593.

47 Lingen, B., Kolter-Jung, D., Dünkelmann, P. et al. (**2003**). *ChemBioChem 4*: 721–726.

48 Yep, A. and McLeish, M.J. (**2009**). *Biochemistry 48*: 8387–8395.

49 Siegert, P., McLeish, M.J., Baumann, M. et al. (**2005**). *Protein Eng. Des. Sel. 18*: 345–357.

50 González, B. and Vicuña, R. (**1989**). *J. Bacteriol. 171*: 2401–2405.

51 Demir, A.S., Pohl, M., Janzen, E., and Müller, M. (**2001**). *J. Chem. Soc. Perkin Trans. 1*: 633–635.

52 Knoll, M., Müller, M., Pleiss, J., and Pohl, M. (**2006**). *ChemBioChem 7*: 1928–1934.

53 Brandt, G.S., Kneen, M.M., Chakraborty, S. et al. (**2009**). *Biochemistry 48*: 3247–3257.

54 Brandt, G.S., Nemeria, N., Chakraborty, S. et al. (**2008**). *Biochemistry 47*: 7734–7743.

55 Zavrel, M., Schmidt, T., Michalik, C. et al. (**2008**). *Biotechnol. Bioeng. 101*: 27–38.

56 Mosbacher, T.G., Mueller, M., and Schulz, G.E. *FEBS J 272*: 6067–6076.

57 Kokova, M., Zavrel, M., Tittmann, K. et al. (**2009**). *Mol. Catal. B-Enzym. 61*: 73–79.

58 Kihumbu, D., Stillger, T., Hummel, W., and Liese, A. (**2002**). *Tetrahedron-Asymm. 13*: 1069–1072.

59 Demir, A.S. and Şeşenoglu, Ö. (**2003**). *Org. Lett. 5*: 2047–2050.

60 Sanchez-Gonzalez, M. and Rosazza, J.P.N. (**2003**). *Adv. Synth. Catal. 345*: 819–824.

61 Demir, A.S., Ayhan, P., Igdir, A.C., and Duygu, A.N. (**2004**). *Tetrahedron 60*: 6509–6512.

62 de María, P.D., Stillger, T., Pohl, M. et al. (**2006**). *Mol. Catal. B-Enzym. 38*: 43–47.

63 de María, P.D., Stillger, T., Pohl, M. et al. (**2008**). *Adv. Syn. Catal.* 350: 165–173.

64 Kurlemann, N., Lara, M., Pohl, M. et al. (**2009**). *Mol. Catal. B-Enzym. 61*: 111–116.

65 Mikolajek, R.J., Spiess, A.C., Pohl, M., and Büchs, J. (**2009**). *Biotecnol. Prog. 25*: 132–138.

66 Brandt, G.S., Kneen, M.M., Petsko, G.A. et al. (**2010**). *J. Am. Chem. Soc. 132*: 438–439.

67 Hischer, T., Gocke, D., Fernández, M. et al. (**2005**). *Tetrahedron 61*: 7378–7383.

68 Muller, C.R. and Perez-Sanchez, M. (**2013**). Domínguez de María, P. *Org. Biomol. Chem. 11*: 2000–2004.

69 Kühl, S., Zehentgruber, D., Pohl, M. et al. (**2007**). *Chem. Eng. Sci.* 62: 5201–5205.

70 Maugeri, Z. (**2014**). Domínguez de María. *P. J. Mol. Catal. B-Enzym. 107*: 120–123.

71 Shanmuganathan, S., Natalia, D., van den Wittenboer, A. et al. (**2010**). Domínguez de María. *P. Green Chem. 12*: 2240–2245.

72 Stillger, T., Pohl, M., Wandrey, C., and Liese, A. (**2006**). *Org. Proc. Res. Dev. 10*: 1172–1177.

73 Mikolajek, R., Spiess, A.C., and Büchs, J. (**2007**). *J. Biotechnol. 129*: 723–725.

74 Mikolajek, R., Spiess, A.C., Pohl, M. et al. (**2007**). *ChemBioChem 8*: 1063–1070.

75 Çalik, P., Yilgör, P., Ayhan, P., and Demir, A.S. (**2004**). *Chem. Eng. Sci. 59*: 5075–5083.

76 Çalik, P., Yilgör, P., Ayhan, P., and Demir, A.S. (**2006**). *Enzym. Microb. Technol. 38*: 617–627.

77 Hischer, T., Steinsiek, S., and Ansorge-Schumacher, M.B. (**2006**). *Biocatal. Biotransform. 24*: 437–442.

78 Kurlemann, N. and Liese, A. (**2004**). *Tetrahedron-Asymm. 15*: 2955–2958.

79 Perez-Sanchez, M. and Müller, C.R. (**2013**). Domínguez de María. *P. ChemCatChem 5*: 2512–2516.

80 Guo, z.; Goswami, A.; Mirfakhrae, K.D.; Patel, R.N (**1999**). *Tetrahedron-Asymm. 10*, 4667-4675.

81 Guo, z.; Goswami, A.; Nanduri, V.B.; Patel, R.N (**2001**). *Tetrahedron-Asymm. 12*, 571-577.

82 Schneider, S., El-Said Mohamed, M., and Fuchs, G. (**1997**). *Arch. Microbiol. 168*: 310–320.

83 Heider, J., Boll, M., Breese, K. et al. (**1998**). *Arch. Microbiol. 170*: 120–131.

84 Debnar-Daumler, C., Seubert, A., Schmit, G., and Heider, J. (**2014**). *J. Bacteriol. 196*: 483–492.

85 Verses, W., Spaepen, S., Vanderleyden, J., and Steyaert, J. (**2007**). *FEBS J. 274*: 2363–2375.

86 Bracco, P., Busch, H., von Langermann, J., and Hanefeld, U. (**2016**). *Org. Biomol. Chem. 14*: 6375–6389.

87 Holt, J. and Hanefeld, U. (**2009**). *Curr. Org. Synth. 6*: 15–37.

88 Purkarthofer, T., Skranc, W., Schuster, C., and Griengl, H. (**2007**). *Appl. Microbiol. Biotechnol. 76*: 309–320.

89 Sharma, M., Sharma, N.N., and Bhalla, T.C. (**2005**). *Enzym. Microb. Technol. 37*: 279–294.

90 Dadashipour, M., Ishida, Y., Yamamoto, K., and Asano, Y. (**2015**). *Proc. Natl. Acad. Sci. U. S. A. 112*: 10605–10610.

91 Hajnal, I., Łyskowski, A., Hanefeld, U. et al. (**2013**). *FEBS J. 280*: 5815–5828.

92 Solís, A., Solís-Oba, M., Pérez, H.I. et al. (**2011**). *Biosci. Biotechnol. Biochem. 75*: 985–986.

93 Dadashipour, M., Yamazaki, M., Momonoi, K. et al. (**2011**). *J. Biotechnol. 153*: 100–110.

94 Ueatrongchit, T., Tamura, K., Ohmiya, T. et al. (**2010**). *Enzym. Microb. Technol. 46*: 456–465.

95 Zagrobelny, M., Bak, S., and Møller, B.L. (**2008**). *Phytochemistry 69*: 1457–1468.

96 Dreveny, I., Andryushkova, A.S., Glieder, A. et al. (**2009**). *Biochemistry 48*: 3370–3377.

97 Gruber, K. and Kratky, C.J. (**2004**). *Polym. Sci. Part A-Polym. Chem. 42*: 479–486.

98 Cui, F.-C., Pan, X.-L., and Liu, J.-Y. (**2010**). *J. Phys. Chem. B 114*: 9622–9628.

99 Costs, D., Wehtje, E., and Adlercreutz, P. (**1999**). *Enzym. Microb. Technol. 25*: 384–391.

100 Griengl, H., Klempier, N., Pöchlauer, P. et al. (**1998**). *Tetrahedron 54*: 14477–14486.

101 Effenberger, F., Eichhorn, J., and Roos, J. (**1995**). *Tetrahedron-Asymm. 6*: 271–282.

102 Van Rantwijk, F. and Stola, A.J. (**2015**). *Mol. Catal. B-Enzym. 114*: 25–30.

103 Hanefeld, U. (**2013**). *Chem. Soc. Rev. 42*: 6308–6321.

104 Cabirol, F.L., Hanefeld, U., and Sheldon, R.A. (**2006**). *Adv. Synth. Catal. 348*: 1645–1654.

105 Effenberger, F. and Heid, S. (**1995**). *Tetrahedron-Asymm. 6*: 2945–2952.

106 Kiljunen, E. and Kanerva, L.T. (**1997**). *Tetrahedron-Asymm. 8*: 1551–1557.

107 Bühler, H., Bayer, A., and Effenberger, F. (**2000**). *Chem. Eur. J. 6*: 2564–2571.

108 Roberge, C., Fleitz, F., Pollard, D., and Devine, P. (**2007**). *Tetrahedron-Asymm. 18*: 208–214.

109 Fechter, M.H., Gruber, K., Avi, M. et al. (**2007**). *Chem. Eur. J. 13*: 3369–3376.

110 Cabirol, F.L., Tan, P.L., Tay, B. et al. (**2008**). *Adv. Synth. Catal. 350*: 2329–2338.

111 Kobler, C., Bohrer, A., and Effenberger, F. (**2004**). *Tetrahedron 60*: 10397–10410.

112 Effenberger, F., Roos, J., Kobler, C., and Buhler, H. (**2002**). *Can. J. Chem. 80*: 671–679.

113 Bhuniya, R. and Nanda, S. (**2012**). *Tetrahedron Lett. 53*: 1990–1992.

114 Dadashipour, M. and Asano, Y. (**2011**). *ACS Catal. 1*: 1121–1149.

115 Avi, M., Gaisberger, R., Feichtenhofer, S., and Griengl, H. (**2009**). *Tetrahedron 65*: 5418–5426.

116 Johnson, D.V., Felfer, U., and Griengl, H. (**2000**). *Tetrahedron 56*: 781–790.

117 Vugts, D.J., Veum, L., al-Mafraji, K. et al. (**2006**). *Eur. J. Org. Chem.* 1672–1677.

118 Bhunya, R., Jana, N., Das, T., and Nanda, S. (**2009**). *Synlett.* 1237–1240.

119 Van den Nieuwendijk, A.M.C.H., Ghisaidoobe, A.B.T., Overkleeft, H.S. et al. (**2004**). *Tetrahedron 60*: 10385–10396.

120 Pscheidt, B., Liu, Z., Gaisberger, R. et al. (**2008**). *Adv. Synth. Catal. 350*: 1943–1948.

121 Scheidt, B., Avi, M., Gaisberger, R. et al. (**2008**). *Mol. Catal. B-Enzym. 52-53*: 183–188.

122 Lu, W., Chen, P., and Lin, G. (**2008**). *Tetrahedron 64*: 7822–7827.

123 Sun, H., Zhang, H., Ang, E.L., and Zhao, H. (**2018**). *Bioorg. Med. Chem. 26*: 1275–1284.

124 Breuer, M., Ditrich, K., Habicher, T. et al. (**2004**). *Angew. Che. Int. Ed. 43*: 788–824.

125 Andexer, J.N., Langermann, J.V., Kragl, U., and Pohl, M. (**2009**). *Trends Biotechnol. 27*: 599–607.

126 Avi, M., Wiedner, R.M., Griengl, H., and Schwab, H. (**2008**). *Chem. Eur. J. 14*: 11415–11422.

127 Gaisberger, R., Weis, R., Luiten, R. et al. (**2007**). *J. Biotechnol. 129*: 30–38.

128 Glieder, A., Weis, R., Skranc, W. et al. (**2003**). *Angew. Chem. Int. Ed. 42*: 4815–4818.

129 Weis, R., Gaisberger, R., Skranc, W. et al. (**2005**). *Angew. Chem. Int. Ed. 44*: 4700–4704.

130 Engleder, M. and Pichler, H. (**2018**). *Appl. Microbiol. Biotechnol. 102*: 5841–5858.

131 Chen, B.-S., Otten, L.G., and Hanefeld, U. (**2015**). *Biotechnol. Adv. 33*: 526,-546.

132 Resch, V. and Hanefeld, U. (**2015**). *Catal. Sci. Technol. 5*: 1385–1399.

133 Wallen, L.L., Benedict, R.G., and Jackson, R.W. (**1962**). *Arch. Biochem. Biophys. 99*: 249–253.

134 Bevers, L.E., Pinkse, M.W.H., Verhaert, P.D.E.M., and Hagen, W.R. (**2009**). *J. Bacteriol. 191*: 5010–5012.

135 Joo, Y.-C., Jeong, K.-W., Yeom, S.-J. et al. (**2012**). *Biochimie 94*: 907–915.

136 Kang, W.-R., Seo, M.-J., Shin, K.-C. et al. (**2017**). *Appl. Environ. Microbiol. 83*: e03351–e03316.

137 Engleder, M., Pavkov-Keller, T., Emmerstorfer, A. et al. (**2015**). *ChemBioChem 16*: 1730–1734.

138 Hirata, A., Kishino, S., Park, S.-B. et al. (**2015**). *J. Lipid Res. 56*: 1340–1350.

139 Volkov, A., Liavonchanka, A., Kamneva, O. et al. (**2010**). *J. Biol. Chem. 285*: 10353–10361.

140 Schmid, J., Steiner, L., Fademrecht, S. et al. (**2016**). *J. Mol. Catal. B-Enzym. 133*: S243–S249.

141 Kim, K.-R. and Oh, D.-K. (**2013**). *Biotechnol. Adv. 31*: 1473–1485.

142 Yang, W., Dostal, L., and Rosazza, J.P.N. (**1993**). *Appl. Environ. Microbiol. 59*: 281–284.

143 Takeuchi, M., Kishino, S., Hirata, A. et al. (**2015**). *J. Biosci. Bioeng. 119*: 636–641.

144 Hudson, A.J., Morvan, B., and Joblin, K.N. (**1998**). *FEMS Microbiol. Lett. 169*: 277–282.

145 Kim, B.N., Joo, Y.-C., Kim, Y.-S. et al. (**2012**). *Appl. Microbiol. Biotechnol. 95*: 929–937.

146 Kim, K.-R., Oh, H.-J., Park, C.-S. et al. (**2015**). *Biotechnol. Bioeng. 112*: 2206–2213.

147 Gocho, S., Tabogami, N., Inagaki, M. et al. (**1995**). *Biosci. Biotech. Biochem. 59*: 1571–1572.

148 Wanikawa, A., Hosoi, K., Takise, I., and Kato, T. (**2000**). *J. Inst. Brew. 106*: 39–43.

149 Kim, B.-N., Yeom, S.-J., and Oh, D.-K. (**2011**). *Biotechnol. Lett. 33*: 993–997.

150 Joo, Y.,.-C., Seo, E.-S., Kim, Y.-S. et al. (**2012**). *J. Biotechnol. 158*: 17–23.

151 Jeon, E.-Y., Lee, J.-H., Yang, K.-M. et al. (**2012**). *Process Biochem. 47*: 941–947.

152 Song, J.-W., Jeon, E.-Y., Song, D.-H. et al. (**2013**). *Angew. Chem. Int. Ed. 52*: 2534–2537.

153 Song, J.-W., Lee, J.-H., Bornscheuer, U.T., and Park, J.-B. (**2014**). *Adv. Synth. Catal. 356*: 1782–1788.

154 van Leeuwen, B.N.M., van der Wulp, A.M., Duijnstee, I. et al. (**2012**). *Appl. Microbiol. Biotechnol. 93*: 1377–1387.

155 Demming, R.M., Otte, K.B., Nestl, B.M., and Hauer, B. (**2017**). *ChemCatChem 9*: 758–766.

156 Kang, W.-R., Seo, M.-J., Shin, K.-C. et al. (**2017**). *Biotechnol. Bioeng. 114*: 74–82.

157 Todea, A., Hiseni, A., Otten, L.G. et al. (**2015**). *Mol. Catal. B-Enzym. 119*: 40–47.

158 Demming, R.M., Fischer, M.-P., Schmid, J., and Hauer, B. (**2018**). *Curr. Opin. Chem. Biol. 43*: 43–50.

159 Brodkorb, D., Gottschall, M., Marmulla, R. et al. (**2010**). *J. Biol. Chem. 285*: 30436–30442.

160 Weidenweber, S., Marmulla, R., Ermler, U., and Harder, J. (**2016**). *FEBS Lett. 590*: 1375–1383.

161 Nestl, B.M., Geinitz, C., Popa, S. et al. (**2017**). *Nat. Chem. Biol. 13*: 275–281.

162 Lüddeke, F. and Harder, J.Z. (**2011**). *Naturforsch C 66*: 409–412.

163 Marmulla, R. and Harder, J. (**2014**). *Front. Microbiol. 5*: 1–14.

164 Kaspera, R., Krings, U., Pescheck, M. et al. (**2005**). *Naturforsch C 60*: 459–466.

165 Duetz, W.A., Bouwmeester, H., van Beilen, J.B., and Witholt, B. (**2003**). *Appl. Microbiol. Biotechnol. 61*: 269–277.

166 Maróstica, M.R. Jr. and Pastore, G.M. (**2007**). *Food Chem. 101*: 345–350.

167 Bicas, J.L., Dionísio, A.P., and Pastore, G.M. (**2009**). *Chem. Rev. 109*: 4518–4531.

168 Bicas, J.L., Fontanille, P., Pastore, G.M., and Larroche, C. (**2010**). *Process Biochem. 45*: 481–486.

169 Bade, A.Z.M., Helmy, S.A., and Morsy, N.F.S. (**2011**). *Food Chem. 126*: 849–854.

170 Adams, A., Demyttenaere, J.C.R., and de Kimpe, N. (**2003**). *Food Chem. 80*: 525–534.

171 Tan, Q. and Day, D.F. (**1998**). *Appl. Microbiol. Biotechnol. 49*: 96–101.

172 Rottava, I., Toniazzo, G., Cortina, P.F. et al. (**2010**). *LWT - Food Sci. Technol. 43*: 1128–1131.

173 Prieto, S., Gloria, A., Perea, V. et al. (**2011**). *Vitae 18*: 163–172.

174 Savithiry, N., Cheong, T.K., and Oriel, P. (**1997**). *Appl. Biochem. Biotechnol. 63*: 213–220.

175 Bicas, J.L., Barros, F.F.C., Wagner, R. et al. (**2008**). *J. Ind. Microbiol. Biotechnol.* *35*: 1061–1070.
176 Onken, J. and Berger, R.G. (**1999**). *J. Biotechnol.* *69*: 163–168.
177 Pesceck, M., Mirata, M.A., Brauer, B. et al. (**2009**). *J. Ind. Microbiol. Biotechnol.* *36*: 827–836.
178 Cadwallader, K.R., Braddock, R.J., and Parish, M.E. (**1992**). *J. Food Sci.* *57*: 241–245.
179 Yogev, O., Naamati, A., and Pines, O. (**2011**). *FEBS J.* *278*: 4230–4242.
180 Woods, S.A., Schwartzbach, S.D., and Guest, J.R. (**1988**). *Biochimica. Biophysica. Acta* *954*: 14–26.
181 Ueda, Y., Yumoto, N., Tokushige, M. et al. (**1991**). *J. Biochem.* *109*: 728–733.
182 Flint, D.H. (**1994**). *Arch. Biochem. Biophys.* *311*: 509–516.
183 Flint, D.H., Emptage, M.H., and Guest, J.R. (**1992**). *Biochemistry* *31*: 10331–10337.
184 Weaver, T., Lees, M., Zaitsev, V. et al. (**1998**). *J. Mol. Biol.* *280*: 431–442.
185 Yang, J., Wang, Y., Woolridge, E.M. et al. (**2004**). *Biochemistry* *43*: 10424–10434.
186 Su, R.-R., Wang, A., Hou, S.-T. et al. (**2014**). *Mol. Biol. Rep.* *41*: 497–504.
187 Rose, I.A. and Weaver, T.M. (**2004**). *Proc. Natl. Acad. Sci. U. S. A.* *101*: 3393–3397.
188 Weaver, T. (**2005**). *Acta Crystallogr. Sect. D* *61*: 1395–1401.
189 Weaver, T. and Banaszak, L. (**1996**). *Biochemistry* *35*: 13955–13965.
190 Weaver, T., Lees, M., and Banaszak, L. (**1997**). *Protein Sci.* *6*: 834–842.
191 Mescam, M., Vinnakota, K.C., and Beard, D.A. (**2011**). *J. Biol. Chem.* *286*: 21100–21109.
192 Yamamoto, K., Tosa, T., Yamashita, K., and Chibata, I. (**1976**). *Eur. J. Appl. Microbiol.* *3*: 169–183.
193 Takata, I., Yamamoto, K., Tosa, T., and Chibata, I. (**1980**). *Enzyme Microb. Technol.* *2*: 30–36.
194 Chibata, I., Tosa, T., and Takata, I. (**1983**). *Trends Biotechnol.* *1*: 9–11.
195 Tosa, T. and Shibatani, T. (**1995**). *Ann. N. Y. Acad. Sci.* *750*: 364–375.
196 Bressler, E., Pines, O., Goldberg, I., and Braun, S. (**2002**). *Biotechnol. Prog.* *18*: 445–450.
197 Ninh, P.H., Honda, K., Yokohigashi, Y. et al. (**2013**). *Appl. Environ. Microbiol.* *79*: 1996–2001.
198 Leuchtenberger, W., Karrenbauer, M., and Ploecker, U. (**1984**). *Ann. N. Y. Acad. Sci.* *434*: 78–86.
199 Cernia, E., Libori, R., Marconi, W., and Soro, S.J. (**1996**). *Mol. Catal. B-Enzym.* *1*: 81–88.
200 Cernia, E., Libori, R., Marconi, W., and Soro, S.J. (**1996**). *Mol. Catal. B-Enzym.* *1*: 89–95.
201 Marcconi, W., Faiola, F., and Piozzi, A.J. (**2001**). *Mol. Catal. B-Enzym.* *15*: 93–99.
202 Teipel, J.W., Hass, G.M., and Hill, R.L. (**1968**). *J. Biol. Chem.* *243*: 5684–5694.

203 Marletta, M.A., Cheung, Y.-F., and Walsh, C. (**1982**). *Biochemistry 21*: 2637–2644.

204 Findeis, M.A. and Whitesides, G.M. (**1987**). *J. Org. Chem. 52*: 2838–2848.

205 Albright, F. and Schroepfer, G.J. Jr. (**1970**). *Biochim. Biophys. Res. Commun. 40*: 661–666.

206 Albright, F. and Schroepfer, G.J. Jr. (**1971**). *J. Biol. Chem. 246*: 1350–1357.

207 Sacks, W. and Jensen, C.O. (**1951**). *J. Biol. Chem. 192*: 231–236.

208 Van der Werf, M.J., van den Tweel, W.J.J., Kamphuis, J. et al. (**1994**). *Trends Biotechnol. 12*: 95–103.

209 Dreyer, J.-L. (**1985**). *Eur. J. Biochem. 150*: 145–154.

210 He, B.-F., Naajima-Kambe, T., Ozawa, T., and Nakahara, T. (**2000**). *Process Biochem. 36*: 407–414.

211 Van der Werf, M.J., van den Tweel, W.J.J., and Hartmans, S. (**1992**). *Appl. Environ. Microbiol. 58*: 2854–2860.

212 Van der Werf, M.J., van den Tweel, W.J.J., and Hartmans, S. (**1993**). *Appl. Environ. Microbiol. 59*: 2823–2829.

213 Van der Werf, M.J., van den Tweel, W.J.J., and Hartmans, S. (**1993**). *Eur. J. Biochem. 217*: 1011–1017.

214 Englard, S., Britten, J.S., and Listowsky, I. (**1967**). *J. Biol. Chem. 242*: 2255–2259.

215 Taggart, J.V., Angielski, S., and Morell, H. (**1962**). *Biochim. Biophys. Acta 58*: 141–144.

216 Cornforth, J.W., Ryback, G., Popjak, G. et al. (**1962**). *Biochem. Biophys. Res. Commun. 9*: 371–375.

217 Rahatekar, H.I., Maskati, F.S., Subramanian, S.S., and Rao, M.R.R. (**1968**). *Indian J. Biochem. 5*: 143–144.

218 Van der Werf, M.J., van den Tweel, W.J.J., Kamphuis, J. et al. (**1994**). *Prog. Biotechnol. 9*: 471–474.

219 Ueda, M., Yamada, H., and Asano, Y. (**1994**). *Appl. Microbiol. Biotechnol. 41*: 215–218.

220 Michielsen, M.J.F., Frielink, C., Wijffels, R.H. et al. (**2000**). *J. Biotechnol. 79*: 13–26.

221 Michielsen, M.J.F., Frielink, C., Meijer, E.A. et al. (**1999**). *Biocatal. Biotransform. 17*: 125–137.

222 Van der Werf, M.J., Huybers, P., van den Tweel, W.J.J., and Hartmans, S. (**1997**). *World J. Microbiol. Biotechnol. 13*: 279–282.

223 Schoemaker, H.E., Boesten, W.H.J., Kaptein, B. et al. (**1996**). *Acta Chim. Scand. 50*: 225–233.

224 Van der Werf, M.J., Hartmans, S., and van den Tweel, W.J.J. (**1995**). *Enzyme Microb. Technol. 17*: 430–436.

225 Hiltunen, J.K. and Qin, Y.-M. (**2000**). *Biochim. Biophys. Acta Mol. Cell Biol. Lipids 1484*: 117–128.

226 Agnihotri, G. and Liu, H.-W. (**2003**). *Bioorg. Med. Chem. 11*: 9–20.

227 Maggio-Hall, L.A. and Keller, N.P. (**2004**). *Mol. Microbiol. 54*: 1173–1185.

228 Bhaumik, P., Koski, M.K., Glumoff, T. et al. (**2005**). *Curr. Opin. Struct. Biol. 15*: 621–628.

229 Holden, H.M., Benning, M.M., Haller, T., and Gerlt, J.A. (**2001**). *Acc. Chem. Res. 34*: 145–157.

230 Bahnson, B.J., Anderson, V.E., and Petsko, G.A. (**2002**). *Biochemistry 41*: 2621–2629.

231 Wu, W.-J., Feng, Y., He, X. et al. (**2000**). *J. Am. Chem. Soc. 122*: 3987–3994.

232 Bell, A.F., Feng, Y., Hofstein, H.A. et al. (**2002**). *Chem. Biol. 9*: 1247–1255.

233 Kasaragod, P., Schmitz, W., Hiltunen, J.K., and Wierenga, R.K. (**2013**). *FEBS J. 280*: 3160–3175.

234 Hisano, T., Tsuge, T., Fukui, T. et al. (**2003**). *J. Biol. Chem. 278*: 617–624.

235 Haataja, T.J.K., Koski, M.K., Hiltunen, J.K., and Glumoff, T. (**2011**). *Biochem. J. 435*: 771–781.

236 Koski, M.K., Haapalainen, A.M., Hiltunen, J.K., and Glumoff, T. (**2004**). *J. Biol. Chem. 279*: 24666–24672.

237 Koski, M.K., Haapalainen, A.M., Hiltunen, J.K., and Glumoff, T. (**2005**). *J. Mol. Biol. 345*: 1157–1169.

238 D'Ordine, R.L., Bahnson, B.J., Tonge, P.J., and Anderson, V.E. (**1994**). *Biochemistry 33*: 14733–14742.

239 Willadsen, P. and Eggerer, H. (**1975**). *Eur. J. Biochem. 54*: 247–252.

240 Hasegawa, J., Ogura, M., Kanema, H. et al. (**1982**). *J. Ferment. Technol. 60*: 501–508.

241 Crosby, J. (**1991**). *Tetrahedron 47*: 4789–4846.

242 Dekishima, Y., Lan, E.I., Shen, C.R. et al. (**2011**). *J. Am. Chem. Soc. 133*: 11399–11401.

243 Cheon, Y., Kim, J.-S., Prk, J.-B. et al. (**2014**). *J. Biotechnol. 182-183*: 30–36.

244 Kaur, B., Kumar, B., Kaur, G. et al. (**2015**). *Appl. Microbiol. Biotechnol. 99*: 3015–3028.

245 Phithakrotchanakoon, C., Champreda, V., Aiba, S.-I. et al. (**2015**). *J. Polym. Environ. 23*: 38–44.

246 Thomson, N.M., Sangiambut, S., Ushimaru, K. et al. (**2017**). *ACS Biomater. Sci. Eng. 3*: 3076–3082.

247 Parmeggiani, F., Weise, N.J., Ahmed, S.T., and Turner, N.J. (**2018**). *Chem. Rev. 118*: 73–118.

248 Nestl, B.M; Hammer, S.C.; Nebel, B.A.; Hauer, B. (**2014**). *Angew. Chem. Int. Ed. 53*, 3070-3095.

249 Heberling, M.M., Wu, B., Bartsch, S., and Janssen, D.B. (**2013**). *Curr. Opin. Chem. Biol. 17*: 250–260.

250 Turner, N.J. (**2011**). *Curr. Opin. Chem. Biol. 15*: 234–240.

251 De Villier, M., Veetil, V.P., Raj, H. et al. (**2012**). *ACS Chem. Biol. 7*: 1618–1628.

252 Veetil, V.P., Fibriansah, G., Raj, H. et al. (**2012**). *Biochemistry 51*: 4237–4243.

253 Tajima, T., Hamada, M., Nakashimada, Y., and Kato, J. (**2015**). *J. Ind. Microbiol. Biotechnol. 42*: 1319–1324.

254 Cárdenas-Fernández, M., Khalikova, E., Korpela, T. et al. (**2015**). *Biochem. Eng. J. 93*: 173–178.

255 Han, C., Yao, P., Yuan, J. et al. (**2015**). *Mol. Catal. B-Enzym. 115*: 113–118.

256 Emery, T.F. (**1963**). *Biochemistry 2*: 1041–1045.

257 Weiner, B., Poelarends, G.J., Janssen, D.B., and Feringa, B.L. (**2008**). *Chem. Eur. J. 14*: 10094–10100.

258 Vogel, A., Schmiedel, R., Hofmann, U. et al. (**2014**). *ChemCatChem 6*: 965–968.

259 Levy, C.W., Buckley, P.A., Sedelnikova, S. et al. (**2002**). *Structure 10*: 105–113.

260 Asuncion, M., Blankenfeldt, W., Barlow, J.N. et al. (**2002**). *J. Biol. Chem. 277* (10): 8306–8611.

261 Barker, H.A., Smyth, R.D., Wawazkiewicz, E.J. et al. (**1958**). *Arch. Biochem. Biophys. 78*: 468–476.

262 Barker, H.A., Smyth, R.D., Wilson, R.M., and Weissbach, H. (**1959**). *J. Biol. Chem. 234*: 320–328.

263 Winkler, M.F. and Williams, V.R. (**1967**). *Biochim. Biophys. Acta 146*: 287–289.

264 Akhtar, M., Cohen, M.A., and Gani, D. (**1986**). *J. Chem. Soc. Chem. Commun.* 1290–1291.

265 Akhtar, M., Cohen, M.A., and Gani, D. (**1987**). *Tetrahedron Lett. 28*: 2413–2416.

266 Akhtar, M., Botting, N.P., Cohen, M.A., and Gani, D. (**1987**). *Tetrahedron 43*: 5899–5908.

267 Gulzar, M.S., Akhtar, M., and Gani, D. (**1997**). *J. Chem. Soc., Perkin Trans. 1*: 649–655.

268 Raj, H., Veetil, V.P., Szymanski, W. et al. (**2012**). *App. Microbiol. Biotechnol. 94*: 385–397.

269 Raj, H., Weiner, B., Veetil, V.P. et al. (**2009**). *Chem BioChem 10*: 2236–2245.

270 Raj, H., Szymanski, W., de Villiers, J. et al. (**2012**). *Nat. Chem. 4*: 478–484.

271 Wang, J. and Zhang, K. (**2015**). *Metab. Eng. 30*: 190–196.

272 Cooke, H.A., Christianson, C.V., and Bruner, S.D. (**2009**). *Curr. Opin. Chem. Biol. 13*: 460–468.

273 MacDonald, M.J. and D'Cunha, G.B. (**2007**). *Biochem. Cell Biol. 85*: 273–282.

274 Kumavath, R.N., Ramana, C.V., Sasikala, C. et al. (**2015**). *Curr. Protein Pept. Sci. 16*: 775–781.

275 Calabrese, J.C., Jordan, D.B., Boodhoo, A. et al. (**2004**). *Biochemistry 43*: 11403–11416.

276 Schwede, T.F., Rétey, J., and Schulz, G.E. (**1999**). *Biochemistry 38*: 5355–5361.

277 Poppe, L. (**2003**). Rétey, *Journal of. Current Organic Chemistry 7*: 1297–1315.

278 Baedeker, M. and Schulz, G.E. (**2002**). *Eur. J. Biochem. 269*: 1790–1797.

279 Schuster, B. and Rétey, J. (**1995**). *Proc. Natl. Acad. Sci. USA 92*: 8433–8437.

280 Bartsch, S. and Bornscheuer, U.T. (**2009**). *Angew. Chem. Int. Ed. 48*: 3362–3365.
281 Yamada, S., Nabe, K., Izuo, N. et al. (**1981**). *Appl. Environ. Microbiol. 42*: 773–778.
282 Renard, G., Guilleux, J.-C., Bore, C. et al. (**1992**). *Biotechnol. Lett. 14*: 673–678.
283 Weise, N.J., Ahmed, S.T., Parmeggiani, F. et al. (**2016**). *Catal. Sci. Technol. 6*: 4086–4089.
284 D'cunha, G.B., Satyanarayan, V., and Nair, P.M. (**1994**). *Enyme Microb. Technol. 16*: 318–322.
285 Hanson, K.R., Havir, E.A., and Ressler, C. (**1979**). *Biochemistry 18*: 1431–1438.
286 Skolaut, A. and Rétey, J. (**2001**). *Arch. Biochem. Biophys. 393*: 187–191.
287 Poppe, L. and Rétey, J. (**2005**). *Angew. Chem. Int. Ed. 44*: 3668–3688.
288 Gloge, A., Zoń, J., Kövári, Á. et al. (**2000**). *Chem. Eur. J. 6*: 3386–3390.
289 Gloge, A., Langer, B., Poppe, L., and Rétey, J. (**1998**). *Arch. Biochem. Biophys. 359*: 1–7.
290 Paizs, C., Katona, A., and Rétey, J. (**2006**). *Chem. Eur. J. 12*: 2739–2744.
291 Paizs, C., Toşa, M.I., Bencze, L.C. et al. (**2011**). *Heterocycles 82*: 1217–1228.
292 Bartsch, S. and Bornscheuer, U.T. (**2010**). *Protein Eng. Des. Sel. 23*: 929–933.
293 Toşa, M.I., Brem, J., Mantu, A. et al. (**2013**). *ChemCatChem 5*: 779–783.
294 Lovelock, S.L. and Turner, N.J. (**2014**). *Bioorg. Med. Chem. 22*: 5555–5557.
295 Bartha-Vári, J.H., Toşa, M.I., Irimie, F.D. et al. (**2015**). *ChemCatChem 7*: 1122–1128.
296 Bartha-Vári, J.H., Bencze, L.C., Bell, E. et al. (**2017**). *Period. Polytechn. Chem. Eng. 61*: 59–66.
297 Weiser, D., Bencze, L.C., Bánóczi, G. et al. (**2015**). *ChemBioChem 16*: 2283–2288.
298 Quinn, A.J., Pickup, M.J., and D'Cunha, G.B. (**2011**). *Biotechnol. Prog. 27*: 1154–1560.
299 Cui, J.D., Zhang, S., and Sun, L.-M. (**2012**). *Appl. Biochem. Biotechnol. 167*: 835–844.
300 Barron, C.C., Sponagle, B.J.D., Arivalagan, P., and D'Cunha, G.B. (**2017**). *Enzyme Microb. Technol. 96*: 151–156.
301 Zhang, Z. (**2008**). *Tetrahedron Lett. 49*: 6468–6470.
302 Shibatani, T., Nishhimura, N., Nabe, K. et al. (**1974**). *Appl. Microbiol. 27*: 688–694.
303 Zhu, L., Zhou, L., Huang, N. et al. (**2014**). *PLoS one 9*: e108586.
304 Parmeggiani, F., Lovelock, S.L., Weise, N.J. et al. (**2015**). *Angew. Chem. Int. Ed. 54*: 4608–4611.
305 Ratnayake, N.D., Theisen, C., Walter, T., and Walker, K.D. (**2016**). *J. Biotechnol. 217*: 12–21.
306 Ratnayake, N.D., Liu, N., Kuhn, L.A., and Walker, K.D. (**2014**). *ACS Catal. 4*: 3077–3090.

307 Weise, N.J., Parmeggiani, F., Ahmed, S.T., and Turner, N.J. (**2015**). *J. Am. Chem. Soc. 137*: 12977-12983.

308 Xiang, L. and Moore, B.S. (**2002**). *J. Biol. Chem. 277*: 32505–32509.

309 Xiang, L. and Moore, B.S. (**2005**). *J J. Bacteriol. 187*: 4286–4289.

310 Wu, B., Szymański, W., de Wildeman, S. et al. (**2010**). *Adv. Synth. Catal. 352*: 1409–1412.

311 Wu, B., Szymański, W., Wybenga, G.G. et al. (**2012**). *Angew. Chem. Int. Ed. 51*: 482–486.

312 Weise, N.J., Ahmed, S.T., Parmeggiani, F., and Turner, N.J. (**2017**). *Adv. Synth. Catal. 359*: 1570–1576.

313 Paizs, C., Katona, A., and Rétey, J. (**2006**). *Eur. J. Org. Chem.* 1113–1116.

314 Parmeggiani, F.; Ahmed, S.T.; Weise, N.J; Turner, N.J. (**2016**). *Tetrahedron 72*, 7256-7262.

315 Ahmed, S.T., Parmeggiani, F., Weise, N.J. et al. (**2016**). *Org. Lett. 18*: 5468–5471.

316 De Lange, B., Hyett, D.J., Maas, P.J.D. et al. (**2011**). *ChemCatChem 3*: 289–292.

317 De Souza, R.O.M.A., Miranda, L.S.M., and Bornscheuer, U.T. (**2017**). *Chem. Eur. J. 23*: 12040–12063.

318 Ahmed, S.T., Parmeggiani, F., Weise, N.J. et al. (**2015**). *ACS Catal. 5*: 5410–5413.

319 Roughly, S.D. and Jordan, A.M. (**2011**). *J. Med. Chem. 54*: 3451–3479.

320 Busto, E., Simon, R.C., and Kroutil, W. (**2015**). *Angew. Chem. Int. Ed. 54*: 10899–10902.

321 Zhou, Y., Wu, S., and Li, Z. (**2016**). *Angew. Chem. Int. Ed. 55*: 11647–11650.

322 Kong, J.-Q. (**2015**). *RSC Adv. 5*: 62587–62603.

323 Katsuyama, Y., Funa, N., and Horinouchi, S. (**2007**). *Biotechnol. J. 2*: 1286–1293.

324 Li, M., Kildegaard, K.R., Chen, Y. et al. (**2015**). *Metab. Eng. 32*: 1–11.

325 Li, M., Schneider, K., Kristensen, M. et al. (**2016**). *Sci. Rep. 6*: 36827.

326 Miyahisa, I., Funa, N., Ohnishi, Y. et al. (**2006**). *Appl. Microbiol. Biotechnol. 71*: 53–58.

327 Trantas, E., Panopoulos, N., and Ververidis, F. (**2009**). *Metab. Eng. 11*: 355–366.

328 Kiddle, G.A., Bennett, R.N., Hick, A.J., and Wallsgrove, R.M. (**1999**). *Plant Cell Environ. 22*: 433–445.

329 Jones, P.R., Manabe, T., Awazuhara, M., and Saito, K. (**2003**). *J. Biol. Chem. 278*: 10291–10296.

330 Yamagata, S., D'andrea, R.J., Fujisaki, S. et al. (**1993**). *J. Bacteriol. 175*: 4800–4808.

331 Steegborn, C., Clausen, T., Sondermann, P. et al. (**1999**). *J. Biol. Chem. 274*: 12675–12684.

332 Nock, L.P. and Mazelis, M. (**1987**). *Plant Physiol. 85*: 1079–1083.

333 Parker, M., Capone, D.L., Francis, I.L., and Herderich, M.J. (**2018**). *J. Agric. Food Chem. 66*: 2281–2286.

334 Kezuka, Y., Yoshida, Y., and Nonaka, T. (**2012**). *Proteins 80*: 2447–2458.

335 Astegno, A., Giorgetti, A., Allegrini, A. et al. (**2013**). *Biomed Res. Int. 2013*: 701536.

336 Toohey, J.I. (**2011**). *Anal. Biochem. 413*: 1–7.

337 Astegno, A., Allegrini, A., Piccoli, S. et al. (**2015**). *Proteins 83*: 78–90.

338 Krupka, H.I., Huber, R., Holt, S.C., and Clausen, T. (**2000**). *Eur. Mol. Biol. Org. J. 19*: 3168–3178.

339 Clause, T., Huber, R., Laber, B. et al. (**1996**). *J. Mol. Biol. 262*: 202–224.

340 Holländer-Czytko, H., Grabowski, J., Sandorf, I. et al. (**2005**). *J. Plant Physiol. 162*: 767–770.

341 Anderson, P.M. and Schultze, M.O. (**1965**). *Arch. Biochem. Biophys. 111*: 593–602.

342 Hiroki, T., Noriko, I., Shigeyasu, I. et al. (**1988**). *Xenobiotica 18*: 1029–1037.

343 Cellini, B., Bertoldi, M., Montioli, R. et al. (**2006**). *Biochemistry 45*: 14140–14154.

344 Bertoldi, M., Cellini, B., Clausen, T., and Voltattorni, C.B. (**2002**). *Biochemistry 41*: 9153–9164.

345 Chu, L., Ebersole, J.L., Kurzban, G.P., and Holt, S.C. (**1999**). *Clin. Infect. Dis. 28*: 442–450.

346 Yoshida, Y., Ito, S., Sasaki, T. et al. (**2008**). *Oral Microbiol. Immunol. 23*: 245–253.

347 Ramírez, E.C. and Whitaker, J.R. (**1998**). *J. Food Biochem. 22*: 427–440.

348 Hamamoto, A. and Mazelis, M. (**1986**). *Plant Physiol. 80*: 702–706.

349 Schwimmer, S. and Kjaer, A. (**1960**). *Biochim. Biophys. Acta 42*: 316–324.

350 Tomisawa, H., Suzuki, S., Ichihara, S. et al. (**1984**). *J. Biol. Chem. 259*: 2588–2593.

351 Tomisawa, H., Hayashi, M., Rukushima, M. et al. (**1993**). *Biochem. Pharmacol. 46*: 1113–1117.

352 Bernstroem, K., Larsen, G.L., and Hammarstroem, S. (**1989**). *Arch. Biochem. Biophys. 275*: 531–539.

353 Iwami, K. and Yasumoto, K. (**1980**). *Agric. Biol. Chem. 44*: 3003–3004.

354 Kumagai, H., Kono, H., Sakurai, H., and Tokimoto, K. (**2002**). *Biosci. Biotechnol. Biochem. 66*: 2560–2566.

355 Morcos, D., Schmier, B.J., Malhotra, A., and Venkatachalam, K.V. (**2015**). *Biochem. Anal. Biochem. 4*: 223.

356 Takada, H., Esaki, N., Tanaka, H., and Soda, K. (**1988**). *Agric. Biol. Chem. 52*: 2897–2901.

357 Naowarojna, N., Cheng, R., Chen, L. et al. (**2018**). *Biochemistry 57*: 3309–3325.

358 Chen, L., Naowarojna, N., Song, H. et al. (**2018**). *J. Am. Chem. Soc. 140*: 4604–4612.

359 Naowarojna, N., Huang, P., Cai, Y. et al. (**2018**). *Org. Lett. 20*: 5427–5430.

360 Hu, W., Song, H., Her, A.S. et al. (**2014**). *Org. Lett. 16*: 5382–5385.

361 Fetzner, S. (**1998**). *Appl. Microbiol. Biotechnol. 50*: 633–657.

362 You, Z.-Y., Liu, Z.-Q., and Zheng, Y.-G. (**2013**). *Appl. Microbiol. Biotechnol. 97*: 9–21.

363 Schallmey, A. and Schallmey, M. (**2016**). *Appl. Microbiol. Biotechnol. 100*: 7827–7839.

364 Hasnaoui-Dijoux, G., Elenkov, M.M., Spelberg, J.H.L. et al. (**2008**). *ChemBioChem 9*: 1048–1051.

365 Castro, C.E. and Bartnicki, E.W. (**1968**). *Biochemistry 7*: 3213–3218.

366 Van Hylckama Vlieg, J.E.T., Tang, L., Lutje Spelberg, J.H. et al. (**2001**). *J. Bacteriol. 183*: 5058–5066.

367 Kavanagh, K.L., J⍰rnvall, H., Persson, B., and Oppermann, U. (**2008**). *Cell. Mol. Life Sci. 65*: 3895–3960.

368 Swanson, P.E. (**1999**). *Curr. Opin. Biotechnol. 10*: 365–369.

369 Kabat, M.M., Daniewski, A.R., and Burger, W. (**1997**). *Tetrahedron-Asymmetry 8*: 2663–2665.

370 Barbachyn, M.R. and Ford, C.W. (**2003**). *Angew. Chem. Int. Ed. 42*: 2010–2023.

371 Assis, H.M.S., Bull, A.T., and Hardman, D.J. (**1998**). *Enzyme Microb. Technol. 22*: 545–551.

372 Jin, H.-X., Hu, Z.-C., Liu, Z.-Q., and Zheng, Y.-G. (**2012**). *Biotechnol. Appl. Biochem. 59*: 170–177.

373 Haak, R.M., Berthiol, F., Jerphagnon, T. et al. (**2008**). *J. Am. Chem. Soc. 130*: 13508.

374 Jerphagnon, T., Haak, R., Berthiol, F. et al. (**2010**). *Top. Catal. 53*: 1002–1008.

375 Wan, N., Tian, J., Wang, H. et al. (**2018**). *Bioorg. Chem. 81*: 529–535.

376 Elenkov, M.M., Hauer, B., and Janssen, D.B. (**2006**). *Adv. Synth. Catal. 348*: 579–585.

377 Chen, J., Zhang, R.-C., Zheng, Y.-G., and Shen, Y.-C. (**2009**). *Adv. Biochemical Eng. Biotechnol. 113*: 33–77.

378 Nakamura, T., Nagasawa, T., Yu, F. et al. (**1994**). *Tetrahedron 50*: 11821–11826.

379 Ma, S.K., Gruber, J., Davis, C. et al. (**2010**). *Green Chem. 12*: 81–86.

380 Fox, R.J., Davis, S.C., Mundorff, E.C. et al. (**2007**). *Nat. Biotechnol. 25*: 338–344.

381 Wan, N.-W., Liu, Z.-Q., Huang, K. et al. (**2014**). *RSC Adv. 4*: 64027–64031.

382 Wan, N.-W., Liu, Z.-Q., Xue, F. et al. (**2015**). *ChemCatChem 7*: 2446–2450.

383 Lutje Spelberg, J.H., van Hylckama Vlieg, J.E.T., Tang, L. et al. (**2001**). *Org. Lett.* 3: 41–43.

384 Lutje Spelberg, J.H., Tang, L., Kellogg, R.M., and Janssen, D.B. (**2004**). *Tetrahedron-Asymmetry 15*: 1095–1102.

385 Campbell-Verduyn, L.S., Szmański, W., Postema, C.P. et al. (**2010**). *Chem. Commun. 46*: 898–900.

386 Szymanski, W., Postema, C.P., Tarabiono, C. et al. (**2010**). *Adv. Synth. Catal. 352*: 2111–2115.

387 Molinaro, C., Guilbault, A.-A., and Kosjek, B. (**2010**). *Org. Lett. 12*: 3772–3775.

388 Elenkov, M.M., Hoeffken, H.W., Tang, L. et al. (**2007**). *Adv. Synth. Catal. 349*: 2279–2285.

389 Elenkov, M.M., Primožič, I., Hrenar, T. et al. (**2012**). *Org. Biomol. Chem.* 5063–5072.

390 Hasnaoui, G., Lutje Spelberg, J.H., de Vries, E. et al. (**2005**). *Tetrahedron-Asymmetry 16*: 1685–1692.

391 Elenkov, M.M., Tang, L., Meetsma, A. et al. (**2008**). *Org. Lett. 10*: 2417–2420.

392 Mikleušević, A., Hameršak, Z., Salopek-Sondi, B. et al. (**2015**). *Adv. Synth. Catal. 357*: 1709–1714.

393 Ager, D.J., Prakash, I., and Schaad, D.R. (**1996**). *Chem. Rev. 96*: 835–875.

394 Lutje Spelberg, J.H., van Hylckama Vlieg, J.E.T., Bosma, T. et al. (**1999**). *Tetrahedron-Asymmetry 10*: 2863–2870.

395 Schrittwieser, J.H., Lavandera, I., Seisser, B. et al. (**2009**). *Eur. J. Org. Chem.* 2293–2298.

396 Chen, S.-Y., Yang, C.-X., Wu, J.-P. et al. (**2013**). *Adv. Synth. Catal. 355*: 3179–3190.

397 Yao, P., Wang, L., Yuan, J. et al. (**2015**). *ChemCatChem 7*: 1438–1444.

398 Jin, H.-X., Liu, Z.-Q., Hu, Z.-C., and Zheng, Y.-G. (**2013**). *Biochem. Eng. J. 74*: 1–7.

399 Fuchs, M., Simeo, Y., Ueberbacher, B.T. et al. (**2009**). *Eur. J. Org. Chem.* 833–840.

6

Organic Synthesis with Isomerases

Many molecules that have the same molecular formula but present in different molecular structures or different three-dimensional arrangements of atoms are called isomers. Accordingly, the variety of isomers can be generally divided into two categories: structural isomers and stereoisomers. For structural isomers, the sequence and/or connectivity of bonds are different from one another. While stereoisomers have the same ordering of individual bonds and the same connection type but differ in their three-dimensional arrangement of bonded atoms. Generally, isomerases are a class of enzymes that convert a molecule from one isomer to another. Specifically, racemases, epimerases, and *cis-trans* isomers carry out the interconversion of stereoisomers, and intramolecular lyases, intramolecular oxidoreductases, and intramolecular transferases catalyze the interconversion of structural isomers. The enzyme classification for isomerases has been assigned in the EC category of EC 5 and further classified into six subclasses. The occurrence of isomerases in nature is rare, while they are important in synthetic organic chemistry by catalyzing "chemical impossible" isomerization reactions.

6.1 Racemases and Epimerases

6.1.1 Amino Acid Racemases and Epimerases

Amino acids play central roles both as building blocks of proteins and as intermediates in the metabolism of organisms, thus are among the most important compounds in nature. With the exception of glycine, all amino acids found in protein have a chiral center at the α-carbon atom that leads to two stereoisomeric forms: the levorotatory (L) and the dextrorotatory (D) form. Although the L-amino acid

Enzyme-Based Organic Synthesis, First Edition. Cheanyeh Cheng.

enantiomers were exclusively found in mammals, various studies demonstrated the presence of D-amino acid enantiomers in some bacteria. D-amino acids (D-AAs) such as D-alanine and D-glutamic acid have long been found in the peptidoglycan of the bacteria cell wall for the protection against peptidase and protease attack [1, 2]. In addition, D-methionine, D-leucine, D-tyrosine, and D-phenylalanine were synthesized in bacteria and incorporated into peptidoglycan to modify it in order to adapt to changing environmental conditions [3]. D-AAs are also constituents of several peptide antibiotics synthesized by bacteria and fungi, including gramicidin D, gramicidin S, polymyxins, penicillin G, cephalosporin, actinomycin D, and bacitracin [2, 4, 5]. Many reports have demonstrated that D-AAs have been found in plants and animals as endogenous molecules either in free form or incorporated form to play different physiological functions. The physiological functions of D-AAs found in mammalian and human tissues are related with aging, signaling, opioid peptides, neuropeptides, and development [2, 5–8].

A number of enzymes can be used for the synthesis of D-AAs including hydrolases, oxidoreductases, D-amino acid aminotransferases, and racemases. D-amino acids are normally produced by the inversion of the configuration of the asymmetric carbon of the respect L-amino acids, which are generally catalyzed by isomerase enzymes. The reversible stereochemical conversion between D- and L-amino acids (L-AAs) is known as "racemization" [9]. Depending on the number of chiral carbons of the amino acid, the nomenclature of isomerases of this kind is referred to as racemases (substrates with one chiral center) and epimerases (substrates with more than one chiral center). Although the basis of amino acid racemization is always the same, that is, the α-hydrogen of an amino acid is removed as a proton to form a planar carbanion transition state followed by the reprotonation to the same carbon atom on either the *re-* or the *si*-face (Scheme 6.1), the reaction mechanism of the array of racemases could be different based on the use of pyridoxal 5'-phosphate (PLP) cofactor. Therefore, amino acid racemases can be divided into two categories: (i) PLP-dependent racemases and (ii) PLP-independent racemases [2, 7].

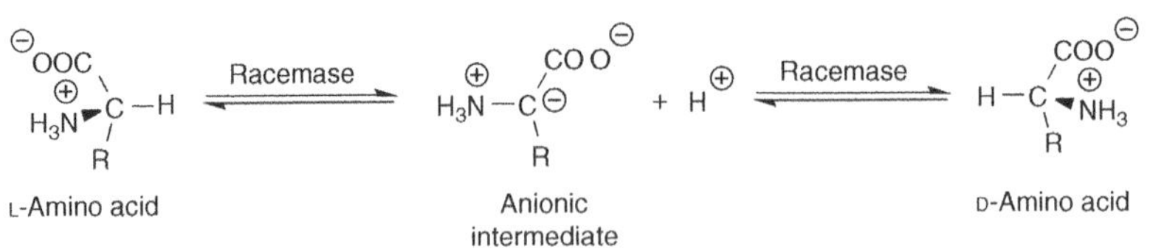

Scheme 6.1 Racemases catalyzed racemization of amino acids.

6.1.1.1 D-Amino Acid Synthesis with PLP-Dependent Amino Acid Racemases

This class of racemases includes alanine racemase (AlaR), serine racemase (SerR), arginine racemase (ArgR), and aspartate racemase (DR). The employment of PLP coenzyme for the racemization is the increase of the acidity of the α-hydrogen of the amino acid. In all these racemases, the PLP coenzyme exists as an internal aldimine (Schiff base) with the ε-amino group of a lysine residue present in the active site. The ε-amino group of lysine can be displaced by the amino moiety of the amino acid substrate to establish a new aldimine (external aldimine), which considerably increases the acidity of the α-hydrogen of the substrate by 15–17 order of magnitude. Thus, the α-hydrogen can be easily abstracted as a proton by a tyrosine residue to form an anionic planar intermediate, which is subsequently reprotonated from either side of this substrate–cofactor complex. The formation of anionic intermediate exhibits the three most significant resonance forms with the quinonoid resonance structure considered the major species responsible for the catalytic power of PLP, in which the electrons produced by the abstraction of the α-hydrogen are neutralized by the positively charged pyridine ring nitrogen [2, 7].

D-alanine generated in bacteria using the ubiquitous AlaR (EC 5.1.1.1) is mainly for the biosynthesis of peptidoglycan, which is a strong and elastic polymer of bacterial cell wall capable of maintaining cell shape, counteracting the osmotic pressure of the cell, and anchoring components of the cell. The presence of D-Ala and D-Glu in the structure of peptidoglycan also provides resistance against peptidase and protease attacks [2, 4]. Producing D-Ala by spore-associated AlaR can be implicated in regulating germination, which ultimately causes germination suppression under unfavorable conditions or sporulation induction in response to the presence of D-AAs [4, 10, 11]. In addition, the cyclic tetrapeptide HC-toxin, an essential virulence determinant for the plant pathogenic fungus *Cochliobolus carbonum* and an inhibitor of histone deacetylase, contains D-Ala and D-Pro that D-Ala is synthesized by the AlaR encoded by the toxG locus of *C. carbonum* [12]. Moreover, cyclosporin A, the widely used immunosuppressant drug in postallogenic organ transplant, is also a D-Ala containing cyclic peptide, which was produced by the AlaR isolated from fungus *Tolypocladium inflatum* [13]. Research study showed that an amino acid racemase found in the fruit-body of a basidiomycetous mushroom, *Lentinus edodes* (Shiitake), is responsible for the existence of D-alanine in the fruit-body. Furthermore, this racemase is able to catalyze the racemization of not only D-alanine but also D-serine, D-homoserine, D-2-aminobutyrate, D-glutamate, and D-asparagine [14].

SerR (EC 5.1.1.18) just like AlaR is a bifunctional enzyme because it catalyzes the racemization of serine and the dehydration of serine to produce pyruvate and ammonia. Thus, the catalysis of the reversible conversion of L- to D-Ser and serine catabolism by α,β-elimination of water by SerR is responsible for the regulation of D-Ser levels in the body [15]. The bacterial serine racemase VanT of *Enterococcus gallinarum* is related to the resistance of the glycopeptide antibiotic vancomycin [16, 17]. Vancomycin inhibits peptidoglycan synthesis by the formation of a complex with the D-Ala-D-Ala residue of the peptidoglycan precursors [18]. Resistance to vancomycin is due to the modification of D-Ala-D-Ala residue to a D-Ala-D-Ser residue by SerR (VanT) that causes a low affinity of the peptidoglycan precursor to vancomycin [19]. VanT serine racemase is also known as the only bacterial racemase to possess a transmembrane domain, which is supposedly involved in the transport of L-Ser from the external medium [20]. Literature report shows that mammalian SerR, the primary enzyme responsible for brain D-Ser production, acts as a D-Asp biosynthetic enzyme in some organs and/or tissues. However, evidences also indicate that there should be some additional enzyme for D-Asp synthesis in mammals [21].

The broad substrate specificity arginine racemase (ArgR, EC 5.1.1.9) was first isolated from *Pseudomonas graveolens* as early as 1971. This PLP-dependent racemase is able to catalyze the racemization of arginine, lysine, ε-*N*-acetyllysine, ornithine, 2,3-diaminopropionate, homoarginine, 2,4-diaminobutyrate, ethionine, citrulline, homo citrulline, δ-*N*-acetylornithine, theanine, glutamine, and methionine, except those hydrophobic, acidic, or aromatic amino acids [22, 23]. Research reports also revealed that ArgR found in *Pseudomonas taetrolens*, *Escherichia coli*, and *Pseudomonas putida* KT2440 presents in the periplasm of the cell, a feature so far unknown for any other amino acid racemase, and is a catabolic enzyme needed for the efficient utilization of D-Lys and D-Arg [24, 25].

PLP-dependent aspartate racemase (DR) has been found in mammals including humans, which converts L-aspartate to D-aspartate [26, 27]. Since D-Asp has been detected in the brain and neuroendocrine tissues, D-Asp plays a crucial role for neurotransmission and neurosecretion in the central neuro system, as well as for the biosynthesis or secretion of hormones in the endocrine glands [2, 28]. Evidences have shown that DR is important in neuronal development, consistent with the high levels of D-Asp in early neonatal stages [2, 26].

6.1.1.2 D-Amino Acid Synthesis with PLP-Independent Amino Acid Racemases

This class of racemases includes proline racemase (ProR), aspartate racemase (AspR), and glutamate racemase (GluR), as well as diaminopimelate epimerase (DAPE). The racemization mechanism of these racemases is through a "two-base" mechanism without involving the PLP coenzyme, which was first proposed in the study of ProR catalyzed interconversion of L-proline and D-proline. Based on the isotope incorporation and enzyme kinetics in D_2O, the "two-base" mechanism demonstrates that one base of the enzyme removes the substrate α-hydrogen

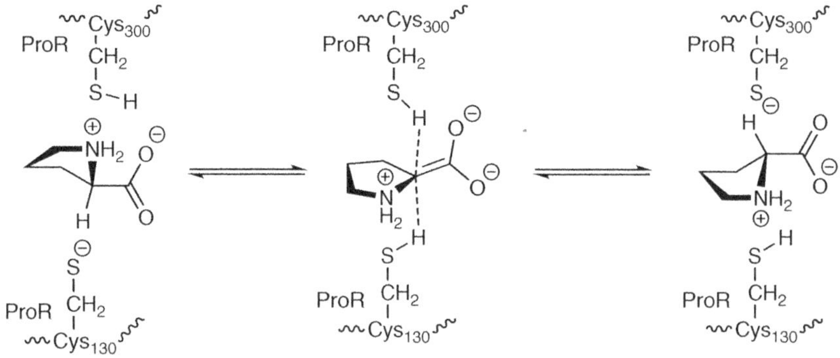

Scheme 6.2 Proline racemase catalyzed interconversion of L-proline to D-proline with a "two-base" mechanism. *Source:* Modified from Conti et al. [2].

as a proton to generate a carbanion intermediate, while the conjugate acid of another base donates a proton to the opposite face of the α-carbon of the intermediate. In this mechanism, no deuterium exchange can be observed from the enzyme–substrate complex indicating that the two bases of the catalytic site are monoprotic. The study also showed that the active site is located at the interface of two identical, or nearly identical, subunits. Each subunit provides one base which is identified as the thiol group of two cysteines (Scheme 6.2) [2]. In addition, there are two forms of ProR; one is able to bind L-proline and the other solely binds D-proline [29, 30].

The first eukaryotic ProR was purified from the human parasite *Trypanosoma cruzi* (*Tc*), which is responsible for the etiological agent of Chagas disease. Later, ProR also has been found in the livestock trypanosomes *Trypanosoma vivax* (*Tv*). The presence of ProR in the genome of *Tc* and *Tv* makes both parasites a way to escape the host immune responses. The substitution of the L-Pro residue with its enantiomer D-Pro into the peptide sequence may increase the half-life, stability, and resistance of *Tc* and *Tv* to host proteases [2, 31].

Glutamate racemase (GluR, EC5.1.1.3) is another PLP-independent racemase that was isolated as early as 1952 in prokaryotes [32]. GluR catalyzes the reversible racemization of L-glutamate and D-glutamate with high specificity and uses a "two-base" mechanism [33]. D-Glutamate as for D-alanine is a key component of the peptidoglycan of the bacterial cell wall in a number of pathogenic organisms that protects them against osmotic lysis, thus is essential for their viability. The rise of antibiotic resistance in pathogenic organisms has led GluR to a target for the development of antibacterial drugs [34, 35].

The PLP-independent aspartate racemase (AspR, EC 5.1.1.13) that catalyzes the racemization of L-aspartate to D-aspartate via a "two-base" mechanism was

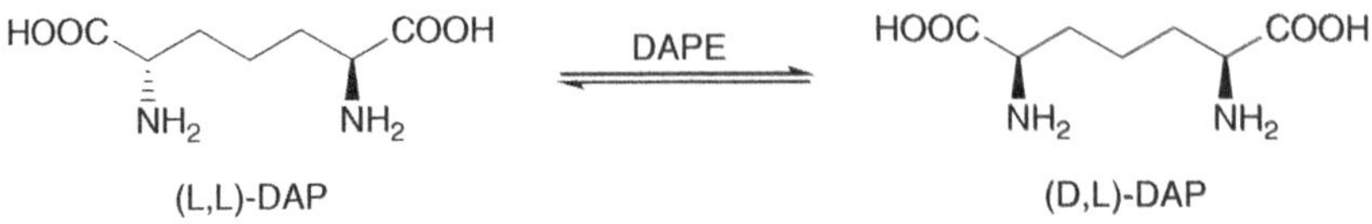

Scheme 6.3 Diaminopimelate epimerase catalyzed stereoconversion of (L,L)-2,6-diaminopimelate to meso-2,6-diaminopimelate. *Source:* Conti et al. [2].

partially purified from *Streptococcus faecalis* in 1972 [2, 36]. Later, the racemization of aspartate was found exclusively in lactic acid bacteria, such as *Streptococcus* species and *Lactobacillus* species, which contain D-Asp in the inter-peptide bridges of their peptidoglycan layers [2, 4]. In the fresh-water cyanobacterium *Microcystis aeruginosa*, the presence of D-Asp in the nonribosomal peptide microcystin was found by the action AspR [4].

Diaminopimelate epimerase (DAPE, EC 5.1.1.7) catalyzes the interconversion of (L,L)-2,6-diaminopimelate (DAP) into meso-2,6-diaminopimelate and is a member of the PLP-independent amino acid racemases (Scheme 6.3) [2]. The meso-2,6-DAP is a precursor of L-lysine and is a key component of the pentapeptide linker in bacterial peptidoglycan [37]. DAPE has been partially purified from *E. coli* in 1957; however, its purification and characterization were implemented until 1984 [38, 39]. DAPE shows its uniqueness by recognizing only the L-configuration at the distal site of the two stereoisomers of (L,L)- and (L,D)-DAP. Although DAPE is involved in the lysine biosynthesis in plants and bacteria, the absence of lysine synthesis in animals makes DAPEs attractive as targets for the rational design of antibiotics.

6.1.1.3 Amino Acid Racemases with Broad Substrate Specificity

In addition to amino acid racemases with high substrate specificities to a single amino acid, such as AlaR, SerR, ProR, GluR, AspR, and DAPE, there is a family of PLP-dependent amino acid racemases (EC 5.1.1.10) exhibit broad substrate specificity that can be collectively called as broad-spectrum racemases (BSRs) [3, 24, 25, 40–44]. The BSR shown in *P. putida* KT2440 catalyzes the racemization of lysine, arginine, alanine, and hydroxyproline, while the BSR in *Pseudomonas taetrolens* shows the racemization ability for lysine, arginine, ornithine, and alanine. D-Amino acids formed by BSR in both *Pseudomonas* sp. can be further catabolized via a unique pathway [24, 41–43]. Noncanonical D-amino acids (i.e. amino acids different from D-Ala and D-Glu in peptidoglycan), such as D-Met and D-Leu, have been found to accumulate at millimolar concentrations in extracellular medium from *Vibrio cholerae*. To date, *V. cholerae* has been proved to be a broad-spectrum amino acid racemase that enables the racemization of aliphatic and basic amino acids as well as several nonproteogenic amino acids including norleucine,

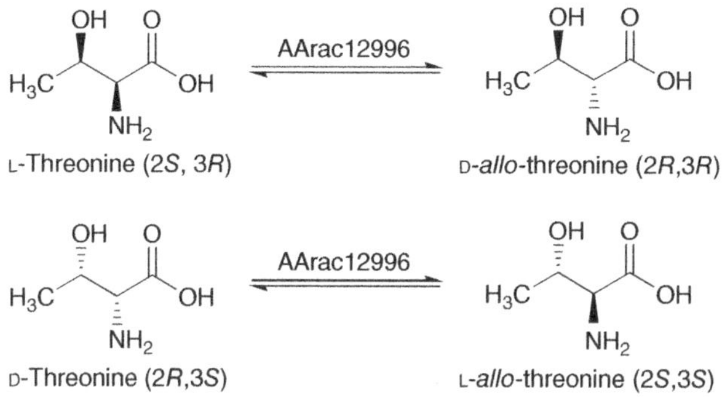

Scheme 6.4 Amino acid racemase catalyzed isomerization of L- or D-threonine at α-C atom to form corresponding diastereomer D-*allo*- or L-*allo*-threonine. *Source:* Based on Würges et al. [46]; Kino et al. [49]; Yagasaki and Ozaki [53].

homoserine, and diaminobutyrate [3, 44]. The roles of the noncanonical D-amino acids in microbial physiology include the modulation of cell-wall structure and the dissolution of biofilms [44, 45].

allo-Threonine (*allo*-Thr) or D-*allo*-Thr is important in the pharmaceutical industry either used as the starting material or used for chemical synthesis of key intermediates to produce potent antibiotics [46]. Although the enzymatic preparation of L-*allo*-Thr using threonine aldolase from *Aeromonas jandaei* DK-39 was described before, the first report for the enzymatic isomerization of threonine using amino acid racemase from *P. putida* ATCC17642 showing threonine-α-epimerase activity (EC 5.1.1.6) was reported in 1993 [47, 48]. Whereas an efficient enzymatic method for the production of very high purity D- and L-*allo*-Thr from inexpensive L- or D-Thr can be obtained using a purified amino acid racemase (AArac12996) from *P. putida* NBRC12996 expressed in *E. coli* (Scheme 6.4) [46, 49]. Crystallization of D- and L-*allo*-Thr was simultaneously coupled to the enzymatic isomerization and performed in a repetitive batch mode to give 30.8 g D-*allo*-Thr and 32.4 g L-*allo*-Thr with very good diastereomeric excess of $de_{D\text{-allo}} > 99.2\%$ and $de_{L\text{-allo}} > 98.4\%$, respectively; the yields were also good with 83% D-*allo*-Thr and 79% L-*allo*-Thr [45]. AArac12996 is a broad-spectrum racemase, which is comparable to those other PLP-dependent amino acid racemases [50].

Isoleucine 2-epimerase (ILEP, EC 5.1.1.21) purified from expressed recombinant *E. coli* of *Lactobacillus buchneri* JCM 1115 is a novel branched-chain amino acid racemase (BCAAR), which catalyzed the racemization of a broad spectrum of nonpolar amino acids, in particular, the epimerization of L-isoleucine to D-*allo*-isoleucine and D-*allo*-isoleucine to L-isoleucine [51]. Further study demonstrated that marked accumulation of D-branched-chain amino acids (D-BCAAs) such as D-valine, D-leucine, and D-*allo*-isoleucine was obtained in the growth medium with the lactic acid bacterium *Lactobacillus otakiensis* JCM 15040. The recombinant BCAAR is a PLP-dependent enzyme that catalyzes the racemization of BCAAs as its main substrate. Research results indicated that a variety of lactic acid bacteria carrying homologs showing 53–60% amino acid sequence identity to the *L. buchneri* ILEP produce D-BCAAs [52].

6.1.1.4 Applications of Amino Acid Racemases and Epimerases

Optically pure D-amino acids are becoming important as intermediates for the production of pharmaceuticals, food additives, and agrochemicals. The industrial manufacturing of D-amino acids is favored by the availability of inexpensive starting material D,L-amino acids or L-amino acids produced by chemical synthesis or fermentation. An efficient one-pot fermentation using enzymatic cascade reactions together with pH control has been developed for the production of D-glutamic acid and D-proline. In this strategy, L-glutamate was converted to the racemic DL-glutamate by the recombinant glutamate racemase of *Lactobacillus brevis* ATCC8287. Subsequently, L-glutamic acid in the racemic mixture was converted to γ-aminobutyric acid (GABA) by the selective decarboxylation using L-glutamate decarboxylase of *E. coli* ATCC11246 (Scheme 6.5) [53]. The same strategy was applied to the production of D-proline. That is, L-proline was first converted to D,L-proline mixture by the recombinant proline racemase of *Clostridium sticklandii* ATCC12262, then, L-proline in the racemic mixture was degraded by *Candida* sp. PRD-234 [53].

The continuous production of enantiopure D-methionine from the racemic D,L-methionine mixture with a yield of 93.5% can be obtained by the integration of an enantioseparation unit of chiral simulated moving bed (SMB) chromatography, a mild enzymatic racemization unit of enzyme membrane reactor, and a nanofiltration unit for recycling concentration of the undesired L-methionine in SMB raffinate prior to racemization. This integrated continuous production of D-methionine ran with an aqueous-organic solvent system to allow dynamic kinetic resolution of racemic D,L-methionine and overcome the 50% yield

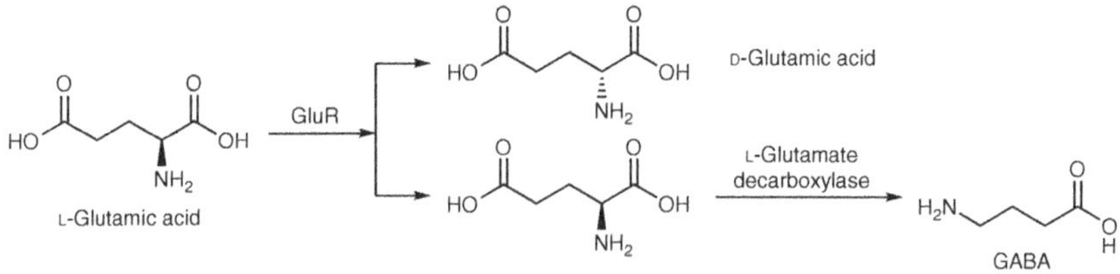

Scheme 6.5 The production of D-glutamic acid from L-glutamic acid by enzymatic cascade reaction using glutamate racemase and L-glutamate decarboxylase. *Source:* Based on Femmer et al. [56]; Schnell et al. [57].

limitation. The amino acid racemase with a broad spectrum of the substrate (EC 5.1.1.10) was from *P. putida* DSM 3263 showing high stability in this implementation [54, 55].

6.1.2 Hydroxy Acid Racemases and Epimerases

6.1.2.1 Lactate Racemase

Lactate racemase (LarA, EC 5.1.2.1) is a microbial enzyme that until now it is only known to catalyze the racemization between D- and L-lactic acid (Scheme 6.6) [56, 57]. The isolation of this enzyme was first reported for two intact cells of *Clostridium* species in 1936 [58]. Later, this enzyme was isolated and characterized from a number of bacteria, such as *Lactobacillus curvatus*, *Lactobacillus sakei*, *Haloferax volcanii*, *Clostridium butylicum*, and *Lactobacillus plantarum* [59–63]. The early studies on the reaction mechanism of lactic acid racemization indicated a direct internal hydride transfer mechanism [61, 62]. Recent studies on the reaction mechanism of lactic acid racemization show a nickel-pincer cofactor dependent proton-coupled hydride transfer (PCHT) mechanism [64–67].

Since the strict substrate specificity of this enzyme, very few applications were implemented with the lactate racemase. The only interesting application with lactic acid production should be its further usage for manufacturing the biodegradable high-quality poly-L-lactic acid (PLA). During the open fermentation of kitchen refuse under nonsterilized conditions for the production of L-lactic acid with L-lactic acid-producing strains, it has been found that the selective proliferation of naturally existing *L. plantarum*, a predominant lactic acid bacteria in the fermented kitchen refuse, catalyzed the racemization of L-lactic acid to D-lactic acid to cause a decreased optical purity in the accumulated lactic acid [68]. However, an engineered *L. plantarum* by double-knockout of the D-lactate dehydrogenases (*ldhD*) gene and the lactate racemase operon (*larA-E*) can completely abolish D-lactic acid production. Then, the Δ*ldhD* Δ*larA-E* mutant can be directly used for the exclusive production of L-lactic acid with high yield and high optical purity. Subsequently expressing α-amylase into the Δ*ldhD* Δ*larA-E* mutant allowed the direct L-lactic acid production from raw corn starch with a yield of 91% g/(g consumed sugar) and optical purity of 98.6% [69].

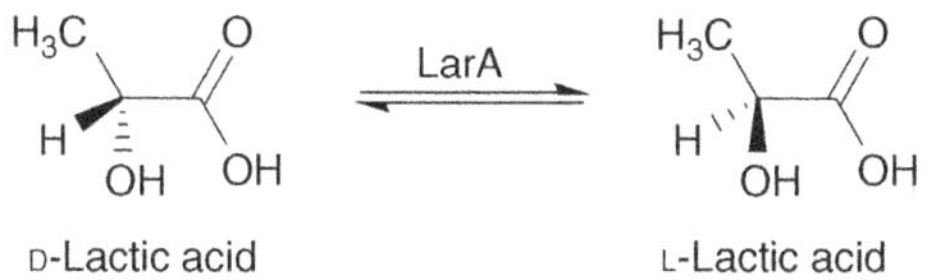

Scheme 6.6 Lactate racemase (LarA) catalyzed racemization of D- and L-lactic acids. *Source:* Based on Schnell et al. [57]; Felfer et al. [70].

6.1.2.2 Mandelate Racemase

Mandelate racemase (EC 5.1.2.2, MR) produced from the soil bacterium *Pseudomonas putida* ATCC 12633 is an inducible divalent metal ion-dependent but a cofactor-independent enzyme that catalyzes the interconversion of (*R*)- and (*S*)-enantiomers of mandelic acid (Scheme 6.7) [57, 70]. The divalent metal ions required for the enzyme activity are Mg^{2+}, Mn^{2+}, Fe^{2+}, Co^{2+}, Ni^{2+}, and Mg^{2+} is the most effective ones [71]. The catalytic mechanism of MR is a so-called "two-base mechanism" as elucidated by X-ray crystal structure and function studies which is feasible through the synergistic actions of chemical operators within the hydrophobic binding active site of MR on the substrate [72–76]. In this mechanism, the abstraction of the α-proton of mandelate by two adjacent enantiomer-specific basic amino acid residues on MR at either side of the chiral α-carbon atom of mandelate, viz., His-297 and Lys-166 for (*R*)- and (*S*)-mandelate, respectively, that one base acts as a base for the deprotonation and the other functions as an acid for protonation from the opposite side to result in an efficient racemization [77, 78]. During the interconversion of mandelate enantiomers, the acidity of the α-proton possesses a p*K*a value of 29 or smaller that is acid enough to induce an abstraction to give the corresponding achiral enolate intermediate [79, 80]. The inducible octameric enzyme (subunit 39 kDa) can be produced in large amounts by a ten-liter batch-cultivation of *P. putida* ATCC 12633 using *rac*-mandelate as an inducer or using a cloned *Pseudomonas aeruginosa* ATCC 15692 cells [81, 82].

Mandelate racemase is a "broad substrate spectrum" catalyst. As early in 1970, it has been found that in addition to its natural substrate mandelate, structure analogous of mandelate with *para*-monosubstituted phenyl groups was also reactive substrates, particularly, the *p*-BrC_6H_4 and *p*-ClC_6H_4 substituents [83]. Later, vinyl glycolate (2-hydroxy-3-butenoic acid) was found a good substrate of MR with a maximal racemization rate of 35% relative to mandelate [84]. This result also indicates the importance of β,γ-unsaturation in promoting MR catalyzed racemization [70]. The enzymatic racemization of mandelic acid derivatives modified at the α-hydroxy acid moiety, such as the mandelic acid amide, was also performed at an acceptable rate using MR. The racemization was significantly

Scheme 6.7 Mandelate racemase catalyzed reversible interconversion of the (*R*)- and (*S*)-enantiomers of mandelate.

enhanced by an electron-withdrawing substituent in the phenyl moiety, which could be explained by steric and electronic reasons [85]. The steric limitations for aryl-substituted mandelate derivatives were particularly striking for *o*-substituents, whereas *m*- and *p*-substituents were freely accepted [86]. Furthermore, the aromatic group of mandelic acid can be extended to a naphthyl group and still with good racemization activity; heteroaromatic analogs, such as thienyl and furyl analogs, are accepted by MR as well [86].

The use of MR for the preparative deracemization of (±)-mandelic acid to overcome the 50% conversion limitation of the kinetic resolution was demonstrated by using a stepwise lipase-mandelate racemase two-enzyme system. The initial step of the process is the *Pseudomonas* sp. lipase-catalyzed enantioselective *O*-acylation of *rac*-mandelic acid in diisopropyl ether using vinyl acetate as the acyl donor to form (*S*)-*O*-acetyl mandelate. After the reaction, the enzyme was filtered off and the organic solvent as well as excess vinyl acetate was removed by evaporation. Then, the remaining unreacted (*R*)-mandelic acid was catalyzed by immobilized MR *in situ* with the (*S*)-*O*-acetyl mandelate in an aqueous buffer. The enzyme and the aqueous buffer were removed by filtration and lyophilization, respectively, and the reaction cycle was resumed (Scheme 6.8). After four cycles, (*S*)-*O*-acetyl mandelic acid obtained was in 80% isolated yield and >98% *e.e.* as the sole product [57, 70, 87].

Similar one-pot enzymatic cascade strategy based on sequential hydrolysis and racemization was employed to improve the enantioselectivity of esterase in order to fulfill the requirement in industrial production of methyl (*R*)-*o*-chloromandelate (*R*-CMM). *R*-CMM is a key intermediate for the production of clopidogrel, a platelet aggregation inhibitor. In this strategy, both the recombinant esterase BioH and mandelate racemase were overproduced and purified from *E. coli* strain BL21. Using the racemic (*R*,*S*)-CMM as the substrate, *S*-CMM is enantioselectively

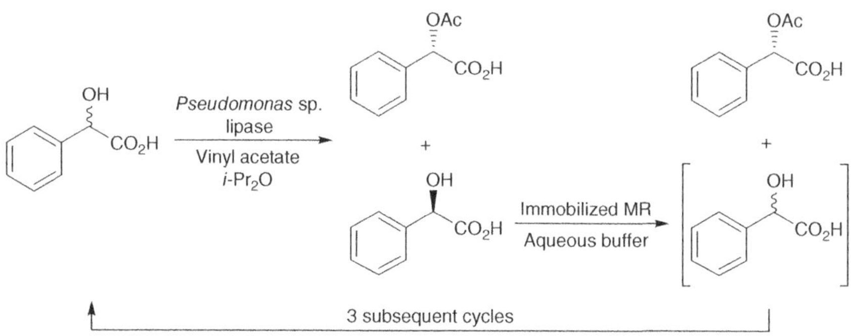

Scheme 6.8 Deracemization of *rac*-mandelate using a stepwise lipase-mandelate racemase two-enzyme system.

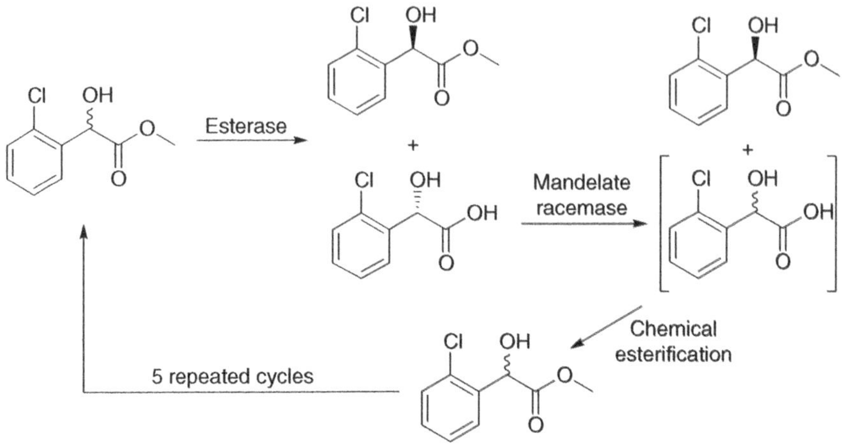

Scheme 6.9 One-pot enzymatic cascade production of methyl (*R*)-*o*-chloromandelate based on sequential hydrolysis and racemization.

hydrolyzed into (*S*)-*o*-chloromandelate (*S*-CM), while the target compound *R*-CMM remains almost unchanged. Subsequently, *S*-CM is catalyzed by the MR to form a racemic (*R*,*S*)-CM mixture. The target compound *R*-CMM was recovered with an *e.e.* value of >97%. The extraction raffinate was then adjusted to pH 2 and extracted with ethyl acetate again to retrieve the racemic (*R*,*S*)-CM mixture, which was further chemically esterified and used as the starting material (Scheme 6.9). When the process was repeated five times, the yield of *R*-CMM can reach 88% [88].

Since enantiopure D-phenylglycine and its derivatives are an important group of chiral chemicals for manufacturing pharmaceuticals and other bioactive compounds such as β-lactam antibiotics and inhibitors, recombinant *E. coli* (LZ110) was engineered to coexpress four enzymes, mandelate racemase, (*S*)-mandelate dehydrogenase (SMDH), D-phenylglycine aminotransferase (DpgAT), and glutamate dehydrogenase (GluDH), to catalyze a one-pot three-step enzymatic cascade transformation for the enantioselective synthesis of D-phenylglycine from racemic mandelic acids (Scheme 6.10). In the third DpgAT catalyzed step, L-glutamate was used as the amine donor to form α-ketoglutarate and was regenerated by GluDH using NH_3. The produced enantiopure D-phenylglycine was in 29.5 g L^{-1} with 99% *e.e.* and 93% conversion. In addition, 12 other D-phenylglycine derivatives were also produced from corresponding mandelic acid derivatives in high conversion (58–94%) and very high *e.e.* (93–99%) using the same enzymatic cascade strategy [89]. This one-pot three-step enzymatic cascade process utilizes cheap, green, and easily available racemic mandelic acid, O_2, NH_3, and glucose as substrate and reagents to produce enantiopure D-phenylglycine in high yield, thus representing one of the most efficient examples of biocatalytic conversion of an

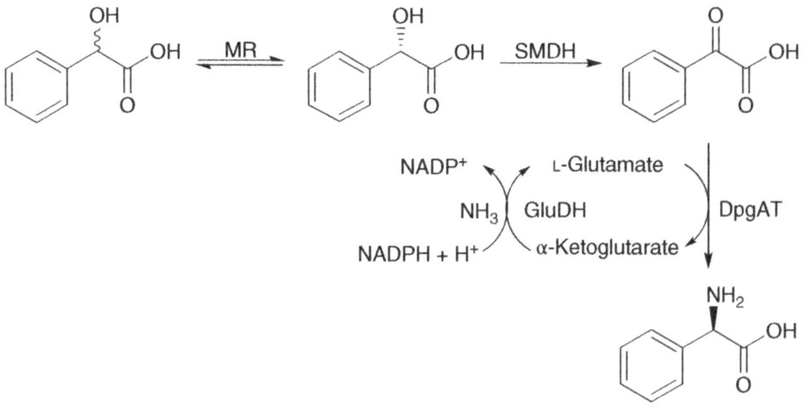

Scheme 6.10 One-pot three-step enzymatic cascade biotransformation for enantioselective synthesis of D-phenylglycine from racemic mandelic acid.

OH group to an NH_2 group. A similar three-step one-pot domino reaction system was established by employing a mandelate racemase from *P. putida*, a novel NAD^+-dependent D-mandelate dehydrogenase from *Lactobacillus brevis*, and a leucine dehydrogenase from *Exiguobacterium sibiricum* for the efficient preparation of L-phenylglycine starting from racemic mandelate acid with internal cofactor recycling to obtain 96.4% conversion, 86.5% isolation yield, >99% *e.e.*, and 50.4 g L^{-1} d^{-1} space-time yield [90].

Benzoylformate (BA) is an important organic synthetic intermediate for thromboxane synthetase inhibitors and antihypertensive agents. Since racemic mandelate is less expensive than BA, the genes encoding mandelate racemase and mandelate dehydrogenase from *P. aeruginosa* strain NUST were cloned and expressed in *E. coli* which was used for producing BA from racemic mandelate. In this process, (*R*)-mandelate is converted to its enantiomer (*S*)-mandelate by MR, then (*S*)-mandelate is specifically oxidized to BA using FMN-dependent *S*-mandelate dehydrogenase. The results showed that the use of whole resting cells of recombinant *E. coli* allowed the conversion of 65.7 mM racemic mandelate to BA at 98.9% yield in 45 h reaction run [91]. One-pot enantioconvergent synthesis of (*R*)-mandelic ethyl ester can be achieved by a novel aqueous/organic two-phase DKR process comprising lipase-catalyzed enantioselective esterification of (*R*)-mandelic acid in the racemic mandelic acid mixture in ethylene dichloride and recombinant *E. coli* cells containing mandelate racemase mediated *in situ* racemization of unreacted enantiomer (*S*)-mandelic acid in the aqueous buffer to (*R*)-mandelic acid within a preparative scale hollow-fiber membrane bioreactor (Scheme 6.11). During the operation of the biocatalytic enantioconvergent process,

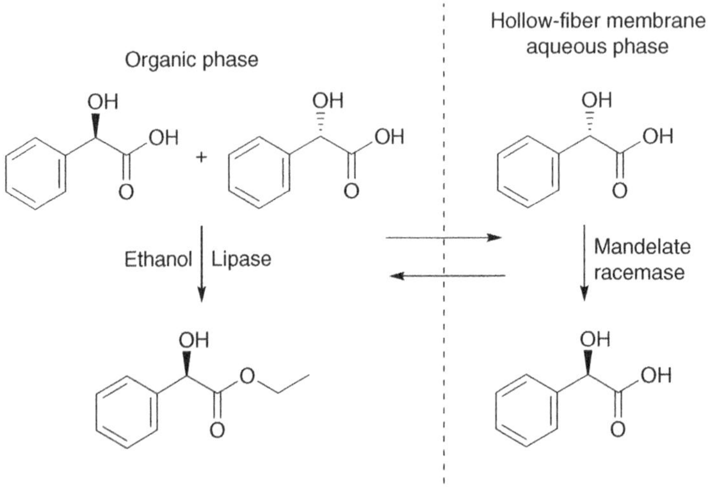

Scheme 6.11 Enantioconvergent synthesis of (*R*)-mandelic ethyl ester catalyzed by lipase-mandelate racemase in aqueous/organic two-phase membrane reactor.

rac-mandelic acid dissolved in an aqueous buffer was circulated through the hydrophilic cellulose membrane lumen side and can be partitioned into the lipase-containing organic phase countercurrent flowing in the shell side of the hollow-fiber bioreactor. In the organic phase, *rac*-mandelic acid was enantioselectively esterified with ethanol by lipase to give ethyl (*R*)-mandelate, which is preferentially soluble and accordingly accumulated in the organic phase. Simultaneously, unreacted (*S*)-mandelic acid was accumulated in the aqueous buffer phase and racemized into *rac*-mandelic acid by recombinant racemase. As a result, the isolated yield of (*R*)-mandelic ethyl ester obtained was 65% with 98% *e.e.* [92, 93].

In order to broaden the applicability of MR and make it more economic, the development of an immobilization method to allow MR recyclable is necessary. The activity of MR from *P. putida* ATCC 12336 was significantly enhanced by immobilization through ionic binding onto DEAE- and TEAE 23-cellulose as compared to the native enzyme. DEAE-cellulose-immobilized MR can be efficiently and repeatedly used for the batch reaction in the racemization of (*R*)-mandelic acid under mild conditions [94]. MR from *P. putida* ATCC 12336 was further immobilized in lyotropic liquid crystal consisting of aqueous buffer, di-*n*-butyl ether, and a nonionic surfactant Brij® 35, which can be activated in organic solvent di-*n*-butyl ether and used for the two-phase batch racemization of (*R*)-mandelic acid [95]. Immobilization of crude MR via covalent bonding onto commercially available support Eupergit CM was used to establish an enzymatic fixed-bed reactor to perform the continuous racemization of (*R*)-mandelate in the racemic mandelate mixture to (*S*)-mandelate. The MR-packed fixed-bed reactor was allowed to couple to an enantioselective preparative chromatographic column for separating the two enantiomers of mandelic acid. Running the combined chiral chromatography and the enzymatic fixed-bed reactor system over several cycles allows the provision of (*S*)-mandelic acid exceeding the 50% yield limitation by conventional kinetic resolution [96, 97].

6.1.3 *N*-Acetyl-D-glucosamine 2-epimerase

In enzymology, the systematic name of *N*-acetyl-D-glucosamine 2-epimerase (AGE) is *N*-acylglucosamine 2-epimerase (EC 5.1.3.8), which is an enzyme that catalyzes the reversible conversion of *N*-acetyl-D-glucosamine (GlcNAc) to *N*-acetyl-D-mannosamine (ManNAc) (Scheme 6.12) and plays a key role in sialic acid metabolism [98–100]. This enzyme has been proved as ATP-dependent epimerase, which ATP is an allosteric activator for its activity [98, 101]. The tertiary structure of an AGE purified from *Anabaena* sp. CH1 solved by X-ray crystallography shows a homodimer having an $(\alpha/\alpha)_6$ barrel core structure. The concerted action of the conserved residues His239 (deprotonation) and His372 (protonation) at the active-site cleft catalyzes the reversible epimerization at the C2 position of GlcNAc with an acid/base ping-pong mechanism [100]. The physiological role of AGE in mammalian cells, such as pigs, rats, and humans, has been related to be the regulation of the renin-angiotensin system [102].

N-Acetyl-D-neuraminic acid (Neu5Ac), the most ubiquitous species among the sialic acids, is widely distributed in the animal kingdom as the terminal positions of glycoproteins and glycolipids. Neu5Ac plays important role in various biological functions by acting as receptors for microorganisms, viruses, toxins, and hormones, by masking receptors, and by regulating of the immune system. Therefore, Neu5Ac has been attracted to the development of new types of influenza drugs that can be used to inhibit neuraminidase of the influenza virus [103]. There are a variety of methods to prepare Neu5Ac. The enzymatic method for preparing Neu5Ac from ManNAc and pyruvate using *N*-acetyl-neuraminate acid lyase (NAL) as a catalyst has been reported [104–106]. While ManNAc is very expensive

Scheme 6.12 Reversible conversion of *N*-acetyl-D-glucosamine to *N*-acetyl-D-mannosamine catalyzed by *N*-acetyl-D-glucosamine 2-epimerase (AGE). *Source:* Based on Altenbuchner et al. [131]; Bommarius et al. [133]; Ishikawa et al. [137].

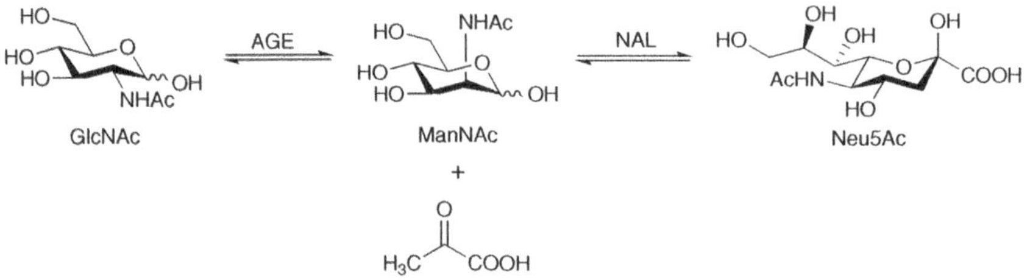

Scheme 6.13 Synthesis of Neu5Ac from GlcNAc with a two-step enzymatic cascade reaction.

and not readily available for large-scale production, an enzymatic cascade reaction using inexpensive GlcNAc and pyruvic acid in the presence of AGE and NAL was developed for the production of Neu5Ac (Scheme 6.13). Therefore, an enzyme membrane reactor with AGE and NAL was developed for the synthesis of Neu5Ac using excess pyruvate to obtain 90% yield but with a large amount of unreacted pyruvate [107]. A fatal defect of this method was the provision of AGE in an efficient amount for an unrestricted scale of the reaction. To solve this difficulty, the gene of AGE from porcine kidney and gene of NAL from *E. coli* K-12 had been cloned and overexpressed in *E. coli*. After feeding appropriate amounts of pyruvate to the reaction mixture, a high conversion of NeuAc from GlcNAc with a 77% conversion rate on a molar base can be obtained and the product was recovered directly using crystallization [108]. The same approach was also carried out by a small-scale stirred glass vessel containing recombinant *E. coli* cells expressing AGE gene from *Anabaena* sp. CH1 and NAL gene from *E. coli* NovaBlue separately and the conversion yield obtained was 33.3% in a 12-h reaction. However, it was found that the recombinant *E. coli* cells can be reused for more than eight cycles with a productivity of >8.0 g Neu5Ac L^{-1} h^{-1} [109]. A similar strategy used to produce Neu5Ac from GlcNAc was performed by overexpressing AGE from a phototrophic cyanobacterium *Synechocystis* sp. PCC6803 and Neu5Ac synthetase from *E. coli* K-1 (ATCC 13027) in *E. coli* NM522. While the necessary substrate phosphoenolpyruvate (PEP) was generated by glucose fermentation of bacterium *Corynebacterium ammoniagenes* DN510 and cofactor ATP were supplied by the activity of *E. coli*. Neu5Ac accumulated after 22 h reaction was at 39.7 mM (12.3 g L^{-1}) using 800 mM GlcNAc and 360 mM glucose initially [110].

Nowadays, a variety of methods have been developed in order to enhance the production of Neu5Ac employing the cascade reactions strategy shown in Scheme 6.13. Genes of AGE and NAL were separately expressed in *E. coli* cells under the control of a temperature-responsive promotor, which is safe compared to chemical-induced systems and coupled for Neu5Ac production from GlcNAc. When 2200 mM pyruvate was used as a carbon source and substrate, 61.9 mM

(19.1 g L^{-1}) Neu5Ac was formed using 200 mM GlcNAc in 36 h [111]. In order to improve the problems of lengthy reaction time and the low conversion rate of the two-step reaction, a novel and simple one-pot synthesis of Neu5Ac from GlcNAc using immobilized *E. coli* coexpressed AGE and NAL on Amberzyme oxirane resin was developed, which gave a conversion rate of ~73% within 24 h. Furthermore, the immobilized enzymes could be used up to five reaction cycles without loss of activity or a significant decrease of the conversion rate [112]. Synthesis of Neu5Ac from GlcNAc and pyruvate was performed by constructing and simultaneously expressing a polycistronic plasmid encoding an AGE gene and a NAL gene in *E. coli* Rosetta (DE3) pLysS. The plasmid, pET-28a-NAL-anAGE with anAGE gene located next to the NAL gene caused the highest amount of Neu5Ac production, which generated 61.3 g L^{-1} in 60 h using whole-cell catalysis without the addition of ATP for AGE activation [113]. Multiple approaches were used to engineer *E. coli* strain to enhance the two-step enzymatic cascade production of Neu5Ac from GlcNAc, first, by increased expression of NAL, second, by knocking out *nan*TEK genes to keep the reactions to the synthetic direction to the final product, and third, improving the performance of the system by promoting substrate transport and optimizing substrate concentrations. The whole-cell biocatalytic reaction performed by the engineered *E. coli* strain produced 240 mM Neu5Ac (74.2 g L^{-1}) with the productivity of 6.2 g Neu5Ac L^{-1} h^{-1} and a conversion yield of 40% from GlcNAc. The engineered strain could even be reused at least five cycles with a productivity >6 g L^{-1} h^{-1} [114].

A novel AGE derived from the cyanobacteria *Anabaena variabilis* ATCC 29143 was recombinantly expressed in *E. coli* and employed to overcome the disadvantages (low specific activity, substantial inhibition by pyruvate, and strong dependence on allosteric activator ATP) of the epimerization of GlcNAc to ManNAc in the one-pot two-enzyme cascade reactions for producing Neu5Ac. The results indicated a high activity of 117 ± 2 U mg^{-1} and a much weaker inhibitory effect of pyruvate as compared to other AGEs. The ATP effect was more on the substrate than on the enzyme activity [115]. Recombinant *E. coli* strains by over expressing AGE from *Synechocystis* sp. and NAL from *E. coli*, respectively, were developed and the activity ratio was optimized by varying recombinant amounts of cell biomass for synthesis Neu5Ac. In addition, substrate concentrations and the ratio of pyruvate and GlcNAc and detergent concentrations were optimized to enhance the production of Neu5Ac. The resulting whole-cell process generated 237.4 mM Neu5Ac with a yield of 40.0% mol/mole GlcNAc. Furthermore, transporter pathways involved in Neu5Ac and GlcNAc were engineered to give the overall whole-cell biocatalytic process a maximum titer of 260.0 mM Neu5Ac (80.4 g L^{-1}) with a conversion yield of 43.3% from GlcNAc, which is suitable to be used for industrial large-scale production of Neu5Ac [116]. A multienzyme complex of NAL from *Corynebacterium glutamicum* ATCC 13032 (CgNAL) and AGE from

Anabaena sp. CH1 (anAGE), polycistronic plasmid with high levels of CgNAL and anAGE expression was constructed by tuning the orders of the genes. The Shine-Dalgarno (SD) sequence and aligned spacing (AS) distance were optimized. The coexpression of both NAL and AGE as multienzyme complex in *E. coli* Rosetta (DE3) by harboring the polycistronic plasmid pET-28a-Sd$_2$-AS$_1$-CgNAL-SD-AS-anAGE increased the production of Neu5Ac by 58.7% to 92.5 g L^{-1} in 36 h using whole-cell catalysis and further increased by 21.9% up to 112.8 g L^{-1} in 24 h with the addition of Triton X-800 [117]. The strategy that combines the expression of the two enzymes, AGE and NAL, balanced by promoter engineering, genes encoding competing pathways by deleting GlcNAc catabolism, and a structure-guided process was used to identify a more efficient NAL and an AGE in *E. coli* mutant with the higher production rate of Neu5Ac. The whole-cell biocatalytic system gave 351.8 mM Neu5Ac with a yield of 58.6% from GlcNAc [118]. Recombinant *E. coli* strain, *E. coli* BL21 (DE3) pET-bage-2nanA, synchronously expressing *Anabaena* sp. CH1 *N*-acetyl-D-glucosamine-2-epimerase (bage), and *E. coli* *N*-acetyl-D-neuraminic acid aldolase (nanA) were used as biocatalysts to produce Neu5Ac from GlcNAc and pyruvate. The reaction generated 412.6 mM Neu5Ac with a GlcNAc yield of 34.4% and a production rate of 15.9 g L^{-1} h^{-1} [119]. A novel AGE from a human gut symbiont *Bacteroides thetaiotaomicron* BT0453 was cloned and characterized, which demonstrates the highest soluble fraction among the AGEs tested. Using GlcNAc and pyruvate as substrates, Neu5Ac production performed by coupling whole cells expressing BT0453 and *E. coli* NAL generated a 53.6% molar yield, 3.6 g L^{-1} h^{-1} productivity, and 42.9 mM titer of Neu5Ac in 36 h [120].

6.1.4 Hydantoin Racemase

The systematic name of the enzyme hydantoin racemase (EC 5.1.99.5, also known as 5'-monosubstituted-hydantoin racemase, HyuA, and HyuE) is D-5-monosubstituted-hydantoin racemase. This enzyme has been purified and characterized in several microorganisms, including *Pseudomonas* sp. strain NS671, *Arthrobacter aurescens* DSM 3747, *Agrobacterium tumefaciens* C58, *Sinorhizobium meliloti* CECT 4114, and *Microbacterium liquefaciens* AJ 3912, which are involved in the production of optically pure D- and L-amino acids [121–126]. Except for *Pseudomonas* sp., the native enzyme found from different microorganisms indicates a homotetramer in structure with a molecular mass of about 100 kDa. Likewise, genomic localization and genetic organization of the genes involved in the production of the amino acids have been reported together with a hydantoinase, a carbamoylase, and a putative hydantoin transport protein [127–130]. Hydantoin racemase catalyzes the reversible transformation of both D- and L-isomers of 5-monosubstituted hydantoins to the corresponding racemic mixtures (Scheme 6.14). The catalytic ability for the racemization of 5-monosubstituted hydantoins has made it very

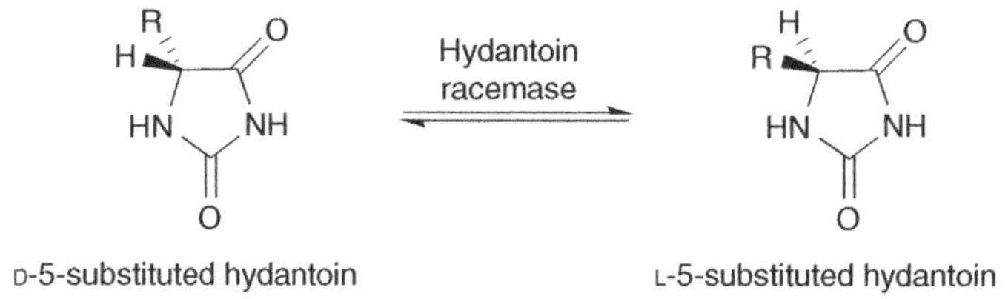

Scheme 6.14 Reversible racemization of D- and L-isomers of 5-monosubstituted hydantoins catalyzed by hydantoin racemase. *Source:* Based on May et al. [142]; Nam et al. [143].

useful in the enzymatic cascade reactions known as hydantoinase process for the production of optically pure natural and unnatural D- or L-amino acids, which are valuable intermediates for the industrial synthesis of antibiotics, sweeteners, pesticides, pharmaceuticals, and biologically active peptides [131–133]. The reaction mechanism of the reversible racemization catalyzed by hydantoin racemase was studied by the three-dimensional crystal structure for the binding properties and specificity of the enzyme toward its substrates [134, 135].

The crystallographic studies of hydantoin racemase suggest a two-base mechanism, that Cys76 (in a thiolate form) in the active site would act as a base and retrieve a proton from the D-isomer of a 5-monosubstituted hydantoin and a plane intermediate of the substrate would be formed. Then, Cys181 in the active site would act as an acid and add a proton to the opposite side of the substrate that produces the L-5-monosubstituted hydantoin. In the same manner, when L-isomer of a 5-monosubstituted hydantoin is available, Cys181 in the active site would bind and act as a base to retrieve a proton, and Cys76 would act as an acid to donate a proton to the putative plane intermediate formed to give the D-5-monosubstituted hydantoin [136].

The "hydantoinase process" is a dynamic kinetic resolution (DKR) process that involves cascade reactions of hydantoinase, *N*-carbamoylase, and hydantoin racemase for the production of 100% yield of optically pure natural and unnatural D- or L-amino acids from cheap and easily available racemic 5-monosubstituted hydantoins (Scheme 6.15) [131, 133, 137]. In this process, hydantoin racemase plays a key role for DKR. Hydantoin racemase is a broad substrate spectrum enzyme: the enzyme from *Pseudomonas* sp. (HyuE) has a preference for hydantoins with aliphatic substituents, such as 5-isopropylhydantoin and 5-(2-methylthioethyl)hydantoin [121, 127], the enzyme from *A. aurescens* shows a preference for hydantoins with arylalkyl substituents, especially 5-benzylhydantoin [122], and the enzyme from *A. tumefaciens* demonstrates the preference for hydantoins with short, for example, 5-ethylhydantoin, rather than long aliphatic side chains or hydantoins with aromatic rings [123, 125].

For the enantioselective preparation of L-amino acid using hydantoin racemase associated "hydantoinase process," a whole-cell *E. coli* coexpressed the genes

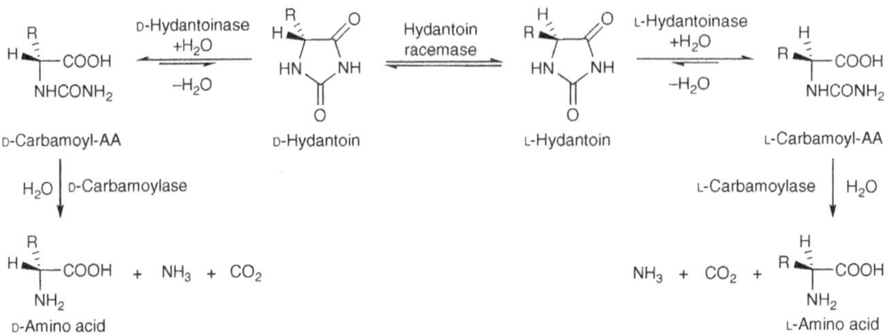

Scheme 6.15 The enzymatic cascade reactions for the total conversion of racemic 5-monosubstituted hydantoins to corresponding D- or L-amino acids.

encoding the L-hydantoinase, the L-*N*-carbamoylase, and the hydantoin racemase from *A. aurescens* DSM 3747 was constructed and used to convert D,L-5-(3'-indolylmethyl)-hydantoin to L-tryptophan. The productivity for L-tryptophan from the corresponding hydantoin was found more than sixfold higher than that achieved by using *A. aurescens* [138]. The three-enzyme "hydantoinase process" was modified by a four-enzyme "double-racemase hydantoinase process" for the production of natural and unnatural L-amino acids from racemic 5-monosubstituted hydantoins. The four enzymes comprising the system include a D-hydantoinase from *A. tumefaciens* BQL9, a hydantoin racemase from *A. tumefaciens* C58, a L-*N*-carbamoylase from *Geobacillus stearothermophilus* CECT43, and a *N*-succinyl-amino acid racemase from *Geobacillus . kaustophilus* CECT4264 (Scheme 6.16). The latter enzyme presents catalytic promiscuity and racemizes *N*-carbamoyl-amino acids to avoid the accumulation of *N*-carbamoyl-D-amino acid in the reaction due to the strict D-enantioselectivity of the hydantoinase. The metal ion cobalt was needed for this process. The enzymatic cascade reactions of this process for producing different optically pure L-amino acids achieved 100% conversion without showing substrate inhibition even at substrate concentration as high as 100 mM [139]. Later, the four enzymes were immobilized separately on different support materials and tested for activity.

The results revealed that support size, weight, and hydrophobicity must be homogeneous to avoid the formation of separate layers of the matrix selected. Therefore, the support IB-350 was used for the immobilization of the four enzymes, even though it may not be the most efficient one for some of the enzymes [140, 141].

Application of the "hydantoinase process" to the optically pure D-amino acids production was first reported in 1976, in which D-phenylglycine and D-*p*-hydroxyphenylglycine were synthesized from corresponding *N*-substituted D,L-5-phenylhydantoin and D,L-5-*p*-hydroxyphenylhydantoin (Scheme 6.17) [142, 143]. D-Phenylglycine and D-*p*-hydroxyphenylglycine are essential building blocks of

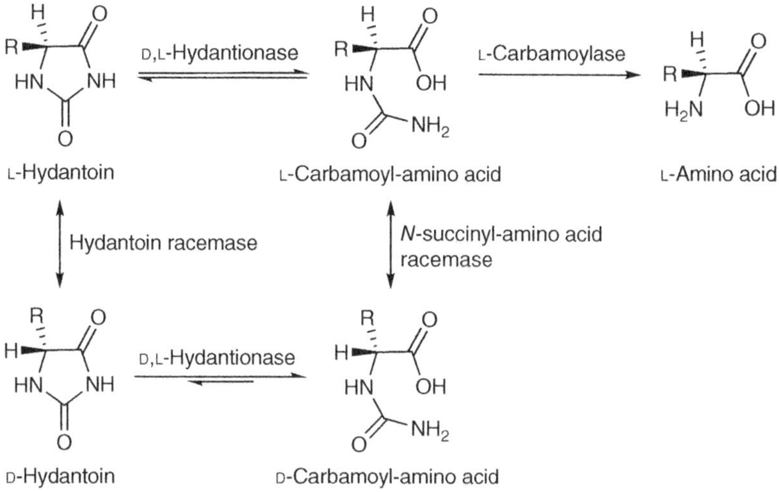

Scheme 6.16 The "double-racemase hydantoinase process" for optically pure L-amino acid production from racemic 5-monosubstituted hydantoins. *Source:* Modified from Kimuraet al. [168].

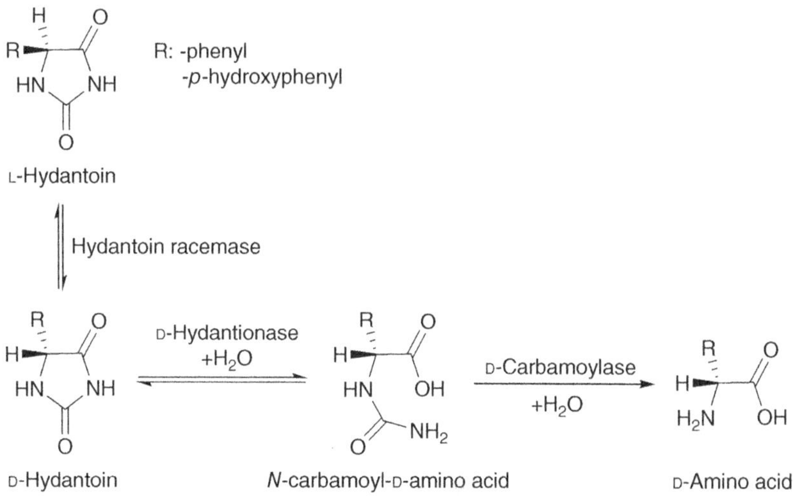

Scheme 6.17 Hydantoinase process used for the production of D-phenylglycine and D-*p*-hydroxyphenylglycine from 5-monosubstituted hydantoins.

the β-lactam antibiotics including amoxicillin, cephalexin, and cefadroxy, commercially produced on a scale > 1000 tons per year today [132, 144]. Nowadays, whole-cell recombinant *E. coli* coexpressing the three genes of D-hydantoinase, D-carbamoylase, and hydantoin racemase from various microorganisms have been widely exploited for the production of optically pure D-amino acids from D,L-5-monosubstituted hydantoins. For example, a reaction system consisting of recombinant *E. coli* whole-cell biocatalysts containing separately expressed D-hydantoinase and D-carbamoylase from *Agrobacterium radiobacter* NRRL B11291 and putative hydantoin racemase from *A. tumefaciens* C58 was capable of producing D-methionine with 100% conversion in 30 min.

The results also showed that the system had high substrate specificity and was effective toward both aliphatic and aromatic D,L-monosubstituted hydantoins [145]. After coexpressing the D-hydantoinase and D-carbamoylase genes from *A. tumefaciens* BQL9 as well as the hydantoin racemase 1 gene ($hyuA1_{At}$) from *A. tumefaciens* C58 in one plasmid in a polycistronic structure in *E. coli*, the whole-cell recombinant reaction system was able to produce 100% conversion and 100% optically pure D-methionine, D-leucine, D-norleucine, D-norvaline, D-valine, D-aminobutyric acid, D-phenylalanine, D-tyrosine, and D-tryptophan from the corresponding hydantoin racemic mixture quickly [146]. Similarly, a whole-cell recombinant reaction system designed by coexpressing D-hydantoinase, *N*-carbamoyl-D-amino acid amidohydrolase, and hydantoin racemase in *E. coli* was applied for the enzymatic production of D-amino acids. The use of this whole-cell system was able to efficiently produce eight amino acids, that is, D-phenylalanine, D-tyrosine, D-tryptophan, *O*-benzyl-D-serine, D-valine, D-norvaline, D-leucine, and D-norleucine, from corresponding D,L-5-monosubstituted hydantoins [147]. The development of a DKR cascade reaction system by combining a newly found D-carbamoylase (*AcHyuC*) from *Arthrobacter crystallopoietes* screened by genome mining with hydantoin racemase from *A. aurescens* (*AaHyuA*) and D-hydantoinase from *A. tumefaciens* (*AtHyuH*) was used for the enantioselective resolution of L-indolylmethylhydantoin into D-tryptophan. L-indolylmethylhydantoin (80 mM) could be totally converted to D-tryptophan with 99.4% yield, > 99.9% *e.e.*, and 36.6 g L^{-1} d^{-1} productivity within 12 h [148].

6.2 Cis-Trans Isomerase

6.2.1 Maleate *Cis-Trans* Isomerase

Maleate *cis-trans* isomerase (EC 5.2.1.2), or simply maleate isomerase (MI), catalyzes the geometric *cis-trans* isomerization of the C2-C3 double bond in maleate to produce fumarate (Scheme 6.18). Maleate isomerase is known to be distributed

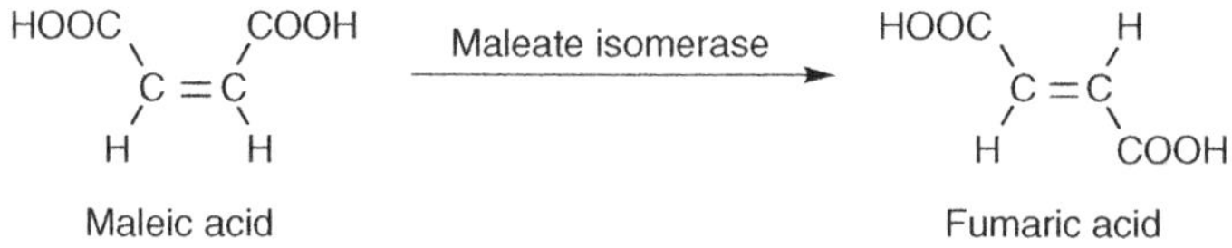

Scheme 6.18 Maleate isomerase catalyzed isomerization of maleic acid to fumaric acid. *Source:* Modified from Bhosale et al. [187].

among a number of maleate-assimilating bacteria, such as *Pseudomonas*, *Arthrobacter*, *Proteus*, *Alcaligenes*, *Bacillus*, *Serratia*, and *Nocardia* species [149–154]. Even more, a putative maleate isomerase gene from *Rhodococcus jostii* was synthesized and expressed in *E. coli*. The protein produced after *N*-terminal truncation efficiently functions the catalytic process of maleate *cis-trans* isomerase for the production of fumaric acid from maleate [155]. The enzyme has a molecular weight of 74 000 and is a member of the Asp/Glu racemase superfamily [149, 155–157]. The structure of MI revealed four molecules in the asymmetric unit, representing two dimers with a flat dimerization surface [155, 158, 159]. Analogous to other superfamily racemase members, each monomer consists of two domains connected by a pseudo-twofold symmetry, with each domain contributes one catalytic cysteine, which is critical to the catalytic activity at the active site of the isomerase [155, 159]. At the active site, a few other amino acid residues help the recognition of the substrate and the stabilization of the reaction intermediate. For instance, in MI from *P. putida* Ser16, Asn17 and Asn169 form hydrogen bonds with the carboxylate group of the maleate distal to Cys82, Tyr139 forms hydrogen bonds with the carboxylate group of the maleate proximal to Cys82, Pro14, and Val84 using van der Waals forces interact with the C2 and C3 carbon atoms of the maleate [155, 159].

Although the catalytic mechanism of MI has not been fully understood, it is considered similarly to the Asp/Glu racemase superfamily. One reaction mechanism revealed from the crystal structure of *Nocardia farcinica* IFM 10152 and its C194A mutant is initiated by nucleophilic attack of the deprotonated sulfur atom in Cys76 to covalently bonding to the C2 atom of the maleate, concomitantly, the thiol proton of Cys194 is transferred to the C3 atom of the maleate to yield a succinyl-cysteine like covalent intermediate. Then, the newly formed C2–C3 single bond is rotated, the single bond between the sulfur atom of Cys76 and the C2 atom is dissociated, and the C3 atom of maleate is deprotonated by Cys194 to form fumarate with the regeneration of a neutral Cys194 [155, 160].

Maleate isomerase catalyzes the *cis-trans* isomerization of maleate to fumarate, which is an important intermediate in a citric acid cycle, and this is biocatalytic reaction step that is involved in butanoate metabolism and nicotinate and nicotinamide metabolism, which is the last step of the metabolic degradation

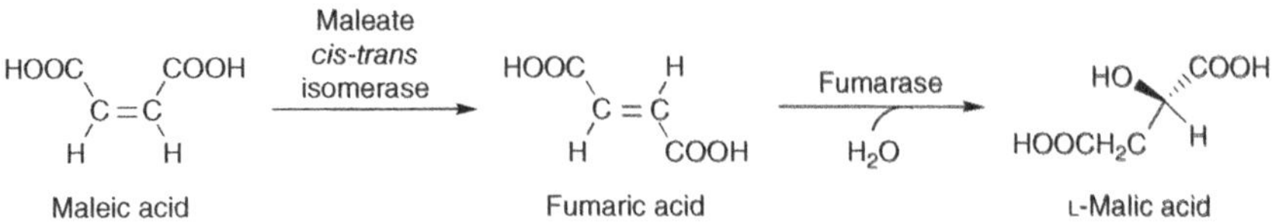

Scheme 6.19 Bioconversion of maleic acid into L-malic acid by consecutive catalysis with maleate *cis-trans* isomerase and fumarase.

pathway of nicotinic acid [161–164]. Fumaric acid is an important starting material for polymerization and esterifications and is used in the food industry and pharmaceutical preparations. A number of industrially useful compounds such as L-aspartic acid, L-alanine, and L-malic acid are produced from fumaric acid via the bioconversion process [153, 162]. Maleic acid is not a natural compound, which is usually produced from maleic anhydride [162]. The conversion of maleic acid into fumaric acid using maleate isomerase is occurred at high temperatures that causes low yields due to by-products formation. In addition, natural maleate isomerase is unstable at moderate temperatures [165]. In order to solve these problems, heat-stable maleate isomerases are engineered and applied [162]. In a study, the heat-treatment of *Pseudomonas alcaligenes* XD-1 can improve the production rate and yield of fumaric acid from maleic acid to as high as 95.0% as a result of inactivating the fumarase of strain XD-1, which converts the fumaric acid produced to L-malic acid [166, 167]. Therefore, an enzymatic process used for the production of L-malic acid from maleic acid by incorporating maleate isomerase and fumarase in *Alcaligenes* sp. T501was performed to give a molar yield of 98.4% (Scheme 6.19) [168].

6.2.2 Linoleate Isomerase

The enzyme linoleate isomerase (LI, EC 5.2.1.5) catalyzes the isomerization reaction of linoleic acid (LA, *cis*-9, *cis*-12-octadecadienoic acid) to produce conjugated linoleic acid (CLA, *cis*-9, *trans*-11-octadecadienoic acid) (Scheme 6.20). The systematic name of linoleate isomerase is linoleate Delta12-*cis*-Delta11-*trans*-isomerase

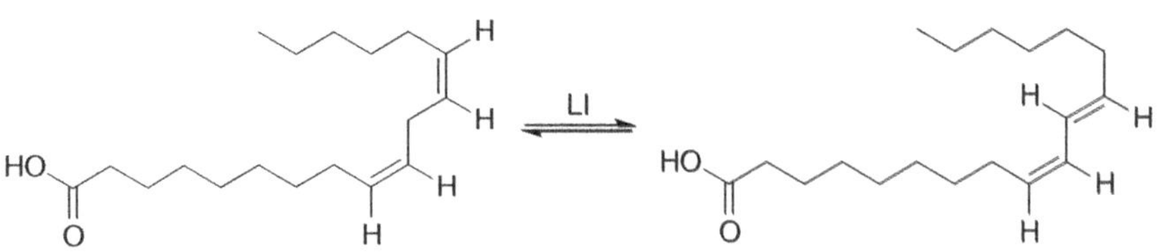

Scheme 6.20 Linoleate isomerase catalyzed isomerization of *cis*-9, *cis*-12-octadecadienoic acid to *cis*-9, *trans*-11-octadecadienoic acid.

and is also called linoleic acid isomerase (LAI). Sources of LAI come from the membrane-bound protein of anaerobic bacteria such as *Butyrivibrio fibrisolvens*, *Lactobacillus acidophilus*, and *Clostridium sporogenes* in rumen or human gut [169–171]. This enzyme involves the metabolism of LA as the first step to convert LA into CLA (mainly *cis*-9, *trans*-11-18:2), then to vaccenic acid (*trans*-11-18:1), and finally to saturated fatty acid stearic acid (18:0). A similar protective metabolic cascade occurs in the rumen, commonly known as biohydrogenation, which eliminates double bonds in polyunsaturated fatty acids (PUFA) [169, 172]. The reaction model based on the crude purification and characterization of LI from *B. fibrisolvens* suggested a concerted mechanism for the isomerization of LA to CLA, which involves a hydrogen abstraction from C11 by a general base and stereospecific transfer of the solvent-derived proton by the enzyme to C13 of the substrate in the pro-*R* configuration [169, 170].

Conjugated linoleic acids are a group of naturally occurring positional and geometric isomers of octadecadienoic fatty acids with conjugated double bonds, for example, *cis*-9, *trans*-11 and *trans*-10, *cis*-12-octadecadienoic acid. These CLAs have been reported to have potential health beneficial effects including anticarcinogenic, antiatherogenic, antidiabetic, antiobese, antioxidative, immunomodulative, antibacterial, cholesterol depressing, and growth-promoting properties. For these reasons, they have been widely studied and intensely worked for efficient preparation. Enzyme extracts prepared from *L. acidophilus* (CCRC14079) and *Propionibacterium freudenreichii* ssp. *shermanii* (CCRC11076) were used to convert LA into a variety of CLAs with *c*9, *t*11-, *t*10, *c*12-, and *c*11, *t*13-CLA being the three major isomers produced. At pH 5, more CLA was formed from *L. acidophilus* treatment than those of *P. freudenreichii* ssp. *shermanii* treatment [173]. Further study using crude enzyme extract from *L. acidophilus* indicated sharp increases in total CLA levels in 50 mg LA treatment and 75 mg LA treatment as enzyme extract level increased from 0 to 50 mg. Forty-eighty percent of CLA isomers in total CLA produced were in *cis,trans*/*trans,cis* form and 14% were in *c*9, *t*11-CLA [174]. Both washed cells of *Lactobacillus delbrueckii* ssp. *bulgaricus* and crude enzyme extract from the culture were also proved to be able to improve CLA production. In addition, the enzyme extract was observed to be capable of oleic and linolenic acid conversions into CLA, demonstrating the possibility of desaturation activity in the enzyme extract [175].

Isolation of different strains of the genus *Bifidobacterium* from the fecal material of neonates was used for assessing their ability to produce the *cis*-9, *trans*-11 CLA isomer from free LA. The species *Bifidobacterium breve* was the most efficient producer that two different strains converted 29 and 27% of the free LA to the *cis*-9, *trans*-11 isomer/μg dry cells, respectively. A strain of *Bifidobacterium bifidum* demonstrated a conversion rate of 18% free LA/μg dry cells [176]. Later experiments demonstrated that there was an increase in *cis*-9, *trans*-11-CLA in the livers of mice and pig after feeding with LA in combination with *B. breve* NCIMB 702258. Moreover, the profile of polyunsaturated FA composition was altered, including higher concentrations of the omega-3 (n − 3) FAs eicosapentaenoic acid and docosahexaenoic acid in adipose tissue. These changes were associated with reductions in the proinflammatory cytokines tumor necrosis factor-α and interferon-γ [177]. A new strain of *B. fibrisolvens* had higher ability to produce CLA from LA and was much more tolerant to LA than other strains. However, CLA reductase causes the conversion of CLA to *trans*-vaccenic acid (*t*-VA). Since fatty acids (FAs) with 18 carbon backbone induce the LI activity and unsaturated FAs induce the CLA reductase, a diet with a low ratio of unsaturated to saturated FAs may benefit CLA production [178]. Another novel strain of *B. fibrisolvens* MDT-5 was isolated from the goat rumen, which exclusively converted LA to CLA with high levels of *c*9, *t*11-CLA. The oral administration of MDT-5 to mice resulted in increased amounts of CLA in the contents of the large intestine as well as in adipose tissue. A prolonged period of MDT-5 administration can further increase the CLA content in body fat with a high-LA diet. Thus, MDT-5 may be useful in pet animals as a probiotic to provide CLA continuously [179].

A screening of lactic acid bacteria (LAB) indicates that *Lactobacillus curvatus, Lactobacillus plantarum, and Lactobacillus sakei* strains displayed LI activity by converting LA and α-linolenic acid (α-LNA) to CLA and conjugated linolenic acid (CLNA), respectively. The conversion rate of CLA production for the three strains was low (2–5%) and CLNA conversion rates varied largely (1–60%). Furthermore, the production of CLA and CLNA is pH and temperature-dependent assessing by *L. sekei* LMG 13558 [180]. Six *L. plantarum* strains isolated from traditional dairy products of minority nationalities were found to be able to convert CLA from free LA *in vitro* using sunflower oil as a substrate of during soy milk fermentation. An increased production of CLA was observed when adding a high concentration of LA to sunflower oil. The strain IMAU60042 gave the highest CLA both in sunflower oil and soy milk. The CLA produced had two isomers: *cis*-9, *trans*-11 CLA and *trans*-10, *cis*-12 CLA with *cis*-9, *trans*-11 CLA the most abundant CLA in the total CLA formed except for the strain P8. During storage, CLA in the fermented soy milk decreases and the decrease of *trans*-10, *cis*-12 CLA is more rapid than *cis*-9, *trans*-11 CLA. The proportion of unsaturated FAs varied after fermentation with different *L. plantarum* strains [181]. The freeze-thawing permeabilization method was employed to modify *L. plantarum* A6-1F cells in order to enhance the LI activity of the cells for producing CLA from LA. It was observed that pH, temperature, substrate concentration, and cell concentration were the main factors to affect the yield of CLA production. Under the optimal conditions, the maximum CLA produced was 275.7 μg mL^{-1} and the main CLA isomer in the reaction mixture was *c*9, *t*11-CLA [182]. The permeabilized *L. acidophilus* cells were also used for the production of CLA from LA through LI in order to overcome the problem of enzyme extraction and poor permeability of cell to LA. The maximum production of CLA could be obtained after the cells were treated with cetyltrimethylammonium bromide. Under optimized batch reaction conditions, the conversion yield of LA to CLA was up to 86.4% corresponding to a yield up to 1.52 g g^{-1} wet permeabilized cells. The permeabilized cells could be reused for at least 10 cycles in the biosynthesis reaction [183].

The ability of producing CLA isomers *cis*-9, *trans*-11-CLA, *trans*-10, *cis*-12-CLA, and *trans*-9, *trans*-11-CLA from free LA was tested for two strains *L. acidophilus* and *Lacticaseibacillus casei* in MRS broth and skim milk supplemented with LA. The results showed maximum production of free CLA from 80.14 to 131.63 μg mL^{-1} was observed at 24 h of incubation in broth containing 0.02% of free LA. For skim milk supplemented with 0.02% free LA, the total amounts of free CLA after 24 h of incubation were in the range from 54.31 to 116.53 μg mL^{-1}. Free CLA in MRS broth or skim milk would likely to be more readily absorbed by the digestive tract than if it were incorporated into the cells [184]. Cells of *Lactobacillus reuteri* ATCC 55739 grown in MRS broth with and without added LA were harvested and washed. The cells grown in the medium without added LA transformed LA into CLA (mainly c9, t11-C18:2) better than did those cells grown with LA. It was also found that glycocholate occurring in humans will not influence the production of CLA with the resting cells [185]. After screening 85 strains of LAB isolated from artisanal cheeses, four *L. plantarum* and two *L. casei* strains were found capable of synthesizing CLA and the highest level of CLA formed in MRS broth for 48h incubation was 19.26 μg mL^{-1}. Some of the screened strains of LAB were able to produce gamma-aminobutyric acid (GABA) [186].

6.3 Intramolecular Oxidoreductases

6.3.1 Aldose-Ketose isomerases

6.3.1.1 Xylose (Glucose) Isomerase

Xylose isomerase (XI, EC 5.3.1.5) in enzymology is an enzyme that catalyzes the reversible conversion between D-xylose and D-xylulose. Xylose isomerase belongs to those intramolecular oxidoreductases in the isomerases family converting aldoses, such as D-xylose, D-glucose, and D-ribose, to corresponding ketoses. The systematic name of XI is D-xylose aldose-ketoseisomerase. Other names include D-xylose isomerase, D-xylose ketoisomerase, and D-xylose ketol-isomerase. Xylose isomerase is also commonly known as glucose isomerase (GI) due to its ability to interconvert glucose and fructose for their extensive industrial application in the production of high fructose corn syrup (HFCS) from glucose. In view of its catalytic ability to catalyze the reversible isomerase of D-glucose and D-xylose to D-fructose and D-xylulose, respectively, it may be the most important of all

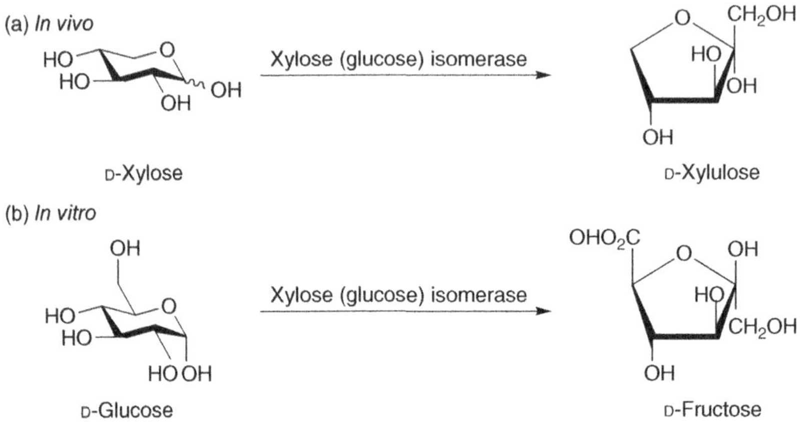

Scheme 6.21 The isomerization of either D-xylose or D-glucose catalyzed by xylose (glucose) isomerase. (a) *In vivo* reaction, (b) *In vitro* reaction.

industrial enzymes (Scheme 6.21) [187]. Generally, the interconversion of xylose and xylulose involves in the xylose metabolism of microorganisms, which serves a nutritional requirement in the saprophytic bacteria for their growth on decaying plant material, and in the meantime aids the ethanol production from hemicellulose [187, 188]. In addition to the above two main applications of XI, recent carbohydrate researches indicated that XI is useful in pharmaceutical industries for the production of rare sugars (20 hexoses and 9 pentoses including xylulose), which can be used to stimulate our immune system, to control diabetes, or as building blocks for anticancer and antiviral drugs [189]. Wide sources of XI (GI) have been reported, including microorganisms, such as bacteria, actinomycetes, and fungi, as well as plants and animals [187, 188, 190].

Xylose (glucose) isomerases purified from different species have similar spatial structures, even if there are certain differences in their primary structure of proteins. Generally, they are all nonglycoproteins with a tetramer or a dimer of similar or identical subunits bound by noncovalent bonds. The molecular mass of XI (GI) is wide from 52 to 191 kDa. Each subunit monomer of the protein is symmetrically distributed and divided into two domains. XI (GI) requires a divalent cation such as Co^{2+}, Mg^{2+}, or Mn^{2+}, or a combination of these cations, for maximum activity. Although both Mg^{2+} and Co^{2+} are essential for enzyme activity, they play different roles. However, Mg^{2+} is superior to Co^{2+} as an activator, Co^{2+} is responsible for the stabilization of the enzyme by holding the ordered conformation, especially the quaternary structure of the enzyme. The structure of XI (GI) consists of eight strains of α/β helical folded structure. Each of the tetramer subunit consists of paired α/β barrels that are nearly coaxial to form two cavities and

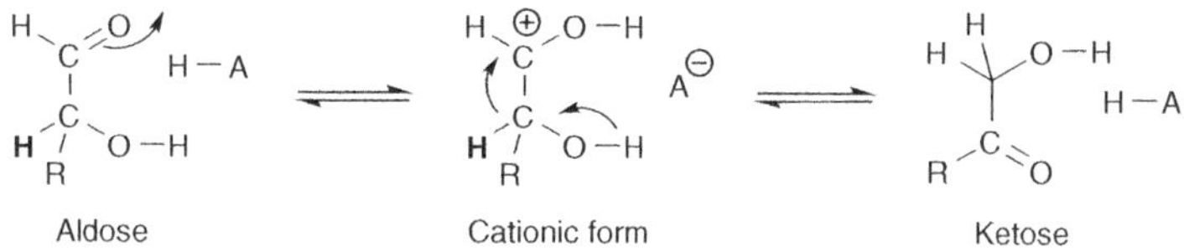

Scheme 6.22 Hydride shift mechanism of xylose (glucose) isomerase.

each of the two cavities can bind the divalent metals. Metal site 1 binds the substrate tightly, while metal site 2 binds the substrate loosely. Both share an acid residue Glu216 of the enzyme to bridge the two cations. Two basic amino acids surround the negatively charged ligands to neutralize them. The second cavity faces the metal cavity and both cavities share the same access route. The second cavity is hydrophobic in nature and has an important residue His53 that is activated by a residue Asp56 that is hydrogen bonded to it. This histidine residue is important in the isomerization of glucose [187, 191–194].

The earlier proposed catalytic mechanism of XI (GI) was assumed to function in a manner similar to sugar-phosphate isomerases that acts as a *cis*-enediol mechanism. Recent studies have attributed the action of XI (GI) to a hydride shift mechanism (Scheme 6.22), which has been studied by the active-site configuration using different approaches including chemical modification, X-ray crystallography, and isotope exchange. The main features of the mechanism proposed for XI (GI) are ring opening of the substrate, isomerization via a hydride shift from C2 to C1, and ring closure of the product. In the isomerization of glucose, His53 is used to catalyze the proton transfer of O1 to O5 via a ring-opening step that the first metal mentioned earlier coordinates to O3 and O4 and is used to dock the substrate. In the isomerization of xylose, xylose sugar binds to the enzyme in an open chain conformation. Metal 1 binds to O2 and O4, and once bound, metal 2 binds to O1 and O2 in the transition state, and these interactions along with a lysine residue help catalyze the hydride shift necessary for isomerization. The transition state consists of a high energy carbonium ion that is stabilized through all the metal interactions with sugar substrate [187, 194–197].

Since the first observation of the isomerization activity for D-xylose to D-xylulose by the bacterium *Lactobacillus pentosus* in 1953 and the isomerization on D-glucose to D-fructose with *Pseudomonas hydrophila* in 1957, a wide range of pentoses and hexoses are found to be suitable substrates for XI (GI) in addition to xylose and glucose, such as D-ribose, D-allose, L-arabinose, L-rhamnose, and 2-deoxyglucose [187, 198–200]. However, the research results indicated that xylose (glucose) isomerase can only catalyze the isomerization for the α-optical isomer of D-glucose or D-xylose, but not for the β-optical isomer [201, 202]. In order to make the carbohydrate conversion reactions of XI (GI) more industrial relevance, a variety of techniques have been developed to increase the yield, thermal stability, or lowering of the pH optimum. Recombinant D-xylose isomerase gene from *L. reuteri* was expressed in the food-grade host *L. plantarum* and the recombinant enzyme was purified and used for the isomerization of D-glucose to D-fructose with a maximum conversion rate of 48% at 60°C [203].

The immobilization of XI (GI) has long been studied since 1971 that GI from *Streptomyces phaeochromogenes* NRRL B-3559 was entrapped in polyacrylamide gel to perform the glucose conversion to fructose at an optimum temperature of 80°C and an optimum pH of about 8 [204]. Similarly, GI from *Streptomyces griseolus* was entrapped in polyacrylamide gel and covalently bound to porous glass, but poor results were achieved for the low activity and the instability of the enzyme prepared by this method [205]. The immobilization by physical entrapment into poly(acrylic acid-co-2-acrylamido 2-methyl propane sulfonic acid) hydrogel polymer networks was applied to GI and employed to carry out the isomerization of glucose to fructose at the optimum pH 7.5 and reaction temperature at 65°C. The immobilized GI retained 81% of its initial activity after being recycled for 15 times, while it retained 87% of its initial activity after being stored at 4°C for 48 days [206]. Instead of using physical entrapment for immobilization, GI from *Streptomyces rubiginosus* was immobilized covalently onto Eupergit C 250L made by the copolymerization of *N,N'*-methylene-*bis*-methacrylamide, glycidyl methacrylate, allyl glycidyl ether, and methacrylamide. The catalytic efficiency of immobilized GI in the isomerization of glucose to fructose was threefold higher than that of free GI. The residual activity of immobilized GI after 18 reuses using a stirred batch reactor was about 85% of its initial activity and the thermal stability was almost the same as that of the free GI at 60°C for 18 h preincubation time. When stored at 5°C and 25°C for 4 weeks, the residual activity was 72 and 69% of the initial activity, respectively [207]. GI from *S. rubiginosus* was also covalently immobilized onto GAMM support (similar to Eupergit C 250L) synthesized by inverse suspension polymerization of glycidyl methacrylate, ally glycidyl ether, *N,N'*-methylene-*bis*(acrylamide), and methacrylamide. The immobilized GI was applied for isomerization of glucose to fructose, which shows better operational stability, thermal stability, and storage stability by retaining 91% of its initial activity after 18 reuses and retaining 97% of its initial activity after being stored at 4°C for 6 weeks [208].

The immobilization of enzyme has been combined with reactor design to perform a continuous type of reaction. In the research report, bilayer GI from *Streptomyces* sp. was immobilized in porous *p*-trimethylamine-polystyrene (TMPS) beads through molecular deposition technique and packed into a column to conduct the continuous production of high fructose syrup (HFS). Under optimized conditions, the maximum catalytic activity of the immobilized bilayer GI in the packed-bed column reactor was 2238 IU g^{-1} dry gel, which retained 68.5% of the initial activity. The half-life of the immobilized GI was approximate 45 d at pH 8.5 and 60°C with 50% w/v glucose. The specific productivity of the immobilized GI was 223 g dry D-glucose/g dry immobilized enzyme per day [209]. In another study, three sizes of XI crystals (20, 40, and 80 μm) from *S. rubiginosus* were prepared and cross-linked. The cross-linked xylose isomerase crystals (CLXIC) of different sizes were separately packed into steel columns to perform the isomerization of rare pentose and hexose sugars. The isomerization of pentose sugars in the plug flow mode CLXIC-reactor was found to concomitantly undergo the product separation, while the concomitant enzymatic isomerization and product separation were weak with hexose sugars. In case of L-arabinose isomerization to L-ribose, the concomitant reaction and separation led to relatively high L-ribose concentration using the CLXIC-packed column reactor, particularly, recycling of saturated L-arabinose solution through the CLXIC-reactor turned out to be more efficient method for the production of L-ribose. Three new substrates, L-lyxose, L-mannose, and D-mannose, also were discovered [210]. The solid–liquid catalyzed isomerization of D-glucose to D-fructose using immobilized GI (Sweetzyme T) prepared from *Suncus murinus* was provided by Novo Nordisk and carried out in a tangential flow two impinging streams reactor (TFTISR). The conversion of glucose using TFTISR was much higher than those expected in conventional reactors [211]. The same type of reaction was studied but performed in a downflow jet loop reactor (DJR). The performance capability for the conversion of glucose to fructose with DIR was found to be higher than that of a conventional batch reactor. The results also revealed that the draft tube diameter has the highest and the ratio of liquid circulation rate to the gas flow rate has the least effects on the conversion of glucose [212]. Furthermore, a novel type impinging streams reactor using immobilized GI gifted by Novo Nordisk was employed to perform the solid-liquid isomerization of glucose to fructose in order to reduce the external mass transfer resistance. This impinging streams reactor shows higher performance capability that 30% conversion of glucose was obtained within 9 s, while the same conversion in packed-bed reactors requires 5 min [213].

A number of methods or their combinations were developed and utilized to solve the thermostability of XI (GI) during the isomerization. Novel thermostable GIs from *Thermoanaerobacterium xylanolyticum* (TxGI), *Thermus oshimai* (ToGI), *Geobacillus thermocatenulatus* (GtGI), and *Thermoanaerobacter siderophilus* (TsGI) were screened via genome mining approach. The results revealed that ToGI had higher catalytic efficiency and superior thermostability toward *in vitro* conversion of D-glucose to D-fructose for producing HFCS among the screened GIs. Its optimum temperature reached 95°C and could retain more than 80% of initial activity in the presence of Mn^{2+} at 85°C for 48 h. The maximum conversion yield of 400 g L^{-1} D-glucose to D-fructose at 85°C was 52.2% [214]. A new hyperthermophilic GI from *Thermoanaerobacter ethanolicus* CCSD1 (TEGI) was identified by genome mining, and then gene encoding the TEGI was synthesized and expressed in *E. coli*, which was further modified through site-directed mutagenesis to improve the activity for the production of HFCS by the isomerization of glucose to fructose. The specific activity, thermostability, and glucose conversion yield of mutant TEGI were 2.3-fold, 1.21-fold (90°C for 24 h), and >55.4% (90°C), respectively, as compared to the wild-type TEGI [215]. Later, whole-cell immobilization of refractory ToGI onto Celite 545 using tris(hydroxymethyl) phosphine (THP) as crosslinker was developed by the same group for producing HFCS at elevated temperature. This THP-immobilized biocatalyst demonstrated an enhanced thermostability with 70.1% residual activity retaining after incubation at 90°C for 72 h and exhibited superior operational stability with 85.8% initial activity retaining after 15 batches of reaction at 85°C [216]. GI from *Thermobifida fusca* was engineered and expressed in *E. coli* supplemented with Co^{2+}. The engineered cells were immobilized with the method of flocculation-cross linking and used for the reversible isomerization glucose to fructose that showed 2.1-fold higher activity and a better operational performance with 43.1% conversion rate at the end of six reaction cycles than cells without the addition of Co^{2+} [217]. The thermostable L-arabinose isomerase of *Bacillus stearothermophilus* US100 (L-AI US100) and the mutant D-glucose isomerase obtained from *Streptomyces* SK were co-expressed in *E. coli* and exploited for the concomitant production of D-tagatose and D-fructose. The recombinant cells were immobilized by entrapment in alginate beads and showed optimal temperatures for D-galactose and D-glucose isomerization of 80 and 85°C, respectively. The two immobilized isomerases showed enhanced acidotolerance and stability at high temperatures. The simultaneous isomerization of D-galactose and D-glucose at 65°C and pH 7.5 using a packed-bed reactor, cells concentration, dilution rate, productivity, and bioconversion rate were 32 g L^{-1}, 2.6 h^{-1}, 3 g L^{-1}·h, and 30%, respectively [218].

Mixed food and beverage (F&B) waste hydrolysate from saccharification is rich in glucose and fructose but also contains trace preservatives, caffeine, colorants, ions, and soluble proteins. Before the enzymatic isomerization of mixed F&B waste hydrolysate to produce HFS, over 99% of the impurities were first removed by adsorptions chromatography and ion-exchange chromatography. After impurities removal, the isomerization of glucose in the mixed F&B waste hydrolysate was performed by using a glass column packed with GI, where glucose was converted to reach 50% of fructose content. This pilot-scale downstream processing was successfully conducted with more than 89% of sugars recovered from hydrolysate and an overall conversion yield of 0.08 kg HFS/kg mixed F&B waste was obtained, which conforms to industrial standards [219]. Hemicellulose is renewable biomass comprising up to 35% of plant materials. D-Xylose is the major product in the hemicellulose hydrolysate. It was found that D-xylulose, an intermediate

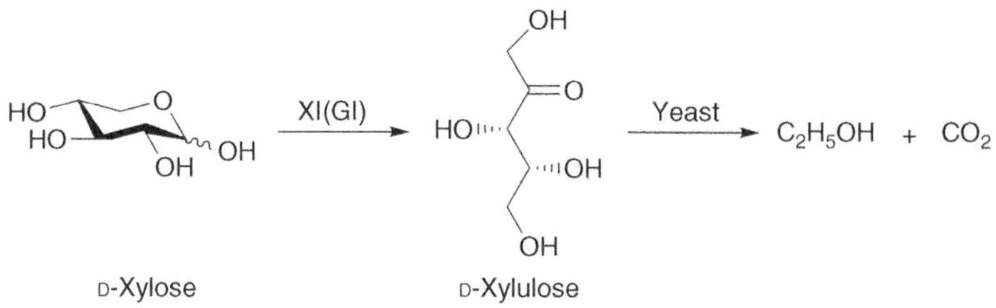

Scheme 6.23 Simultaneous isomerization and fermentation of D-xylose for ethanol production. *Source:* Modified from Makkee et al. [224].

of D-xylose catabolism, can be fermented to ethanol and carbon dioxide in a yield greater than 80% by yeasts. Therefore, ethanol can be produced from D-xylose in a yield greater than 80% by a two-step process, in which D-xylose is converted to D-xylulose by XI, and then D-xylulose is fermented to ethanol by yeasts (Scheme 6.23). In some yeasts, xylitol is a by-product formed during the conversion of D-xylulose into ethanol. Thus, the simultaneous isomerization and fermentation of D-xylose to ethanol by isomerase and yeasts provide an additional dimension in the conversion of waste biomass such as hemicellulose to liquid fuels, especially ethanol [220]. Ethanol production from xylose by the simultaneous isomerization and fermentation was also modified by anaerobic fermentation using an anaerobic yeast strain *Schizosaccharomyces pombe*. The results show that the simultaneous isomerization and fermentation of 6% xylose can produce 2.1% w/v ethanol within 2 days at temperature 39°C [221].

D-Psicose has been already used as a substrate for the preparation of allitol and other rare sugars. A method for the preparation of D-psicose from inexpensive glucose was developed by the direct coupling of immobilized D-xylose isomerase (Sweetzyme T) and immobilized D-tagatose 3-epimerase (D-TE) from *Pseudomonas* sp. ST-24 on Chitopearl beads BCW 2503 (Scheme 6.24).

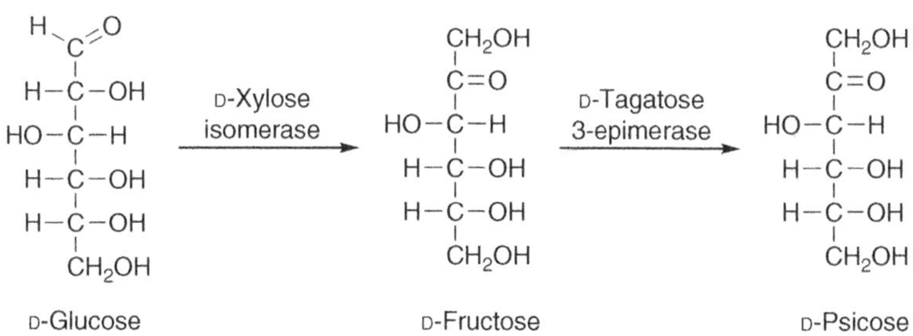

Scheme 6.24 Conversion of D-glucose to D-psicose by coupling D-xylose isomerase and D-tagatose 3-epimerase.

A continuous production of D-psicose from D-fructose was also performed by a packed-bed column filled with immobilized D-TE. About 90 g of D-psicose was obtained from 500 g D-fructose after a 10-day continuous reaction using this column at 45°C [222].

A novel method to convert cellulose into 5-hydroxymethylfurfural (HMF) with a high yield of 46.1% was developed by integrating sequential enzyme cascade reactions in an aqueous system with solid acid chemocatalysis in an organic solvent system. Magnetic Fe_3O_4 nanoparticles of approximately 20 nm were first synthesized, which were then added to the precursor used for mesoporous silica nanoparticles (MSNs) synthesis. Subsequently, the synthesized Fe_3O_4-loaded MSN (Fe_3O_4-MSN) was used as hosts for the immobilization of cellulase (cellulase-Fe_3O_4@MSN) and glucose isomerase (isomerase-Fe_3O_4@MSN). The sulfonic-acid-functionalized MSNs (HSO_3-MSN) were also prepared for use as a solid acid catalyst. These synthesized biocatalysts, that is, cellulase-Fe_3O_4@MSN and isomerase-Fe_3O_4@MSN, and chemical catalyst, that is, HSO_3-MSN, were separately applied in the sequential cellulose-to-glucose, glucose-to-fructose, and fructose-to-HMF conversion, respectively, across both aqueous- and organic–solvent systems after the optimized reaction conditions (Scheme 6.25). This method could be an effective and economically friendly process for various catalytic applications [223].

The crystalline of D-mannitol is nonhygroscopic and has no teeth decaying effects. It also will not cause laxatives with a daily intake up to 10–20 g. Therefore, D-mannitol is used as a sweet builder in "sugar-free" chewing gum and in pharmaceutical preparation to mask the unpleasant taste as well as to give a higher mechanical strength and stability of drugs. The industrial large-scale production of D-mannitol, 1:1 mixtures of D-glucose/D-fructose are reduced by high-pressure hydrogenation using Raney-Nickle as the catalyst to yield D-mannitol and D-glucitol (D-sorbitol) concomitantly, which leads to a low D-mannitol yield of 25–30%.

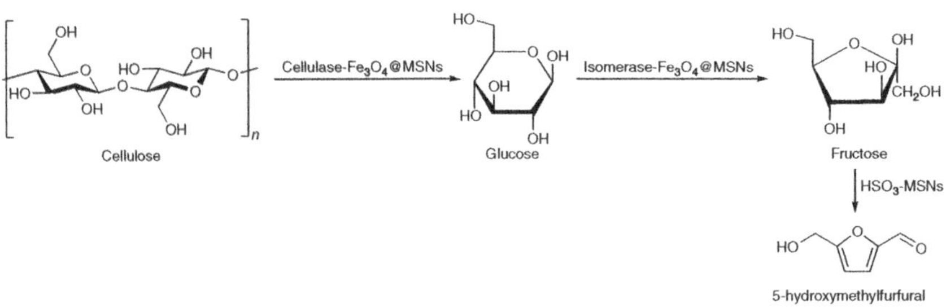

Scheme 6.25 Integrated enzyme cascade-chemocatalytic conversion of cellulose into HMF with enzyme immobilized on magnetic MSN and acid functionalized MSN. *Source:* Based on Lim and Oh [251]; Bhuiyan et al. [252].

During the hydrogenation, D-glucitol and D-mannitol are formed from D-fructose, while D-glucose is reduced to D-glucitol. Thereby, expensive downstream processing of the product is required. The source of D-glucose/D-fructose mixture can be acquired either by enzymatic hydrolysis of sucrose using invertase or the enzymatic isomerization of D-glucose using GI. Consequently, the development of alternative procedures for producing D-mannitol is needed and is interested by many researchers. A survey of the investigations indicates that among the variety of procedures, a large part involves the combination of enzymatic and chemocatalytic reactions, and can be divided into four categories, based on the species exploited toward D-mannitol: 1. D-glucose, 2. D-fructose, 3. D-mannose, and 4. D-mannose + D-fructose. However, enzymatic isomerization of D-glucose using GI is found an important process and used in many of the combined procedures to give high D-mannitol yield. One example is the process based on the enzymatic interconversion of D-glucose into D-fructose using GI with simultaneous and preferential hydrogenation of D-fructose using Cu/SiO_2 catalyst, which results in a 65% yield of D-mannitol (Scheme 6.26) [224].

In addition to the chemocatalytic hydrogenation or combined enzymatic catalysis/chemocatalytic hydrogenation for the production of D-mannitol from inexpensive sugars, whole-cell biotransformation was also developed for the formation

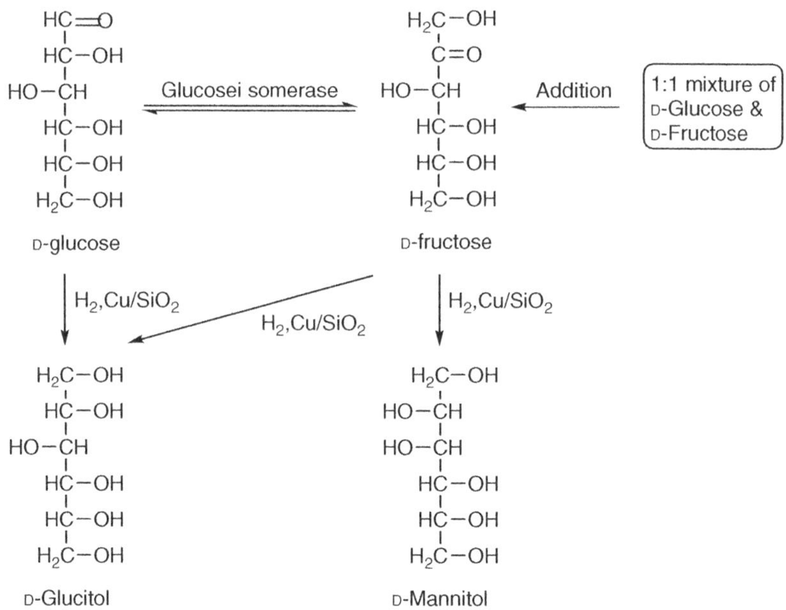

Scheme 6.26 D-Mannitol production by simultaneous enzymatic isomerization and hydrogenation.

of D-mannitol [224]. Additionally, the supplementation of the recombinant *E. coli* strain with extracellular glucose isomerase from *S. rubiginosus* was employed to convert 1000 mM D-glucose to yield 800 mM D-mannitol. Coexpression of a glucose isomerase (*xylA*) gene of *E. coli* in the biotransformation strain resulted in a D-mannitol concentration of 420 mM from 1000 mM D-glucose [225].

6.3.1.2 L-Rhamnose Isomerase

L-Rhamnose isomerase (L-RI, EC 5.3.1.14), an aldose-ketose isomerase, catalyzes the reversible isomerization between L-rhamnose (6-deoxy-L-mannose) and L-rhamnulose (6-deoxy-L-fructose) (Scheme 6.27) and initiates microbial catabolism of L-rhamnose, an essential component of mycobacterial cell wall. The systematic name of L-RI is L-rhamnose aldose-ketose isomerase and other common names in use include rhamnose isomerase and L-rhamnose keto-isomerase. L-RI is found in many microorganisms due to its essential role for L-rhamnose metabolism, that is, a sequential conversion: L-rhamnose → L-rhamnulose → L-rhamnulose 1-phosphate → dihydroxyacetone phosphate plus L-lactaldehyde. The former product enters the tricarboxylic acid (TCA) cycle for glycolysis. In *E. coli*, L-lactaldehyde is converted to L-lactate and further converted to pyruvate under aerobic conditions, while it is reduced to L-1,2-propanediol to be secreted out of the cell [226–228].

L-Rhamnose isomerase has been isolated, purified, and characterized from *E. coli* [229], *Pseudomonas* sp. strain LL172 [230], *Pseudomonas stutzeri* [231, 232], *Bacillus pallidus* Y25 [233], *Thermoanaerobacterium saccharolyticum* NTOU1 [234], *Thermotoga maritima* ATCC 43589 [235], *Caldicellulosiruptor saccharolyticus* ATCC 43494 [236], *Bacillus halodurans* ATCC BAA-125 [237, 238], *Mesorhizobium loti* Tono [239], *Dictyoglomus turgidum* DSMZ 6724 [240], *Bacillus subtilis* ATCC 23587 [241], and *Bacillus subtilis* 168 [242]. In summary, except one showing a homodimer all of L-RIs are in tetramer configuration that gives the subunit molecular mass in the range 42–49 kDa and the total molecular mass

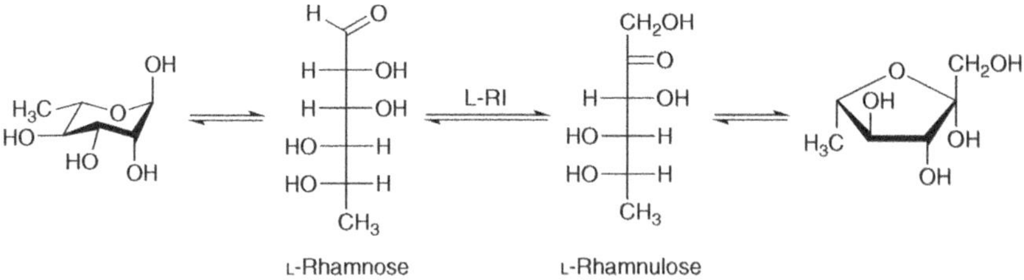

Scheme 6.27 L-Rhamnose isomerase catalyzed reversible isomerization between L-rhamnose and L-rhamnulose. *Source:* Based on Shommpoosang et al. [265]; Wen et al. [266].

between 184 kDa and 194 kDa (except for the dimer 121 kDa). The optimal pH and temperature for the activity of this enzyme are between 7.0 and 9.0, as well as between 60°C and 90°C, respectively. The crystal structure of L-RI indicates a tight tetramer of four $(\beta/\alpha)_8$-barrels, which is most similar to xylose isomerase. The crystal structure also confirmed the presence of two metal ion-binding sites: a structural metal ion-binding site for substrate binding, and a catalytic metal ion-binding site used for catalyzing the isomerization by a metal-mediated hydride-shift mechanism for substrates in a linear form, like D-xylose isomerase [232, 237, 243, 244]. In general, the metal ion required for the mechanism is Mn^{2+} or Co^{2+} as a divalent metal cofactor for activation, although Zn^{2+} also appears to bind at the structural binding site [234, 236, 243–246]. The structural analysis of the metal ion-bound catalytic site revealed Asp237 to be responsible for the ring opening of furanose, while a catalytic water molecule acting as an acid/base catalyst in the isomerization reaction is likely to be involved in the pyranose-ring opening. Furthermore, a newly found substrate subbinding site in the vicinity of the catalytic site may recognize different anomers of substrates [245, 246].

L-RI shows very broad substrate specificity toward a variety of aldoses and ketoses and thus displays a great potential in biological production of many expensive rare sugars. According to the definition by the International Society of Rare Sugars, rare sugars are monosaccharides, and their derivatives exist in nature in very limited quantities [226, 247, 248]. L-Rhamnulose is a precursor of the strawberry aroma furaneol and has been used in the flavor industry. Free L-RI from *B. pallidus* Y25 produces L-rhamnulose from L-rhamnose with a turnover ratio of 45% [226, 233]. L-Rhamnulose was also continuously produced by immobilized L-RI from *D. turgidum* DSMZ 6724 in a packed bed bioreactor with a conversion yield 43% [240]. L-Rhamnulose was prepared from L-rhamnose by a convenient, efficient, and cost-effective one-pot two-enzyme two-step strategy using L-RI and L-rhamnulose kinase. In the first reaction step, L-rhamnose was isomerized and phosphorylated to form L-rhamnulose 1-phosphate in the presence of ATP to increase the conversion ratio. In the second step, sugar phosphates were purified by silver nitrate precipitation, followed by the hydrolysis of the phosphate group of L-rhamnulose 1-phosphate to produce L-rhamnulose. The purity of L-rhamnulose produced exceeds 99% and the conversion yield is more than 80% [249]. In order to improve the thermostability of L-RI, a thermostable recombinant L-RI from the hyperthermophile *Caldicellulosiruptor obsidiansis* OB47 was used for catalyzing the isomerization of L-rhamnose to produce L-rhamnulose. The activity of the recombinant L-RI was maximum at pH 8.0 and 85°C in the presence of Co^{2+} and the conversion ratio was 44% [250]. In addition to L-rhamnulose, L-RI is also able to catalyze the isomerization reaction of other aldoses or ketoses, including the isomerization of D-allose/D-allulose, D-gulose/D-sorbose, L-mannose/L-fructose, L-talose/L-tagatose, L-galactose/L-tagatose, and

L-lyxose/L-xylulose. Recently, rare sugars have been proven to be of paramount significance in the food industry, nutraceuticals, pharmaceuticals, and other applications.

The rare sugar D-allose, an aldohexose, has beneficial pharmaceutical activities including anticancer, antitumor, anti-inflammatory, antioxidative, antihypertensive, cryoprotective, and immunosuppressant activities. In addition, D-allose is a noncaloric sugar with no toxicity, thus it has potential as a functional ingredient in formulated food. D-Allose exists in several cyclic forms, consisting of β-D-allo-1,5-pyranose (77.5%), α-D-allo-1,5-pyranose (14%), β-D-allo-1,4-furanose (5%), and α-D-allo-1,4-furanose (3.5%). They are white odorless crystalline solid and are soluble in water and insoluble in alcohol [251]. The first report about the production of D-allose was from D-psicose using immobilized L-RI on Chitopearl beads of BCW 2603 from *P. stutzeri* LL172. The conversion yield of the batch reaction after a 20-day reaction period was 40%. However, the reaction substrate D-psicose was in turn obtained from D-fructose using immobilized D-tagatose 3-epimerase (D-TE) (Scheme 6.28) [252]. Applying the same strategy, 25% of D-psicose was isomerized to D-allose with 8% by-product D-altrose by crude recombinant L-RI crosse-linked with glutaraldehyde within 30 days. The D-allose in the product mixture containing unreacted substrate D-psicose was easily separated and crystallized by ethanol. The operative half-life of the cross-linked enzyme was 2 months after repeated usage [253]. Next, large-scale production of D-allose from D-psicose using recombinant L-RI partially purified from *E. coli* and immobilized on BCW-2510 Chitopearl beards was performed by a continuous bioreactor. When 50% w/w D-psicose was employed in the column, approximately 30% D-psicose was isomerized to D-allose in a period of 17 days [254]. In addition, L-RIs obtained from other microorganisms, such as *B. pallidus* Y25, *T. saccharolyticum* NTOU1, *C. saccharolyticus* ATCC 43494, and *B. subtilis* str. 168, were also applied for the enzymatic production of D-allose from D-psicose without any by-product, and the maximal conversion yield was 35, 34, 33, and 37.5%,

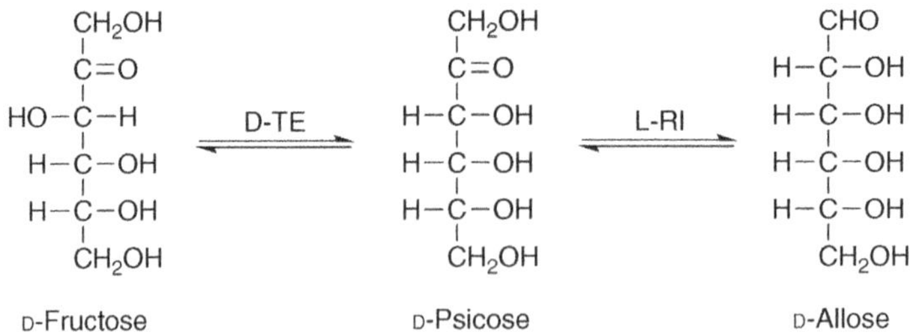

Scheme 6.28 Production of D-allose from D-fructose via D-psicose by using D-tagatose 3-epimerase and L-rhamnose isomerase in sequential.

respectively [233, 234, 236, 242, 255]. As a matter of fact, D-allose can be synthesized from D-glucose via a three-step enzyme reaction, that is, the applying of D-xylose isomerase for converting D-glucose to D-fructose as the first step in the two-step enzymatic process shown in Scheme 6.28 [251].

The isomerization of D-allulose (D-psicose) for the production of D-allose using a thermostable recombinant L-RI from *Thermobacillus composti* KWC4 as the biocatalyst was activated in the presence of Mn^{2+} at optimal pH 7.5 and temperature 65°C, and D-allose was the sole product obtained from D-allulose [256]. A recombinant L-RI from the thermophilic bacterium *Clostridium stercorarium* was also reported to have a high isomerization activity in the presence of Mn^{2+} to convert D-allulose to D-allose without any by-product. A conversion yield of 33% was achieved under the optimized conditions of pH 7 and 70°C [257]. The thermostability and catalytic activity of wild-type L-RI from *C. obsidiansis* OB47 on D-allulose were improved by rationally designed site-directed mutagenesis in order to meet the requirements of industrial production of D-allose. Compared with that of the wild-type enzyme, the relative activities of two mutants toward D-allulose were increased by 105.6 and 134.1%, respectively, and the thermostability of another mutant was also enhanced by raising the half-life to 7.7 and 1.1 h at 70 and 80°C, respectively [258].

Gulose is an aldohexose sugar. It is very rare in nature, but has been found in archaea, bacteria, and eukaryotes. It also exists as a syrup with a sweet taste. It is soluble in water and slightly soluble in methanol. Studies revealed that it can be used as a drug-formulation agent and food additive. Owing to its scarcity in nature, it is very expensive, therefore, it must be obtained by synthesis. A simple enzymatic production of D-gulose from D-sorbose has been developed using immobilized L-RI. L-Rhamnose isomerase from *P. stutzeri* was immobilized on BCW 2603 Chitopearl beads. At equilibrium, the production yield of D-gulose was 10% [259].

L-Lyxose is generally the second favored aldose substrate of L-RI, because of its high structural similarity with L-rhamnose [226]. The thermostable L-RI from *T. maritima* shows much higher specific activity toward L-lyxose than L-rhamnose. Under the optimum conditions of pH 8.0 and 85°, and in the presence of Mn^{2+}, 500 g L^{-1} L-xylulose was converted to 225 g L^{-1} L-lyxose without any by-product in 3 h using 100 U enzyme mL^{-1} [235]. When 100 g L^{-1} of L-xylulose was utilized for the isomerization by the recombinant L-RI from *B. subtilis* under optimal conditions (pH 8.0 and 60°C), 40 g L^{-1} of L-lyxose was solely produced in 60 min [241]. L-Lyxose could be prepared from ribitol by a method comprising sequential reactions with microbial cells and immobilized enzymes. In the first reaction step, the complete transformation of ribitol to L-ribulose was attained by using washed cells of *Acetobacter aceti* IFO 3281. Then, the L-ribulose produced was employed as the substrate for the subsequent production of L-xylulose by the

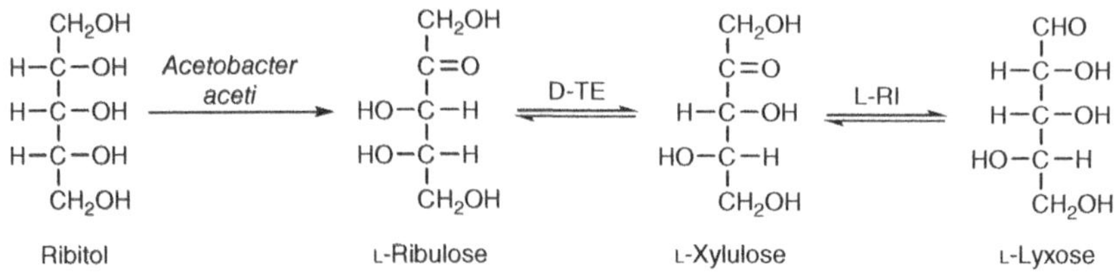

Scheme 6.29 Production of L-lyxose from ribitol using sequential microbial and enzymatic reactions.

immobilized D-tagatose 3-epimerase (D-TE) of recombinant *E. coli* JM 105, and followed by the conversion of L-xylulose to L-xylose using immobilized L-RI of *Pseudomonas* sp. strain LL172 in a one-pot reaction strategy (Scheme 6.29). The conversion yield of L-xylose from L-ribulose was about 60% and overall about 5.0 g L-xylose can be recovered from 10.0 g ribitol in a flask reaction [260]. Xylitol can also be used as a raw material for the production of L-lyxose and L-xylose by a one-pot coupled microbial-immobilized enzymatic reaction strategy, that is, xylitol was first converted to L-xylulose using NAD-dependent xylitol dehydrogenase in *Alcaligenes* 701B strain with a yield of 34%, then the produced L-xylulose was converted to L-xylose and L-lyxose using immobilized L-RI. The final equilibrium mass ratio between L-xylulose, L-xylose, and L-lyxose was 53:26:21 [261].

The first report of the enzymatic production of L-mannose from L-fructose was through an immobilized L-RI from *P. stutzeri*. The immobilized enzyme has a broad pH range (pH 5-11) for optimal activity that gave a production yield 30% at the optimal temperature 60°C [262]. Later, a recombinant L-RI from a facultative thermophilic bacterium *B. pallidus* Y25 was used for catalyzing L-fructose to L-mannose. The conversion yield was 25% when the enzyme was at its maximum activity (pH 7.0 and 65°C) [233]. In addition, free L-RIs either in recombinant form or native form from *B. subtilis* or *T. maritima*, respectively, were also found to be employed to perform the production of L-mannose from L-fructose. For recombinant L-RI from *B. subtilis*, 25 g L^{-1} L-mannose was produced from 100 g L^{-1} L-fructose for 80 min under the optimum conditions of pH 8.0 and 60°C [241]. For purified L-RI from *T. maritima*, 500 g L^{-1} L-fructose was converted to 175 g L^{-1} L-mannose under the optimum conditions of pH 8.0 and 85°C, and in the presence of Mn^{2+} [235].

The rare sugar L-talose was first enzymatically prepared from L-tagatose using immobilized L-RI of *Pseudomonas* sp. strain LL172 to afford a production yield of 12% [259]. Recombinant L-RI from *M. loti* was immobilized and applied to the production of L-talose from L-tagatose. At equilibrium, the L-tagatose to L-talose conversion ratio was 90:10 [239]. Both immobilized L-RIs from *Pseudomonas* sp. LL172 and *M. loti* gave no by-product, while recombinant L-RI from *P. stutzeri* produced L-galactose as a by-product when it was applied to L-talose production from L-tagatose [263].

When L-RI of *P. stutzeri* immobilized on BCW 2510 was utilized for the conversion of L-tagatose to L-galactose, L-talose was formed as a by-product. At equilibrium, the conversion ratio of L-tagatose/L-galactose/L-talose was 60:30:10. A novel two-step enzymatic route using immobilized D-TE from *Pseudomonas cichorii* mutant and recombinant L-RI of *P. stutzeri* for preparing L-galactose from L-sorbose was developed. At equilibrium, an overall yield of 7.5% for L-galactose production was obtained with some isomerization by-product L-talose [226, 264].

Deoxy sugars are monosaccharide derivatives in which one or more of the hydroxyl groups of the common monosaccharide has been replaced by a hydrogen atom. They are important carbohydrates that play roles as potentially active components of natural products. 6-Deoxy sugars are building blocks of a variety of natural products, including antifungals, antibiotics, and anticancer agents. 6-Deoxy-L-psicose is a 6-deoxy rare sugar that is used for the preparation of 6-deoxy-L-glucose, 6-deoxy-L-altrose, and 6-deoxy-L-allose. The enzymatic production of 6-deoxy-L-psicose was performed by incubating L-rhamnose (6-deoxy-L-mannose) with recombinant L-RI and recombinant D-TE. The final equilibrium mixture containing L-rhamnose, L-rhamnulose, and 6-deoxy-L-psicose in a ratio of 55:35:15 was obtained. After purification, the conversion yield of 6-deoxy-psicose with respect to L-rhamnose was only 8.7%. Therefore, a more efficient and convenient two-step strategy of "isomerization → epimerization → phosphorylation → dephosphorylation" cascade reaction was developed to prepare 6-deoxy-L-psicose from L-rhamnose, that is, L-rhamnose was first isomerized to L-rhamnulose catalyzed by L-RI, then L-rhamnulose was epimerized to 6-deoxy-L-psicose catalyzed by D-TE, and finally 6-deoxy-L-psicose was selectively phosphorylated with ATP at C-1 position to produce 6-deoxy-L-psicose 1-phosphate catalyzed by fructose kinase (FK). In the second reaction step, the purified 6-deoxy-L-psicose 1-phosphate was hydrolyzed to eliminate the phosphate group using acid phosphatase (AphA) to produce 6-deoxy-L-psicose in a high yield of 81% with respect to L-rhamnose. (Scheme 6.30) [265, 266]. The rare deoxy sugar 6-deoxy-L-talose can be directly produced from L-fucose (6-deoxy-L-galactose) via the intermediate L-fuculose by a novel coupling reaction of immobilized D-arabinose isomerase (D-AI) and immobilized L-RI. In this one-pot synthesis, the equilibrium ratio of L-fucose, L-fuculose, and 6-deoxy-L-talose was 80:9:11 that corresponds to a production yield of 7.1% with respect to L-fucose [267].

6.3.1.3 L-Fucose Isomerase

The L-fucose isomerase (L-FI, EC 5.3.1.25) is an enzyme that catalyzes the aldose L-fucose into the corresponding ketose L-fuculose using Mn^{2+} as a cofactor (Scheme 6.31). In addition, this enzyme can also act on D-arabinose (L-fucose

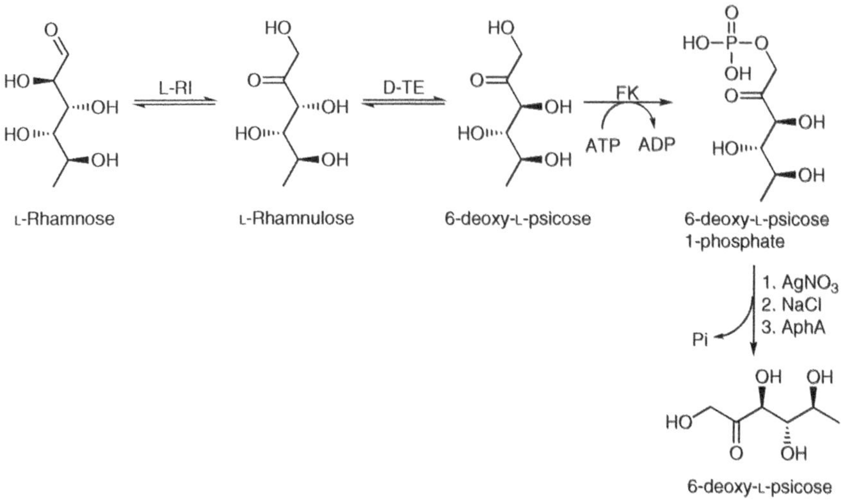

Scheme 6.30 Two-step enzymatic synthesis of 6-deoxy-L-psicose from L-rhamnose.

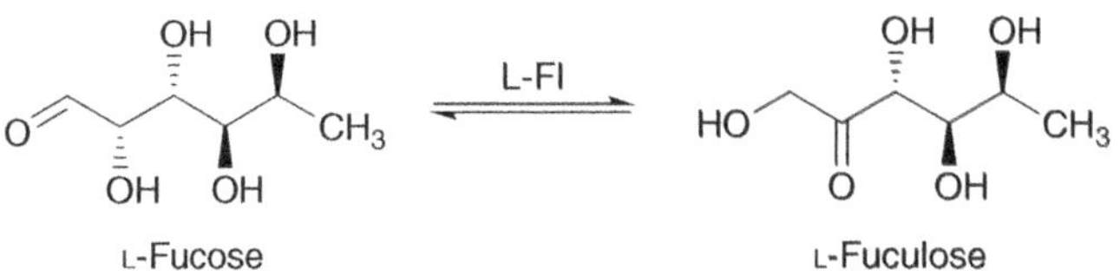

Scheme 6.31 The isomerization reaction of L-fucose catalyzed by L-fucose isomerase. *Source:* Modified from Wong et al. [277].

without the 6-methyl group) to convert it into D-ribulose, thus is known as D-arabinose isomerase (L-AI, EC 5.3.1.3) as well in literature [268–272]. The enzyme existed as a hexamer, which isolated from *E. coli* forms the largest structurally known ketol isomerase of 65 kDa per subunit and shows neither sequence nor structural similarity with other ketol isomerases. The structure was solved with an L-fucitol molecule bound to the catalytic center to mimic a bound L-fucose molecule in its open-chain form. The protein environment strongly suggests an enol-diol type reaction mechanism [273, 274]. However, the native L-FI purified from *C. saccharolyticus* is a trimer with the molecular mass 204 kDa [275].

L-Fucose isomerase shows a broad spectrum of substrate specificity. However, the isomerization activity of the enzyme with aldose substrates displays activity in a decreasing order for L-fucose, D-arabinose, D-altrose, and L-galactose, which are converted to the ketoses L-fuculose, D-ribulose, D-psicose, and L-tagatose, respectively [274, 275]. L-Rhamnulose and L-fuculose are two crucial rare deoxy ketoses that play important roles in sugar metabolism. In bacteria, L-fucose must be

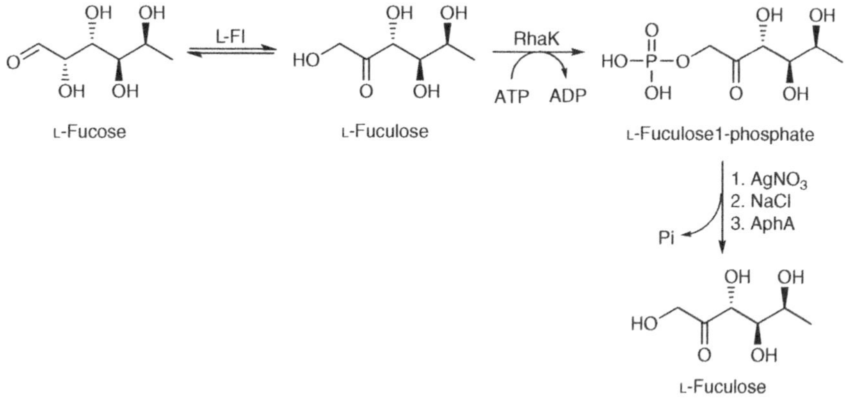

Scheme 6.32 Two-step enzymatic synthesis of L-fuculose from L-fucose.

converted to their ketose 1-phosphate forms, which are later split into dihydroxy acetone phosphate and L-lactaldehyde by L-fuculose 1-phosphate aldolase (FucA) to facilitate further metabolic function, therefore FucA is a powerful biocatalyst and has been widely used in synthetic chemistry to prepare rare ketoses or their derivatives. In addition, L-fuculose can also be directly isomerized or epimerized into other rare sugars [249]. Based on the metabolism of L-fucose in bacteria, an efficient preparation of L-fuculose from L-fucose was developed by a two-step enzymatic synthesis strategy. The first reaction step of this strategy involves the one-pot isomerization of L-fucose to L–fuculose and subsequently phosphorylation of L-fuculose to L-fuculose 1-phosphate in the presence of ATP as the phosphate donor by L-FI and L-rhamnulose kinase (L-RhaK), respectively. The by-products (ATP and ADP) were selectively separated by the silver nitrate precipitation method. In the second reaction step, the phosphate group of L-fuculose 1-phosphate was removed by hydrolysis using acid phosphatase (AphA) (Scheme 6.32). The overall yield (gram scale) of L-fuculose obtained with regard to L-fucose was 84% and a purity exceeding 99% [249].

The coupled enzyme cascade reaction strategy involving L-FI was exploited for the synthesis of unnatural low-calorie sweeteners L-glucose and L-fructose. The reaction was carried out in a step-by-step synthesis manner (Scheme 6.33) using rhamnulose-1-phosphate aldolase (RhuA) first to perform the stereospecific aldol condensation of dihydroxyacetone phosphate (DHAP) and L-glyceraldehyde to yield L-fructose 1-phosphate. After isolation, the barium salt of L-fructose 1-phosphate was converted to sodium salt followed by neutralization with NaOH. Then L-fructose 1-phosphate was subsequently hydrolyzed by acid phosphatase to remove the phosphate group and produce L-fructose as a single product. Finally, L-fructose was isomerized to L-glucose with L-FI to give a 40:60

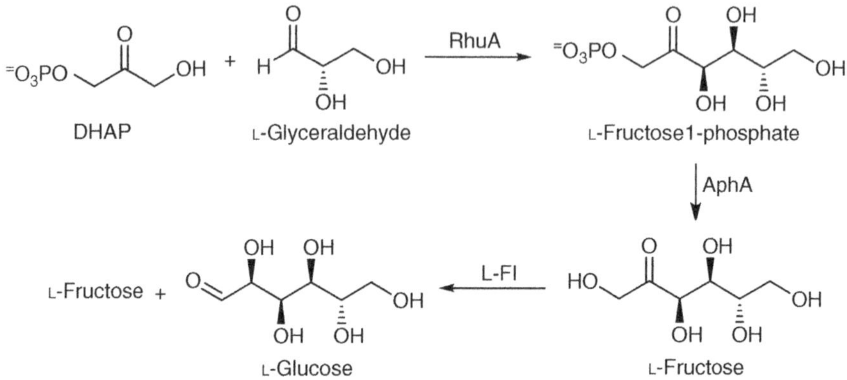

Scheme 6.33 A step-by-step coupled enzymatic synthesis of L-glucose and L-fructose.

(L-glucose:L-fructose) mixture, which was further chromatographed to yield pure products. This method uses relatively inexpensive cell-free extracts prepared from commercially available recombinant *E. coli* strains and shows the advantages of shorter synthesis step and higher yields [276]. This strategy was further applied for the synthesis of L-fucose and analogs.

For the preparation of L-fucose, the starting materials used were DHAP and DL-lactaldehyde and the enzymes used in sequence were L-fuculose-1-phosphate aldolase (L-FucA), AphA, and L-FI. Moreover, DL-lactaldehyde was prepared from DL-lactaldehyde dimethyl acetal and used directly without further purification. Both the stepwise and the one-pot sequential procedures were successfully used for the synthesis of L-fucose with a yield of 38.6% and 55.2%, respectively. For the preparation of L-fucose analogs, the combined enzymatic process was carried out by either a stepwise approach or sequential approach (Scheme 6.34) [277]. The same strategy using the same three enzymes, which L-FucA and L-FI were from

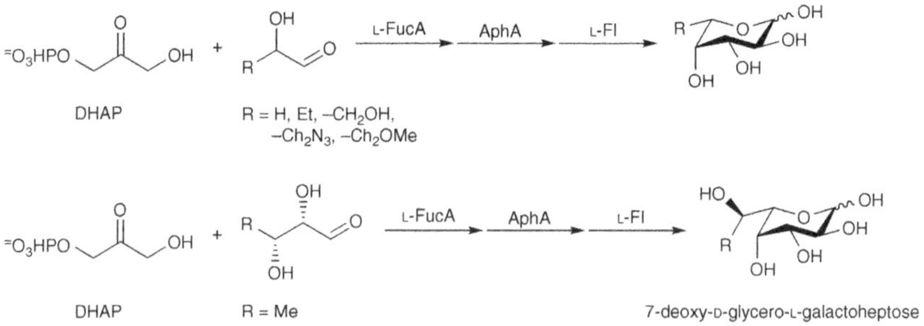

Scheme 6.34 Enzyme cascade reactions for the synthesis of L-fucose analogs.

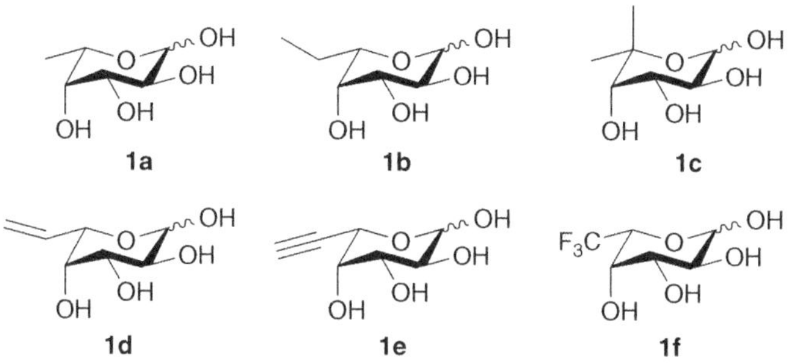

Scheme 6.35 Structures of L-fucose and L-fucose analogs modified at chain termini. *Source:* Modified from Moustafa et al. [292].

recombinant *E. coli*, the L-fucose analogs modified at the nonpolar methyl terminus of L-fucose (**1a**) were also reported.

Starting from the appropriate hydroxy aldehyde precursors and dihydroxyacetone phosphate, those fucose derivatives bearing extended linear (**1b**) and branched (**1c**) saturated, or various unsaturated aliphatic chains (**1d, 1e**) and the 6,6,6-trifluoro derivative (**1f**) have been prepared by a sequential method (Scheme 6.35). An overall yield of up to 30% has been acquired for these enantiomerically pure L-fucose analogs [278].

In addition to the isomerization activity of aldoses to ketoses catalyzed by L-FI, the report has shown that L-xylulose can be used as a raw material for the production of L-xylose with a recombinant *E. coli* L-FI as the catalyst. The enzyme had a very alkaline pH (over 10.5) and the half-lives determined for the enzyme at 35°C and 45°C were 6 h 50 min and 1 h 30 min, respectively. The equilibrium ratio between L-xylulose and L-xylose was 15:85 at 35°C, which shows a favorable production of L-xylose. The results also suggest that L-FI is a more suitable catalyst for L-xylose production from L-xylulose than L-RI, since the by-product L-lyxose was not detected in the reaction catalyzed by the former enzyme. Since xylitol appears to inhibit L-xylulose isomerization only to some extent, the L-xylose production from L-xylulose is economically feasible because L-xylulose production from xylitol can be combined with the isomerization of L-xylulose to L-xylose [279].

6.4 Intramolecular Transferases

6.4.1 Phosphopentomutase

Phosphopentomutases (PPMs, EC 5.4.2.7) reversibly catalyze the interconversion between α-D-ribose 5-phosphate (ribose 5-phosphate, R5P) and α-D-ribose 1-phosphate (ribose 1-phosphate, R1P) (Scheme 6.36), which bridges glucose metabolism

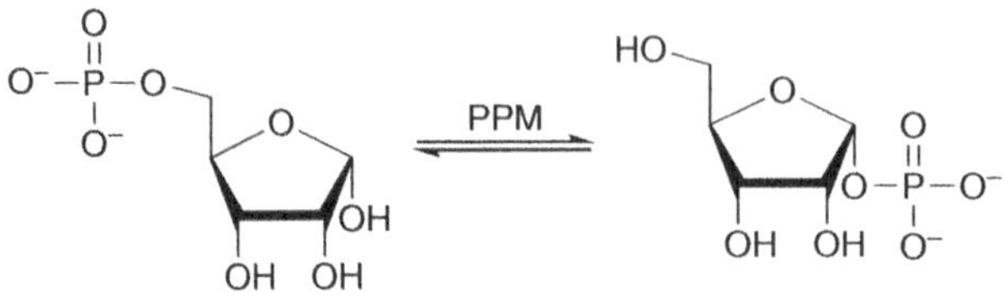

Scheme 6.36 Phosphopentomutase catalyzed reversible transfer of a phosphate group within a molecule.

and RNA biosynthesis [280]. PPM has been purified from both prokaryotes and eukaryotes, while the sequence and structure of PPMs from the two sources are unrelated [281–284]. The molecular identity of the mammalian PPM is still not ascertained. Bacterial PPMs are alkaline phosphatase superfamily members, which is a large and diverse family of dimetalloenzymes catalyzing phosphoryl or sulfonyl transfer via a conserved reaction cycle [283, 285]. The catalytic cycle for the interconversion of R5P and R1P proposed according to PPM from *Bacillus cereus* is a six-step mechanism: in step 1, the enzyme is phosphorylated and active; in step 2, the substrate, R5P encounters the active enzyme; in step 3, a phosphoryl group is transferred from the enzyme to the 1-OH of R5P, resulting in a ribose 1,5-bisphosphate (R1,5bP) chemical intermediate and an unphosphorylated enzyme intermediate; in step 4, the R1,5bP is reoriented in the enzyme active site; in step 5, the reoriented R1,5bP encounters the unphosphorylated enzyme; in step 6, a phosphoryl group is transferred from R1,5bP back to the enzyme, which results in the R1P product and an active phosphorylated enzyme [286, 287]. Study results also indicated that PPMs catalyze the reversible transfer of an intramolecular phosphate group between C1 and C5 carbon atoms of deoxyribose [288, 289].

The modified nucleosides are used as antiviral and antitumoral agents, thus they have an important impact on human health. The chemical preparation of nucleoside analogs often requires difficult and time-consuming multistep processes, particularly, the challenge of stereoselective glycosylation. Therefore, an alternative biotechnological preparation of nucleosides was employed instead of the traditional chemical synthesis. In a chemoenzymatic strategy, ribose, arabinose, and deoxyribose 5-phosphates were first chemoenzymatically synthesized from corresponding ribose, arabinose, and deoxyribose. These furanose 5-phosphates were used as starting materials to prepare natural and modified purine and pyrimidine nucleosides by coupling PPM and purine nucleoside phosphorylases (PNPs) overexpressed in *E. coli* [290]. Since out of the natural environment, PPM is unstable and its ability is affected by parameters such as pH and temperature. In order to stabilize PPM, PPM overexpressed in *E. coli* was irreversibly immobilized on several supports. The results showed that PPM in 10% (v/v) glycerol solution

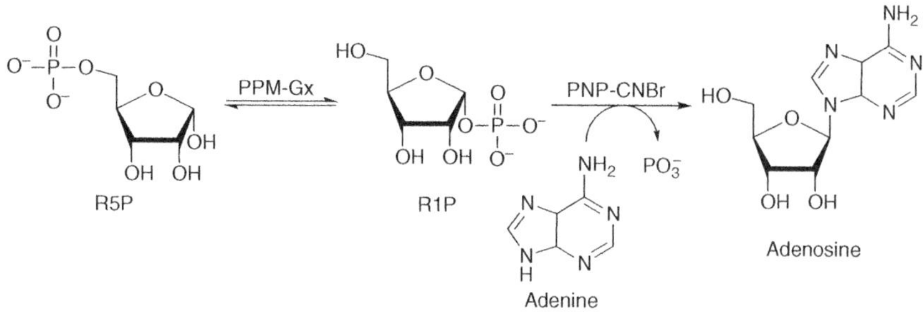

Scheme 6.37 Biosynthesis of adenosine from ribose 5-phosphate by immobilized PPM and stabilized PNP multienzymatic system. *Source:* Modified from Valino et al. [300].

covalently immobilized on glyoxyl-agarose (PPM-Gx) was most stable and activity evaluated for adenosine biosynthesis using PPM-Gx and stabilized PNP (PNP-CNBr) multienzymatic system from R5P gave a high yield (Scheme 6.37). The immobilized PPM was further used for the biosynthesis of ribavirin, a guanosine analog with a broad spectrum of antiviral activity, from R5P [291].

An *in vitro* non-natural enzymatic pathway to utilize the chemical energy stored in xylooligosaccharides from biomass to split water to produce a nearly theoretical yield of H_2 was proposed. This green hydrogen production pathway through water splitting at low temperatures was constructed on the basis of the novel enzymatic cascade activities of PPM catalyzing the conversion of D-xylose 1-phosphate into D-xylose 5-phosphate and of ribose 5-phosphate isomerase (R5P isomerase) catalyzing the conversions of D-xylose 5-phosphate and D-xylulose 5-phosphate (Scheme 6.38) [292].

2'-Deoxyribonucleoside is used as a building block of promising antisense drugs for cancer therapy, a synthesis intermediate for antiviral agents such as azidothymidine, and also a precursor of an indispensable material 2'-deoxyribonucleoside triphosphate for widespread PCR applications. Since PPM shows activity for the

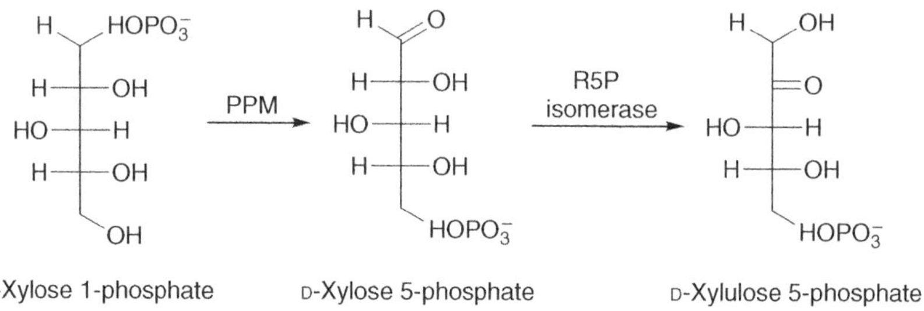

Scheme 6.38 Isomerization of D-xylose 1-phosphate into D-xylulose 5-phosphate catalyzed by enzymatic cascade reactions. *Source:* Based on Aberhart et al. [301]; Frey [302]; and Frey and Reed [303].

reversible transfer of an intramolecular phosphate group between C1 and C5 carbon atoms of deoxyribose, PPM was exploited for the biochemical production of 2'-deoxyribonucleoside via 2-deoxyribose 5-phosphate (DR5P) synthesis from glucose and acetaldehyde. In this multienzymatic cascade strategy, three enzymes involved are deoxyriboaldolase (DERA), which catalyze the condensation of acetaldehyde and glyceraldehyde 3-phosphate (G3P) to DR5P, PPM, which catalyzes intermolecular phosphate transfer from DR5P to 2-deoxyribose 1-phosphate (DR1P), and nucleoside phosphorylase (NP), which catalyzes 2'-deoxyribonucleoside formation from a nucleobase and DR1P. The production of DR5P was through the glycolysis of glucose using *Klebsiella pneumoniae* B-4-4 and DERA-overexpressing *E. coli* cells in the presence of ATP. Under the optimum conditions, a molar yield of 84% for DR5P was obtained in 2 h from acetaldehyde and dihydroxyacetone phosphate (DHAP) [293, 294].

The microbial production of 2'-deoxyribonucleosides from glucose, acetaldehyde, and a nucleoside was in essence through the reverse reactions of 2'-deoxyribonucleoside degradation and the glycolytic pathway. However, the glycolysis of glucose performed by baker's yeast (*Saccharomyces cerevisiae*) yields fructose 1,6-diphosphate (FDP) with ATP generated from AMP through alcoholic fermentation of the yeast. FDP is further transformed to DR5P in the presence of acetaldehyde by DERA-expressing *E. coli* cells via D-G3P. Subsequently, DR5P and a nucleobase can be transformed into 2'-deoxyribonucleosides via DR1P through the reverse degradation reactions of 2'-deoxyribonucleoside by *E. coli* transformants expressing PPM and NP (Scheme 6.39). The molar yield of

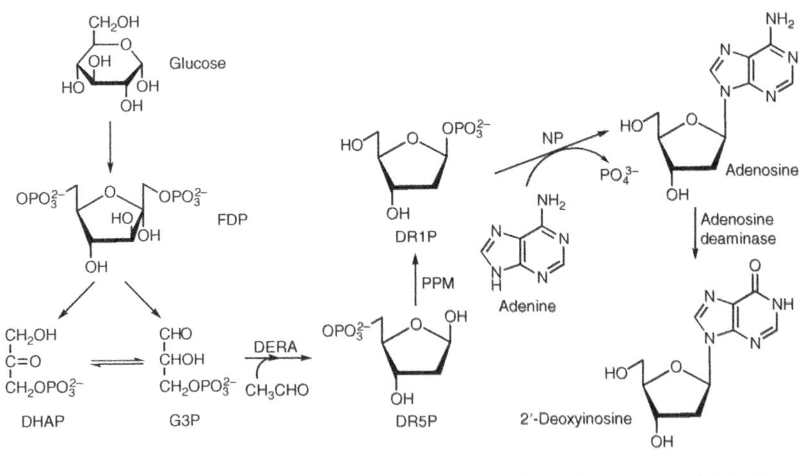

Scheme 6.39 Enzymatic synthesis of 2'-deoxyinosine from glucose, acetaldehyde, and adenine through the glycolytic pathway and the reverse reaction of 2'-deoxyribonucleoside degradation.

2'-deoxyinosine produced from glucose, acetaldehyde, and adenine as to glucose was 17.8% [295]. Based on the same method, a one-pot enzymatic synthesis of 2'-deoxyinosine from 100 mM glucose, 333 mM acetaldehyde, and 100 mM adenine was performed and about 33 mM 2'-deoxyinosine was produced in 24 h. The production of 2'-deoxyinosine from adenine was due to the adenosine deaminase activity of *E. coli* transformants [296]. A novel PPM and a DERA were found in hyperthermophilic archaeon, *Thermococcus kodakarensis*, which clearly indicate the presence of a metabolic link between central carbon metabolism and pentose biosynthesis and catabolism in *T. kodakarensis* [297]. An acetaldehyde-tolerant and phosphorylated compound-tolerant PPM was also found in *Bacillus sphaericus*, which is an excellent catalyst for the production of 2'-deoxyribonucleoside in the presence of acetaldehyde and phosphorylated compounds such as FDP and G3P derived from the glycolysis of glucose with yeast cells [298].

One of the two approaches used for the synthesis of 2-chloro-9-(2-deoxy-2-fluoro-β-D-arabinofuranosyl)adenine (clofarabine) starts with the cascade one-pot enzymatic transformation of 2-deoxy-2-fluoro-D-arabinose into 2-deoxy-2-fluoro-α-D-arabinofuranose-1-phosphate, followed by its condensation with 2-chloroadenine thereby affording clofarabine in ca. 48% yield in 24 h. In this approach, the sequential conversion of 2-deoxy-2-fluoro-D-arabinose into 2-deoxy-2-fluoro-α-D-arabinofuranose-1-phosphate was carried out by the following recombinant *E. coli* enzymes in sequence: ribokinase, PPM (no 1,6-diphosphates of D-hexoses as cofactor required), and PNP. The substrate activities of D-arabinose, D-ribose, and D-xylose in the similar cascade syntheses of the relevant 2-chloroadenine nucleosides were compared with the activities of 2-deoxy-2-fluoro-D-arabinose. The results indicated that D-ribose showed the best substrate activity (90% yield of 2-chloroadenosine in 30 min) [299].

Purine nucleoside analogs such as abacavir, fludarabine, and amdoxovir as well as pyrimidine ones such as stavudine, zidovudine, and lamivudine are widely used in clinical treatments for HIV. Since the preparation of nucleoside 5'-diphosphate by classical methodologies is complex, one-pot multistep enzymatic synthesis of pyrimidine nucleosides diphosphates from readily available reagents such as glucose was developed. For thymidine 5'-diphosphate synthesis, two consecutive one-pot multistep enzymatic processes were employed. In the first step, 2'-deoxyribose 5-phosphate (DR5P) was obtained from glucose by *Erwinia carotovora* whole cells containing DR5P aldolase (DERA) as a catalyst for the aldol addition between acetaldehyde and D-glyceraldehyde 3-phosphate (G3P), which is *in situ* generated by the glycolytic pathway, followed by the action of PPM and thymine phosphorylase (TP) affording thymidine in 85% conversion yield relative to DR5P. In the next step, the thymidine was converted to

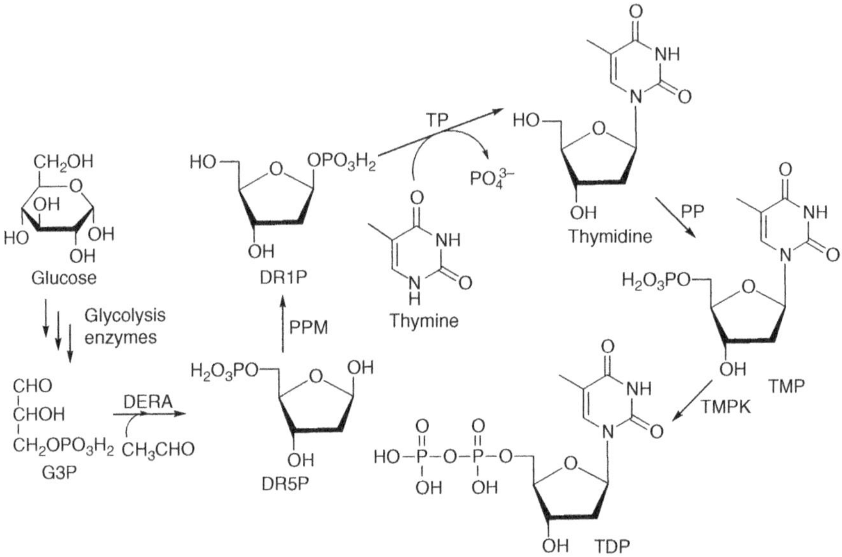

Scheme 6.40 One-pot multistep enzymatic synthesis of thymidine 5'-diphosphate from glucose. *Source:* Baumann et al. [320].

thymidine 5'-diphosphate (TDP) via thymidine 5'-monophosphate (TMP) by the combined action of nonspecific acid phosphatase (PP) in *E. coli* and thymidine monophosphate kinase (TMPK) in *S. cerevisiae* (Scheme 6.40) [300].

6.4.2 Lysine 2,3-Aminomutase

Lysine 2,3-aminomutase (2,3-LAM, EC 5.4.3.2) catalyzes the reversible interconversion of L-lysine and L-β-lysine, a reaction that involves the intramolecular migration of the α-amino group to the β-carbon with the concomitant transfer of the β-hydrogen to the α-carbon (Scheme 6.41) [301–303]. The enzyme contains PLP and [4Fe-4S] cluster and is activated by *S*-adenosylmethionine (SAM). This enzyme in *Clostridium subterminale* SB4 and in other anaerobic bacteria was the

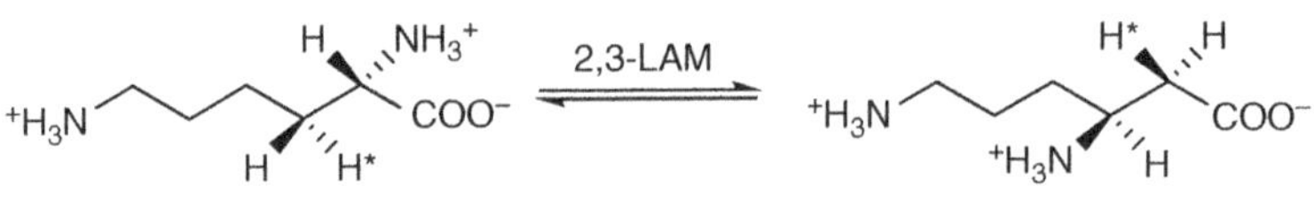

Scheme 6.41 Lysine 2,3-aminomutase catalyzes the interconversion of L-lysine and L-β-lysine. *Source:* Based on Frey et al. [303]; Wu et al. [324]; Morley and Stsdtman [325]; Morley and Stsdtman [326]; Morley and Stsdtman [327]; Morley and Stsdtman [328].

first aminomutase discovered and purified during investigations of lysine metabolism in *Clostridia* [304]. The enzyme purified from *C. subterminale* SB4 exhibits an apparent subunit molecular weight of 48 kDa as detected by sodium dodecyl sulfate polyacrylamide gel electrophoresis (SDS-PAGE), and the native enzyme has an apparent molecular weight of 285 kDa as detected by electrophoresis and gel filtration chromatography that shows a hexameric structure [304–306]. The enzyme contains cobalt as the major metal cofactor with small amounts of zinc and copper [307]. The structure of 2,3-LAM was determined by X-ray crystallography to 2.1-Å resolution. The unit cell contains a dimer of hydrogen-bonded, domain-swapped dimers, the subunits of which adopt a fold that contains all three cofactors in a central channel defined by six β/α structural units. Zinc coordination links the domain-swapped dimers [308]. In biosynthesis, β-lysine is a precursor for several antibiotics in *Streptomyces*, such as streptothricin F and viomycin [303, 309].

The generalized nonpolar, free radical reaction mechanism for the interconversion of L-lysine and L-β-lysine by 2,3-LAM from *C. subterminale* SB4 takes place in five steps: (1) L-lysine is bound as the external PLP-aldimine by transaldimination with the internal Lys337-aldimine of PLP; (2) SAM is assumed to be cleaved to generate the 5'-deoxyadenosyl radical in a reversible process, and the resultant radical abstracts the 3-pro-*R* hydrogen from the lysyl side chain to generate the substrate-related radical 1 and 5'-deoxyadenosine, which remains bound to the enzyme; (3) radical 1 undergoes isomerization to the azacyclopropylcarbinyl radical 2 and then to the L-β-lysine-related radical 3; (4) hydrogen abstraction from 5'-deoxyadenosine generates the external PLP-aldimine of L-β-lysine and regenerates SAM; (5) transaldimination with Lys337 releases the product and regenerates the internal Lys337-aldimine of PLP [303]. In addition, LAM was identified to be encoded by the *yodO* gene of *B. subtilis*, which similarly employing the above-described mechanism catalyzes the interconversion of L-lysine and L-β-lysine [310, 311]. There was also a report to show that, in the absence of chemical reductants, 2,3-LAM from *Porphyromonas gingivalis* can be activated using a reducing protein such as flavodoxin-$NADP^+$ reductase, flavodoxin, or ferredoxin [312].

The stereochemistry of the reversible interconversion of L-α-lysine and L-β-lysine catalyzed by 2,3-LAM from *C. subterminale* SB4 has been elucidated and the corresponding equilibrium constants as well as binding energy have been studied in detail as well [301, 313–315]. While research evidence indicates that 2,3-LAM from *E. coli* catalyzes (*S*)-α-lysine (L-α-lysine) to the product (*R*)-β-lysine (D-β-lysine), an opposite stereochemistry exhibited by 2,3-LAM from *C. subterminale* SB4. The radical in the reaction of *E. coli* 2,3-LAM has the (*R*)-configuration at the active site that has been confirmed by electron paramagnetic resonance (EPR) studies. Therefore, even though (*S*)-β-lysine is not a substrate of *E. coli* 2,3-LAM, it undergoes hydrogen abstraction to form an (*S*)-β-lysine-related radical with the same stereochemistry of hydrogen transfer from C2 of (*S*)-β-lysine to the 5'-deoxyadenosyl radical as in the action of the *Clostridium* 2,3-LAM. The β-lysyl radical formed possesses an (*R*)-configuration, which is different from that at the active site of the *Clostridium* 2,3-LAM [316].

Streptothricin F is a representative of a large family of ubiquitous nitrogen-containing antibiotics, which includes the β-lysine moiety in its chemical structure. The incorporation of the β-lysine moiety into streptothricin F has been reported by a chemoenzymatic synthetic method. In this method, DL-lysine was the source of β-lysine, which was chemically synthesized from (bromopropyl)phthalimide with 9% overall yield. Then, DL-lysine and pure helianthate salt were microbial transformed to produce streptothricin F using *Streptomyces* L-1689-23 as the catalyst. It was shown that only L-α-lysine was utilized for the formation of the β-lysine moiety in streptothricin F [309, 317].

The tuberactinomycins are a family of nonribosomal peptide antituberculosis drugs to treat the bacterium *Mycobacterium tuberculosis*. Tuberactinamine A, tuberactinomycin A, and tuberactinomycin B (viomycin) are members of this antibiotic family with structural difference shown in Scheme 6.42. The general structural difference between tuberactinamine A and tuberactinomycins is the β-lysine side chain present in tuberactinomycins but not in tuberactinamine A. It has been proved that the β-lysine side chain is derived from L-lysine by the action of 2,3-LAM. The fermentation of *Streptomyces griseoverticillatus* var. *tuberacticus* NRRL 3482 was used to produce predominantly tuberactinomycin A with minor

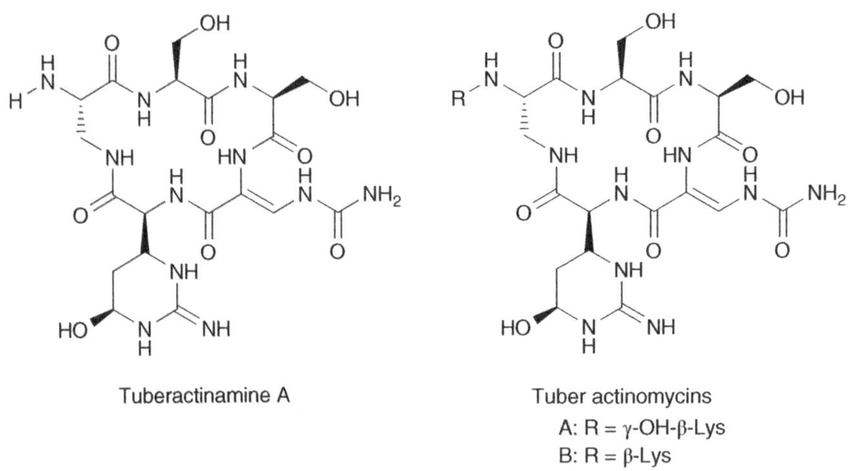

Scheme 6.42 Chemical structures of tuberactinamine A, tuberactinomycin A, and tuberactinomycin B (viomycin). *Source:* Based on Kreimeyer et al. [329]; Berkovitch et al. [333]; Sandal et al. [337].

viomycin and other tuberactinomycins. In accordance, the inhibitor of 2,3-LAM, (*S*)-2-aminoethyl-L-cysteine, was demonstrated for the efficient production of tuberactinamine A during the fermentation of *Streptomyces griseoverticillatus* var. *tuberacticus* NRRL 3482 [318]. Later, viomycin was also produced by the heterologous biosynthesis of *Streptomyces lividans* 1326 [319].

The strategy used by the majority of living cells is the accumulation of small, soluble, organic molecules that are termed compatible solutes. The unique compatible solute N^{ε}-acetyl-β-lysine found in methanogenic archaea is produced at high salinities to cope with the salt stress. The 2,3-LAM and β-lysine acetyltransferase genes, *ablA* and *ablB*, were identified and characterized from the methanogenic archaea *Methanosarcina mazei* Gö1, *Methanosarcina acetivorans*, *Methanosarcina barkeri*, *Methanococcus jannaschii*, and *Methanococcus maripaludis*, which were proved to be essential for the biosynthesis of N^{ε}-acetyl-β-lysine via L-lysine (Scheme 6.43) [320]. Furthermore, a mutant of *M. mazei* Gö1 by deletion of the genes that encode the 2,3-LAM (*ablA*) and the β-lysine acetyltransferase (*ablB*) could still grow well at low to intermediate salinities and the growth was only slowed down at high salt concentration but not impaired. Analysis results revealed an increased glutamate pool in the mutant and a novel solute, alanine, was produced, which indicating alanine is used as a compatible solute by *M. mazei* Gö1 [321]. The production of the archaeal osmolyte N^{ε}-acetyl-β-lysine also can be performed by homologous overexpression of the *yodP-kamA* genes, that encode a *ablA*-related 2,3-LAM and *ablB*-related putative acetyltransferase, in *B. subtilis*. The recombinant *B. subtilis* synthesized considerable amounts (0.28 μmol mg^{-1} protein) of N^{ε}-acetyl-β-lysine [322]. The 2,3-LAM (*ablA*) genes from the marine *M. mazei* N2M9705, halotolerant *Methanocalculus chunghsingensis* K1F9705b, and halophilic *Methanohalophilus portucalensis* FDF1 were cloned by PCR southern hybridization. *AblAs* from methanoarchaea are

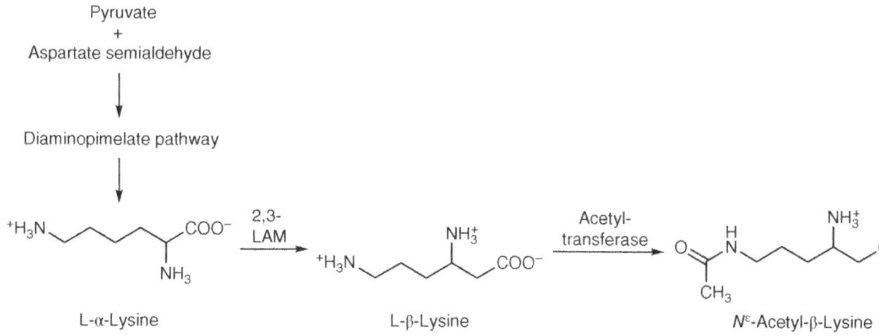

Scheme 6.43 Pathway of N^{ε}-acetyl-β-lysine synthesis in methanogenic archaea. *Source:* Modified from Wu et al. [324].

characterized and identified as 2,3-LAM, which were then applied for the synthesis of β-lysine *in vivo* and *in vitro* [323].

6.4.3 Lysine 5,6-Aminomutase

A PLP and adenosylcobalamin (coenzyme B_{12}, AdoCbl)-dependent aminomutase lysine 5,6-aminomutase (5,6-LAM, EC 5.4.3.3) was initially discovered in the studies of lysine metabolism in anaerobic bacteria, such as *C. sticklandii* and *Porphyromonas gingivalis*, that catalyzes the reversible interconversion of (*S*)-β-lysine (L-β-lysine) and 3,5-diaminohexanoic acid as well as (*R*)-lysine (D-lysine) and 2,5-diaminohexanoic acid (Scheme 6.44) [303, 324–328]. 5,6-LAM from *C. sticklandii*, a gram-positive anaerobe, participates in the second step of the fermentation pathway of lysine in which lysine is transformed to acetic acid, ammonia, and butyric acid [329].

The purified native 5,6-LAM comprises two protein components, the core enzyme E_1 is a 170 kDa heterotetramer composed of two α- (~52 kDa) and two β- (~32 kDa) subunits and formulated as $(\alpha/\beta)_2$, whereas the auxiliary activating protein E_2 has an approximate molecular mass of 80 kDa. The crystal structure of the PLP and adenosylcobalamin bound 5,6-LAM complex studied by X-ray crystallography shows that the α-subunit harbors a PLP-binding triosephosphate isomerase-like (TIM) barrel with PLP embedded and liked to the β-subunit through the aldimine bond to Lys144β and the β-subunit incorporates an AdoCbl-binding Rossmann-like domain. The crystal structure shows adenosylcobalamin cleaved to cobalamin and 5'-deoxyadenosine [303, 324, 328, 330–333]. The reaction mechanism of 5,6-LAM is similar to that of 2,3-LAM by substituting the adenosylcobalamin cofactor for SAM as the source of the 5'-deoxyadenosyl radical and the ε-amino group of the substrate forming the external aldimine with PLP and undergoing migration between C6 and C5 so that the substrate (*S*)-β-lysine is

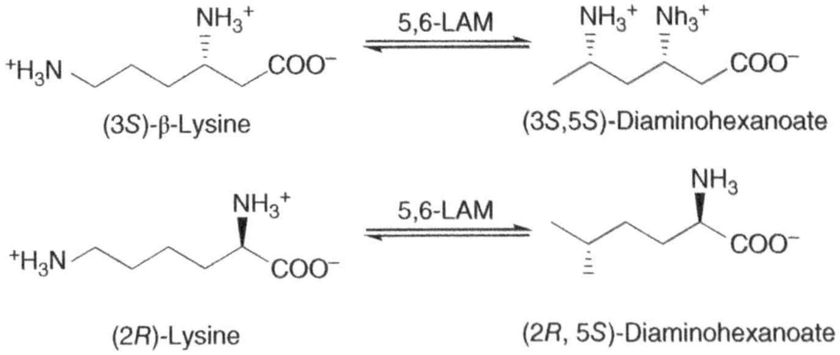

Scheme 6.44 Stereochemical reactions catalyzed by lysine 5,6-aminomutase. *Source:* Based on Cooke and Bruner [345]; Lin et al. [346] and Huang et al. [347].

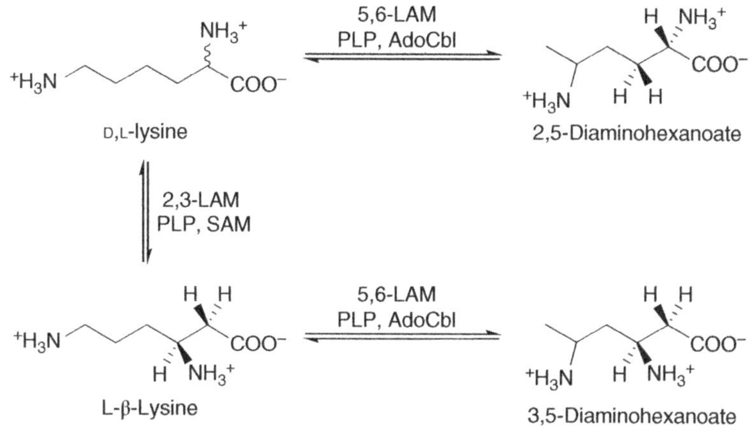

Scheme 6.45 Reaction of racemic lysine catalyzed by lysine 2,3-aminomutase and/or lysine 5,6-aminomutase. *Source:* Based on Rachid et al. [348]; Rachid et al. [349]; Krug and Müller [350].

catalyzed to produce (3*S*,5*S*)-diaminohexanoate [333–336]. The reaction of 5,6-LAM is believed to follow an intramolecular radical rearrangement mechanism involving the formation of a cyclic azacyclopropylcarbinyl radical intermediate. The arrangement of PLP and adenosylcobalamin binding sites in internal aldimine is presented as a locking mechanism preventing adventitious cleavage of the cobalt-carbon bone, with the potential complications of exposing the enzyme to the destructive effects of radical side reactions in the absence of the substrate [303].

As early as the 1950s, it has been demonstrated that L-α-lysine was catabolized to acetate, butyrate, and ammonia by *C. sticklandii* and by two strains of *E. coli*. The first two steps of the fermentation pathway are catalyzed by 2,3-LAM and 5,6-LAM, respectively, which are responsible for the transformation of L-α-lysine to the intermediate (3*S*, 5*S*)-diaminohexanoate (Scheme 6.45). The *in vitro* combination of 2,3-LAM and 5,6-LAM also provides a method for the simultaneous production of 2,5-diaminohexanoate and 3,5-diaminohexanote (Scheme 6.45) [329, 333, 337].

6.4.4 Tyrosine 2,3-Aminomutase

In enzymology, tyrosine 2,3-aminomutase (TAM, EC 5.4.3.6) is a 4-methylideneimidazole-5-one (MIO) cofactor dependent aminomutase that catalyzes the amine shift isomerization of L-tyrosine to (*S*)-β-tyrosine (Scheme 6.46) [324]. This enzyme is also called tyrosine α,β-amino mutase, which was first isolated from the

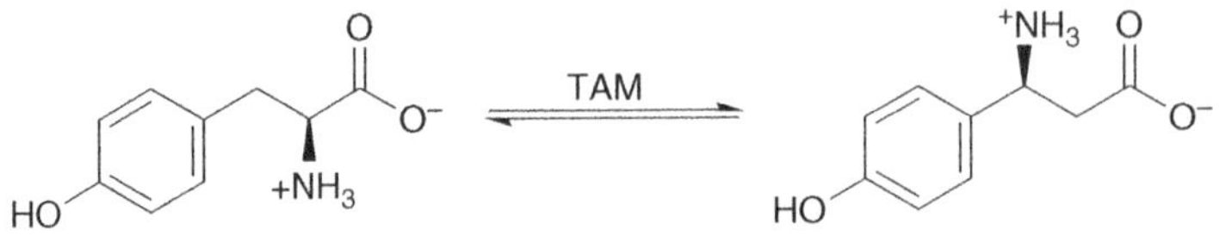

Scheme 6.46 Tyrosine 2,3-aminomutase catalyzed isomerization of L-tyrosine to (*S*)-β-tyrosine. *Source:* Spiteller et al. [354].

bacterial cells of *Bacillus brevis* Vm4 and characterized with a molecular mass of 75 kDa [338]. This enzyme is also found in the bacterium *Streptomyces globisporus*, which was termed *Sg*TAM and was used for the biosynthesis of C-1027, an enediyne antitumor antibiotic of chromoprotein family [339, 340]. The X-ray crystal structure of *Sg*TAM was solved at 2 Å resolution, which provides strong evidence to an MIO-containing aminomutase with structural homology to ammonia lyases. However, the crystal structure shows that *Sg*TAM has a closed active site well suited to retain ammonia and minimize the formation of lyase elimination products [341]. The mechanism of this enzyme has been extensively studied that employs the cofactor MIO in the active site as the potent electrophile formed by the self-condensation of the tripeptide sequence Ala-Ser-Gly in the protein backbone. The α-amine of L-tyrosine adds into the electrophilic moiety of MIO via a conjugate addition. Formation of this covalent adduct facilitates deprotonation of the β-hydrogen and elimination of ammonia to produce *p*-hydroxycinnamate. The enzyme-bound ammonia and *p*-hydroxycinnamate in the active site are then allowed readdition at the β-position of the substrate. Ultimately, the product (*S*)-β-tyrosine is released to regenerate the prosthetic group MIO [342, 343].

The cultures of *B. brevis* Vm4 were used for the production of two peptide antibiotics, edeine A and edeine B, that contain a group of novel amino acids including β-tyrosine. In the cultures, β-tyrosine is formed by the isomerization of L-α-tyrosine catalyzed by TAM [338, 344]. The enediyne antitumor antibiotic C-1027 is a natural secondary metabolite produced by *S. globisporus*. C1027 is composed of an apoprotein (CagA) and a reactive enediyne chromophore, which has four distinct chemical moieties, including a fascinating and unique β-amino acid moiety, (*S*)-3-chloro-4,5-dihydroxy-β-phenylalanine. The biosynthesis of the β-amino acid moiety starts from L-tyrosine requiring five enzymes in which the tyrosine aminomutase *Sg*TAM is the first enzyme employed for the formation of (*S*)-β-tyrosine in the biosynthetic pathway. Then, a sixth enzyme catalyzes the incorporation of this β-amino acid moiety into C-1027 (Scheme 6.47) [345–347].

The highly cytotoxic chondramides are mixed nonribosomal peptide/polyketide secondary metabolites produced by the myxobacterium *Chondromyces crocatus* Cm c5, which exhibit striking structural similarity to the antifungal marine depsipeptides jaspamides characterized by an unusual (*R*)-β-tyrosine moiety.

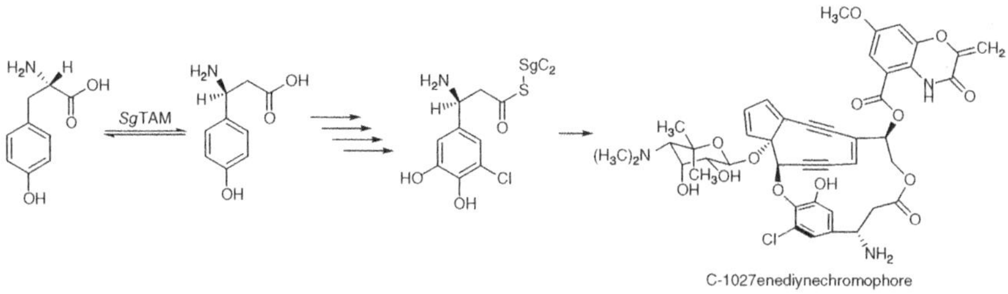

Scheme 6.47 Biosynthetic pathway for the β-amino acid moiety of the C-1027 enediyne chromophore from L-tyrosine. *Source:* Based on Wu et al. [324]; Walker et al. [355]; Steele et al. [356].

Chondramides belong to a family of molecules, which potently affect the function of the actin cytoskeleton. (*R*)-β-tyrosine in chondramides was demonstrated to be produced directly from L-tyrosine by the tyrosine aminomutase CmdF in *C. crocatus* (*Cc*TAM) through heterologous gene expression and site-directed mutagenesis. The biosynthetic pathway of chondramides from L-tyrosine to (*R*)-β-tyrosine and its incorporation into the natural product was also elucidated through sequenced chondramides gene cluster identification and characterization for polyketide synthases (PKSs) and nonribosomal polypeptide synthetases (NRPSs) (Scheme 6.48) [348–350]. The cryptic stereochemistry of the CcTAM mechanism was evaluated by deuterium-labeled isotopomers of α-tyrosine. The results revealed that the α-amino group was transferred to Cβ of the phenylpropanoid skeleton with retention of configuration and the pro-(3*S*) proton exchanges with protons from the bulk media during its migration to Cα during catalysis. The migratory proton attached to Cα of the product also with retention of configuration. The enantioselectivity of *Cc*TAM yields 85% (*R*)-β-tyrosine yet also forms 15% (*S*)-β-tyrosine through inversion of configuration. Interestingly, the production of (*S*)-β-tyrosine increased with solvent pH [351] In addition to bacteria, the

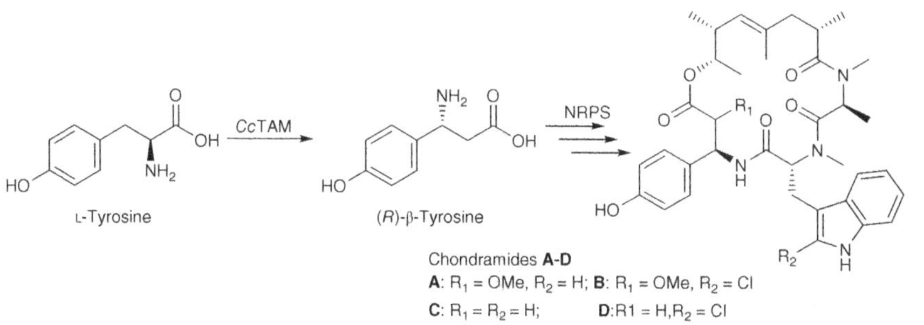

Scheme 6.48 Biosynthetic pathway for the synthesis of chondramides from L-tyrosine via (*R*)-β-tyrosine. *Source:* Cox et al. [373].

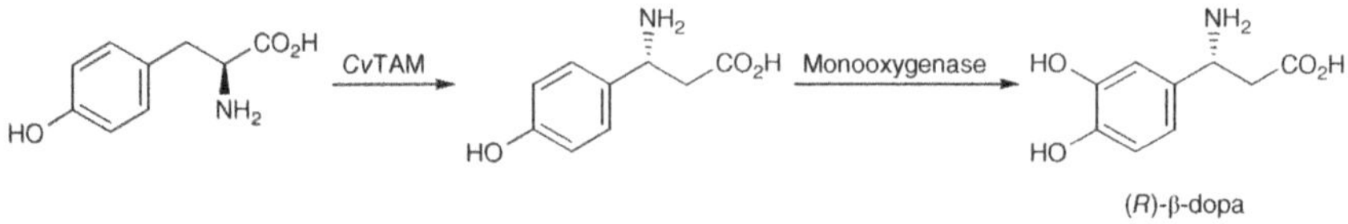

Scheme 6.49 Biosynthesis of (*R*)-β-dopa in *C. violaceous* with L-tyrosine as precursor. *Source:* Wu et al. [374].

MIO-dependent tyrosine aminomutase was also discovered from Japanese cultivated rice *Oryza sativa* (*Os*TAM). *Os*TAM converts L-tyrosine to a mixture of (*R*)- and (*S*)-β-tyrosines with unique high enantioselectivity and retention of configuration to 94% *e.e.* in (*R*)-β-tyrosine, which does not change with pH [352, 353].

The new natural β-amino acid (*R*)-3,4-dihydroxy-β-phenylalanine ((*R*)-β-dopa) was found present in the mushroom *Cortinarius violaceus* as the iron(III)-catechol complex to give the blue-purple color of their fruit bodies. The biosynthesis of (*R*)-β-dopa in *C. violaceus* was proved to proceed through (*R*)-β-tyrosine using L-tyrosine as the substrate, which is catalyzed by the tyrosine 2,3-aminomutase (*Cv*TAM) and (*R*)-β-tyrosine was subsequently hydroxylated by a monooxygenase (Scheme 6.49) [354]. This study also showed that *rac*-3-fluorotyrosine was able to produce 5-fluoro-β-dopa with traces of 3-fluoro-β-tyrosine. The investigation of the substrate specificity of *Cv*TAM using both (*S*)-3-fluorotyrosine and racemic mixture was performed with *C. violaceus* that revealed a strictly stereospecific to give only (*R*)-enantiomer 3-fluoro-β-tyrosine for both experiments. Therefore, *Cv*TAM transforms exclusively (*S*)-3-fluorotyrosine. Likewise, the hydroxylation of 3-fluoro-β-tyrosine is strictly stereospecific to (*R*)-5-fluoro-β-dopa [354].

6.4.5 Phenylalanine Aminomutase

6.4.5.1 L-β-Phenylalanine Forming

The phenylalanine aminomutase from *Taxus chinensis* (*Tc*PAM, EC 5.4.3.10) catalyzes the reversible isomerization of L-α-phenylalanine ((2*S*)-α-phenylalanine) to L-β-phenylalanine ((3*R*)-β-phenylalanine), which is the first committed step in the biosynthesis of the C-13 side chain of the widely used anticancer alkaloid Taxol (paclitaxel) (Scheme 6.50) [324, 355, 356]. *Taxus* PAM also has low phenylalanine ammonia lyase (PAL) activity to catalyze the α,β-elimination of ammonia from the α-isomer to form *trans*-cinnamic acid. However, the first detection of this kind of aminomutase was from the cell-free extracts of *Taxus brevifolia* tissue [357]. As PAM cDNA from *Taxus cuspidata* was cloned and heterologously expressed in *E. coli*, the enzyme was confirmed virtually identical to the recombinant enzyme from *T. chinensis* involved in Taxol biosynthesis [355]. Purified PAM

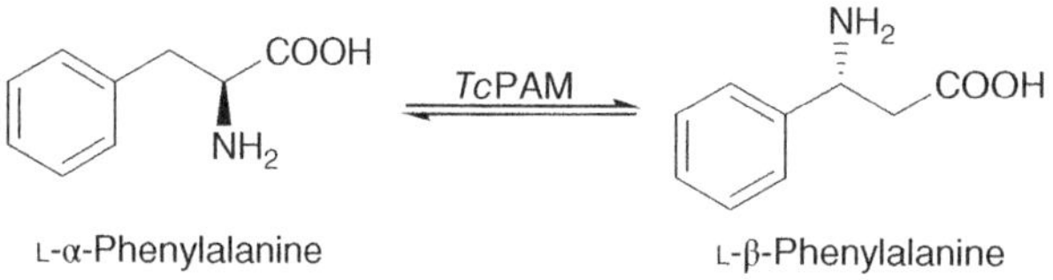

Scheme 6.50 Phenylalanine aminomutase from *Taxus chinensis* catalyzed L-α-phenylalanine to (*R*)-β-phenylalanine. *Source:* Kajani et al. [375]. Licensed under CC BY 3.0.

from *T. chinensis* cell cultures shows a K_m of 1.1 mM, a V_{max} of 110.1 μm min^{-1} mg^{-1} protein, a pH optimum of 7.5–8.0, and the PAM cDNA deduced molecular mass of 75.3 kDa [356].

The first X-ray crystal structure of a PAM from *Taxus canadensis* (*Tca*PAM) was determined at 2.4 Å resolution. *Tca*PAM is a tetramer with protomers in the same relative orientation as observed for the PAL and *Sg*TAM, and each monomer has a molecular mass of 76 kDa. The active site of the *Tc*PAM contains the cofactor 4-methylidene-imidazole-5-one (MIO) prosthesis, observed in all catalysts of the class I lyase-like family. The active site of the *Tc*PAM structure also is complex with (*E*)-cinnamate [358]. The later found crystal structure of PAM from *T. chinensis* (*Tc*PAM) also shows a tetramer structure with four monomers assembling into a dimer of dimers. The *Tc*PAM monomer has an elongated shape and contains three domains, the tail domain, the body domain, and the head domain. The head domain contains the MIO group and the lid loop, which covers the active site [359]. The reaction mechanism studied by experimental and computational methods revealed that the MIO-dependent *Taxus* sp. PAM catalyzes the isomerization of (*S*)-α-phenylalanine to (*R*)-β-phenylalanine through the exchange of the amino group from C2 of the substrate to C3 of the product and the pro-C3*S* hydrogen of the substrate to C2 of the product via a *trans*-cinnamic acid intermediate as well as an aminated-methylidene imidazolone (NH_2-MIO) adduct with both retention of configuration [358–366].

A range of arylalanine substrates with various substituents on the phenyl ring can be catalyzed by *Taxus* sp. PAM to the corresponding (*R*)-β-arylalanines [367]. *Para*-fluoro-phenylalanine had the best activity whereas some substrates had very poor activity. While the results indeed show that the native PAM is able to accept some substrates with substituents on the aromatic ring, no clear trend in the catalytic rate emerged in relating to electron-withdrawing or electron-donating substituents. Since *trans*-cinnamic acid (*t*-CA) is an intermediate in the mutase catalyzed isomerization reaction, PAM was used for catalyzing the addition of ammonia to substituted cinnamic acid to produce enantiopure α- and β-phenylalanine. Due to the high stereoselectivity of PAM, only (*R*)-α-phenylalanine and (*S*)-β-phenylalanine are formed. As in the isomerization reaction, PAM also accepts a variety of (*E*)-cinnamates with different substituents, such as alkyl, alkoxy, halogen, nitro, on the phenyl ring [368, 369]. The regioselectivity of the conversion indicated from the results demonstrates that substrates with electron-donating groups on the phenyl ring are preferably converted to β-amino acids, whereas cinnamic acids with electron-withdrawing groups on the phenyl ring produce mainly α-amino acids. Further studies have shown that the regioselectivity of the MIO-dependent *Tc*PAM can be tailored toward β-regioselectivity. An engineering redesigned *Tc*PAM mutant allows the synthesis of almost pure (*R*)-β-phenylalanine and its derivatives by one-step asymmetric amination of cinnamic acid [370]. Similarly, two amino acid residues (Asn458 and Leu108) in active sites of *Tc*PAM were mutated to yield variant *Tc*PAM-Asn458Phe/Leu108Glu, which improves the β-selectivity of the hydroamination of *t*-CA to β-phenylalanine approximately 5.2-fold higher than the wild-type *Tc*PAM and decreases the α-selectivity to 68%. The percentage of β-phenylalanine in the product mixture increased from 42 to 83% [371]. The mutant was also applied to synthesize substituted β-arylalanines with various substituent groups on the phenyl ring using substituent *t*-CA as a substrate. The regioselectivity was affected by the electronic properties and position of the substituent groups, and the 4-methoxy and methyl substituent *t*-CA were transferred into their corresponding β-arylalanines with conversion rates larger than 90%. Furthermore, an MIO-dependent *Tca*PAM was repurposed to irreversibly catalyze the conversion of ring-substituted *trans*-cinnamate epoxide racemates to their corresponding arylserines. This catalyst also converted each cinnamate epoxide analog to its corresponding isoserine, which play a key role in building the pharmacophore seen in anticancer and protease inhibitor drugs [372].

A dynamic kinetic resolution process was developed by coupling *Tca*PAM with a broad-spectrum alanine racemase from *P. putida* in order to achieve a yield greater than 50% for the racemic α-arylalanines to the corresponding (*R*)-β-arylalanines. The use of a range of different arylalanines as substrates was investigated as well. The inclusion of a biocatalytic racemization step alongside the PAM-catalyzed reaction moderately increased the overall yield of the enantiomerically pure β-arylalanines between 4 and 19% (depending on the arylalanine), which corresponded to as much as a 63% increase compared to the turnover with the aminomutase reaction alone (Scheme 6.51) [373].

A simple and efficient enzymatic tandem reaction was employed for preparing a series of enantiomerically pure (*S*)-β-phenylalanine and its analogs from the corresponding racemates. In this approach, an efficient kinetic resolution of (*R*)-β-phenylalanine in the racemates is initially enantioselective isomerization to (S)-α-phenylalanine by *Tc*PAM. The latter is then deaminated *in situ* to the corresponding cinnamic acids by phenylalanine ammonia lyase from *Rhodosporidium*

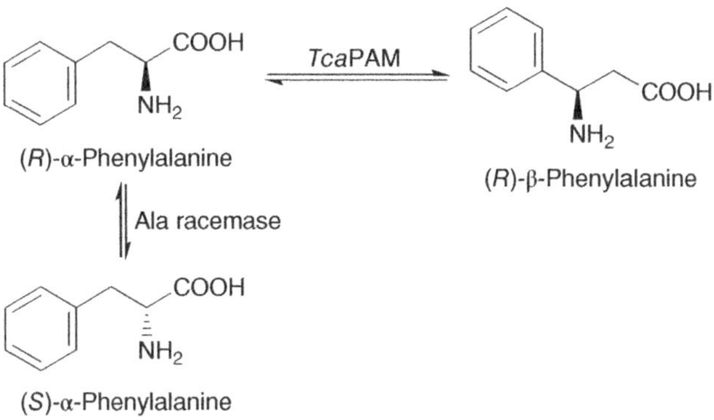

Scheme 6.51 Dynamic kinetic resolution of racemic α-phenylalanine to yield (*R*)-β-phenylalanine using combined enzymes.

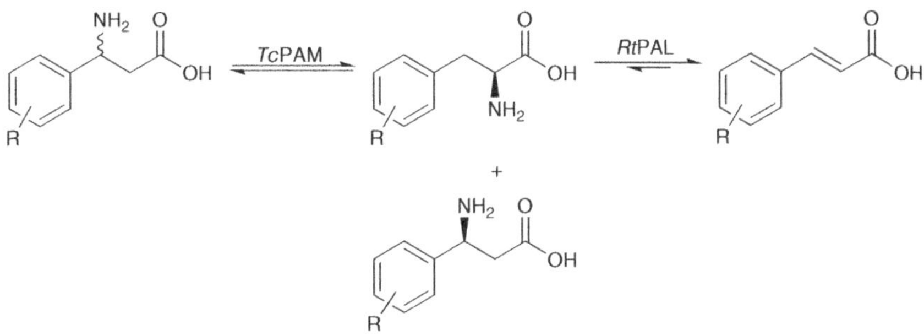

Scheme 6.52 Kinetic resolution of racemic β-phenylalanine and its derivatives using coupled PAM and PAL. *Source:* Csuka et al. [384].

toruloides (*Rt*PAL) (Scheme 6.52) [374]. The use of the latter enzyme overcomes the unfavorable equilibrium to convert a range of aromatic (*S*)-β-amino acids with excellent enantioselectivity of 99% *e.e.* Small scale preparative reactions resulted in *ca.* 50% yield for the aromatic (*S*)-β-amino acids.

A broad range of natural diterpenoids known as toxoids (taxanes) are produced by *Taxus* species and approximately 350 of them have been identified. Among the identified toxoids, taxol has been proven to be one of the most important anticancer drugs with powerful effects against a range of cancers. In general, the molecular structure of taxol consists of two components: the taxane ring moiety (10-deacetylbaccatin III) derived from the terpenoid pathway and C-13 side chain derived from the phenylpropanoid pathway (Scheme 6.53) [375]. There are about 20 distinct enzymatic steps for the taxol biosynthesis, while the taxane ring moiety is much more readily available, the low content of taxol should be related to the side chain

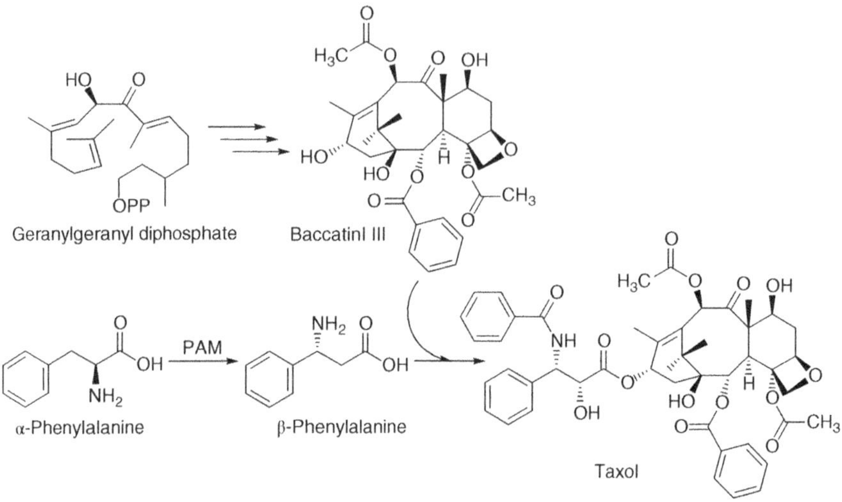

Scheme 6.53 Conversion of α-phenylalanine to β-phenylalanine catalyzed by PAM and its attachment to the C-13 side chain of Baccatin III in the biosynthesis of Taxol. *Source:* Based on Jin et al. [377]; Magarvey et al. [378]; Ratnayake et al. [385].

assembly. The first step in the side chain assembly of taxol involves the conversion of α-phenylalanine to β-phenylalanine catalyzed by PAM, which has been first demonstrated in the cell-free extracts of *T. brevifolia* [357]. PAM gene from *T. cuspidata* and *T. chinensis* has been cloned and functionally expressed in *E. coli* and the corresponding PAM has been characterized and employed for the taxol's β-phenylalanine side chain biosynthesis [355, 356]. In addition, the PAM gene from *Taxus baccata L.*, which contains higher content of baccatin III ring moiety of taxol, was cloned, expressed, and used for the taxol side chain synthesis [375].

6.4.5.2 D-β-Phenylalanine Forming

The MIO-dependent phenylalanine aminomutase from the bacterium *Pantoea agglomerans* (*Pa*PAM, EC 5.4.3.11) catalyzes the isomerization of L-phenylalanine ((2*S*)-α-phenylalanine) and D-β-phenylalanine ((3*S*)-β-phenylalanine) (Scheme 6.54), which is the enantiomer of the product made by the mechanistically similar aminomutase *Tc*PAM [376]. *Pa*PAM activity was first reported and designated as AdmH in the antibiotic andrimid production by *Pantoea agglomerans* (formerly *Erwinia herbicola*) using gene cluster cloned into *E. coli*, which functions the formation of β-Phe [377]. The activity of aminomutase AdmH for its specific conversion from L-Phe to (*S*)-β-Phe was further confirmed in the study concerned about the early stages of the andrimid biosynthetic assembly line [378]. *Pa*PAM is a member of the class I lyase family that includes aminomutase PAMs and TAMs, as well as ammonia lyase PALs, TALs, and HALs.

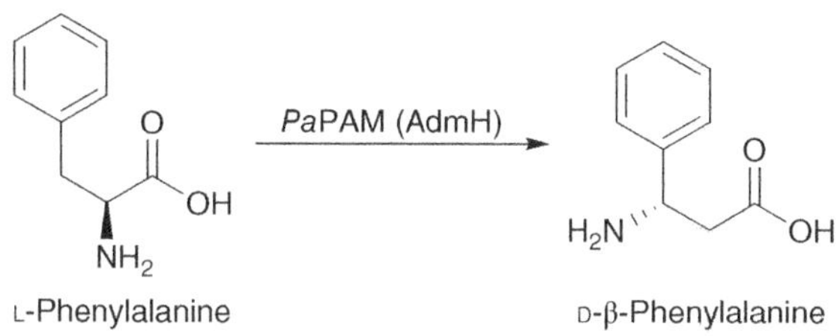

Scheme 6.54 Isomerization of L-phenylalanine to (*S*)-β-phenylalanine catalyzed by *Pa*PAM.

The X-ray crystal structure of *Pa*PAM was determined at 1.7 Å resolution with cinnamate included in the crystallization buffer, which shows similar architecture to the class of MIO-dependent aminomutases and ammonia lyases. *Pa*PAM is a tetramer with the configuration of dimer of dimers and each subunit contains an active site. The inner-loop region is just above the active site and is packed toward MIO. The *Pa*PAM MIO is made autocatalytically from three amino-acid residues Thr-Ser-Gly instead of from the common Ala-Ser-Gly [379]. The crystal structure of *Pa*PAM was solved as α- and β-phenylpropanoid adducts, which provide that *Pa*PAM reacts by an alkylamine elimination pathway (such as a Hoffmann-type or E2 elimination process) as demonstrated previously for *Sg*TAM. The carbon skeleton of the (*S*)-phenylalanine substrate remains in one rotameric conformation, while the exocyclic C–N bond of the amine-MIO adduct rotates into position below the α- and β-carbon atoms to complete the isomerization reaction and results in an inversion of configuration at each migration terminus [379]. Later investigation of the reaction mechanism using MS and NMR for recombinant *Pa*PAM also shows that the isomerization reaction is performed through the intramolecular exchange of NH_2 with *pro*-3*R* hydrogen of α-phenylalanine with inversion of configuration. The *Pa*PAM shuttles the α-NH_2 of α-phenylalanine to β site to replace the *pro*-3*R* hydrogen and the *pro*-3*R* hydrogen is simultaneously shifted to α site to produce β-phenylalanine. In addition, Phe455 in the active site is the key residue to control the exchange way by changing the binding orientation of the carboxyl group of the intermediate *trans*-cinnamic acid to control the NH_2-H pair exchange [380]. Moreover, factors that affect MIO-dependent mutase and lyase activity were explored for enzyme AdmH. The results reveal that the enzyme AdmH displays thermal bifunctionality, with mutase activity predominant at lower temperature and lyase activity preferred at a higher temperature, which is associated with the higher-energy barrier of the movement of the active-site loops. In addition, the MIO-dependent β-lyase activity is consistent with the amino-MIO adduct mechanism as well [381].

Since *Pa*PAM has a broad spectrum of substrate specificity, several commercially available (*S*)-arylalanine substrates were selected to survey the substrate specificity of *Pa*PAM for making non-natural β-amino acids. Compared to phenylalanine, substrates containing substituents that were either electron-withdrawing or -donating through resonance or inductive effects affected the k_{cat} of PAM. Generally, the turnover and catalytic efficiency of *Pa*PAM for the *meta*-isomers were better than for the corresponding *para*- and *ortho*-isomers, with some exceptions. *Pa*PAM principally synthesizes the β-amino acids at >90% and the cinnamate by-products at <10% for 11 of the 19 productive substrates. The yield for those less productive substrates was 14–65% of the cinnamate analog. The computational chemistry was also used to gain further insights on the substrate selectivity of *Pa*PAM that the steric barriers created by specific active-site residue interactions with the substituted aryl portion of the substrate significantly influence the substrate selectivity [382]. Substrate specificity of *Pa*PAM was further surveyed with a wide range of racemic α- and β-arylalanines. Both α- and β-arylalanines were accepted as substrates when the aryl moiety was relatively small, like phenyl, 2-, 3-, 4-flurophenyl or thiophen-2-yl. While 2-substituted α-arylalanines bearing bulky electron-withdrawing substituents did not react, the corresponding substituted β-aryl analogs were converted rapidly. Conversion of 3- and 4-substituted α-arylalanines happened smoothly, while conversion of the corresponding β-arylalanines was poor or nonexistent. In the range of pH 7–9, there was no significant influence on the conversion of racemic α- and β-(thiophene-2-yl)alanines, whereas increasing the concentration of ammonia inhibited the isomerization progressively and decreased the amount of the by-product ((*E*)-3-(thiophen-2-yl)acrylic acid). Except for (*S*)-2-nitro-α-phenylalanine produced from its β-isomer with a 92% *e.e.*, all of the tested substrates showed high *e.e.* values of the products that indicated excellent enantioselectivity and stereospecificity of the isomerization [383].

Pseudomonas fluorescens R124 strain isolated from a nutrient-limited orthoquartzite sandstone cave encodes three different MIO-dependent enzymes, including a histidine ammonia-lyase (HAL), a tyrosine/phenylalanine/histidine ammonia-lyase (XAL), and a phenylalanine aminomutase (*Pf*PAM). Identification and characterization of *Pf*PAM show a 51% overall sequence identity to *Pa*APAM and to the full identity of the conserved catalytically important motifs. Therefore, *Pf*PAM showed similar stereoselectivity to that of *Pa*PAM and catalyzed the conversion of either (*S*)-Phe or (*R*,*S*)-Phe to (*S*)-β-Phe in up to 99% *e.e.* When *Pf*PAM was used to catalyze the reverse reaction for (*R*,*S*)-β-Phe, (*S*)-Phe was formed with 11.8% conversion from the consumed (*S*)-β-Phe. In addition, the product mixture of *Pf*PAM-catalyzed biotransformation of (*S*)-Phe contained 55% (*S*)-β-Phe, 25% (*S*)-Phe, and 20% (*E*)-cinnamic acid revealing that *Pf*PAM also exhibits ammonia-lyase activity. The kinetic study demonstrated that the rate of conversion from (*S*)-Phe to (*S*)-β-Phe proceeded at a higher rate than that of the reverse reaction (Scheme 6.55) [384].

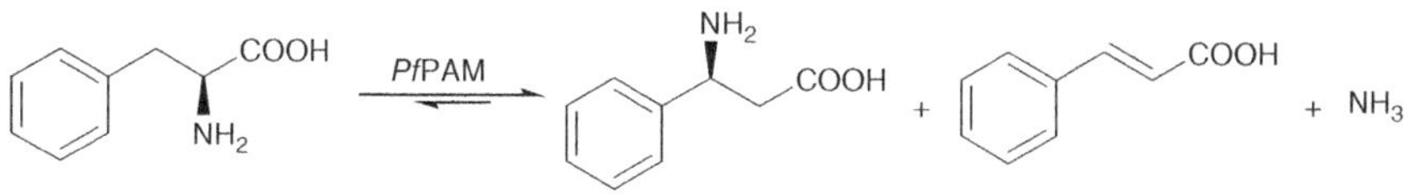

Scheme 6.55 Isomerization of L-phenylalanine catalyzed by PAM from *P. fluorescens* to (*S*)-β-phenylalanine and (*E*)-cinnamic acid. *Source:* Christianson [388].

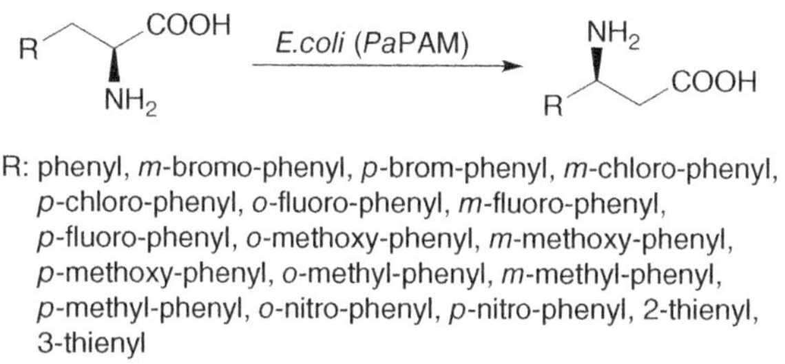

Scheme 6.56 Whole-cell biocatalyst catalyzed substituted α-phenylalanines to their corresponding (*S*)-β-aryl-β-amino acids.

Whole-cell *E. coli* was engineered to express *Pa*PAM, which was employed for the *in vivo* conversion of 18 variously substituted α-phenylalanines to their unnatural (3*S*)-β-aryl-β-amino acids at >99% *e.e.* (Scheme 6.56) [385]. The small-scale fermentation conversion yields of the whole-cell biocatalyst biosynthesis of 18 β-arylalanines were ranged from 4% (*p*-nitro-β-phenylalanine) to 96% (*m*-bromo-β-phenylalanine), corresponding to the production levels between 8.5 and 235 mg L^{-1} over 6 h, respectively. Notably, *E. coli* cells can be reused at least five reaction cycles without noticeable loss of activity and cell viability.

A newly discovered PAM from *Bacillus* sp. strain LM 4-2 (PabH) was found with (*S*)-selectivity for L-phenylalanine to produce (S)-β-phenylalanine, which could be linked to the secondary metabolism of a pyloricidin-like secondary metabolite. However, PabH was shown to display better β-lyase activity in many cases. Despite the PabH, the kinetic resolution strategy by combining an (*S*)-selective aminomutase EncP from *Streptomyces maritimus* (*Sm*PAM) and a strict α-lyase from *Anabaena variabilis* (*Av*PAL) was applied to several racemic substituted β-phenylalanines to acquire better conversion yield of enantiopure substituted (*R*)-β-phenylalanine. The results showed that high conversion for five racemic substrates with >98% (*R*) enantiomeric excess was obtained using a one-pot preparative-scale reaction [386].

The antibiotic andrimid is a polyketide/non-ribosomal peptide that inhibits the bacterial β-subunit of acetyl coenzymeA carboxylase (ACC). Since ACC converts acetyl-CoA to malonyl-CoA, which is used for fatty acid biosynthesis, the

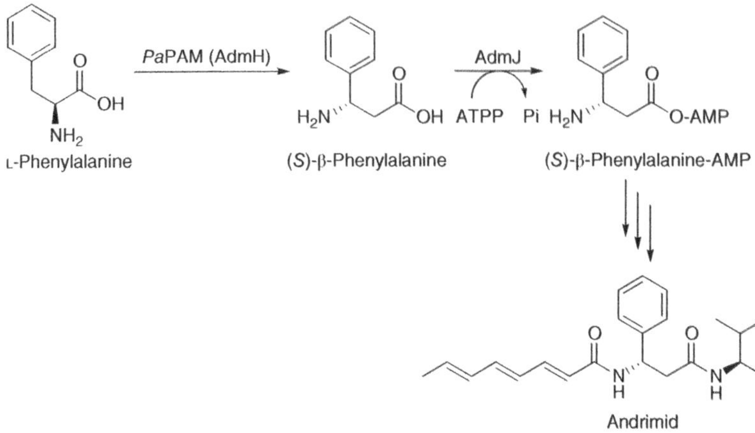

Scheme 6.57 L-Phenylalanine to (*S*)-β-phenylalanine catalyzed by the enzyme AdmH (*Pa*PAM) involved in the biosynthetic pathway of andrimid by *P. agglomerans*. *Source:* Modified from Su et al. [407].

inhibition of ACC prevents cell growth. The structure of andrimid contains three amino acids, phenylalanine, valine, and glycine, while the phenylalanine moiety is actually (*S*)-β- Phe. Andrimid biosynthesis is widely distributed in the bacterial realm, while *P. agglomerans* readily produces andrimid by converting L-phenylalanine to (*S*)-β-phenylalanine using the enzyme AdmH (*Pa*PAM). (*S*)-β-Phenylalanine, which in turn is activated by AdmJ and ATP to form (*S*)-β-Phe-aminoacyl-AMP and followed by a sequence of steps to form andrimid (Scheme 6.57) [377, 378].

Hitachimycin, also known as stubomycin, isolated from *Streptomyces scabrisporus* is a macrolactam antibiotic showing antiprotozoal and antitumor activity, as well as having potent antimicrobial activity against *B. subtilis* and *Micrococcus luteus*. The main structural features of hitachimycin are the unique β-phenylalanine starter unit of its polyketide skeleton and a bicyclic structure with an internal five-membered carbocycle. The biosynthetic gene cluster for hitachimycin in *S. scabrisporus* was identified by genome mining, including a putative phenylalanine-2,3-aminomutase (HitA), five polyketide synthases, four β-amino-acid-carrying enzymes, and a characteristic amidohydrolase. Incorporation studies illustrated that the β-Phe starter unit is derived from α-Phe, presumably by the action of PAM. The enzymatic activity of HitA PAM was confirmed by HitA knock mutant. The results suggest that the identified gene cluster is responsible for the biosynthesis of hitachimycin (Scheme 6.58). A plausible biosynthetic pathway for hitachimycin, containing a unique polyketide skeletal transformation mechanism, was proposed [387].

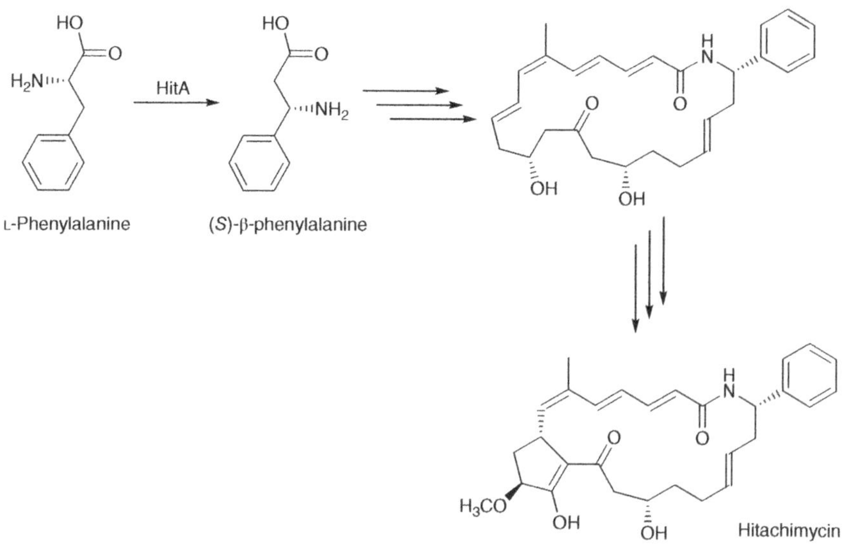

Scheme 6.58 PAM HitA catalyzed conversion of L-phenylalanine to (*S*)-β-phenylalanine for hitachimycin biosynthesis. *Source:* Modified from Zhou et al. [416].

6.5 Intramolecular Lyases

6.5.1 *ent*-Copalyl Diphosphate Synthase

The enzyme, *ent*-copalyl diphosphate synthase (CPS, EC 5.5.1.13), is an enzyme that catalyzes the cyclization of all *trans* geranylgeranyl diphosphate (GGPP) to copalyl diphosphate (CDP), a labdane-related diterpenoid (Scheme 6.59) [388]. This reaction is the first step in the gibberellin (GA) biosynthetic pathway. The systematic name of this enzyme class is *ent*-copalyl-diphosphate lyase (decyclizing) and other names such as *ent*-kaurene synthase A and *ent*-kaurene synthetase A are also common in use. *ent*-Copalyl diphosphate synthase belongs to class II

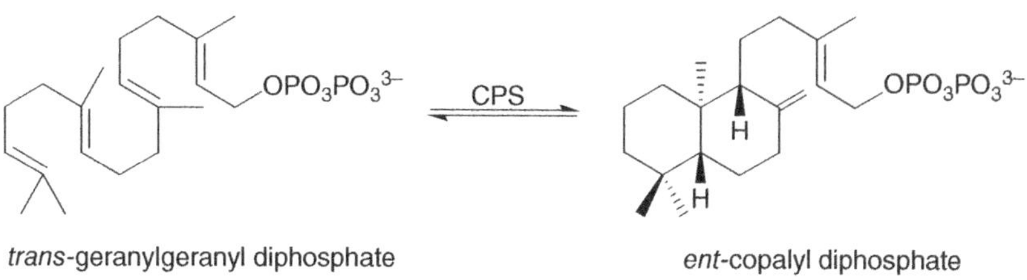

Scheme 6.59 Cyclization of all-*trans* geranylgeranyl diphosphate to *ent*-copalyl diphosphate catalyzed by *ent*-copalyl diphosphate synthase.

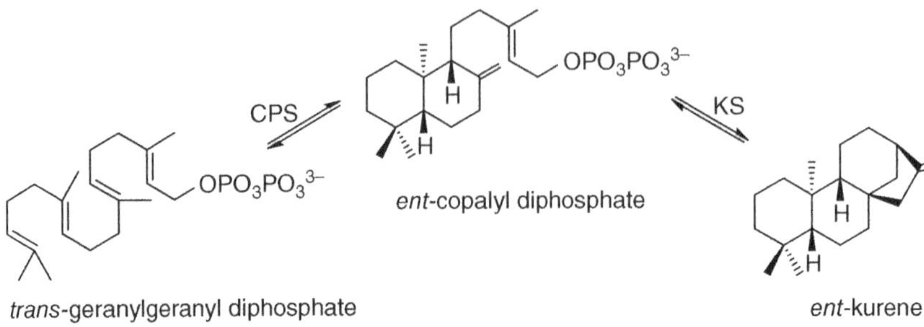

Scheme 6.60 Conversion from geranylgeranyl diphosphate to *ent*-kaurene catalyzed by *ent*-copalyl diphosphate synthase and *ent*-kaurene synthase in sequence. *Source:* Modified from Mafu et al. [420].

terpenoid cyclases and is found in plants, fungi, and bacteria to serve various functions [388].

CPSs from fungi and mosses also have a distinct *ent*-kaurene synthase (KS) activity associated with the same protein molecule, which catalyzes the next step from *ent*-CDP to *ent*-kaurene in the gibberellin biosynthesis [389–394]. However, two separated proteins, one for CPS and one for KS (EC 4.2.3.19), are found in higher plants, even though they are probably associated as weakly bound dimers or enzymes complexes [391, 395–397]. Therefore, the labdane diterpenoid *ent*-CDP is an intermediate in the plant biosynthetic pathway leading to the formation of *ent*-kaurene and ultimately to the gibberellin phytohormones (Scheme 6.60), which are important for plant growth and development [388]. Initially, CPS was designated as *ent*-kaurene synthetase A when first cloned from *Arabidopsis thaliana* [398]. Only one form of *ent*-copalyl diphosphate synthase was isolated from eubacterial *Streptomyces* sp. strain KO-3938 to convert GGPP into *ent*-CDP [399]. The bifunctional enzymes from lower plants are larger (106–107 kDa) than the monofunctional enzymes from higher plants (90–98 kDa).

The crystal structures of CPS complexed with substrate analog (*S*)-15-aza-14,15-dihydrogeranylgeranyl thiolodiphosphate at 2.25 Å resolution and product analog 13-aza-13,14-dihydrocopalyl diphosphate at 2.75 Å resolution were obtained from *A. thaliana*, which reveals three α-helical domains (α, β, and γ) [388, 400]. The catalytic general acid is the central aspartate in the characteristic sequence motif DXDD [401]. The active site of a class II terpenoid cyclase is located at the interface of the β and γ domains in a protein with overall βγ or αβγ domain architecture. A βγ enzyme is a monofunctional class II cyclase, whereas an αβγ enzyme can be a monofunctional class II cyclase or a bifunctional class I-class II cyclase. Class II diterpene cyclases initiate catalysis by protonation of the terminal π bond of GGPP, likely in concert with C10-C15 bond formation and C6-C11 bond formation to yield a bicyclic tertiary carbocation intermediate. As with cyclization reaction catalyzed by class I terpenoid cyclases, the resulting carbocation intermediate can undergo further reactions, hydride transfers, methyl migrations, etc. before termination of the reaction by proton elimination or addition of a solvent molecule [402–404]. The first crystal structure of a bacterial class II diterpene cyclase at 1.80 Å reveals the βγ domain architecture [405].

Gibberellins (GAs) are *ent*-kaurene-derived diterpenoid phytohormones produced by plants, fungi, and bacteria sharing a common tetracyclic 6-5-6-5 fused hydrocarbon ring (gibberellane) skeletal structure, which form an important group of plant hormones with various functions in different species and at different stages of the plant's lifetime. Disorders in GA biosynthesis commonly show themselves as growth disorders, particularly as dwarfism, and some of those can be traced to reduced *ent*-CPS activity [397, 406]. The roots and rhizomes of *Salvia miltiorrhiza* Bunge are highly valued in traditional Chinese medicine because its bioactive diterpenoid tanshinones have been reported to have many pharmaceutical activities. Previous studies also demonstrated four different diterpenoid biosynthetic pathways from the universal diterpenoid precursor GGPP in *Salvia miltiorrhiza*. One study clarifies the functional characterization of CPS, KS, and kaurene oxidase (KO) in the GA biosynthetic pathway that CPS catalyzes the cyclization of GGPP to *ent*-CDP, which is converted to *ent*-kaurene by KS. Then, KO catalyzes the three-step oxidation of *ent*-kaurene to *ent*-kaurenoic acid, which finally leads to the production of GAs by a series of cytochrome P450s in fungi and, additional 2-oxoglutarate dependent dioxygenases (2ODDs) in plants (Scheme 6.61) [407]. The results also reveal that the fused enzyme KS-CPS increases *ent*-kaurene production by several fold compared with separate expression of CPS and KS in yeast strains. Tanshinones are abietane-type norditerpenoid

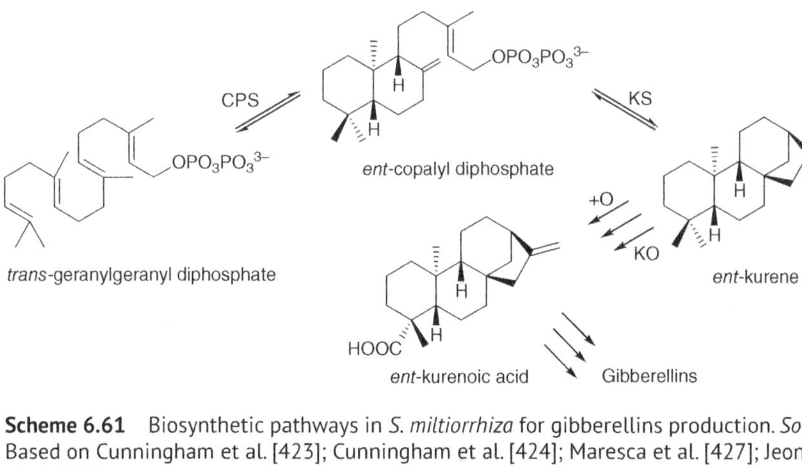

Scheme 6.61 Biosynthetic pathways in *S. miltiorrhiza* for gibberellins production. *Source:* Based on Cunningham et al. [423]; Cunningham et al. [424]; Maresca et al. [427]; Jeong et al. [428].

quinone natural products. The initial results from a functional genomics-based investigation of tanshinone biosynthesis, specifically the functional identification of the relevant diterpene synthases from *S. miltiorrhiza*, show that the cyclohexa-1,4-diene arrangement of the distal ring poises the resulting miltiradiene for the ensuing aromatization and hydroxylation to ferruginol [408]. In bacterium *Bradyrhizobium japonicum*, an operon encodes separate *ent*-CPS and *ent*-KS, which together with a ferredoxin, a short-chain alcohol dehydrogenase, three P450s, and a GGPP synthase are involved in GA synthesis [409]. Since *ent*-kaurene is also a dedicated precursor pool and is responsible for synthesizing natural sweeteners such as steviol glycosides, two *ent*-kaurene genes encoding *ent*-CPS and *ent*-KS from *Stevia rebaudiana* (a typical plant producing steviol glycosides) were molecularly constructed and expressed in *E. coli* for *ent*-kaurene production [410]. The results demonstrated that *E. coli* strain MG1655 coexpressed synthetic CPS-KS module and geranylgeranyl diphosphate synthase (GGPPS) from *Rhodobacter sphaeroides* gave an enhanced *ent*-kaurene production of approximately 41.1 mg L^{-1}. The *ent*-kaurene production was further increased up to 179.6 mg L^{-1} by overexpression of the three key enzymes for isoprenoid precursor, 1-deoxyxylulose-5-phosphate synthase, farnesyl diphosphate synthase, and isopentenyl diphosphate isomerase from *E. coli* to obtain the highest titer of *ent*-kaurene (578 mg L^{-1}) with a specific yield of *ent*-kaurene of 143.5 mg g^{-1} dry cell weight in 1 L bioreactor containing 20 g L^{-1} glycerol.

A mixture of at least eight different glycosides derived from the tetracyclic diterpene steviol is accumulated in *S. rebaudiana* Bertoni leaves, which taste intensely sweet and have similar biosynthetic origins to those of gibberellic acid (GiA). To date, 136 different gibberellic acids (GiAs) have been identified, but the most biologically active ones are GiA1, GiA3, GiA4, and GiA7. For GiA production, an efficient CRISPR/Cas9-based genome editing tool for the filamentous fungus *Fusarium fujikuroi* was developed to engineer the metabolic pathways responsible for the accumulation of a series GiAs in *F. fujikuroi*. The established technology successfully changed its GiA product profile from GiA3 to tailor-made GiA4 and GiA7 mixtures, which are more efficient for agricultural use, especially for fruit growth, and the developed strains will greatly improve industrial GiA production [411]. Since both the CPS and KS genes from *S. rebaudiana* have been isolated and found that recombinant CPS and KS were catalytically active, steviol biosynthesis uses the same mechanism as for GiA biosynthesis. While both CPS and KS participate in the synthesis of steviol glycosides, a secondary metabolic pathway from a highly regulated primary route previously committed to hormone gibberellins biosynthesis [412].

Labdane-related diterpenoids (LRDs) comprise a large group of over 5000 known natural products defined as minimally containing the fused bicyclic hydrocarbon structure found in the labdane family of diterpenoids. Eight different hydrocarbon skeletal structures, with one as distinct double bond isomers, were produced with three pGGxC constructs, proving the utility of our modular approach for facile biosynthesis of labdane-related diterpenes (abieta-7,13-diene, *ent*-kaur-16-ene, *ent*-pimara-8(14),15-diene, *ent*-isokaur-15-ene, *ent*-cassa-12,15-diene, *ent*-sandaracopimaradiene, *syn*-pimara-7,15-diene, *syn*-stemar-13-ene, and *syn*-stemod-13(17)-ene). These diterpenes are precursors to literally thousands of known compounds that establish a foundation for the biosynthetic production of elaborate LRD natural products [413]. The structural diversity of LRDs critically depends on the hydrocarbon skeletal structures generated, in large part, by class I diterpene synthases. In plant kingdom, the relevant class I diterpene synthases are clearly derived from the *ent*-KS and are often termed KS-like (KSL). A further investigation of the alleles of a KSL from rice (*Oryza sativa*), OsKSL5, led to the discovery of a nearby residue that tunes product outcome. Substitution for this newly identified residue is additionally shown to exert an epistatic effect in KSs, altering product distribution only if combined with the replacement of the key isoleucine. Interestingly, the pair of residues was found to exert additive effects on the product outcome mediated by distantly related KSLs from the eudicot castor bean. Therefore, the use of a rational combination of substitutions for this pair of residues to engineer significantly increase (dominant) selectivity for novel 8α-hydroxy-*ent*-pimar-15-ene product outcome in the KS from the dicot *A. thaliana* [414].

Engineering biosynthetic pathways in heterologous microbial host organisms offer an elegant approach to pathway elucidation via the incorporation of putative biosynthetic enzymes and characterization of resulting novel metabolites. Therefore, the incorporation of either a heterologous mevalonate pathway (MEV) or through the enhancement of the endogenous methylerythritol phosphate pathway (MEP) with a facile modular metabolic engineering system was developed. With MEP pathway enhancement, pyruvate supplement media and simultaneous overexpression of three genes (*idi*, *dxs*, and *dxr*) resulted in the greatest increase in diterpene yield, which indicates distributed metabolic control within this pathway. Incorporation of a heterologous MEV pathway in bioreactor grown cultures resulted in significantly higher yields than MEP pathway enhancement. Diterpene production levels ranging from 10 to >100 mg L^{-1} of *E. coli* culture have been established with suitable growth conditions, which enables characterization of enzymatic products and pathway elucidation. The results represent an up to >1000-fold improvement in diterpene production from the facile modular platform, with MEP pathway enhancement offering a cost-effective alternative with reasonable yield [415]. Miltiradiene is the product of the stepwise conversion of GGPP catalyzed by *Sm*CPS and *Sm*KSL from *S. miltiorrhiza*, which has been identified as the precursor to tanshinones. The modular pathway engineering (MOPE) strategy was applied for rapid assembling synthetic miltiradiene pathways in the

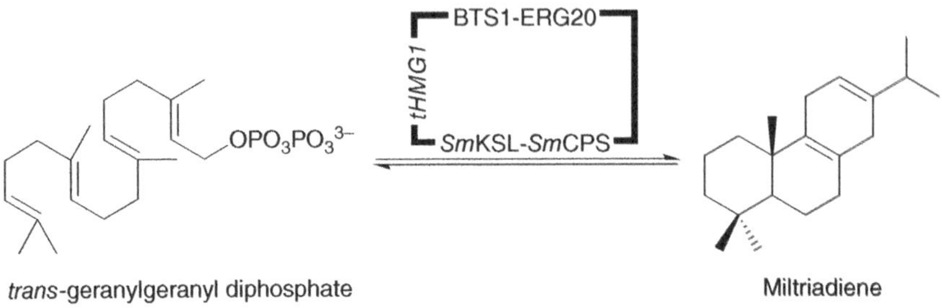

Scheme 6.62 Miltiradiene production from GGPP catalyzed by the fusion of *Sm*CPS and *Sm*KSL as well as the fusion of BTS1 and ERG20. *Source:* Based on Howitt et al. [431]; Jensen et al. [432].

yeast *S. cerevisiae*. The active sites of *Sm*CPS and *Sm*KSL were engineered into proximity for enhanced metabolic flux channeling to miltiradiene biosynthesis by constructing protein fusions. The fusion of *Sm*CPS and *Sm*KSL, as well as the fusion of BTS1 (GGPP synthase) and ERG20 (farnesyl diphosphate synthase), led to significantly improved miltiradiene production and reduced by-product accumulation (Scheme 6.62) [416]. As a result, the best pathway facilitated by the MOPE strategy with the diploid strain YJ2X reached miltiradiene titer of 365 mg L^{-1} in a 15-L bioreactor.

A medicinal plant, *Tripterygium wilfordii*, exhibits impressive and effective anti-inflammatory, immunosuppressive, and antitumor activities, and the main active ingredients are diterpenoids and triterpenoids, such as triptolide and celastrol, respectively. A suspension cell culture of *T. wilfordii* produced triptolide via the relevant abietane olefin precursor miltiradiene in an inducible fashion (Scheme 6.63). Eight putative (di)terpene synthases were then identified and characterized for their potential involvement in triptolide biosynthesis using the transcriptome data. This included not only biochemical studies, revealing the expected presence of class II diterpene cyclases that produce the intermediate CDP, along with the finding of an atypical class I (di)terpene synthase that acts on CDP to produce the abietane olefin miltiradiene, but also their subcellular localization and genetic analysis. Particularly, RNA interference targeting either both of the CDP synthases, *Tw*TPS7v2 and *Tw*TPS9v2, or the subsequently acting miltiradiene synthase, *Tw*TPS27v2, led to decreased production of triptolide [417].

Platensimycin (PTM) isolated from *Streptomyces platensis* MA7327 and platencin (PTN) isolated from *S. platensis* MA7339 in two different soil samples are potent and selective inhibitors of bacterial and mammalian fatty acid synthases and have emerged as promising drug leads for both antibacterial and antidiabetic therapies. Comparative study of the PTM and PTN biosynthetic machineries in *S. platensis* MA7327 and MA7339 revealed that the divergence of PTM and PTN

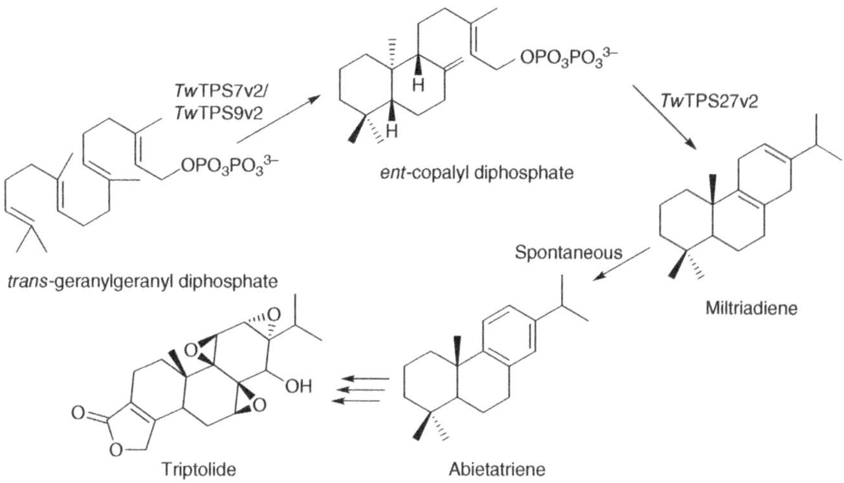

Scheme 6.63 Triptolide production catalyzed by *Tripterygium wilfordii*. *Source:* Krügel et al. [440].

biosynthesis is controlled by dedicated *ent*-kaurene and *ent*-atiserene synthases, thereby latter of which represents a new pathway for diterpenoid biosynthesis [418]. The PTM and PTN biosynthetic machineries provide a rare glimpse at how secondary metabolic pathway evolution increases natural product structural diversity and support the wisdom of applying combinatorial biosynthesis methods for the generation of novel PTM and/or PTN analogs, thereby facilitating drug development efforts based on these privileged natural product scaffolds.

Terpenoids are a major component of maize (*Zea mays*) chemical defenses that mediate responses to herbivores, pathogens, and other environmental challenges. In one study, *ent*-CPS is produced by maize plants in response to the attack by *Fusarium* fungi, which is used to form a precursor for producing defensive phytoalexins in plant [419]. The biosynthesis and production of a class of maize defense-related diterpenoids, named dolabralexins, involve the sequential activity of two diterpene synthases, *ent*-CPS (*Zm*AN2) and KSL4 (*Zm*KSL4), which lead to form the diterpene hydrocarbon dolabradiene. Furthermore, a cytochrome P450 monooxygenase, *Zm*CYP71Z16, was characterized, which catalyzes the oxygenation of dolabradiene to yield the epoxides 15,16-epoxydolabrene (epoxydolabrene) and 3β-hydroxy-15,16-epoxydolabrene (epoxydolabranol) (Scheme 6.64) [420]. In addition, much of the epoxydolabranol is further converted into 3β,15,16-trihydroxydolabrene (trihydroxydolabrene), which is the predominant diterpenoid of the field-grown maize root tissues. Epoxydolabranol significantly inhibited the growth of both fungal pathogens *Fusarium*

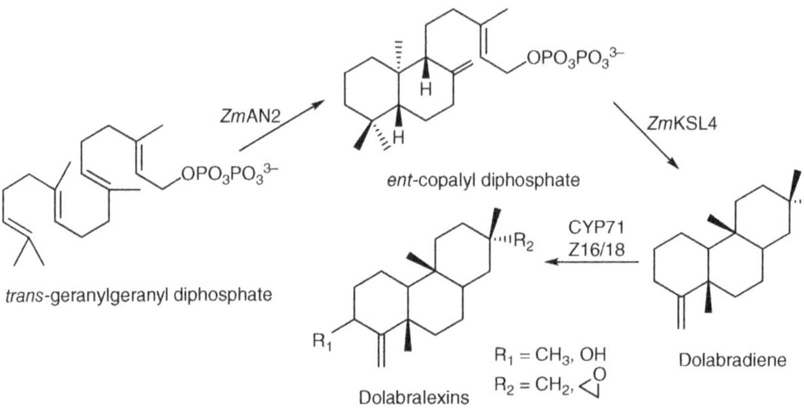

Scheme 6.64 The biosynthesis of defense-related dolabralexins in maize. *Source:* Modified from Dambek et al. [448].

verticillioides and *Fusarium graminearum in vitro* at 10 μg mL^{-1}, while trihydroxydolabrene-mediated inhibition was specific to *F. verticillioides*.

The *Isodon* plants (Lamiaceae) have been used in traditional Chinese medicine to alleviate suffering from inflammations and cancers, which is attributed to the pharmacologically active *ent*-kaurane diterpenoids such as eriocalyxin B and oridonin. The *Isodon eriocalyx* (Dunn) Kudô species native to southwest China can accumulate a particularly high content of *ent*-kaurane diterpenoids (~1.5% w/w of dried leaves). In addition to two *ent*-copalyl diphosphate synthases (*Ie*CPS1 and *Ie*CPS2), another three other diterpene synthase genes (*IeCPS3*, *IeKS1*, and *IeKSL1*) were mined from the transcriptome of *I. eriocalyx* plant. Their functional characterization reveals that *IeCPS3* is an *ent*-CDP synthase, and *IeKS1* catalyzes the production of *ent*-kaurene from *ent*-CDP and thus an *ent*-KS. In *I. eriocalyx*, *ent*-kaurene is further metabolized to form eriocalyxn B. When *IeKSL1* was combined with *Ie*CPS2 or *Ie*CPS3, no product was detected. Biochemical and genetic evidence suggests that both *IeCPS2* and *IeKS1* possess dual roles in the biosynthesis of gibberellins and *Isodon ent*-kaurene diterpenoids [421].

6.5.2 Lycopene β-Cyclase

Lycopene β-cyclases (β-LCY, EC 5.5.1.19, *CrtL*, *Crt-b*, *CrtY*) are a class of enzymes that convert the acyclic carotenoid lycopene to the bicyclic molecule β-carotene by catalyzing the formation of two beta-rings at each end of the linear carotene structure [422–428]. This enzyme requires NAD(P)H as the coenzyme for the catalytic reaction [429]. This reaction takes place as a two-step reaction in which both sides of the lycopene molecule are cyclized in sequence into β-carotene rings via the

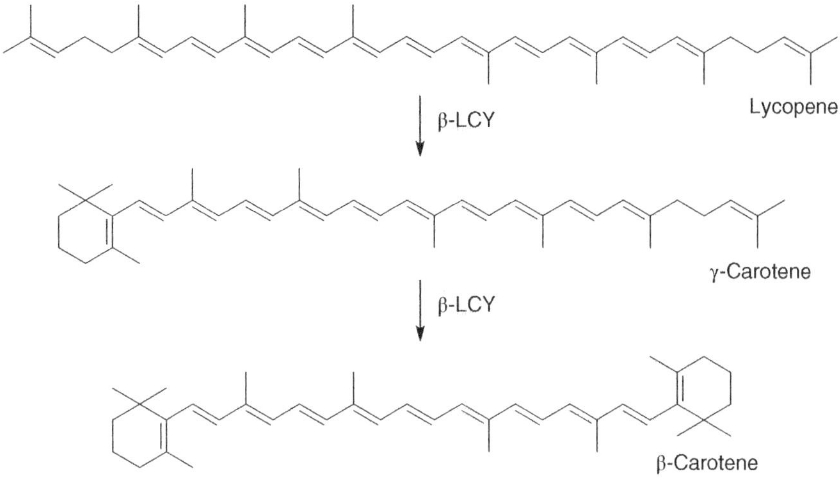

Scheme 6.65 Lycopene β-cyclase catalyzed conversion of lycopene to β-carotene.

monocyclic γ-carotene as an intermediate (Scheme 6.65) [423, 424, 427, 428]. This enzyme is widely distributed in nature including bacteria, fungi, algae, and plants. β-LCY obtained from cyanobacterium *Synechococcus* sp. Strain PCC7942 has a molecular mass of 46 125 with a polypeptide of 411 amino acid residues, while β-LCY from marine bacterium *Paracoccus haeundaensis* expressed in *E. coli* exhibits a molecular mass of approximately 43 kDa characterized with 386 amino acid residues [424, 428]. The mechanism of the cyclization reaction at each end of the lycopene molecule to form the β-ring in β-carotene was elucidated by ^{2}H isotope labeling [429, 430]. During the cyclization, a hydrogen atom from water is introduced stereospecifically at C2 of the folded acyclic end group of the lycopene molecule, which is in a required shape by the cyclase enzyme, to initiate the reaction. The electrophilic attack at C2 is accompanied with the concerted ring closure of the C1–C6 bond to form the ring. This would give a transient C5 carbocation which is presumably stabilized by the enzyme. Finally, H^+ at C6 is eliminated to generate the β-ring end group.

Carotenoids are the second most abundant pigment in nature, which exist over 700 members in the carotenoid family to carry out important functions as antioxidants, hormone precursors, colorants, and essential components of the photosynthetic apparatus. Most carotenoid pigments are C40 polyenes and all are derived from phytoene. The colors of carotenoids range from yellow, orange to red with variations of brown and purple, for example: β-carotene from carrots, lycopene from tomatoes, and lutein from marigold flowers. For many years due to the importance of carotenoids to plants and humans, the carotenoid biosynthetic pathway in higher plants is well established and it takes place in plastids by

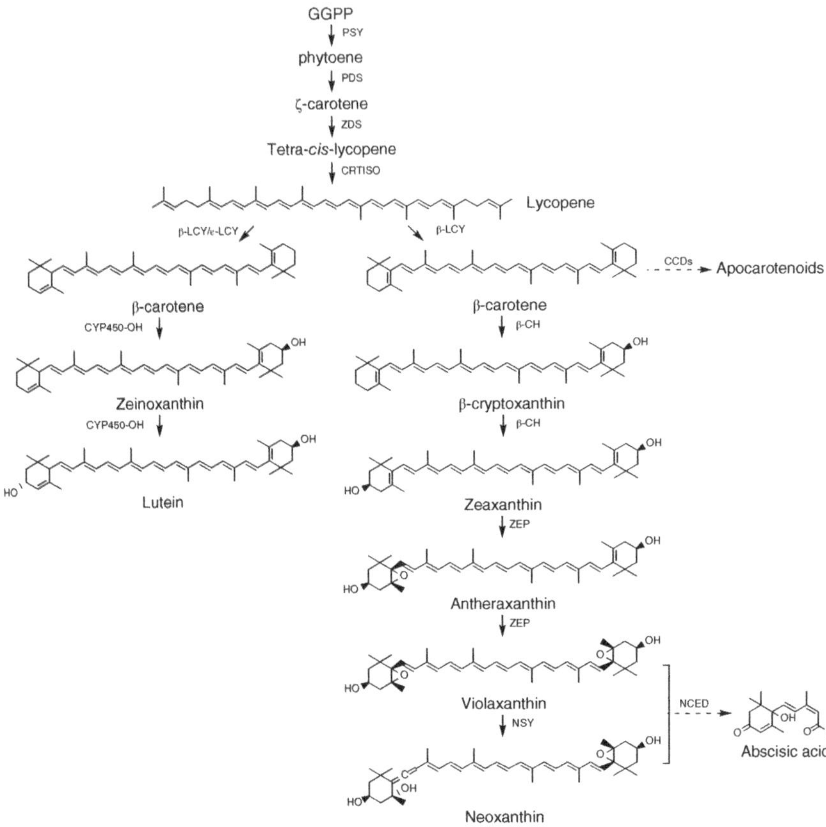

Scheme 6.66 Carotenoid biosynthetic pathway in higher plants.

nuclear-encoded enzymes (Scheme 6.66) [431, 432]. Carotenoid biosynthesis starts with the isoprenoid pathway (mevalonate-dependent or mevalonate-independent) to generate the C5 isoprene unit, isopentenyl pyrophosphate (IPP). IPP is condensed with its isomer dimethylallyl pyrophosphate (DMAPP) to C10 geranyl pyrophosphate (GPP) and elongated to C15 farnesyl pyrophosphate (FPP). FPP is further catalyzed by GGPP synthase (*CrtE*) to form GGPP [422]. Then, the first committed step in the biosynthesis of carotenoids is the condensation of two molecules of GGPP catalyzed by phytoene synthase (PSY, *CrtB*) to form the colorless carotenoid 15-*cis*-phytoene (phytoene), which does not usually accumulate in tissues. In plants, four desaturation reactions catalyzed by phytoene desaturase synthase (PDS) and ζ-carotene desaturase (ZDS), respectively, then convert the colorless phytoene into the red colored tetra-*cis*-lycopene via di-*cis*-ζ-carotene, by introducing two symmetric double bonds. The tetra-*cis*-lycopene is further catalyzed by carotenoid isomerase (CRTISO) to form all-*trans*-lycopene (lycopene), which is the preferred substrate for the cyclase. Cyclization of lycopene is a branch point in the carotenoid biosynthesis pathway, the β,β-branch leads to β-carotene and its derivatives, and the β,ε-branch leads to α-carotene and its derivatives [431]. The formation of β-carotene is catalyzed by β-LCY, while the formation of α-carotene requires β-LCY and lycopene ε-cyclase (ε-LCY, EC 5.5.1.18).

α-Carotene and β-carotene are further modified to produce xanthophyll's, which are enzymatically formed oxidation products [431]. In the ε,β-branch, the biosynthesis of lutein from α-carotene requires the sequential action of two separate carotenoid hydroxylases belonging to the cytochrome P450-type monooxygenase family, β-ring hydroxylase (CYP450-βOH) and ε-ring hydroxylase (CYP450-εOH). In the β,β-branch of the carotenoid pathway, hydroxylation of β-carotene by β-carotene hydroxylase (β-CH) in two steps produces β-cryptoxanthin and then zeaxanthin, a 3,3'-dihydroxy-α-carotene. Zeaxanthin also can be further epoxidized by zeaxanthin epoxidase (ZEP) in a two-step reaction to produce violaxanthin via antheraxanthin. Violaxanthin is converted to neoxanthin by neoxanthin synthase (NSY), which is the last carotenoid of the β,β-branch of the carotenoid pathway in higher plants. Both the 9-*cis*-neoxanthin and 9-*cis*-violaxanthin can be cleaved by 9-*cis*-epoxycarotenoid dioxygenase (NCED). The cleaved products (called apocarotenoids) can then be further modified to produce abscisic acid (ABA). In addition to the transcriptional regulation, factors such as post-translational regulation, chromoplast biogenesis, epigenetic mechanisms, and enzymatic degradation of carotenoids to apocarotenoids by the carotenoid cleavage dioxygenases (CCDs) can have a profound role on the fruit carotenoid metabolism. In fact, some apocarotenoid compounds are recognized as important signaling, aroma, flavor, and pigment components in fruits and berries.

β-Carotene is the main dietary precursor of vitamin A and therefore also referred to as provitamin A. β-LCY genes from the eubacterium *Erwinia herbicola* and the higher plant daffodil (*Narcissus pseudonarcissus*) were introduced into the tomato plastid genome. While expression of the bacterial enzyme did not strongly alter the carotenoid composition, expression of the plant enzyme efficiently converted lycopene into β-carotene. In green leaves of the transplastomic tomato plants, more lycopene was channeled into the β-branch of carotenoid biosynthesis, resulting in increased accumulation xanthophyll cycle pigments and correspondingly reduced accumulation of the α-branch xanthophyll lutein. In fruits, most of the lycopene was converted into β-carotene with provitamin A levels reaching 1 mg per g dry weight [433]. A β-LCY gene (*TaL*CYB) was cloned from the hexaploid wheat cultivar Chinese Spring (*Triticum aestivum* L.), which was expressed differently in different tissues of wheat. Although *TaL*CYB had a higher expression level in the later stages of grain development, the β-carotene content still showed a decreasing tendency. The expression of *TaL*CYB in leaves was dramatically induced by strong light and the β-carotene content variation corresponded with changes in *TaLCYB* expression [434]. An engineered photosynthetic *R. sphaeroides* producing β-carotene was constructed by optimizing the promoter of the *crtY*, *zwf* knockout, and *dxs* integration to improve β-carotene. The engineered RS-C3 has the highest β-carotene content of 14.93 mg g^{-1} dry cell weight (DCW) among all of the reported photosynthetic bacteria. Preliminary studies of producing β-carotene under light conditions by the engineered bacteria were also conducted to lead a β-carotene content reached 3.34 mg g^{-1} DCW [435].

Bilberry was found to be a good source of carotenoids among fruits and berries. Eight carotenoid biosynthetic genes including PSY, PDS, ZDS, CRTISO, β-LCY, ε-LCY, β-CH, and CYP450-βOH, as well as a carotenoid cleavage dioxygenase CCD1 were isolated from bilberry (*Vaccinium myrtillus* L.) and expressed in bilberry fruit at five different development stages and different light quality conditions [436]. The most abundant carotenoids throughout the berry development were lutein and β-carotene, as well as accompanied by lower amount of 9*Z*-β-carotene, violaxanthin, neoxanthin, zeaxanthin, antheraxanthin, and β-cryptoxanthin with the expression of eight biosynthetic genes, which indicates a metabolic flux toward β-branch of the carotenoid pathway. However, the carotenoid levels decreased in both the β-branch and ε,β-branch toward bilberry fruit ripening along with increased CCD1 expression, similarly to NCED1, indicating enzymatic carotenoid cleavage and degradation. Intense white light conditions increased the expression of the carotenoid biosynthetic genes but also the expression of the cleavage genes CCD1 and NCED1, especially in unripe fruits. Instead, mature bilberry fruits responded specifically to red/far-red light wavelengths by inducing the expression of both the carotenoid biosynthetic and the cleavage genes indicating tissue and developmental stage-specific regulation of apocarotenoid formation by light quality. In another report, the effects of UV-B radiation and $CaCl_2$ on the enhancement of carotenoid content in germinated yellow corn (cultivar: Suyu 29) kernels were investigated [437]. The results showed that the contents of predominant carotenoids in germinated corn kernels were significantly improved by 0.375 μW m^{-2} UV-B radiation, but their germination was inhibited. The contents of lutein, zeaxanthin, and α-cryptoxanthin were further enhanced by addition of 5 mmol L^{-1} $CaCl_2$. Meanwhile, $CaCl_2$ treatment could promote the germination of corn kernels and alleviate the damage of UV-B radiation. UV-B radiation and $CaCl_2$ treatments could effectively increase superoxide dismutase and peroxidase activities. Furthermore, the essential genes involved in the carotenoid biosynthesis pathway showed different patterns in UV-B radiation and $CaCl_2$ treatments, which reveals that UV-B radiation can increase carotenoid content and antioxidant enzyme activity. Moreover, $CaCl_2$ can further improve carotenoid content and reduce photooxidative damage caused by UV-B radiation.

Full-length cDNA encoding ZDS, ε-LCY, β-CH, and ZEP, as well as partial-length cDNA encoding CYP450-εOH were isolated in oriental melon Chamoe (*Cucumis melo* L. var. *makuwa*), which were expressed in the peel, pulp, and stalk of chamoe cultivars Ohbokggul and Gotgam. Most of the carotenoid biosynthetic genes were expressed at their highest levels in the stalk, whereas carotenoids were highly distributed in the peel. The expression levels of all carotenoid biosynthetic genes in fruits of the native cultivar Gotgam chamoe were higher than those in the cultivar Ohbokggul chamoe, consistent with the abundant carotenoid accumulation in Gotgam chamoe fruits and trace carotenoid content of Ohbokggul chamoe fruit. Lutein and β-carotene were the dominant carotenoids in the peel of Gotgam chamoe with 278.05 $\mu g\ g^{-1}$ and 112.02 $\mu g\ g^{-1}$ dry weight, respectively [438].

Canthaxanthin is directly and indirectly applied for food coloring. When combined with β-carotene, it is also useful in cosmetology and pharmacology as a dermal photoprotector. A carotenoid biosynthesis gene cluster containing five genes identified as *crtE*, *certY*, *crtI*, *crtB*, and *crtW* involved in canthaxanthin production was isolated from the photosynthetic *Bradyrhizobium* sp. strain ORS278. The five genes in the cluster are organized in at least two operons. The functional assignment of each open reading frame was confirmed by complementation studies. It is recognized that the four genes *crtE*, *crtB*, *crtI*, and *crtY*, encoding geranylgeranyl diphosphate synthase, phytoene synthase, phytoene desaturase, and lycopene β-cyclase, respectively, involve in the carotenoid biosynthesis for β-carotene formation from farnesyl pyrophosphate (FPP), whereas the gene *crtW* encoding β-carotene ketolase is responsible for the production of canthaxanthin from β-carotene [439].

The aromatic carotene isorenieratene is a carotenoid light-harvesting pigment. The biosynthesis of isorenieratene is restricted to green photosynthetic bacteria and a few actinomycetes. Five genes out of seven of two adjacent operons in one cluster could be identified from *S. griseus*, which are sufficient for the synthesis of isorenieratene. Stepwise deletions of these genes demonstrated their coding for geranylgeranyl pyrophosphate synthase (*crtE*), phytoene desaturase (*crtI*), phytoene synthase (*crtB*), and lycopene β-cyclase (*crtY*). The novel gene *crtU* was assigned to encode a unique desaturase responsible for the conversion of β-carotene via β-isorenieratene to isorenieratene by a desaturation/methyltransferation mechanism (Scheme 6.67) [440]. The sequence of *crtU* comprises a motif typically found in methyltransferases. The two remaining genes, *crtT* and *crtV*, of the cluster left the carotenoid biosynthetic pathway unaffected.

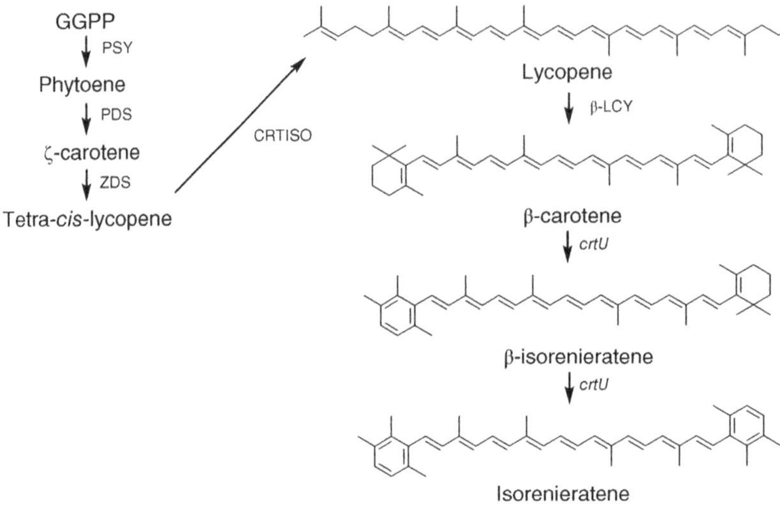

Scheme 6.67 Biosynthesis of isorenieratene form β-carotene via β-isorenieratene.

Zeaxanthin is a commercially important high-value carotenoid that is used in nutraceuticals, cosmetics, food, and animal feed industries. A 14 kb genomic DNA fragment of the Gram-negative bacterium *Flavobacterium* sp. strain R1534 was cloned and a 5.1 kb piece containing the carotenoid biosynthesis genes was sequenced. Functional assignment for the five genes (*crtE*, *crtB*, *crtY*, *crtI*, and *crtZ*) of the carotenoid biosynthetic pathway necessary for the synthesis of zeaxanthin was done in *E. coli* transformed with deletion mutants of the original plasmid carrying all five genes [441]. A model unicellular freshwater microalga *Chlamydomonas reinhardtii* was employed to gain knowledge about the influence of light on the carotenoid composition and the expression of the gene encoding the main steps of the carotenoid biosynthetic pathway [442].

The results showed that there is an activation of the xanthophyll cycle not only in response to high light but also in response to other stress conditions, such as nitrogen starvation. Measurements of carotenoid content in the presence of inhibitors of protein and carotenoid synthesis suggest that only one of the two possible routes for the synthesis of zeaxanthin upon transference to high light, either the *de novo* synthesis of carotenoids or the interconversion between violaxanthin and zeaxanthin, is dependent on protein synthesis. The high increase in the transcript levels of the cytochrome-dependent carotene β- and ε-hydroxylases in response to high light suggests an important role of these enzymes in regulation of xanthophyll synthesis upon light stress. Because of the increasing demand for zeaxanthin, a molecularly engineered *E. coli* by optimizing the synthesis of zeaxanthin from lycopene using fusion protein-mediated substrate channeling as well as by the introduction of tunable intergenic regions was developed to serve alternative sources other than the chemical synthesis or a natural product purified from microorganisms. The tunable intergenic regions approach was more efficient compared with protein fusion for coordinating expression of β-LCY gene *crtY* and β-CH gene *crtZ*. Gene *crtZ* from *Pantoea ananatis* is used to generate a recombinant strain of *E. coli* BETA-1 to produce higher amounts of 11.95 ± 0.21 mg g^{-1} DCW zeaxanthin [443].

A carotenoid biosynthesis gene cluster containing five carotenogenic genes designated as *crtW* (β-carotene ketolase), *crtZ* (β-CH), *crtY* (β-LCY), *crtI* (PDS), and *crtB* (PSY), which is responsible for the production of astaxanthin, was isolated from the marine bacterium *Agrobacterium aurantiacum* by using an *E. coli* host that synthesizes β-carotene via the *Erwinia uredovora crt* genes. The functions of the five *crt* genes of *A. aurantiacum* were determined through the analyses of the pigments accumulated in some *E. coli* transformants carrying various combinations of the *E. carotene desaturase uredovora* and *A. aurantiacum* carotenogenic genes. As a result, the astaxanthin biosynthetic pathway is proposed for the first time at the level of the biosynthesis genes. The *crtW* and *crtZ* gene products, which mediated the oxygenation reactions from β-carotene to astaxanthin, were found to have low substrate specificity. This allowed the production of many presumed intermediates of astaxanthin, that is, adonixanthin, phoenicoxanthin (adonirubin), canthaxanthin, 3'-hydroxyechinenone, and 3-hydroxyechinenone [444].

Pepper (*Capsicum* spp.) is a worldwide popular vegetable and spices valued for heat, nutrition, and rich pigment content. The most abundant carotenoids in ripe pepper fruits are β-carotene, capsanthin, and capsorubin produced by the carotene and xanthophyll biosynthetic pathway in *Capsicum* (Scheme 6.68). The

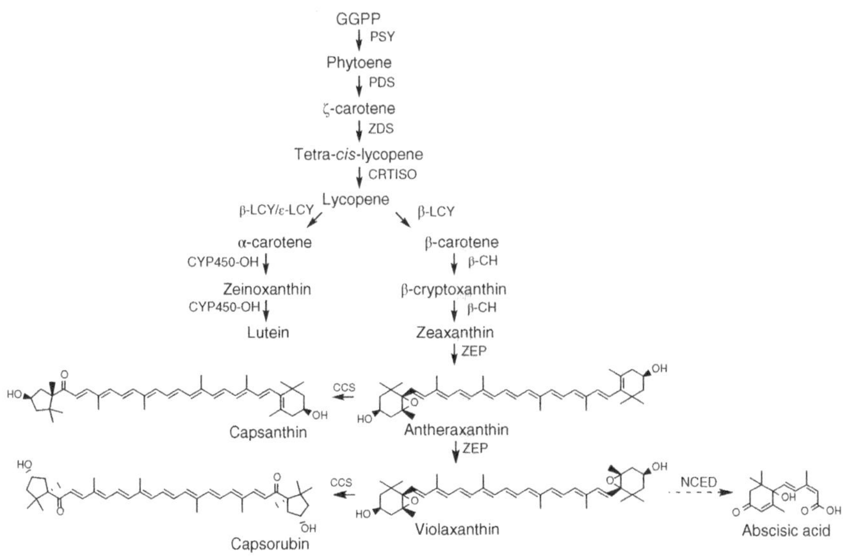

Scheme 6.68 Biosynthetic pathway of carotene and xanthophyll in *Capsicum*.

biosynthetic pathway of *Capsicum* species uniquely exhibits capsanthin-capsorubin synthase (Ccs) that synthesizes two red pigments, capsanthin and capsorubin. In a report, the carotenoid composition of orange fruited *Capsicum* lines was defined along with the allelic variability of the biosynthetic enzymes. The carotenoid content present in seven orange pepper varieties was determined by a UPLC method, which indicated the orange appearance of the fruit was due either to the accumulation of β-carotene or in two cases, due to only the accumulation of red and yellow carotenoids. Four carotenoid biosynthetic genes, *Psy*, *Lcyb*, *CrtZ*-2, and *Ccs* were cloned and sequenced from these cultivars. The data showed that the coding regions within *Psy* and *CrtZ*-2 did not change in any of the lines, the genomic sequence contained introns not previously reported. *Lcyb* and *Ccs* contained no introns but did exhibit polymorphisms resulting in amino acid changes; a new *Ccs* variant was found. When selectively breeding for high provitamin A levels, phenotypic recurrent selection based on fruit color is not sufficient. Based on these results, specific alleles are candidate molecular markers for the selection of orange pepper lines with high β-carotene and thus high provitamin A levels [445]. Since abscisic acid (ABA) is an important plant hormone that plays an important role in stress responses and can also accelerate ripening in climacteric and non-climacteric fruit, the effects of ABA treatment on capsanthin accumulation in pepper fruit and on the expression of key genes involved in the capsanthin biosynthetic pathway was evaluated. Thus, the pepper fruit was treated with ABA at green mature stage. The results indicate that ABA treatment increased capsanthin content in pepper fruit, with the best result obtained with 150 mg L^{-1} ABA solution. Application of exogenous ABA also increased the expression levels of the capsanthin synthesis genes phytoene synthase (*Psy*), lycopene β-cyclase (*Lcyb*), β-carotene hydroxylase (*CrtZ*), and capsanthin/capsorubin synthase (*Ccs*), likely explaining the significant capsanthin content increase in pepper fruit [446].

Furthermore, the carotenoid formation in a red-fruited discovery panel of *Capsicum annuum* (chili pepper) has been characterized. The data revealed that color intensity correlated with the amount of capsanthin and its esters, along with transcript levels of the 1-deoxy-D-xylulose 5-phosphate synthase (DXS) and phytoene synthase-1 (PSY-1) genes. Quantification of carotenoids through development and ripening suggested the presence of separate biosynthesis and accumulation phases. Subplastid fractionation demonstrated the differential sequestration of pigments in high- and low-intensity lines and indicated the PSY protein to be most active in the membrane fractions when abundance was highest in the fibril fractions. Carotenoid accumulation was associated with the esterification of xanthophyll, expression of a putative carotenoid acyltransferase, and increased fibril content within the plastid. The analysis of subplastid fractions suggests that the plastoglobuli are likely to be the progenitor of the characteristic fibrils found in pepper fruit [447].

Fucoxanthin is the second most abundant algal carotenoid and is responsible for *in vivo* absorption at about 500–600 nm. Thus, fucoxanthin acts as an antenna and transfers excitation energy to chlorophyll. Diadinoxanthin is the major carotenoids in phaeophyceae and bacillariophyceae (diatom) and is a light-harvesting carotenoid. The biosynthesis pathway to diadinoxanthin and fucoxanthin in *Phaeodactylum tricornutum* was elucidated by a two-step strategy. For the initial part to β-carotene, genes were selected from the genomic database and the function of several of them was identified by genetic pathway complementation in *E. coli*. They included genes encoding PSY, PDS, ZDS, and β-LCY. In order to establish the reaction sequences of the final steps leading to fucoxanthin and diadinoxanthin, intermediates of the pathway present only in trace amounts were accumulated by TLC and identified as violaxanthin and neoxanthin. Neoxanthin is a branching point for the synthesis of both diadinoxanthin and fucoxanthin and the mechanisms for their formation were proposed (Scheme 6.69) [448]. A single isomerization of one of the allenic double bonds in neoxanthin yields diadinoxanthin. Two reactions, hydroxylation at C8 in combination with a keto-enol tautomerization and acetylation of the 3'-OH group, result in the formation of fucoxanthin.

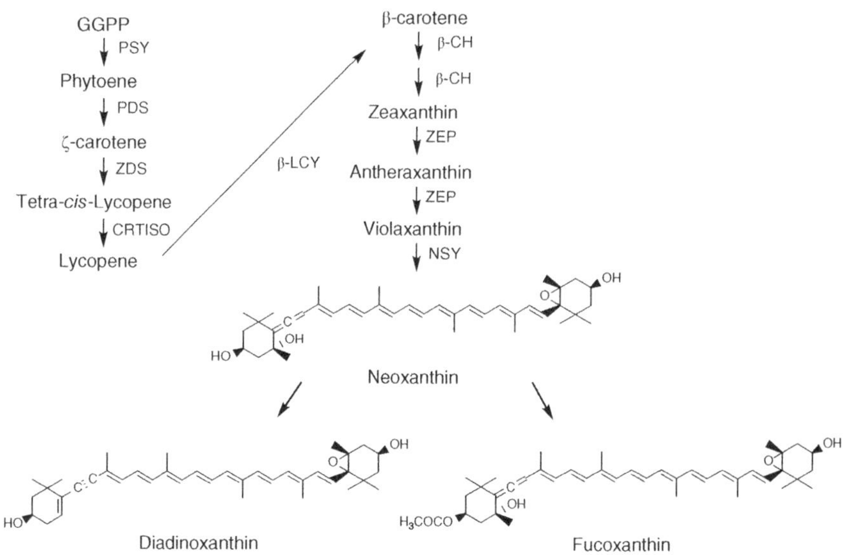

Scheme 6.69 Biosynthetic pathway in *P. tricornutum* from β-carotene to diadinoxanthin and fucoxanthin.

References

1 Höltje, J.-V. (**1998**). *Microbiol. Mol. Biol. Rev. 62*: 181–203.
2 Conti, P., Tamborini, L., Pinto, A. et al. (**2011**). *Chem. Rev. 111*: 6919–6946.
3 Lam, H., Oh, D.-C., Cava, F. et al. (**2009**). *Science 325*: 1552–1555.
4 Hernández, S.B. and Cava, F. (**2016**). *Environ. Microbiol. 18*: 1673–1685.
5 Gao, X., Ma, Q., and Zhu, H. (**2015**). *Appl. Microbiol. Biotechnol. 99*: 3341–3349.
6 Fujii, N. and Saito, T. (**2004**). *Chem. Rec. 4*: 257–278.
7 Yoshimura, T. and Esaki, N. (**2003**). *J. Biosci. Bioeng. 96*: 103–109.
8 Ohide, H., Miyoshi, Y., Maruyama, R. et al. (**2011**). *J. Chromatogr. B 879*: 3162–3168.
9 Radkov, A.D. and Moe, L.A. (**2014**). *Appl. Microbiol. Biotechnol. 98*: 5363–5374.
10 O'Connor, K.A. and Zusman, D.R. (**1997**). *Mol. Microbiol. 24*: 839–850.
11 Redmond, C., Baillie, L.W.J., Hibbs, S. et al. (**2004**). *Microbiology 150*: 355–363.
12 Cheng, Y.-Q. and Walton, J.D. (**2000**). *J. Biol. Chem. 275*: 4906–4911.
13 Di Salvo, M.L., Florio, R., Paiardini, A. et al. (**2013**). *Arch. Biochem. Biophys. 529*: 55–65.
14 Watanabe, A., Yamaguchi, S., Urabe, K., and Asada, Y.J. (**2003**). *Mol. Catal. B-Enzym. 23*: 379–387.
15 Foltyn, V.N., Bendikov, I., Miranda, J.D. et al. (**2005**). *J. Biol. Chem. 280*: 1754–1763.
16 Arias, C.A., Martin-Martinez, M., Blundell, T.L. et al. (**1999**). *Mol. Microbiol. 31*: 1653–1664.
17 Arias, C.A., Weisner, J., Blackburn, J.M., and Reynolds, P.E. (**2000**). *Microbiology 146*: 1727–1734.
18 Yoshimura, T. and Goto, M. (**2008**). *FEBS J. 275*: 3527–3537.
19 Grohs, P., Gutmann, L., Legrand, R. et al. (**2000**). *J. Bacteriol. 182*: 6228–6232.
20 Arias, C.A., Peña, J., Panesso, D., and Reynolds, P. (**2003**). *J. Antimicrob. Chemother. 51*: 557–564.
21 Ito, T., Hayashida, M., Kobayashi, S. et al. (**2016**). *J. Biochem. 160*: 345–353.
22 Soda, K., Yorifuji, T., and Ogata, K. (**1971**). *J. Biol. Chem. 246*: 5085–5092.
23 Soda, K., Yorifuji, T., and Misono, H. (**1971**). *J. Biol. Chem. 246*: 5093–5101.
24 Matsui, D., Oikawa, T., Arakawa, N. et al. (**2009**). *Appl. Microbiol. Biotechnol. 83*: 1045–1054.
25 Radkov, A.D. and Moe, L.A. (**2018**). *Front. Microbiol. 9*: 1343.
26 Kim, P.M., Duan, X., Huang, A.S. et al. (**2010**). *PNAS 107*: 3175–3179.
27 Sakai, K., Homma, H., Lee, J.-A. et al. (**1998**). *Arch. Biochem. Biophys. 351*: 96–105.
28 Topo, E., Soricelli, A., D'Aniello, A. et al. (**2009**). *Reprod. Biol. Endocrinol. 7*: 120.
29 Cardinal, G.J. and Abeles, R.H. (**1968**). *Biochemistry 7*: 3970–3978.
30 Rudnick, G. and Abeles, R.H. (**1975**). *Biochemistry 14*: 4515–4522.

31 Chamond, N., Goytia, M., Coatnoan, N. et al. (**2005**). *Mol. Microbiol. 58*: 46–60.
32 Ayengar, P. and Roberts, E. (**1952**). *J. Biol. Chem.* 453–460.
33 Gall, K.A., Tanner, M.E., and Knowles, J.R. (**1993**). *Biochemistry 32*: 3991–3997.
34 Vance, N.R., Witkin, K.R., Rooney, P.W. et al. (**2018**). *ChemMedChem 13*: 2514–2521.
35 Pal, M. and Bearne, S.L. (**2014**). *Bioorg. Med. Chem. Lett. 24*: 1432–1436.
36 Lamont, H.,.C., Staudenbauer, W.L., and Strominger, J.L. (**1972**). *J. Biol. Chem. 247*: 5103–5106.
37 Schleifer, K.H. and Kandler, O. (**1972**). *Bacteriol. Rev. 36*: 407–477.
38 Antia, M., Hoare, D.S., and Work, E. (**1957**). *J. Biochem. 65*: 448–459.
39 Wiseman, J.S. and Nichols, J.S. (**1984**). *J. Biol. Chem. 259*: 8907–8914.
40 Lim, Y.H., Yoshimura, T., Kurokawa, Y. et al. (**1998**). *J. Biol. Chem. 273*: 4001–4005.
41 Revelles, O., Wittich, R.-M., and Ramos, J.L. (**2007**). *J. Bacteriol. 189*: 2787–2792.
42 Radkov, A.D. and Moe, L.A. (**2013**). *J. Bacteriol. 195*: 5016–5024.
43 Matsui, D. and Oikawa, T. (**2010**). *Chem. Biodivers. 7*: 1591–1602.
44 Espaillat, A., Carrasco-López, C., Bernardo-García, N. et al. (**2014**). *Acta Cryst. D70*: 79–90.
45 Cava, F., Lam, H., de Pedro, M.A., and Waldor, M.K. (**2011**). *Cell. Mol. Life Sci. 68*: 817–831.
46 Würges, K., Mackfeld, U., Pohl, M. et al. (**2011**). *Adv. Synth. Catal. 353*: 2431–2438.
47 Kataoka, M., Wada, M., Ikemi, M. et al. (**1998**). *Ann. N. Y. Acad. Sci. 864*: 318–322.
48 Lim, Y.-H., Yokoigawa, K., Esaki, N., and Soda, K. (**1993**). *J. Bacteriol. 175*: 4213–4217.
49 Kino, K., Sato, M., Yoneyama, M., and Kirimura, K. (**2007**). *Appl. Microbiol. Biotechnol. 73*: 1299–1305.
50 Ebbers, E.J., Ariaans, G.J.A., Houbiers, J.P.M. et al. (**1997**). *Tetrahedron 53*: 9417–9476.
51 Mutaguchi, Y., Ohmori, T., Wakamatsu, T. et al. (**2013**). *J. Bacteriol. 195*: 5207–5215.
52 Mutaguchi, Y., Kasuga, K., and Kojima, I. (**2018**). *Front. Microbiol. 9*: 1540.
53 Yagasaki, M. and Ozaki, A.J. (**1998**). *Mol. Catal. B-Enzym. 4*: 1–11.
54 Fuereder, M., Femmer, C., Storti, G. et al. (**2016**). *Chem. Eng. Sci. 152*: 649–662.
55 Bechtold, M., Makart, S., Reiss, R. et al. (**2007**). *Biotechnol. Bioeng. 98*: 812–824.
56 Femmer, C., Bechtold, M., Roberts, T.M., and Panke, S. (**2016**). *Appl. Microbiol. Biotechnol. 100*: 7423–7436.
57 Schnell, B., Faber, K., and Kroutil, W. (**2003**). *Adv. Synth. Catal. 345*: 653–666.
58 Tatum, E.L., Peterson, W.H., and Fred, E.B. (**1936**). *Biochem. J. 30*: 1892–1897.
59 Hiyama, T., Fukui, S., and Kitahara, K. (**1968**). *J. Biochem. 64*: 99–107.
60 Oren, A. and Gurevich, P. (**1995**). *Can. J. Microbiol. 41*: 302–307.

61 Cantwell, A. and Dennis, D. (**1974**). *Biochemistry 13*: 287–291.
62 Pepple, J.S. and Dennis, D. (**1976**). *Biochim. Biophys. Acta 429*: 1036–1040.
63 Desguin, B., Goffin, P., Bakouche, N. et al. (**2015**). *J. Bacteriol. 197*: 219–230.
64 Rankin, J.A., Mauban, R.C., Fellner, M. et al. (**2018**). *Biochemistry 57*: 3244–3251.
65 Yu, M.-J. and Chen, S.-L. (**2017**). *Chem. Eur. J. 23*: 7545–7557.
66 Xu, T., Wodrich, M.D., Scopelliti, R. et al. (**2017**). *Proc. Natl. Acad. Sci. U.S.A. 114*: 1242–1245.
67 Desguin, B., Goffin, P., Viaene, E. et al. (**2014**). *Nat. Commun. 5*: 3615.
68 Sakai, K., Fujii, N., and Chukeatirote, E. (**2006**). *J. Biosci. Bioeng. 102*: 227–232.
69 Okano, K., Uematsu, G., Hama, S. et al. (**2018**). *Biotechnol. J. 13*: e1700517.
70 Felfer, U., Goriup, M., Koegl, M.F. et al. (**2005**). *Adv. Synth. Catal. 347*: 951–961.
71 Fee, J.A., Hegeman, G.D., and Kenyon, G.L. (**1974**). *Biochemistry 13*: 2528–2532.
72 Neidhart, D.J., Howell, P.L., Petsko, G.A. et al. (**1991**). *Biochemistry 30*: 9264–9273.
73 Kenyon, G.L., Gerlt, J.A., Petsko, G.A., and Kozarich, J.W. (**1995**). *Acc. Chem. Res. 28*: 178–186.
74 Kallarakal, A.T., Mitra, B., Kozarich, J.W., and Gerlt, J.A. (**1995**). *Biochemistry 34*: 2788–2797.
75 Mitra, B., Kallarakal, A.T., Kozarich, J.W., and Gerlt, J.A. (**1995**). *Biochemistry 34*: 2777–2787.
76 St. Maurice, M. and Bearne, S.L. (**2004**). *Biochemistry 43*: 2524–2532.
77 Powers, V.M., Koo, C.W., and Kenyon, G.L. (**1991**). *Biochemistry 30*: 9255–9263.
78 Maggio, E.T., Kenyon, G.L., Mildvan, A.S., and Hegeman, G.D. (**1975**). *Biochemistry 14*: 1131–1139.
79 Gerlt, J.A. and Kozarich, J.W. (**1991**). *J. Am. Chem. Soc. 113*: 9667–9669.
80 Kenyon, G.L. and Hegeman, G.D. (**1970**). *Biochemistry 9*: 4036–4043.
81 Stecher, H., Felfer, U., and Faber, K. (**1997**). *J. Biotechnol. 56*: 33–40.
82 Tsou, A.Y., Ransom, S.C., and Gerlt, J.A. (**1989**). *Biochemistry 28*: 969–975.
83 Hegeman, G.D., Rosenberg, E.Y., and Kenyon, G.L. (**1970**). *Biochemistry 9*: 4029–4036.
84 Li, R., Powers, V.M., Kozarich, J.W., and Kenyon, G.L. (**1995**). *J. Org. Chem. 60*: 3347–3351.
85 Goriup, M., Strauss, U.T., Felfer, U. et al. (**2001**). *Mol. Catal. B-Enzym. 15*: 207–221.
86 Felfer, U., Strauss, U.T., Kroutil, W. et al. (**2001**). *Mol. Catal. B-Enzym. 15*: 213–222.
87 Strauss, U.T. and Faber, K. (**1999**). *Tetrahedron-Asymmetry 10*: 4079–4081.
88 Gu, J., Ye, L., Guo, F. et al. (**2015**). *Tetrahedron Lett. 56*: 1489–1491.
89 Zhou, Y., Wu, S., and Li, Z. (**2017**). *Adv. Synth. Catal. 359*: 4305–4316.
90 Fan, C.-W., Xu, G.-C., Ma, B.-D. et al. (**2015**). *J. Biotechnol. 195*: 67–71.
91 Li, D., Zeng, Z., Yang, J. et al. (**2013**). *Biosci. Biotechnol. Biochem. 77*: 1236–1239.

92 Choi, W.J., Lee, K.Y., Kang, S.H., and Lee, S.B. (**2007**). *Sep. Purif. Technol. 53*: 178–182.

93 Ei Gihani, M.T. and Williams, J.M.J. (**1999**). *Curr. Opin. Chem. Biol. 3*: 11–15.

94 Strauss, U.T., Kandelbauer, A., and Faber, K. (**2000**). *Biotechnol. Lett. 22*: 515–520.

95 Bauer, C., Boy, M., Baber, K. et al. (**2001**). *Mol. Catal. B-Enzym. 16*: 91–100.

96 Wrzosek, K., Harriehausen, I., and Seidel-Morgenstern, A. (**2018**). *Org. Process Res. Dev. 22*: 1761–1771.

97 Wrzosek, K., García Rivera, M.A., Bettenbrock, K., and Seidel-Morgenstern, A. (**2016**). *Biotechno.l J 11*: 453–463.

98 Datta, A. (**1970**). *Biochemistry 9*: 3363–3370.

99 Luchansky, S.J., Yarema, K.J., Takahashi, S., and Bertozzi, C.R. (**2003**). *J. Biol. Chem. 278*: 8035–8042.

100 Ito, S. (**2010**). *J. Appl. Glycosci. 57*: 1–6.

101 Liao, H.-F., Kao, C.-H., Lin, W.-D. et al. (**2012**). *Process Biochem. 47*: 948–952.

102 Schmitz, C., Gotthardt, M., Hinderlich, S. et al. (**2000**). *J. Biol. Chem. 275*: 15357–15362.

103 Babu, Y.S., Chand, P., Banita, S. et al. (**2000**). *J. Med. Chem. 43*: 3482–3486.

104 Bednarski, M.D., Chenault, H.K., Simon, E.S., and Whitesides, G.M. (**1987**). *J. Am. Chem. Soc. 109*: 1283–1285.

105 Kim, M.-J., Hennen, W.J., Sweers, H.M., and Wong, C.-H. (**1988**). *J. Am. Chem. Soc. 110*: 6481–6486.

106 Mahmoudian, M., Noble, D., Drake, C.S. et al. (**1997**). *Enzyme Microb. Technol. 20*: 393–400.

107 Kragl, U., Kittelmann, M., Ghisalba, O., and Wandrey, C. (**1995**). *Ann. NY Acad. Sci. 750*: 300–305.

108 Maru, I., Ohnishi, J., Ohta, Y., and Tsukada, Y. (**1998**). *Carbohydr. Res. 306*: 575–578.

109 Lee, Y.-C., Chien, H.-C.R., and Hsu, W.-H. (**2007**). *J. Biotechnol. 129*: 453–460.

110 Tabata, K., Koizumi, S., Endo, T., and Ozaki, A. (**2002**). *Enzyme Microb. Technol. 30*: 327–333.

111 Zhang, Y., Tao, F., Du, M. et al. (**2010**). *Appl. Microb. Biotechnol. 86*: 481–489.

112 Hu, S., Chen, J., Yang, Z. et al. (**2010**). *Appl. Microb. Biotechnol. 85*: 1383–1391.

113 Sun, W., Ji, W., Li, N. et al. (**2013**). *Bioresour. Technol. 130*: 23–29.

114 Lin, B.-X., Zhang, Z.-J., Liu, W.-F. et al. (**2013**). *Appl. Microb. Biotechnol. 97*: 4775–4784.

115 Klermund, L., Groher, A., and Castiglione, K. (**2013**). *J. Biotechnol. 168*: 256–263.

116 Zhou, J., Chen, X., Lu, L. et al. (**2016**). *Mol. Catal. B-Enzym. 125*: 42–48.

117 Wang, Z., Zhuang, W., Cheng, J. et al. (**2017**). *J. Agric. Food Chem. 65*: 7467–7475.

118 Chen, X., Zhou, J., Zhang, L. et al. (**2018**). *Metab. Eng. 47*: 374–382.

119 Kao, C.-H., Chen, Y.-Y., Wang, L.-R., and Lee, Y.-C. (**2018**). *Mol. Biotechnol. 60*: 427–434.

120 Gao, X., Zhang, F., Wu, M. et al. (**2019**). *J. Agric. Food Chem. 67*: 285–6291.

121 Watabe, K., Ishikawa, T., Mukohara, Y., and Nakamura, H. (**1992**). *J. Bacteriol. 174*: 7989–7995.

122 Wise, A., Pietzsch, M., Syldatk, C. et al. (**2000**). *J. Biotechnol. 80*: 217–230.

123 Las Heras-Vázquez, F.J., Martínez-Rodríguez, S., Mingorance-Cazorla, L. et al. (**2003**). *Biochem. Biophys. Res. Commun. 303*: 541–547.

124 Martínez-Rodríguez, S., Las Heras-Vázquez, F.J., Mingorance-Cazorla, L. et al. (**2004**). *Appl. Environ. Microbiol. 70*: 625–630.

125 Martínez-Rodríguez, S., Las Heras-Vázquez, F.J., Clemente-Jiménez, J.M., and Rodríguez-Vico, F. (**2004**). *Biochimie 86*: 77–81.

126 Suzuki, S., Onishi, N., and Yokozeki, K. (**2005**). *Biosci. Biotechnol. Biochem. 69*: 530–536.

127 Watabe, K., Ishikawa, T., Mukohara, Y., and Nakamura, H. (**1992**). *J. Bacteriol. 174*: 3461–3466.

128 Hils, M., Münch, P., Altenbuchner, J. et al. (**2001**). *Appl. Microbiol. Biotechnol. 57*: 680–688.

129 Wise, A., Syldatk, C., Mattes, R., and Altenbuchner, J. (**2001**). *Arch. Microbiol. 176*: 187–196.

130 Suzuki, S., Takenaka, Y., Onishi, N., and Yokozeki, K. (**2005**). *Biosci. Biotechnol. Biochem. 69*: 1473–1482.

131 Altenbuchner, J., Siemann-Herzberg, M., and Syldatk, C. (**2001**). *Curr. Opin. Biotechnol. 12*: 559–563.

132 Ogawa, J. and Shimizu, S.J. (**1997**). *Mol. Catal. B-Enzym. 2*: 163–176.

133 Bommarius, A., Schwarm, M., and Drauz, K.J. (**1998**). *Mol. Catal. B-Enzym. 5*: 1–11.

134 Martínez-Rodríguez, S., González-Ramírez, L.A., Clemente-Jiménez, J.M. et al. (**2008**). *Acta Cryst. F64*: 50–53.

135 Andújar-Sánchez, M., Martínez-Rodríguez, S., Las Heras-Vázquez, F.J. et al. (**2006**). *Biochim. Biophys. Acta 1764*: 292–298.

136 Martínez-Rodríguez, S., Andújar-Sánchez, M., Neira, J.L. et al. (**2006**). *Protein Sci. 15*: 2729–2738.

137 Ishikawa, T., Watabe, K., Mukohara, Y., and Nakamura, H. (**1997**). *Biosci. Biotechnol. Biochem. 61*: 185–187.

138 Wilms, B., Wiese, A., Syldatk, C. et al. (**2001**). *J. Biotechnol. 86*: 19–30.

139 Rodríguez-Alono, M.J., Clemente-Jiménez, J.M., Rodríguez-Vico, F., and Las Heras-Vázquez, F.J. (**2015**). *Biochem. Eng. J. 101*: 68–76.

140 Rodríguez-Alono, M.J., Rodríguez-Vico, F., Las Heras-Vázquez, F.J., and Clemente-Jiménez, J.M. (**2017**). *Catalysts 7*: 192.

141 Rodríguez-Alono, M.J., Rodríguez-Vico, F., Las Heras-Vázquez, F.J., and Clemente-Jiménez, J.M. (**2016**). *J. Chem. Technol. Biotechnol. 91*: 1972–1981.

142 May, O., Verseck, S., Bommarius, A., and Drauz, K. (**2002**). *Org. Process Res. Dev. 6*: 452–457.

143 Nam, S.-H., Park, H.-S., and Kim, H.-S. (**2005**). *Chem. Rec. 5*: 298–307.

144 Syldatk, C., May, O., Altenbuchner, J. et al. (**1999**). *Appl. Microbiol. Biotechnol. 51*: 293–309.

145 Martínez-Rodríguez, S., Las Heras-Vázquez, F.J., Clemente-Jiménez, J.M. et al. (**2002**). *Biotechnol. Prog. 18*: 1201–1206.

146 Martínez-Gómez, A.I., Martínez-Rodríguez, S., Clemente-Jiménez, J.M. et al. (**2007**). *Appl. Environ. Microbiol. 73*: 1525–1531.

147 Nozaki, H., Takenaka, Kira, I., Watanabe, K., and Yokozeki, K. (**2005**). *J. Mol. Catal. B-Enzym. 32*: 213–218.

148 Liu, Y., Xu, G., Han, R. et al. (**2018**). *Bioresour. Technol. 249*: 720–728.

149 Scher, W. and Jakoby, W.B. (**1969**). *J. Biol. Chem. 244*: 1878–1882.

150 Kato, Y., Yamagishi, J., and Asano, Y. (**1995**). *J. Ferment. Bioeng. 80*: 610–612.

151 Hatakeyama, K., Asai, Y., Uchida, Y. et al. (**1997**). *Biochem. Biophys. Res. Commun. 239*: 74–79.

152 Hatakeyama, K., Goto, M., Uchida, Y. et al. (**2000**). *Biosci. Biotechnol. Biochem. 64*: 569–576.

153 Hatakeyama, K., Goto, M., Kobayashi, M. et al. (**2000**). *Biosci. Biotechnol. Biochem. 64*: 1477–1485.

154 Jiménez, J.I., Canales, Á., Jiménez-Barbero, J. et al. (**2008**). *Proc. Natl. Acad. Sci. U. S. A. 105*: 11329–11334.

155 Fisch, F., Fleites, C.M., Delenne, M. et al. (**2010**). *J. Am. Chem. Soc. 132*: 11455–11457.

156 Liu, X., Zhao, Q., Ren, J. et al. (**2013**). *Mol. B-Enzym. 93*: 44–50.

157 Ruzheinikov, S.N., Taal, M.A., Sedelnikova, S.E. et al. (**2005**). *Structure 13*: 1707–1713.

158 Ohtaki, A., Nakano, Y., Iizuka, R. et al. (**2008**). *Proteins 70*: 1167–1174.

159 Chen, D., Tang, H., Lv, Y. et al. (**2013**). *Mol. Microbiol. 87*: 1237–1244.

160 Dokainish, H.M., Ion, B.F., and Gauld, J.W. (**2014**). *Phys. Chem. Chem. Phys. 16*: 12462–12474.

161 Otsuka, K. (**1961**). *Agric. Biol. Chem. 25*: 726–730.

162 Engel, C.A.R., Straathof, A.J.J., Zijlmans, T.W. et al. (**2008**). *Appl. Microbiol. Biotechnol. 78*: 379–389.

163 Behrman, E.J. and Stanier, R.Y. (**1957**). *J. Biol. Chem. 228*: 923–945.

164 Tang, H., Yao, Y., Wang, L. et al. (**2012**). *Sci. Rep. 2*: 377.

165 Takamura, Y., Takamura, T., Soejima, M., and Uemura, T. (**1969**). *Agric. Biol. Chem. 33*: 718–728.

166 Nakajima-Kambe, T., Nozue, T., Mukouyama, M., and Nakahara, T. (**1997**). *J. Ferment. Bioeng. 84*: 165–168.

167 Ichikawa, S., Iino, T., Sato, S. et al. (**2003**). *Biochem. Eng. J. 13*: 7–13.

168 Kimura, T., Kawabata, Y., and Sato, E. (**1986**). *Agric. Biol. Chem. 50*: 89–94.

169 Liavonchanka, A. and Feussner, I. (**2008**). *ChemBioChem 9*: 1867–1872.

170 Kepler, C.R. and Tove, S.B. (**1967**). *J. Biol. Chem. 242*: 5686–5692.

171 Peng, S.S., Deng, M.-D., Grund, A.D., and Rosson, R.A. (**2007**). *Enzyme Microb. Technol. 40*: 831–839.

172 Devillard, E., McIntosh, F.M., Duncan, S.H., and Wallace, R.J. (**2007**). *J. Bacteriol. 189*: 2566–2570.

173 Lin, T.Y., Lin, C.W., and Wang, Y.J. (**2002**). *J. Food Sci. 67*: 1502–1505.

174 Lin, T.Y., Lin, C.-W., and Wang, Y.-J. (**2003**). *Food Chem. 83*: 27–31.

175 Lin, T.Y. (**2006**). *Food Chem. 94*: 437–441.

176 Rosberg-Cody, E.; Ross, R.P.; Hussey, S.; Ryan, C.A; Murphy, B.P.; Fitzgerald, G.F.; Devery, R.; Stanton, C. (**2004**). *Appl. Environ. Microbiol. 70*, 4635-4641.

177 Wall, R., Ross, R.P., Shanahan, F. et al. (**2009**). *Am. J. Clin. Nutr. 89*: 1393–1401.

178 Fukuda, S., Furuya, H., Suzuki, Y. et al. (**2005**). *J. Gen. Appl. Microbiol. 51*: 105–113.

179 Fukuda, S., Suzuki, Y., Murai, M. et al. (**2006**). *App. Microbiol. 100*: 787–794.

180 Gorisson, L., Weckx, S., Vlaeminck, B. et al. (**2011**). *J. Appl. Microbiol. 111*: 593–606.

181 Zhao, H.-W., Lv, J.-P., and Li, S.-R. (**2011**). *Biotechnol. Biotechnol. Equip. 25*: 2266–2272.

182 Li, H., Liu, Y., Bao, Y. et al. (**2012**). *J. Food Sci. 77*: M330–M336.

183 Wei, M., Ding, X.-L., Xue, Z.-L., and Zhao, S.-G.J. (**2014**). *Mol. Catal. B-Enzym. 108*: 59–63.

184 Alnoso, L., Cuesta, E.P., and Gilliland, S.E. (**2003**). *J. Dairy Sci. 86*: 1941–1946.

185 Roman-Nunez, M., Cuesta-Aloson, E.P., and Gilliland, S.E. (**2007**). *J. Food Sci. 72*: M140–M143.

186 Renes, E., Linares, D.M., González, L. et al. (**2017**). *Funct. Food. 34*: 340–346.

187 Bhosale, S.H., Rao, M.B., and Deshpande, V.V. (**1996**). *Microbiol. Rev. 60*: 280–300.

188 Van Maris, A.J.A., Winkler, A.A., Kuyper, M. et al. (**2007**). *Adv. Biochem. Eng. Biotechnol. 108*: 179–204.

189 Beerens, K., desmet, T., and Soetaert, W. (**2012**). *J. Ind. Microbiol. Biotechnol. 39*: 823–834.

190 Lobanok, A.G., Sapunova, L.I., Dikhtievski, Y.O., and Kazakevich, I.O. (**1998**). *World J. Microbiol. Biotechnol 14*: 259–262.

191 Takasaki, Y., Kosugi, Y., and Kanbayashi, A. (**1969**). *Agric. Biol. Chem. 33*: 1527–1534.

192 Yamanaka, K. and Takahara, N. (**1977**). *Agric. Biol. Chem. 41*: 1909–1915.

193 Chanitnun, K. and Pinphanichakarn, P. (**2012**). *Braz. J. Microbiol.* 1084–1093.

194 Jenkins, J., Janin, J., Rey, F. et al. (**1992**). *Biochemistry 31*: 5449–5458.

195 Makkee, M., Kieboom, A.P.G., and van Bekkum, H. (**1984**). *Recl. Trav. Chim. Pays-Bas-J. Roy. Neth. Chem. Soc. 103*: 361–364.

196 Lee, C., Bagdasarian, M., Meng, M., and Zeikus, J.G. (**1990**). *J. Biol. Chem. 265*: 19082–19090.

197 Katz, A.K., Li, X., Carrell, H.L. et al. (**2006**). *Proc. Natl. Acad. Sci. U.S.A. 103*: 8342–8347.

198 Mitsuhashi, S. and Lampen, J.O. (**1953**). *J. Biol. Chem. 204*: 1011–1018.

199 Marshall, R. and Kooi, E. (**1957**). *Science 125*: 648–649.

200 Meng, M., Lee, C., Bagdasarian, M., and Zeikus, J.G. (**1991**). *Proc. Natl. Acad. Sci. U.S.A. 88*: 4015–4019.

201 Bock, K., Meldal, M., Meyer, B., and Wiebe, L. (**1983**). *Acta Chem. Scand. B 37*: 101–108.

202 Lee, H.S. and Hong, J. (**2000**). *J. Biotechnol. 84*: 145–153.

203 Staudigl, P., Haltrich, D., and Peterbauer, C.K. (**2014**). *J. Agric. Food Chem. 62*: 1617–1624.

204 Strandberg, G.W. and Smiley, K.L. (**1971**). *Appl. Microbiol. 21*: 588–593.

205 Giovenco, S., Morisi, F., and Pansolli, P. (**1973**). *FEBS Lett. 36*: 57–60.

206 Kamal, H., Hegazy, E.-S.A., Sharada, H.M. et al. (**2014**). *J. Radiat. Res. Appl. Sci. 7*: 154–162.

207 Tükel, S.S. (**2008**). AlagöD. *Food Chem. 111*: 658–662.

208 Yu, H., Guo, Y., Wu, D. et al. (**2011**). *Mol. Catal. B-Enzym. 72*: 73–76.

209 Ge, Y., Zhou, H., Kong, W. et al. (**1998**). *Appl. Biochem. Biotechnol. 69*: 203–215.

210 Jokela, J., Pastinen, O., and Leisola, M. (**2002**). *Enzyme Microb. Technol. 31*: 67–76.

211 Sohrabi, M. and Marvast, M.A. (**2000**). *Ind. Eng. Che. Res. 39*: 1903–1910.

212 Salehi, Z., Sohrabi, M., Kaghazchi, T., and Bonakdarpour, B. (**2005**). *Process Biochem. 40*: 2455–2460.

213 Fatourehchi, N., Sohrabi, M., Dabir, B., and Royaee, S.J. (**2014**). *J. Chem. Technol. Biotechnol. 89*: 1918–1923.

214 Jia, D.-X., Zhou, L., and Zheng, Y.-G. (**2017**). *Enzyme Microb. Technol. 99*: 1–8.

215 Liu, Z.-Q., Zheng, W., Huang, J.-F. et al. (**2015**). *J. Ind. Microbiol. Biotechnol. 42*: 1091–1103.

216 Jia, D.-X., Wang, T., Liu, Z.-J. et al. (**2018**). *J. Biosci. Bioeng. 126*: 176–182.

217 Zhang, F., Duan, X., Chen, S. et al. (**2013**). *Process Biochem. 48*: 1502–1508.

218 Rhimi, M., Messaoud, E.B., Borgi, M.A. et al. (**2007**). *Enzyme Microb. Technol. 40*: 1531–1537.

219 Kwan, T.H., Ong, K.L., Haque, M.A. et al. (**2018**). *Food Bioprod. Process. 111*: 141–152.

220 Gong, C.-S., Chen, L.-F., Flickinger, M.C. et al. (**1981**). *Appl. Environ. Microbiol. 41*: 430–436.

221 Lastick, S.M., Tucker, M.Y., Beyette, J.R. et al. (**1989**). *Appl. Microbiol. Biotechnol. 30*: 574–579.

222 Itoh, H., Sato, T., and Izumori, K. (**1995**). *J. Ferment. Bioeng. 80*: 101–103.

223 Lee, Y.-C., Dutta, S., and Wu, K.C.-W. (**2014**). *ChemSusChem 7*: 3241–3246.

224 Makkee, M., Kieboom, A.P.G., and Van Bekkum, H. (**1985**). *Starch-Stärke 37*: 136–141.

225 Kaup, B., Bringer-Meyer, S., and Sahm, H. (**2005**). *Appl. Microbiol. Biotechnol. 69*: 397–403.

226 Xu, W., Zhang, W., Zhang, T. et al. (**2016**). *Appl. Microbiol. Biotechnol. 100*: 2958–2992.

227 Levinson, S.L. and Krulwich, T.A. (**1976**). *J. Gen. Microbiol. 95*: 277–286.

228 Badia, J., Gimenez, R., Baldoma, L. et al. (**1991**). *J. Bacteriol. 173*: 5144–5150.

229 Tagaki, Y. and Sawada, H. (**1964**). *Biochim. Biophys. Acta 92*: 10–17.

230 Bhuiyan, S.H., Itami, Y., and Izumori, K. (**1997**). *J. Ferment. Bioeng. 84*: 319–323.

231 Leang, K., Takada, G., Ishimura, A. et al. (**2004**). *Appl. Environ. Microbiol. 70*: 3298–3304.

232 Yoshida, H., Takeda, K., Izumori, K., and Kamitori, S. (**2010**). *Protein Eng. Des. Sel. 23*: 919–927.

233 Poonperm, W., Takata, G., Okada, H. et al. (**2007**). *Appl. Microbiol. Biotechnol. 76*: 1297–1307.

234 Lin, C.-J., Tseng, W.-C., Lin, T.-H. et al. (**2010**). *J. Agric. Food Chem. 58*: 10431–10436.

235 Park, C.-S., Yeom, S.-J., Lim, Y.-R. et al. (**2010**). *Biotechnol. Lett. 32*: 1947–1953.

236 Lin, C.-J., Tseng, W.-C., and Fang, T.-Y. (**2011**). *J. Agric. Food Chem. 59*: 8702–8708.

237 Doan, T.-N.-T., Prabhu, P., Kim, J.-K. et al. (**2010**). *Acta Cryst. F66*: 677–680.

238 Prabhu, P., Doan, T.T.N., Jeya, M. et al. (**2011**). *Appl. Microbiol. Biotechnol. 89*: 635–644.

239 Takata, G., Uechi, K., Taniguchi, E. et al. (**2011**). *Biosci. Biotechnol. Biochem. 75*: 1006–1009.

240 Kim, Y.-S., Shin, K.-C., Lim, Y.-R., and Oh, D.-K. (**2013**). *Biotechnol. Lett. 35*: 259–264.

241 Park, C.-S. (**2014**). *Biotechnol. Bioproc. Eng. 19*: 18–25.

242 Bai, W., Shen, J., Zhu, Y. et al. (**2015**). *Food Sci. Technol. Res. 21*: 13–22.

243 Korndörfer, I.P., Fessner, W.-D., and Matthew, B.W. (**2000**). *J. Mol. Biol. 300*: 917–933.

244 Prabhu, P., Doan, T.-N.-T., Tiwari, M. et al. (**2014**). *FEBS J. 281*: 3446–3459.

245 Yoshida, H., Yamaji, M., Ishii, T. et al. (**2010**). *FEBS J. 277*: 1045–1057.

246 Yoshida, H., Yoshihara, A., Teraoka, M. et al. (**2013**). *FEBS Open Bio. 3*: 35–40.

247 Fang, Z., Zhang, W., Zhang, T. et al. (**2018**). *Appl. Microbiol. Biotechnol. 102*: 7283–7292.

248 Wu, R., Xie, H., Mo, Y., and Cao, Z. (**2009**). *J. Phys, Chem. 113*: 11595–11603.

249 Wen, L., Zang, L., Huang, K. et al. (**2016**). *Bioorg. Med. Chem. Lett. 26*: 969–972.

250 Chen, Z., Xu, W., Zhang, W. et al. (**2018**). *J. Sci. Food Agric. 98*: 2184–2193.

251 Lim, Y.-R. and Oh, D.-K. (**2011**). *Appl. Microbiol. Biotechnol. 91*: 229–235.

252 Bhuiyan, S.H., Itami, Y., Rokui, Y. et al. (**1998**). *J. Ferment. Bioeng. 85*: 539–541.

253 Menavuvu, B.T., Poonperm, W., Leang, K. et al. (**2006**). *J. Biosci. Bioeng. 101*: 340–345.

254 Morimoto, K., Park, C.-S., Ozaki, M. et al. (**2006**). *Enzyme Microbial Technol. 38*: 855–859.

255 Mu, W., Yu, L., Zhang, W. et al. (**2015**). *Appl. Microbiol. Biotechnol. 99*: 6571–6584.

256 Xu, W., Zhang, W., Tian, Y. et al. (**2017**). *Process Chem. 53*: 153–161.

257 Seo, M.-J., Choi, J.-H., Kang, S.-H. et al. (**2018**). *Biotechnol. Lett. 40*: 325–334.

258 Chen, Z., Chen, J., Zhang, W. et al. (**2018**). *J. Agric. Food Chem. 66*: 12017–12024.

259 Bhuiyan, S.H., Itami, Y., Takada, G., and Izumori, K. (**1999**). *J. Biosci. Bioeng. 88*: 567–570.

260 Bhuiyan, S.H., Ahmed, Z., Utamura, M., and Izumori, K. (**1998**). *J. Ferment. Bioeng. 86*: 513–516.

261 Granström, T.B., Takata, G., Morimoto, K. et al. (**2005**). *Enzyme Microb. Technol. 36*: 976–981.

262 Bhuiyan, S.H., Itami, Y., and Izumori, K. (**1997**). *J. Ferment. Bioeng. 84*: 558–562.

263 Leang, K., Takata, G., Fukai, Y. et al. (**2004**). *Biochim. Biophys. Acta 1674*: 68–77.

264 Leang, K., Maekawa, K., Menavuvu, B.T. et al. (**2004**). *J. Biosci. Bioeng. 97*: 383–388.

265 Shommpoosang, S., Yoshihara, A., Uechi, K. et al. (**2014**). *Biosci. Biotechnol. Biochem. 78*: 317–325.

266 Wen, L., Huang, K., Zheng, Y. et al. (**2016**). *Tetrahedron Lett. 57*: 3819–3832.

267 Shompoosang, S., Yoshihara, A., Uechi, K. et al. (**2016**). *J. Biosci. Bioeng. 121*: 1–6.

268 Green, M. and Cohen, S.S. (**1956**). *J. Biol. Chem. 219*: 557–568.

269 Eagon, R.G. (**1961**). *J. Bacteriol. 82*: 548–550.

270 Eugene, O. and Mortlock, R.P. (**1971**). *J. Bacteriol. 108*: 293–299.

271 Izumori, K. and Yamanaka, K. (**1974**). *Agric. Biol. Chem. 38*: 267–273.

272 Takeda, K., Yoshida, H., Izumori, K., and Kamitori, S. (**1804**). *Biochim. Biophys. Acta* **2010**: 1359–1368.

273 Seemann, J.E. and Schulz, G.E. (**1997**). *J. Mol. Biol. 273*: 256–268.

274 Hong, S.-H., Lim, Y.-R., Kim, Y.-S., and Oh, D.-K. (**2012**). *Biochimie 94*: 1926–1934.

275 Ju, Y.-H. and Oh, D.-K. (**2010**). *Biotechnol. Lett. 32*: 299–304.

276 Alajarín, R., García-Junceda, E., and Wong, C.-H. (**1995**). *J. Org. Chem. 60*: 4294–4295.

277 Wong, C.-H., Alajarín, R., Morís-Vars, F. et al. (**1995**). *J. Org. Chem. 60*: 7360–7363.

278 Fessner, W.-D., Goβe, C., Jaeschke, G., and Eyrisch, O.E. (**2000**). *J. Org. Chem.* 125–132.

279 Usvalampi, A., Turunen, O., Valjakka, J. et al. (**2012**). *Enzym. Microb. Technol. 50*: 71–76.

280 Tozzi, M.G., Camici, M., Mascia, L. et al. (**2006**). *FEBS J. 273*: 1089–1101.

281 Hammer-Jespersen, K. and Munch-Petersen, A. (**1970**). *Eur. J. Biochem. 17*: 397–407.

282 Kammen, H.O. and Koo, R. (**1969**). *J. Biol. Chem. 244*: 4888–4893.

283 Galperin, M.Y., Bairoch, A., and Koonin, E.V. (**1998**). *Protein Sci. 7*: 1829–1835.

284 Maliekal, P., Sokolova, T., Vertommen, D. et al. (**2007**). *J. Biol. Chem. 282*: 31844–31851.

285 Panosian, T.D., Nannemann, D.P., Bachmann, B.O., and Iverson, T.M. (**2010**). *Acta Cryst. F66*: 811–814.

286 Panosian, T.D., Nannemann, D.P., Watkins, G.R. et al. (**2011**). *J. Biol. Chem. 286*: 8043–8054.

287 Iverson, T.M., Panosian, T.D., Birmingham, W.R. et al. (**2012**). *Biochemistry 51*: 1964–1975.

288 Barsky, D.L. and Hoffee, P.A. (**1983**). *Biochim. Biophys. Acta 743*: 162–171.

289 Hamamoto, T., Noguchi, T., and Midorikawa, Y. (**1998**). *Biosci. Biotechnol. Biochem. 62*: 1103–1108.

290 Tavern-Porro, M., Bouvier, L.A., Pereira, C.A. et al. (**2008**). *Tetrahedron Lett. 49*: 2642–2645.

291 Rivero, C.W., De Benedetti, E.C., Gallego, F.L. et al. (**2017**). *J. Biotechnol. 249*: 34–41.

292 Moustafa, H.M.A., Kim, E.-J., Zhu, Z. et al. (**2016**). *ChemCatChem 8*: 2898–2902.

293 Horinouchi, N., Ogawa, J., Sakai, T. et al. (**2003**). *Appl. Environ. Microbiol. 69*: 3791–3797.

294 Ogawa, J., Saito, K., Sakai, T. et al. (**2003**). *Biosci. Biotechnol. Biochem. 67*: 933–936.

295 Horinouchi, N., Ogawa, J., Kawano, T. et al. (**2006**). *Appl. Microbiol. Biotechnol. 71*: 615–621.

296 Horinouchi, N., Ogawa, J., Kawano, T. et al. (**2006**). *Biotechnol. Lett. 28*: 877–881.

297 Rashid, N., Imanaka, H., Fukui, T. et al. (**2004**). *J. Bacteriol. 186*: 4185–4191.

298 Horinouchi, N., Kawano, T., Sakai, T. et al. (**2009**). *New Biotech. 26*: 75–82.

299 Fateev, IV.; Antonov, K.V.; Konstantinova, I.D.; Muravyova, T.I.; Seela, F.; Esipov, R.S.; Miroshnikov, A.I.; Mikhailopulo, I.A. (**2014**). *Beilstein J. Org. Chem. 10*, 1657-1669.

300 Valino, A.L., Iribarren, A.M., and Lewkowicz, E.J. (**2015**). *Mol. Catal. B-Enzym. 114*: 58–64.

301 Aberhart, D.J., Lin, H.-J., and Weiller, B.H. (**1981**). *J. Am. Chem. Soc. 103*: 6750–6752.

302 Frey, P.A. (**1993**). *FASEB J. 7*: 662–670.

303 Frey, P.A. and Reed, G.H. (**1814**). *Biochim. Biophys. Acta* **2011**: 1548–1557.

304 Chirpich, T.P., Zappia, V., Costilow, R.N., and Barker, H.A. (**1970**). *J. Biol. Chem. 245*: 1778–1789.

305 Zappia, V. and Baker, H.A. (**1970**). *Biochim. Biophys. Acta 207*: 505–513.

306 Song, K.B. and Frey, P.A. (**1991**). *J. Biol. Chem. 266*: 7651–7655.

307 Petrovich, R.M., Ruzicka, F.J., Reed, G.H., and Frey, P.A. (**1991**). *J. Biol. Chem. 266*: 7656–7660.

308 Lepore, B.W., Ruzicka, F.J., Frey, P.A., and Ringe, D. (**2005**). *Proc. Natl. Acad. Sci. U.S.A. 102*: 13819–13824.

309 Gould, S.J. and Thiruvengadam, T.K. (**1981**). *J. Am. Chem. Soc. 103*: 6752–6754.

310 Chen, D., Ruzicka, F.J., and Frey, P.A. (**2000**). *Biochem. J. 348*: 539–549.

311 Chen, D. and Frey, P.A. (**2001**). *Biochemistry 40*: 596–602.

312 Brazeau, B.J., Gort, S.J., Jessen, H.J. et al. (**2006**). *Appl. Environ. Microbiol. 72*: 6402–6404.

313 Aberhart, D.J., Gould, S.J., Lin, H.-J. et al. (**1983**). *J. Am. Chem. Soc. 105*: 5461–5470.

314 Chen, D., Tanem, J., and Frey, P.A. (**2007**). *Biochim. Biophys. Acta 1774*: 297–302.

315 Wang, S.C. and Frey, P.A. (**2007**). *Biochemistry 46*: 12889–12895.

316 Behshad, E., Ruzicka, F.J., Mansoorabadi, S.O. et al. (**2006**). *Biochemistry 45*: 12639–12646.

317 Thiruvengadam, T.K., Gould, S.J., Aberhart, D.J., and Lin, H.-J. (**1983**). *J. Am. Chem. Soc. 105*: 5470–5476.

318 Morse, B.K., Brown, M.S., Gagane, J.W. et al. (**1997**). *Antibiotics 50*: 698–670.

319 Barkei, J.J., Kevany, B.M., Felnagle, E.A., and Thomas, M.G. (**2009**). *ChemBioChem 10*: 366–376.

320 Baumann, K.P., Gottschalk, G., Lin, W. et al. (**2003**). *Appl. Environ. Microbiol. 69*: 6047–6055.

321 Saum, R., Mingote, A., Santos, H., and Müller, V. (**2009**). *Environ. Microbiol. 11*: 1056–1065.

322 Müller, S., Hoffmann, T., Santos, H. et al. (**2011**). *Appl. Microbiol. Biotechnol. 91*: 689–697.

323 Hung, C.-C. and Lai, M.-C. (**2013**). *J. Microbiol. Immunol. Infect. 46*: 1–10.

324 Wu, B., Szymański, W., Heberling, M.M. et al. (**2011**). *Trends Biotechnol. 29*: 352–362.

325 Morley, C.G.D. and Stsdtman, T.C. (**1970**). *Biochemistry 9*: 4890–4900.

326 Morley, C.G.D. and Stsdtman, T.C. (**1971**). *Biochemistry 12*: 2325–2329.

327 Morley, C.G.D. and Stsdtman, T.C. (**1972**). *Biochemistry 11*: 600–605.

328 Morley, C.G.D. and Stsdtman, T.C. (**1973**). *Biochemistry 12*: 1054–1063.

329 Kreimeyer, A., Perret, A., Lechaplais, C. et al. (**2007**). *J. Biol. Chem. 282*: 7191–7197.

330 Maity, A.N., Chen, Y.-H., and Ke, S.-C. (**2014**). *Int. J. Mol. Sci. 15*: 3064–3087.

331 Chang, C.H. and Frey, P.A. (**2000**). *J. Biol. Chem. 275*: 106–114.

332 Tang, K.-H., Harms, A., and Frey, P.A. (**2002**). *Biochemistry 41*: 8767–8776.

333 Berkovitch, F., Behshad, E., Tang, K.-H. et al. (**2004**). *Proc. Natl. Acad. Sci. U.S.A. 101*: 15870–15875.

334 Kunz, F., Retey, J., Arigoni, D. et al. (**1978**). *Helv. Chim. Acta 61*: 1139–1145.

335 Retey, J., Kunz, F., Arigoni, D., and Stadtman, T.C. (**1978**). *Helv. Chim. Acta 61*: 2989–2998.

336 Wetmore, S.D., Smith, D.M., and Radom, L. (**2001**). *J. Am. Chem. Soc. 123*: 8678–8689.

337 Sandal, G.M., Smith, D.M., and Radom, L. (**2006**). *J. Am. Chem. Soc. 128*: 16004–16005.

338 Kurylo-Borowska, Z. and Abramsky, T. (**1972**). *Biochim. Biophys. Acta 264*: 1–10.

339 Christenson, S.D., Liu, W., Toney, M.D., and Shen, B. (**2003**). *J. Am. Chem. Soc. 125*: 6062–6063.

340 Christenson, S.D., Wu, W., Spies, M.A. et al. (**2003**). *Biochemistry 42*: 12708–12718.

341 Christianson, C.V., Montavon, T.J., Van Lanen, S.G. et al. (**2007**). *Biochemistry 46*: 7205–7214.

342 Christenson, C.V., Montavon, T.J., Festin, G.M. et al. (**2007**). *J. Am. Chem. Soc. 129*: 15744–15745.

343 Montavon, T.J., Christianson, C.V., Festin, G.M. et al. (**2008**). *Bioorg. Med. Chem. Lett. 18*: 3099–3102.

344 Parry, R.J. and Kurylo-Borowska, Z. (**1980**). *J. Am. Chem. Soc. 102*: 836–837.

345 Cooke, H.A. and Bruner, S.D. (**2010**). *Biopolymers 93*: 802–810.

346 Lin, S., Van Lanen, S.G., and Shen, B. (**2008**). *J. Am. Chem. Soc. 130*: 6616–6623.

347 Huang, S.-X., Lohamn, J.R., Huang, T., and Shen, B. (**2013**). *Proc. Natl. Acad. Sci. U.S.A. 110*: 8069–8074.

348 Rachid, S., Frug, D., Kunze, B. et al. (**2006**). *Chem. Biol. 14*: 667–681.

349 Rachid, S., Krug, D., Weissman, K.J., and Müller, R. (**2007**). *J. Biol. Chem. 282*: 21810–21817.

350 Krug, D. and Müller, R. (**2009**). *ChemBioChem 10*: 741–750.

351 Wanninayake, U. and Walker, K.D. (**2013**). *J. Am. Chem. Soc. 135*: 11193–11204.

352 Yan, J., Aboshi, T., Teraishi, M. et al. (**2015**). *Plant Cell 27*: 1265–1278.

353 Walter, T., King, Z., and Walker, K.D. (**2016**). *Biochemistry 55*: 1–4.

354 Spiteller, P., Rüth, M., von Nussbaum, F., and Steglich, W. (**2000**). *Angew. Chem. Int. Ed. 39*: 2754–2756.

355 Walker, K.D., Klettke, K., Akiyama, T., and Croteau, R. (**2004**). *J. Biol. Chem. 279*: 53947–53954.

356 Steele, C.L., Chen, Y., Dougherty, B.A. et al. (**2005**). *Archiv. Biochem. Biophys. 438*: 1–10.

357 Walker, K.D. and Floss, H.G. (**1998**). *J. Am. Chem. Soc. 120*: 5333–5334.

358 Feng, L., Wanninayake, U., Strom, S. et al. (**2011**). *Biochemistry 50*: 2919–2930.

359 Wybenga, G.G., Szymanski, W., Wu, B. et al. (**2014**). *Biochemistry 53*: 3187–3198.

360 Mutate, W., Klettke, K.L., Foster, C., and Walker, K.D. (**2007**). *Biochemistry 46*: 9785–9794.

361 Cooke, H.A., Christianson, C.V., and Bruner, S.D. (**2009**). *Curr. Opin. Chem. Biol. 13*: 460–468.

362 Turner, N.J. (**2011**). *Curr. Opin. Chem. Biol. 15*: 234–240.

363 Wanninayake, U., DePorre, Y., Ondari, M., and Walker, K.D. (**2011**). *Biochemistry 50*: 10082–10090.

364 Wanninayake, U. and Walker, K.D. (**2012**). *Biochemistry 51*: 5226–5228.

365 Wang, K., Hou, Q., and Liu, Y. (**2013**). *J. Mol. Graph. Model. 46*: 65–73.

366 Heberling, M.M., Masman, M.F., Bartsch, S. et al. (**2015**). *ACS Chem. Biol. 10*: 989–997.

367 Klettke, K.L., Sanyal, S., Mutatu, W., and Walker, K.D. (**2007**). *J. Am. Chem. Soc. 129*: 6988–6989.

368 Szymanski, W., Wu, B., Weiner, B. et al. (**2009**). *J. Org. Chem. 74*: 9152–9157.

369 Wu, B., Szymanski, W., Wietzes, P. et al. (**2009**). *ChemBioChem 10*: 338–344.

370 Wu, B., Szymanski, W., Wybenga, G.G. et al. (**2012**). *Angew. Chem. Int. Ed. 51*: 482–486.

371 Zhu, L., Ge, F., Li, W. et al. (**2018**). *Enzyme Microb. Technol. 114*: 22–28.

372 Shee, P.K., Ratnayake, N.D., Walter, T. et al. (**2019**). *ACS Catal. 9*: 7418–7430.

373 Cox, B.M., Bilsborrow, J.B., and Walker, K.D. (**2009**). *J. Org. Chem. 74*: 6953–6959.

374 Wu, B., Szymanski, W., de Wildeman, S. et al. (**2010**). *Adv. Synth. Catal. 352*: 1409–1412.

375 Kajani, A.A., Mofid, M.R., and Saeid, K.A. (**2013**). *Iran. J. Biotechnol. 11*: 96–103.

376 Ratnayaki, N.D., Wanninayake, U., Geiger, J.H., and Walker, K.D. (**2011**). *J. Am. Chem. Soc. 133*: 8531–8533.

377 Jin, M., Fischbach, M.A., and Clardy, J. (**2006**). *J. Am. Chem. Soc. 128*: 10660–10661.

378 Magarvey, N.A., Fortin, P.D., Thomas, P.M. et al. (**2008**). *ACS Chem. Biol. 3*: 542–554.

379 Strom, S., Wanninayake, U., Ratnayake, N.D. et al. (**2012**). *Angew. Chem. Int. Ed. 51*: 2898–2902.

380 Zhu, L., Yang, J., Feng, G. et al. (**2020**). *Enzyme Microb. Technol. 132*: 109428.

381 Chesters, C., Wilding, M., Goodall, M., and Micklefield, J. (**2012**). *Angew. Chem. Int. Ed. 51*: 4344–4348.

382 Ratnayake, N.D., Liu, N., Kuhn, L.A., and Walker, K.D. (**2014**). *ACS Catal. 4*: 3077–3090.

383 Varga, A., Bánóczi, G., Nagy, B. et al. (**2016**). *RSC Adv. 6*: 56412–56420.

384 Csuka, P., Juhász, V., Kohári, S. et al. (**2018**). *ChemBioChem 19*: 411–418.

385 Ratnayake, N.D., Theisen, C., Walter, T., and Walker, K.D. (**2016**). *J. Biotechnol. 217*: 12–21.

386 Weise, N.J., Ahmed, S.,.T., Parmeggiani, F., and Turner, N.J. (**2017**). *Adv. Synth. Catal. 359*: 1570–1576.

387 Kudo, F., Kawamura, K., Uchino, A. et al. (**2015**). *ChemBioChem 16*: 909–914.

388 Christianson, D.W. (**2017**). *Chem. Rev. 117*: 11570–11648.

389 Fall, R.R. and West, C.A. (**1971**). *J. Biol. Chem. 246*: 6913–6928.

390 Kawaide, H., Imai, R., Sassa, T., and Kamiya, Y. (**1997**). *J. Biol. Chem. 272*: 21706–21712.

391 Kawaide, H., Sassa, T., and Kamiya, Y. (**2000**). *J. Biol. Chem. 275*: 2276–2280.

392 Toyomasu, T., Kawaide, H., Ishizaki, A. et al. (**2000**). *Bisci. Biotechnol. Biochem 64*: 660–664.

393 Xu, M., Hillwig, M.L., Tiernan, M.S., and Peters, R.J. (**2017**). *J. Nat. Prod. 80*: 328–333.

394 Hayashi, K.-I., Kawaide, H., Notomi, M. et al. (**2006**). *FEBS Lett. 580*: 6175–6181.

395 Frost, R.G. and West, C.A. (**1977**). *Plant Physiol. 59*: 22–29.

396 Duncan, J.D. and West, C.A. (**1981**). *Plant Physiol. 68*: 1128–1134.

397 Keeling, C.I., Dullat, H.K., Yuen, M. et al. (**2010**). *Plant Physiol. 152*: 1197–1208.

398 Sun, T.-P. and Kamiya, Y. (**1994**). *Plant Cell 6*: 1509–1518.

399 Ikeda, C., Hayashi, Y., Itoh, N. et al. (**2007**). *J. Biochem. 141*: 37–45.

400 Koeksal, M., Hu, H., Coates, R.M. et al. (**2011**). *Nat. Chem. Biol. 7*: 431–433.

401 Prisic, S., Xu, J., Coates, R.M., and Peters, R.J. (**2007**). *ChemBioChem 8*: 869–874.

402 Rudolf, J.D., Dong, L.-B., Cao, H. et al. (**2016**). *J. Am. Chem. Soc. 138*: 10905–10915.

403 Potter, K., Criswell, J., Zi, J. et al. (**2014**). *Angew. Chem. Int. Ed. 53*: 7198–7202.

404 Zerbe, P., Chiang, A., and Bohlmann, J. (**2012**). *Phytochemistry 74*: 30–39.

405 Potter, K.C., Zi, J., Hong, Y.J. et al. (**2016**). *Angew. Chem. Int. Ed. 55*: 634–638.

406 Prisic, S., Xu, M., Wilderman, P.R., and Peters, R.J. (**2004**). *Plant Physiol. 136*: 4228–4236.

407 Su, P., Tong, Y., Cheng, Q. et al. (**2016**). *Sci. Rep. 6*: 23057.

408 Gao, W., Hillwig, M.L., Huang, L. et al. (**2009**). *Org, Lett 11*: 5170–5173.

409 Morrone, D., Chambers, J., Lowery, L. et al. (**2009**). *FEBS Lett. 583*: 475–480.

410 Kong, M.K., Kang, H.-J., Kim, J.H. et al. (**2015**). *J. Biotechnol. 214*: 95–102.

411 Richman, A.S., Gijzen, M., Starratt, A.N. et al. (**1999**). *Plant J. 19*: 411–421.

412 Shi, T.-Q., Gao, J., Wang, W.-J. et al. (**2019**). *ACS Synth. Biol. 8*: 445–454.

413 Cyr, A., Wilderman, P.R., Determan, M., and Peters, R.J. (**2007**). *J. Am. Chem. Soc. 129*: 6684–6685.

414 Jia, M., Zhou, K., Tufts, S. et al. (**2017**). *ACS Chem. Biol. 12*: 862–867.

415 Morrone, D., Lowry, L., Determan, M.K. et al. (**2010**). *Appl. Microbiol. Biotechnol. 85*: 1893–1906.

416 Zhou, Y.J., Gao, W., Rong, Q. et al. (**2012**). *J. Am. Chem. Soc. 134*: 3234–3241.

417 Su, P., Guan, H., Zhao, Y. et al. (**2018**). *Plant J. 93*: 50–65.

418 Smanski, M.J., Yu, Z., Casper, J. et al. (**2011**). *Proc. Natl. Acad. Sci. U.S.A. 108*: 13498–13503.

419 Harris, L.J., Saparno, A., Johnston, A. et al. (**2005**). *Plant Mol. Biol. 59*: 881–894.

420 Mafu, S., Ding, Y., Murrphy, K.M. et al. (**2018**). *Plant Physiol. 176*: 2677–2690.

421 Du, G., Gong, H.-Y., Feng, K.-N. et al. (**2019**). *Phytochmistry 158*: 96–102.

422 Cheng, Q. (**2006**). *J. Ind. Microbiol. Biotechnol. 33*: 552–559.

423 Cunningham, F.X. Jr., Chamovitz, D., Misawa, N. et al. (**1993**). *FEBS Lett. 328*: 130–138.

424 Cunningham, F.X. Jr., Sun, Z., Chamovitz, D. et al. (**1994**). *Plant Cell 6*: 1107–1121.

425 Hugueney, P., Badillo, A., Chen, H.-C. et al. (**1995**). *Plant J. 8*: 417–424.

426 Pecker, I., Gabbay, R., Cunningham, F.X. Jr., and Hirschberg, J. (**1996**). *Plant Mol. Biol. 30*: 807–819.

427 Maresca, J.A., Graham, J.E., Wu, M. et al. (**2007**). *Proc. Natl. Acad. Sci. U.S.A. 104*: 11784–11789.

428 Jeong, T.H., Ji, K., and Kim, Y.T. (**2013**). *J. Microbiol. Biotechnol. 23*: 144–148.

429 Hornero-Méndez, D. and Britton, G. (**2002**). *FEBS Lett. 515*: 133–136.

430 Cunningham, F.X. Jr., Pogson, B., Sun, Z. et al. (**1996**). *Plant Cell 8*: 1613–1626.

431 Howitt, C.A. and Pogson, B.J. (**2006**). *Plant Cell Environ. 29*: 435–445.

432 Jensen, S.L., Hegge, E., and Jackman, L.M. (**1964**). *Acta Chem. Scand. 18*: 1703–1718.

433 Apel, W. and Bock, R. (**2009**). *Plant Physiol. 151*: 59–66.

434 Zeng, J., Wang, C., Chen, X. et al. (**2015**). *BMC Plant Biol. 15*: 112.

435 Qiang, S., Su, A.P., Li, Y. et al. (**2019**). *J. Agric. Food Chem. 67*: 9560–9568.

436 Karppinen, K., Zoratti, L., Sarala, M. et al. (**2016**). *BMC Plant Biol. 16*: 95.

437 He, W., Wang, Y., Dai, Z. et al. (**2019**). *Food Chem. 278*: 509–514.

438 Tuan, P.A., Lee, J., Park, C.H. et al. (**2019**). *Food 8*: 77.

439 Hannibal, L., Lorquin, J., D'Ortoli, N.A. et al. (**2000**). *J. Bacteriol. 182*: 3850–3853.

440 Krügel, H., Krubasik, P., Weber, K. et al. (**1999**). *Biochim. Biophys. Acta 1439*: 57–64.

441 Pasamontes, L., Hug, D., Tessier, M. et al. (**1997**). *Gene 185*: 35–41.

442 Couso, I., Vila, M., Vigara, J. et al. (**2012**). *Eur. J. Physiol. 47*: 223–232.

443 Li, X.-R., Tian, G.-Q., and Shen, H.-J. (**2015**). *J. Ind. Microbiol. Biotechnol. 42*: 627–636.
444 Misawa, N., Satomi, Y., Kondo, K. et al. (**1995**). *J. Bacteriol. 177*: 6575–6584.
445 Guzman, I., Hamby, S., Romero, J. et al. (**2010**). *Plant Sci. 179*: 49–59.
446 Tian, S.-L., Li, L., Tian, Y.-Q. et al. (**2016**). *J. Am. Soc. Hort. Sci. 141*: 609–616.
447 Berry, H.M., Rickett, D.V., Baxter, C.J. et al. (**2019**). *J. Exp. Bot. 70*: 2637–2650.
448 Dambek, M., Eilers, U., Breitenbach, J. et al. (**2012**). *J. Exp. Bot. 63*: 5607–5612.

7

Organic Synthesis with Ligases

In enzymology, a ligase, also called *synthetase*, is any one of a class of about 50 enzymes that involve in the reactions of conservation of chemical energy and provide a link between energy-demanding synthetic processes and energy-yielding breakdown reactions. In general, they catalyze the joining of two large molecules by forming a new chemical bond, usually with accompanying hydrolysis of a small pendant chemical group on one of the larger molecules or the linking together of two compounds, for example, enzymes that catalyze the joining of two molecules with the formation of C–O, C–S, C–N, C–C, and phosphate ester bonds. The needed energy is usually derived from the cleavage of an energy-rich phosphate bond (in many cases, by the simultaneous conversion of adenosine triphosphate [ATP] to adenosine diphosphate [ADP]). The enzyme classification for ligases has been assigned in the EC category of EC 6 and further classified into six subclasses. The applications of this class of enzymes in organic synthesis are not as wide as enzymes from class EC 1 to EC 5 [1, 2].

7.1 Ligases for Carbon–Oxygen Bonds Formation

7.1.1 Olefin β-Lactone Synthetase

Olefin β-lactone synthetase (OβLS, EC 6.1.3.1), also known as *oleC* (gene name), involved in olefin biosynthesis is an ATP-dependent enzyme that catalyzes the conversion of (2*R*,3*S*)-2-alkyl-3-hydroxyalkanoic acid to the β-lactone *cis*-3-alkyl-4-alkyloxetan-2-one (Scheme 7.1) [3–5]. The systematic name of this enzyme is (2*R*,3*S*)-2-alkyl-3-hydroxyalkanoate ligase, which belongs to the acyl-CoA synthetase, nonribosomal peptide synthetase (NRPS), and luciferase (ANL) enzyme superfamily. This enzyme, found in certain bacterial species, participates in the biosynthetic pathway for the production of olefins by its action to form the

Enzyme-Based Organic Synthesis, First Edition. Cheanyeh Cheng.

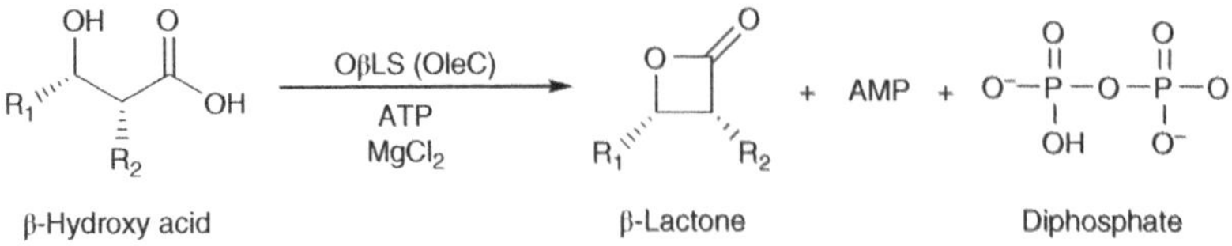

Scheme 7.1 ATP-dependent olefin β-lactone synthetase catalyzed β-lactone formation from corresponding β-hydroxy acid. *Sources:* Based on the Enzyme List [3], Raushel [4], and Robison et al. [5].

intermediate β-lactone. The OleC protein from *Stenotrophomonas maltophilia* was purified and crystalized. The crystal structure was analyzed by synchrotron X-ray diffraction at 3.4 Å resolution to show a trigonal space group $P3_121$ or $P3_221$, with unit-cell parameters $a = b = 98.8$, $c = 141.0$ Å. The crystal of OleC protein presents a 60 kDa monomer per asymmetric unit. More detailed molecular structure of OleC protein is under way [6]. The OleC protein from *Xanthomonas campestris* was also purified in a recombinant form from separate *Escherichia coli* expression host lines by gel filtration chromatography, which shows a monomer with a molecular mass of 58.5 kDa by the size exclusion chromatography [7].

Previously, it has been reported that OleC from *S. maltophilia* catalyzed the β-hydroxy acid to directly generate the *cis*-olefin through the decarboxylative dehydration reaction [8]. However, the study of the reaction catalyzed by the enzyme OleC from four different bacteria using ^{1}H nuclear magnetic resonance (NMR) demonstrates that thermally labile *cis*- and *trans*-β-lactones were produced from *syn*- and *anti*-β-hydroxy acids, respectively, in an ATP-dependent reaction; no alkenes are observed [9]. The results are also consistent with that OleC is a member of the ATP-dependent ANL superfamily, which catalyzes the ATP-dependent ligation of substrates via the formation of an adenylate intermediate [10]. Therefore, MgATP is likely required to activate the β-hydroxyl group or the carboxylate of the substrate prior to β-lactone formation. Accordingly, a reaction mechanism was proposed that the most probable intermediate would occur by initial attack of the substrate carboxylate on the α-phosphoryl group of ATP [4]. A detailed investigation of the reaction mechanism of the OleC-catalyzed β-lactone formation from β-hydroxy acids was performed by a novel, continuous assay for β-lactone synthetase activity using synthetic β-hydroxy acid substrates with alkene or alkyne moieties. The proposed mechanism from the kinetic study suggests that OleC from *X. campestris* performs carboxylic acid adenylation followed by intramolecular nucleophilic attack to facilitate β-lactone ring closure (Scheme 7.2) [5].

The structure of β-lactones has a strained four-membered heterocycle, which is the core scaffold in many natural products isolated from diverse organisms including sea sponges, butterflies, and microbial plant pathogens. β-Lactone is an

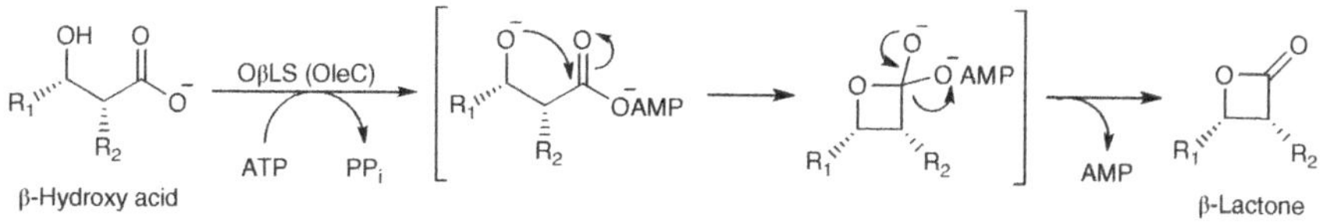

Scheme 7.2 Proposed reaction mechanism for the formation of β-lactone from β-hydroxy acid via a carboxy-adenylate intermediate and intramolecular ring closure. *Source:* Robison et al. [5].

intermediate during the olefin biosynthesis pathway found in a variety of bacteria, thus the olefin biosynthetic pathway is important in organic synthesis. The olefin biosynthetic pathway of bacteria is encoded by a four-gene cluster, *oleABCD*, which starts from the "head-to-head" coupling of CoA-activated fatty acids (FAs) to unstable β-keto acids catalyzed by OleA, then the enzyme OleD couples the reduction of β-keto acids with NADPH oxidation to yield stable β-hydroxy acids, and finally, the enzyme OleC catalyzes the ring closure to form β-lactones and is further reacted by enzyme OleB to produce *cis*-olefins [11–16]. The application of β-lactones formed from corresponding β-hydroxy acids catalyzed by enzyme OleC was applied for pharmaceutical industry to synthesize lipstatin, which is an inhibitor of pancreatic lipase and is the precursor of the FDA-approved anti-obesity drug, orlistat (Scheme 7.3) [17], or the synthesis of ebelactone A, a commercially available esterase inhibitor, which contains the β-lactone ring and is formed spontaneously from the final, enzyme-linked, β-hydroxy-thioester intermediate (Scheme 7.3) [18].

7.2 Ligases for Carbon–Sulfur Bonds Formation

7.2.1 Acetate-CoA Ligase (AMP-forming)

Acetate-CoA ligase or acetyl-CoA synthetase (ACS, EC 6.2.1.1, adenosine monophosphate [AMP]-forming) or short-chain fatty acyl-CoA synthetase belongs to the acyl-adenylate-forming enzyme superfamily including acyl- and aryl-CoA synthetases, firefly luciferase, and NRPSs, which catalyzes the formation of acetyl-CoA and AMP from acetate, ATP, and CoA (Scheme 7.4) [19, 20]. This enzyme presents in all organisms from bacteria to humans, where it is located in various cell compartments such as cytosol, smooth endoplasmic reticulum (SER), mitochondria, chloroplast, and peroxisomes and exhibits wide tissue distribution. Acetyl-CoA is an essential metabolic intermediate in various metabolic pathways including tricarboxylic acid (TCA) cycle, FA biosynthesis, and cholesterol biosynthesis [21–23].

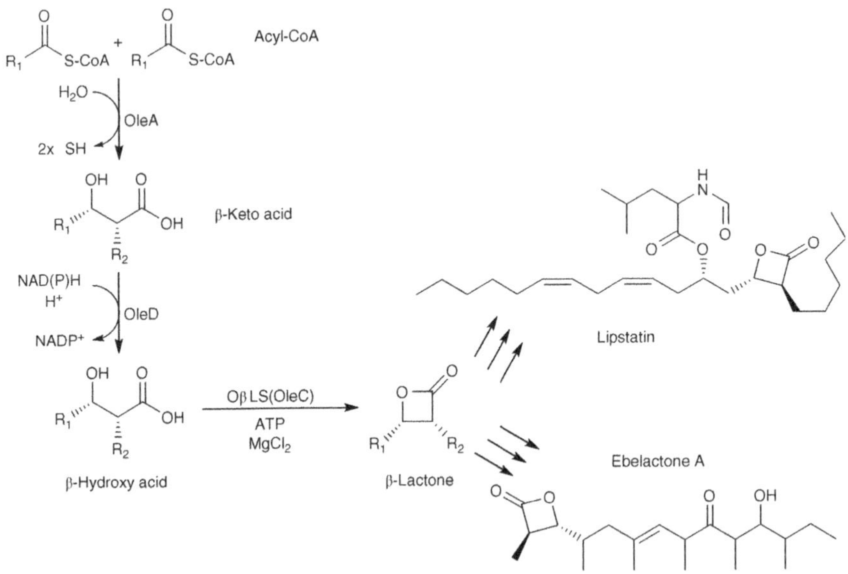

Scheme 7.3 β-Lactones formation from β-hydroxy acids catalyzed by ATP-dependent enzyme OleC and used for the synthesis of lipstatin or ebelactone A. *Sources:* Based on Weibel et al. [17] and Wyatt et al. [18].

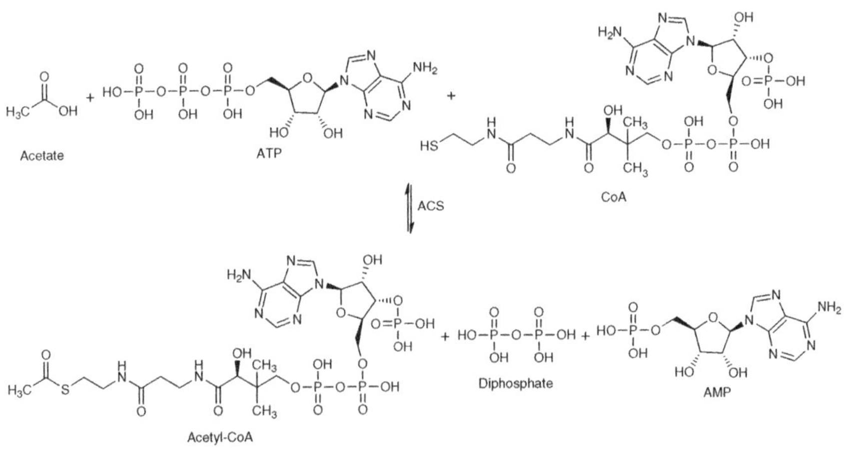

Scheme 7.4 Formation of acetyl-CoA from acetate, ATP, and CoA catalyzed by acetyl-CoA synthetase. *Sources:* Based on Knights [19] and Liedvogel and Stumpf [20].

$$CH_3COO^- + ATP \xrightarrow{ACS} CH_3O\text{–}AMP + PP_i$$

$$CH_3CO\text{–}AMP + CoA \xrightarrow{ACS} CH_3CO\text{–}CoA + AMP$$

Scheme 7.5 Two-step reaction for the conversion of acetate into acetyl-CoA catalyzed by acetyl-CoA synthetase.

ACS was first identified in pig and rabbit heart muscle, which is active only with acetate and propionate [24]. Partially purified ACS from rat liver has a molecular weight of 58 kDa and shows higher affinity for acetate than other short-chain FAs [25, 26]. While ACS purified from bovine and murine heart exhibits a molecular weight of 72 kDa with acetate specifically utilized as substrate, the intact molecule contains 30% α-helix and 30% β-structure [27, 28]. The gene encoding *Bradyrhizobium japonicum* ACS was cloned and expressed in *E. coli*. The purified enzyme comprises 648 amino acid residues with a calculated molecular mass of 71 996 Da. The kinetic studies for the site-directed mutants suggest ACS catalyzes a two-step reaction for the formation of acetyl-CoA from acetate with an acetyl-AMP intermediate (Scheme 7.5). Gly-266 and Lys-269 of the active site participate in the formation of acetyl-AMP, whereas Glu-414 may play a role in acetate binding [29].

The X-ray crystal structure of yeast *Saccharomyces cerevisiae* ACS in a binary complex with AMP was determined at 2.3 Å resolution. The crystal structure contains a large N-terminal domain and a small C-terminal domain, where AMP is bound at the interface between the two domains. The conformation of the structure is competent for catalyzing the first step of the reaction that Lys675 located in the active site is critical for this step. A rotation of 140° in the small domain is needed for the binding of CoA and the catalysis of the second step. In contrast to the monomeric bacterial enzyme, yeast ACS is a stable trimer [30]. Furthermore, through the investigation of the *Salmonella enterica* ACS structure, modeling of the *Methanothermobacter thermautotrophicus MT*ACS1 on the *S. cerevisiae* ACS1 and *S. enterica* ACS structures showed similar active-site architecture, and alignment of the amino acid sequences of proven ACSs identified the four residues, Ile312, Thr313, Val388, and Trp416, that form the putative acetate binding pocket. Replacements at these positions significantly alter the enzyme's affinity for acetate as well as the range of acyl substrates that can be utilized [31].

Studies using mutations of *S. enterica* ACS to test the ability of the enzymes to catalyze the complete reaction and the adenylation half-reaction were performed that showing the substitution of Lys609 with alanine results in the suppression of the adenylation reaction, while the substitution of Gly524 to leucine is unable to catalyze the complete reaction yet catalyzes the adenylation half-reaction. In addition, the steady-state kinetic data of putative substrate-binding residues demonstrate that no single residue plays a dominant role in dictation CoA binding. Results also demonstrate that mutations in the active site can alter the acyl substrate specificity [32]. To investigate whether *Mycobacterium tuberculosis* *Mt*ACS is regulated by lysine acetylation as reported by *S. enterica* ACS, the recombinant *Mt*ACS in *E. coli* was purified and characterized and its enzyme activity was determined to reveal that Lys617 is critical for its function. Results also demonstrate that *Mt*ACS undergoes auto-acetylation with acetate but not with acetyl-CoA as the acetyl donor due to the inhibition of CoA [33]. The two-step reaction mechanism of acetyl-CoA formation from acetate and the regulatory role of Lys617 for the first half acetylation reaction were also proved by *Mt*ACS via ^{31}P NMR. Furthermore, the chemical mechanism of the enzyme relies on a conformational change controlled by the protonation state of Asp525 was demonstrated as well [34].

The investigation of the recombinant *MT*ACS1 indicates the highly conserved Tyr in the first position of motif III plays a key role in the active-site architecture through the hydrogen bond interaction with a highly conserved active-site Gln. The invariant Asp in the fifth position plays a critical role in ATP binding and catalysis through interaction with the 2′- and 3′-OH groups of the ribose moiety of ATP [35]. The kinetics and thermodynamics parameters of *E. coli* ACS were characterized to study the adenylating step of the two-step acetyl-CoA formation. The results showed that four amino acid residues, T264, K270, D500, and K609, play important role in the catalysis and ATP binding [36]. On the other hand, parasite *Leishmania donovani* ACS (*Ld*ACS) was identified in recombinant *E. coli*. The biochemical and biophysical characterization of this AMP-forming *Ld*ACS revealed the presence of the conserved PX_4GK motif, which plays key role in governing substrate specificity rather than structural integrity except glycine at sixth position [37]. Further studies showed that the role of leucine 684 residue, a conserved residue of *Ld*ACS, is involved in substrate recognition and the catalysis of acetylation of the enzyme [38].

The AMP-forming ACS gene (*acs*) of anaerobic ammonium-oxidizing bacteria *Kuenenia stuttgartiensis* was heterologously expressed in *E. coli* and employed to investigate the substrate specificity and kinetic properties of the putative *Ks*ACS. Enzyme activity was observed toward several short-chain organic acids, that is, acetate, propionate, and butyrate, but was highest with acetate [39]. The specificity of ACS from *Arabidopsis thaliana* (*At*ACS) in catalyzing acetyl-CoA generation was for only acetate. However, the carboxylate binding pocket of *At*ACS was engineered to successfully switch the highly specific enzyme for only using acetate, to be equally specific for using longer linear (up to hexanoate) or branched chain (methylvalerate) carboxylate substrates. Hence, engineering *At*ACS to utilize novel carboxylate substrates would generate alternative acyl-CoA substrates for FA or polyketide biosynthesis [40]. Similarly, the substrate selectivity of the $Trp^{416}Gly$ mutant of *MT*ACS ($Trp^{416}Gly$*MT*ACS) showed activities for saturated straight chain carboxylic acids from C_2 to C_8, for ω-alkenyl straight chain carboxylic acids from C_4 to C_7 and for ω-alkynyl straight chain carboxylic acids from C_5 to C_7 [41].

Except for the most well-known function of the product acetyl-CoA from ACS in the use of TCA cycles as well as in the production of FA, ACS plays a variety of other functions in organisms. Experiments showed that acetyladenylate during the acetyl-CoA formation from acetate catalyzed by ACS plays a role in controlling the direction of flagellar rotation of *E. coli* cells [42]. Yeast acetyl-CoA synthetases (ACS1p and 2p) not only are found important for carbon metabolism but also are shown to supply the acetyl-CoA for histone acetylation by histone acetyltransferases (HATs). Since yeast mitochondrial and nucleocytosolic acetyl-CoA pools are biochemically isolated, acetyl-CoA metabolism is directly linked to chromatin regulation and may affect diverse cellular processes in which acetylation metabolism intersect, such as disease states and aging [43]. Research studies have provided evidence for the central and conserved role for ACS in regulating life span in yeast and flies by a mechanism involving autophagy, which is essential for the maintenance of cellular homeostasis during periods of stress [44]. In adult mice, downregulation of the *acs* gene expression in hippocampal region impairs the long-term spatial memory, a cognitive process reliant on histone acetylation, of the animal. The results again reveal a link between cellular metabolism, gene regulation, and cognitive function [45]. This enzyme is also an emerging key enzyme for cancer metabolism, which supplies acetyl-CoA for tumor cells by capturing acetate as a carbon source under stressed conditions. When the gene is present, the cells are able to take acetate as a food source to convert it to acetyl-CoA during stressed conditions. In the cases of advanced carcinoma tumors, this enzyme was downregulated and indicated a poor five-year survival rate. Thus, this enzyme shows an interesting biomarker for the presence of tumors in colorectal carcinoma [46]. Green microalgae *Dunaliella tertriolecta* contains large amounts of lipid (36–42%), thus a potential biodiesel feedstock. Genes of ACS from *D. tertriolecta* were isolated, characterized, and expressed in *E. coli* (*Dt*ACS). It has been found that the expression levels of *Dt*ACS were increased under nitrogen starvation stress, indicating that ACS activity may be related to the lipid accumulation in nitrogen-deficient condition [47].

While ACS's activity plays important role in the metabolic pathways associated with many biological functions, ACS is also employed in organic synthetic chemistry for producing useful industrial chemicals. Since CoA and the catalytic product acetyl-CoA in the ACS-catalyzed reaction are expensive, the recycle of CoA and acetyl-CoA is important to make the organic synthesis related with ACS practical. For example, in the enzyme-catalyzed aldol condensation of acetyl-CoA with oxaloacetic acid (OAA) to form citric acid catalyzed by citrate synthase (CS,

EC 2.3.3.1), one method used for regenerating acetyl-CoA is through the coupling of ACS-catalyzed acetylation of CoA with acetate, which continuously furnishes the acetyl-CoA used in the aldol condensation by supplying inexpensive acetate and ATP (Scheme 7.6) [48]. The recycling number of CoA can be as high as 1000 on a multi-millimole scale. This coupled-enzyme strategy can be further considered to apply for the production of expensive enantiopure L-carnitine, which is used in the clinical treatment of lipid storage myopathy.

Poly(hydroxyalkanoic acid) (PHA) is a kind of biodegradable thermoplastics, which is found to accumulate in a wide range of bacteria as a carbon and energy storage material and deposits as insoluble granules in the cells. A three-enzymatic cascade reaction system was constructed by enzymes ACS, recombinant poly(hydroxyalkanoic acid) synthase (PHAS) from *Chromatium vinosum*, and propionyl-coenzyme-A transferase (PCT) from *Clostridium propionicum* for the *in vitro* biosynthesis of poly(3-hydroxybutyric acid) (P[3HBA]) using free D-(−)-3-hydroxybutyric acid (3HBA) as substrate. In this three-enzyme-coupled reaction system, both acetate and CoA supplied for producing acetyl-CoA and used in subsequent polymerization reaction can be recycled (Scheme 7.7a) [49]. The energy required for the *in vitro* synthesis of P(3HBA) was provided by the ATP hydrolyses resulting in acetyl-CoA synthesis catalyzed by ACS. The established CoA recycling P(3HBA) biosynthesis system may provide a useful strategy for the synthesis of diverse PHAs on a semipreparative scale. Similarly, a novel coupled-enzyme system capable of recycling CoA and regenerating NADPH was established using enzymes β-ketothiolase (β-KT), NADPH-dependent acetoacetyl-CoA reductase (AACR), PHAS, ACS, and glucose dehydrogenase (GDH), for the synthesis of P(3HBA) from acetate (Scheme 7.7b), which is based on the P(3HBA) biosynthetic pathway in *Raistonia eutropha* [50]. The *in vitro* synthesis yield of P(3HBA) from acetate was 5.6 mg in a 5-mL reaction mixture, and the weight-averaged molecular weight as well as polydispersity were 6.64×10^6 and 1.36, respectively.

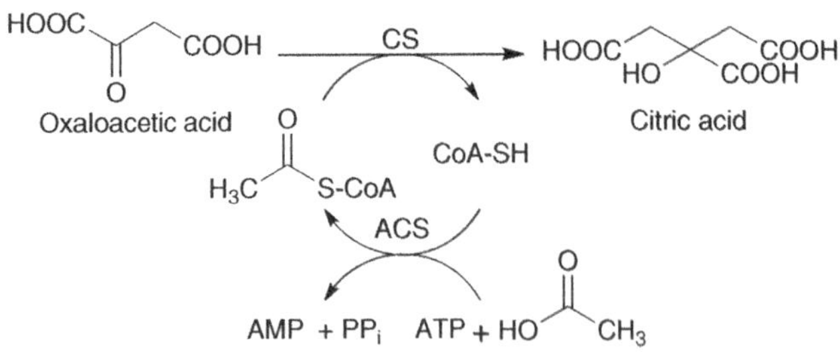

Scheme 7.6 Coupled-enzymatic reactions for citric acid production from oxaloacetic acid with recycling of Co-A. *Source:* Patel et al. [48].

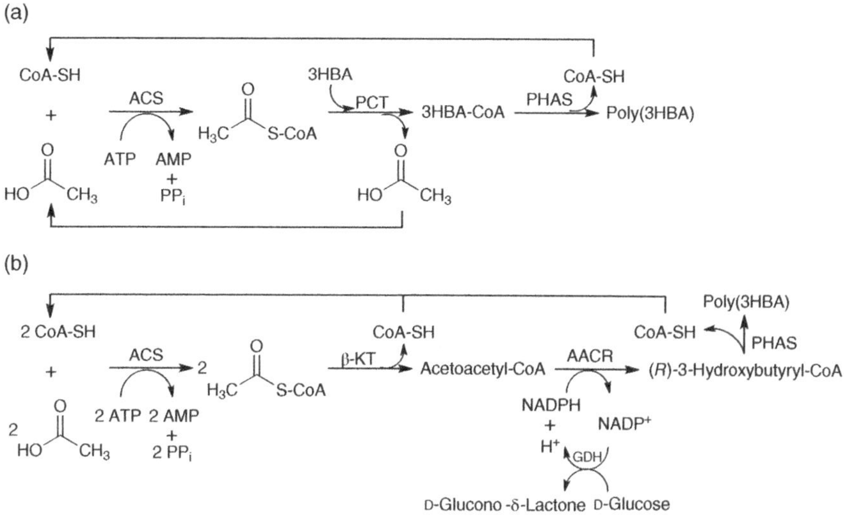

Scheme 7.7 *In vitro* biosynthesis of poly(3-hydroxybutyric acid) using cascade enzymatic reaction system: (a) recycling of acetate and CoA; (b) recycling of CoA and NADPH regenerating. *Source:* (a) Modified from Jossek and Steinbüchel [49].

CoA can be reused at least 26 times; thus NADPH also can be regenerated at least 26 times during a 24-hour reaction.

FAs, promising precursors for advanced biofuels, can be prepared from inexpensive acetic acid (HAc) using engineered *E. coli* with pH-coupled fed-batch fermentation. The genetic engineering strategies include overexpression of ACS (*acs*) and acyl-acyl carrier protein [ACP] thioesterase (TE, *tesA*) genes and knockout of acyl-CoA dehydrogenase gene (*fadE*) to avoid FAs degradation in an *E. coli* strain. The fermentation metabolic pathways of FAs biosynthesis from acetic acid in an *E. coli* strain is shown in Scheme 7.8 [51]. The microbial production of ~$1\,g\,L^{-1}$ FAs from HAc with ~20% theoretical yield. When the engineered strain was cultured with HAc-rich liquid wastes, the yield was ~$0.43\,g\,L^{-1}$ FAs using waste streams from dilute acid hydrolysis of lignocellulosic biomass and ~$0.17\,g\,L^{-1}$ FAs using effluent from anaerobic-digested sewage sludge. This study opens the opportunities for coupling the waste treatment with the biosynthesis of advanced biofuel using genetically engineered microbial species.

FAs are composed of long alkyl chains, which is a primary metabolite in cells serving both chemical and energy storage functions and represents nature's "petroleum." FAs and its extended metabolite lipids (triacylglycerol, TAG) are isolated from plant and animal oils for a diverse set of products ranging from fuels to oleochemicals. In consideration of energy costs and environmental protection,

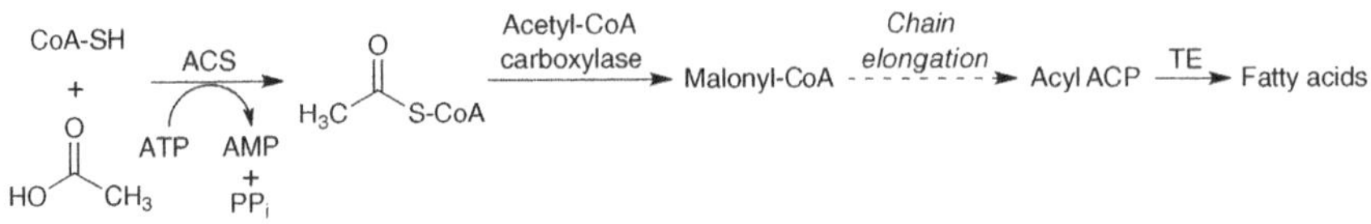

Scheme 7.8 Simplified metabolic pathways for fatty acids biosynthesis from acetic acid. *Source:* Modified from Xiao et al. [51].

the production of FAs and TAG is today focused on the microbial production from renewable resources such as simple sugars and acetic acid [52]. Since acetyl-CoA serves as a central precursor metabolite in multiple biological pathways, it is thought to play a key role in building FAs and lipids. Thus, engineered *E. coli* has been used to produce structurally tailored fatty esters (biodiesels), fatty alcohols, and waxes directly from simple sugars [52]. It has been found that microalgal species including *Scenedesmus obliquus, Ourococcus multisporus, Chlamydomonas pitschmannii, Micractinium reisseri, Botryococcus braunii,* and *Chlamydomonas reinhardtii* have high lipid content, thus they are used for biomass/biodiesel production and advanced wastewater treatment. Application of lipidomics and transcriptomics to manipulate the lipid-associated gene ACS in microalgae improves the accumulative lipid content [53, 54].

Cholesterol is an essential component of cell membranes and the precursor for the synthesis of steroid hormones and bile acids. The synthesis of this molecule occurs partially in a membranous world (especially the last steps), where the enzymes, substrates, and products involved tend to be extremely hydrophobic. The importance of cholesterol has increased in the past half-century because of its association with cardiovascular diseases, which are considered one of the leading causes of death worldwide. The enzymes involved in cholesterol biosynthesis start from the initial two-carbon acetyl-CoA building block, which is formed from acetate, CoA, and ATP catalyzed by ACS, an essential metabolite in all kinds of orgnisms [55].

Since ACS is an important precursor molecule for the synthesis of natural and unnatural chemicals, the development of a reusable and a more robust entity of this enzyme is urgently needed for industrial applications. One approach to attain these goals is the immobilization of ACS. Therefore, ACS was adsorbed on poly(*N*-isopropylacrylamide)-poly(ethylenimine) microgel. The adsorption of ACS on microgel showed high binding of enzymes, with a maximum binding capacity of 286 $\mu g\, mg^{-1}$ of microgel for ACS was achieved. The immobilization of the enzyme resulted in improved catalytic efficiency over free enzymes; in addition, the immobilized enzyme showed no significant changes in the enzyme structure. This enzyme bioconjugate was further immobilized on an ultrathin membrane to assess the two-step reaction for acetyl-CoA formation (Scheme 7.5) in flow-through condition. Bioconjugate was covalently immobilized by crosslinking on a thin layer of preformed microgel support upon polyethylene terephthalate (PET) track etched membrane. The membrane reactor constructed with a dead-end filtration device showed consistent operational stability and maintained >70% of initial activity after seven consecutive operation cycles [56].

7.2.2 Medium-chain Acyl-CoA Ligase

The systematic name of medium-chain acyl-CoA ligase (or synthetase) (MACS, EC 6.2.1.2) is medium-chain FA:CoA ligase (AMP-forming), which also has many other common names, such as butyryl-CoA synthetase, acyl-activating enzyme, short-chain acyl-CoA synthetase, medium-chain acyl-CoA synthetase, etc. This enzyme belongs to the ANL adenylating enzymes superfamily including acyl- and aryl-CoA synthetases, firefly luciferase, and adenylation domains of the modular NRPSs and preferentially acts on medium-chain fatty acids (MCFAs) from C4 to C12 in the initial reaction of FA metabolism, which characterizes a two-step reaction with the initial adenylation of a carboxylate to form an acyl-AMP intermediate, followed by a second thioester-forming reaction to produce corresponding acyl-CoA (Scheme 7.9) [57].

The first identification of this enzyme reported in 1953 was prepared by comminuting heart, liver, or kidney of hogs or cattle but was obtained in highly purified form from beef liver, which catalyzed a wide variety of saturated fatty acids (SFAs) with chain lengths from C2 to C20 but preferentially from C4 to C12. The molecular weight of the enzyme was comparatively low from 30 to 60 kDa [58]. Later, a butyryl-CoA synthetase was purified from bovine heart mitochondria, which shows activity toward butyrate and appreciable activity toward propionate, valerate, and caproate [59]. Two maurine MACSs, MACS1 and Sa protein, were overexpressed in COS cells, purified to homogeneity, and characterized to show a molecular mass of ~62 kDa. Among C4–C16 FAs, MACS1 preferentially utilizes octanoate, whereas isobutyrate is the most preferred FA among C2–C6 FAs for Sa protein. Both enzymes, mainly detected in the mitochondrial matrix of liver and

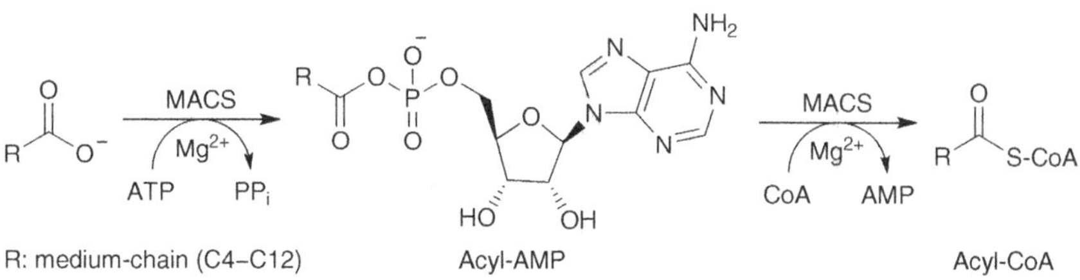

Scheme 7.9 MACS-catalyzed two-step reaction of medium-chain carboxylate to form corresponding acyl-CoA. *Source:* Modified from Gulick [57].

kidney, are utilized mainly for oxidation [60]. Liver and kidney contain MACSs that are capable of forming the CoA thioester of not only short- and MCFAs but also numerous carboxylic acid xenobiotics. These enzymes are referred to as xenobiotic/medium-chain FA:CoA ligases (XM-ligase). Two XM-ligases, HMXA and HXMB, were isolated from human liver mitochondria with a molecular mass of 64 230 Da, which bears 56.2% amino acid homology to the MACS1 enzyme, 58.7% homology to the bovine XL-III XM-ligase, and 81.5% homology to the bovine XL-I XM-ligase. While the enzyme expressed in COS cells had greater benzoate activity than phenylacetate activity, which is consistent with the known substrate specificity of HXMA [61].

The first X-ray MACS crystal structure of human ACSM2A in a series of substrate/product/cofactor complexes was determined at 1.6–2.6 Å resolutions. The data present a substantial rearrangement between the N- and C-terminal domains, driven purely by the identity of the bound ligand in the active site. The determined structures allowed the identification of the presence or absence of the ATP pyrophosphates as the conformational switch and elucidated new mechanistic details, including the role of invariant Lys557 and a divalent Mg^{2+} in coordinating the ATP pyrophosphates, as well as the involvement of a Gly-rich P-loop and the conserved Arg472-Glu365 salt bridge in the domain rearrangement [62]. The three-dimensional X-ray crystal structure of an MACS from *Methanosarcina acetivorans* was determined at 2.1 Å resolution. The cavity volume of the binding pocket of this acyl-CoA synthetase is comparable to the enzyme 4-chlorobenzoyl-CoA ligase. While the acyl-binding pocket of this enzyme shows a much more shallow pocket than other previously identified medium-chain carboxylate ACSs, the results emphasize the high sequence and structural diversity among this family in their acyl-binding pockets [63]. The MACS of *M. acetivorans* uses 2-methylbutyrate as its preferred substrates, which indicates the interaction between the sidechain of Cys298 and Lys256 of this enzyme is important for the catalytic activity based on the computational studies of the binding mechanism for the second reaction step conformation [64, 65]. Early studies of the FAs activation reaction of MACS based on the overall reaction of butyrate activation demonstrate a Bi Uni Uni Bi Ping Pong model, in which two interchangeable conformational states of the enzyme were proposed to explain the opposite effects of ATP on the partial reactions, as well as the inhibition by CoA and ATP of ATP formation and by PPi of the butyl-AMP-dependent CoA disappearance [66]. Recent biochemical and structural evidence has provided insights into the active site and catalytic mechanisms used by the ANL superfamily enzymes for the two-step reactions to activate a carboxylate with ATP and to form acyl-CoA. The enzymes use a 140° domain rotation to present opposing faces of the dynamic C-terminal domain to the active site for the different partial reactions [57].

In addition to the activation of MCFAs in the first step of FA metabolism through the transfer of acyl-CoA, MACSs also play many other important functions in cells. A novel *o-macs* gene expressed in the rat olfactory epithelium (OE) encodes a medium-chain acyl-CoA synthetase termed O-MACS, which demonstrates substrate preference toward straight and saturated-FAs with chain length of C6–C12 and plays important roles in processing odorants in a zone-specific manner, or the zonal patterning of the OE during development [67]. The anaerobic β-oxidation pathway is distinct from the aerobic FA degradation pathway in a number of properties. However, research studies reported that *E. coli* can utilize FAs as sole carbon sources during growth under anaerobic conditions. Two gene clusters, *yfcYX* and *ydiQRSTD* (*ydiD*), in *E. coli* encoding two enzymes FadIJ and FadK, respectively, were reported. FadIJ is required for hydration, oxidation, and thiolytic cleavage of the acyl chain, while FadK is indeed a medium-chain FA ACS that is primarily active on short-chain-length FAs (<10 C atoms). Expression of FadK is repressed during aerobic growth and is maximally expressed under anaerobic conditions in the presence of the terminal electron acceptor, fumarate [68]. Association studies in Japanese Suita cohort of MACS polymorphisms and various phenotypes revealed the contribution of the Leu513Ser polymorphism in medium-chain acyl-CoA synthetase 2 (MACS2) to multiple risk factors of the metabolic syndrome. Further research results also suggested an involvement of the MACS2 Leu513Ser polymorphism in the development of the metabolic syndrome in Caucasian population. In addition, the higher triglyceride and glucose levels after an oral metabolic tolerance test support a possible functional impact of the polymorphism *in vivo* [69]. To understand the mechanism of cancer metastasis, the expression, clinical significance, and biological functions of acyl-CoA medium-chain synthetase 3 (ACSM3) were investigated by tissue microarray and hepatocellular carcinoma (HCC) clinical samples. The results indicated ACSM3 was downregulated in HCC tissues. HCC patients with low expression of ACSM3 exhibited poor prognosis. Overexpression of ACSM3 attenuated migration and invasion of HCC cells *in vitro* and *in vivo* and downregulated the phosphorylation of WNK1 (Thr60) and AKT (Thr308). These findings reveal ACSM3 is a novel prognostic marker and a potential therapeutic target for HCC [70].

Medium-chain ACSs not only act in the FAs metabolism but also participate in the glycine conjugation pathway in the detoxification of xenobiotics such as benzoic acid and endogenous carboxylic acids. Various endogenous and xenobiotic toxic metabolites must be conjugated to glycine to more hydrophilic conjugates, which are less toxic than the parent compounds and be excreted to the urine. The first reaction of glycine conjugation pathway is the activation of xenobiotic carboxylic acids and FAs catalyzed by MACS to form corresponding acyl-CoAs, which can be subsequently conjugated to glycine by glycine *N*-acyltransferase (GLYAT, EC 2.3.1.13) to produce acylglycines (Scheme 7.10)

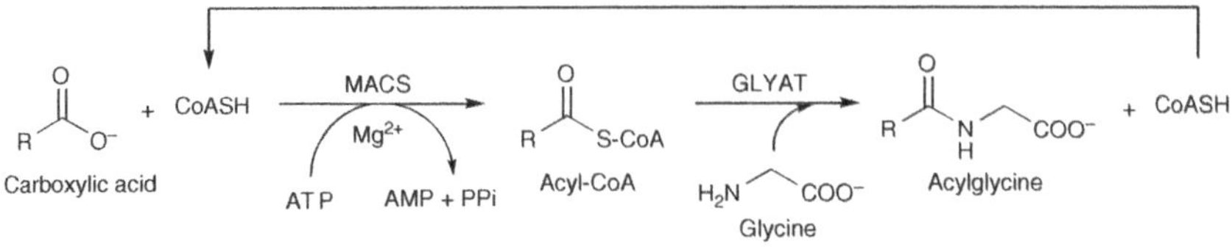

Scheme 7.10 Glycine conjugation of carboxylic acids catalyzed by MACS and GLYAT in sequence. *Sources:* Based on Badenhorst et al. [71] and Knights [19].

[19, 71]. An initial study of substrate specificity for MACS demonstrates that the purified enzyme accepts not only straight medium-chain FAs but also aromatic acids. Among the arylacetic acids, activity can be obtained with naphthylacetic acids, whereas a loss of activity was observed by the introduction of a methyl group at the α-position. High activity was also observed with cyclohexanoic acid [72]. Further study of the molecular specificity demonstrates that MACS purified from mouse kidney mitochondria showed highest activity with hexanoic acid and high activities for benzoic acid derivatives with alkyl and alkoxyl groups in the *para*- and *meta*-positions of the benzene ring. An *ortho*-substituted acid exhibited no activity. The enzyme also exhibited less activity with valproic acid, tranexamic acid, indomethancin, and ketoprofen. Diflunisal, 2-hydroxydodecanoic acid, and salicylic acid inhibited the enzyme. While the chlorine and methyl substitutions in the para- and meta-positions of the benzene ring yielded an increase in glycine conjugation. The acids with a methoxy group showed a low degree of glycine conjugation. The acids with nitro or amino groups were conjugated to a slight extent with glycine. Salicylic acid was *in vitro* conjugated with glycine in kidney but not in liver [73]. Similarly, MACS purified from mouse liver mitochondria catalyzed the first reaction of amino acid conjugation for straight medium-chain FAs as well as aromatic and arylacetic acids with maximal activity for hexanoic acid. High activities were observed with benzoic acid having methyl, pentyl, and methoxy groups in the para- or meta-positions of the benzene ring [74].

The activation of free fatty acids (FFAs) into acyl-CoA thioesters catalyzed by ACS can be further catabolized and reduced into derivatives like alcohols and alkanes. Alcohols and alkanes derived from MCFAs (C6–C12) are potential biofuels. Thus, MCFAs are important precursors to fuel-like compounds and industrial chemicals. Because MACS (FadD) from *E. coli* shows low activity on MCFAs, *E. coli* was engineered to generate mutations in *fadD* for enhancing the MACS activity on MCFAs. Data demonstrated that FadD mutations result in an enhancement of *E. coli* growth rate on the MCFAs hexanoate, octanoate, and decanoate. Partially purified wild-type FadD and mutant variants showed an increased activity on octanoate and decanoate [75].

Polyhydroxyalkanoates are polyesters of various (*R*)-hydroxyalkanoates synthesized by bacteria such as *Pseudomonas* sp. from renewable plant-derived carbon feedstocks. They are carbon and energy storage material in bacteria and have physical properties similar to those of petroleum-based plastic polyethylene but are biodegradable. Therefore, they have acquired attention as alternatives to more common petroleum-based polymers and recognized as candidate materials for sustainable development. It has been found that (*R*)-3-hydroxyacyl-CoA, the precursor for PHA synthesis, is thought to be derived from FA biosynthesis when the microorganism is grown on unrelated carbon sources, such as glucose and glycerol, and β-oxidation when the microorganism is grown on related carbon sources, such as FAs [76–78]. Therefore, enoyl-CoA, 3-ketoacyl-CoA, (*S*)-3-hydroxyacyl-CoA, and (*R*)-3-hydroxyacyl-ACP serve as major precursor pool for medium-chain-length-(*R*)-3-hydroxyacyl-CoAs [79, 80]. An application focuses on the utilization of 2-alkenoic acids of acyl chain lengths ranging from C6 to C12 for the biosynthesis of medium-chain-length polyhydroxyalkanoates (mcl-PHAs) using recombinant β-oxidation-defective mutants of *E. coli*. When metabolically engineered cells were fed with 2-alkenoic acids and glucose, broad substrate specificity MACS is able to catalyze the ligation of 2-alkenoic acids and CoA to generate enoyl-CoAs. Subsequently, (*R*)-3-hydroxyalkanoate-CoAs (3HA-CoAs) are produced from enoyl-CoAs by the action of (*R*)-specific enoyl-CoA hydratase (PhaJ) in the absence of β-oxidation. Then, mcl-PHAs can be synthesized from (*R*)-3HA-CoAs by the action of PHA synthase (PhaC). Thus, novel type of PHA consisting of 3-hydroxydecanoate (3HD) was synthesized from 2-decenoic acid. Solvent cast film of poly(3HD) was transparent and exhibited thermal property similar to that of polycaprolactone. In addition, PHAs from various carbon chain lengths 2-alkenoic acids (C6–C12) were with over 95 mol% of the corresponding single monomer units. Since the pyrolysis product of poly(3HD) was dominantly 2-decenoic acid, it demonstrates the recycling feasibility of PHA via 2-alkenoic acids (Scheme 7.11) [81].

Reports have shown that the physical properties of PHA copolymers consisting primarily of one short-chain-length (SCL) repeating unit and a small concentration of medium-chain-length (MCL) repeating units resemble the petroleum-based plastic polyethylene. Moreover, the SCL-*co*-MCL PHA copolymers can be used in biomedical applications such as tissue engineering and drug delivery [82, 83]. The metabolic pathways for these two classes of PHAs are different – SCL PHAs are synthesized from acetyl-CoA and MCL PHAs are synthesized either from FA *de novo* biosynthesis or β-oxidation [80]. Research results have shown that a key enzyme PhaG from *Pseudomonas putida*, which together with a CoA ligase, AlkK, is essential to MCL PHAs production from glucose [79]. The co-expression of PhaG, AlkK, and an engineered PHA synthase enabled *E. coli* to produce MCL PHA from glucose [80]. Since a much lower yield of SCL-*co*-MCL

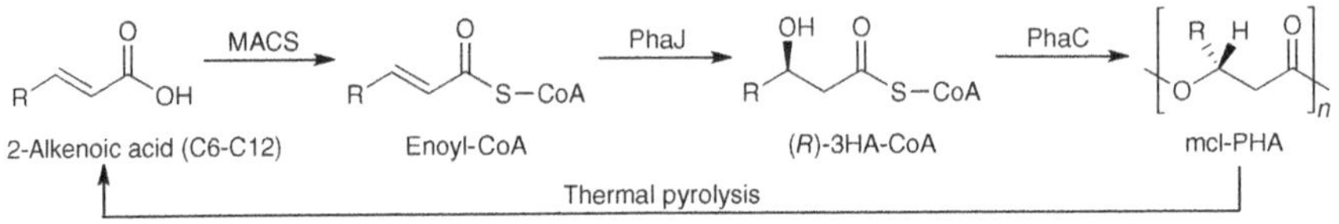

Scheme 7.11 Biosynthetic pathway for mcl-PHAs synthesis from 2-alkenoic acids. *Source:* Sato et al. [81].

PHA copolymers is often obtained from bacterial production as compared to SCL PHA biosynthesis produced from glucose, a new versatile route was designed for SCL-*co*-MCL PHA copolymer production from glucose (Scheme 7.12). An engineered *E. coli* strain (*E. coli* LS5218-STQKABGK) was introduced into the two separate biosynthetic pathways for SCL and MCL PHAs and employed for the copolymer production using batch fermentation. A highest yield of SCL-*co*-MCL PHA copolymers produced from glucose was reported, which the PHA copolymers produced consisted of repeating units with 4,6, 8, 10, and 12 carbons at mol% concentrations similar to that of other SCL-*co*-MCL PHA copolymers reported with desirable physical properties [84]. Levulinic acid (LA), a five-carbon γ-keto acid that can be readily obtained as a non-enzymatic acid hydrolysis product of lignocellulosic biomass, can serve as the sole carbon source for some bacteria such as *P. putida* to catabolize to acetyl-CoA and propionyl-CoA [85]. Thus, in addition to glucose, LA is a renewable substrate used for PHA copolymer synthesis.

It has been known that only the (*R*)-enantiomers of 2-aryloxypropanoic acids exhibit herbicide activity. In addition, of those 2-aryloxypropanoic acids, (*R*)-2-(4-chlorophenoxy)propanoic acid (CPPA) is capable of lowering the level of serum cholesterol and prevents platelet aggregation, but the (*S*)-antipode leads to muscle irritability and spasms by inhibiting chloride channels [86–88]. By screening microorganisms for deracemizing CPPA, only *Mycobacterium smegmatis* KU1047, *Brevibacterium ketoglutamicum*, and *Pseudomonas aeruginosa* KU1097 were found to show deracemization activity. The enantioselectivity of these microorganisms was found to be similar such that racemic CPPA was deracemized to the (*R*)-enantiomer. In addition, the cell-free extract of *B. ketoglutamicum* also displayed deracemization activity when the necessary cofactors (ATP, CoASH, and Mg^{2+}) for the acyl-CoA synthetase catalyzed formation of acyl-CoA thioester were supplied to the reaction mixture, but no activity was detected when any one of the above cofactors was absent. Therefore, a new enzyme, α-methyl carboxylic deracemizing enzyme 1 (MCAD1), was purified and cloned from *B. ketoglutamicum* KU1073 to carry out the (*S*)-enantioselective thioesterification reaction of CPPA (Scheme 7.13). Since MCAD1 shows highest activity toward MCL (C8 carbon atom) FAs, the substrate specificity indicates that MCAD1 is an MCL acyl-CoA synthetase [89].

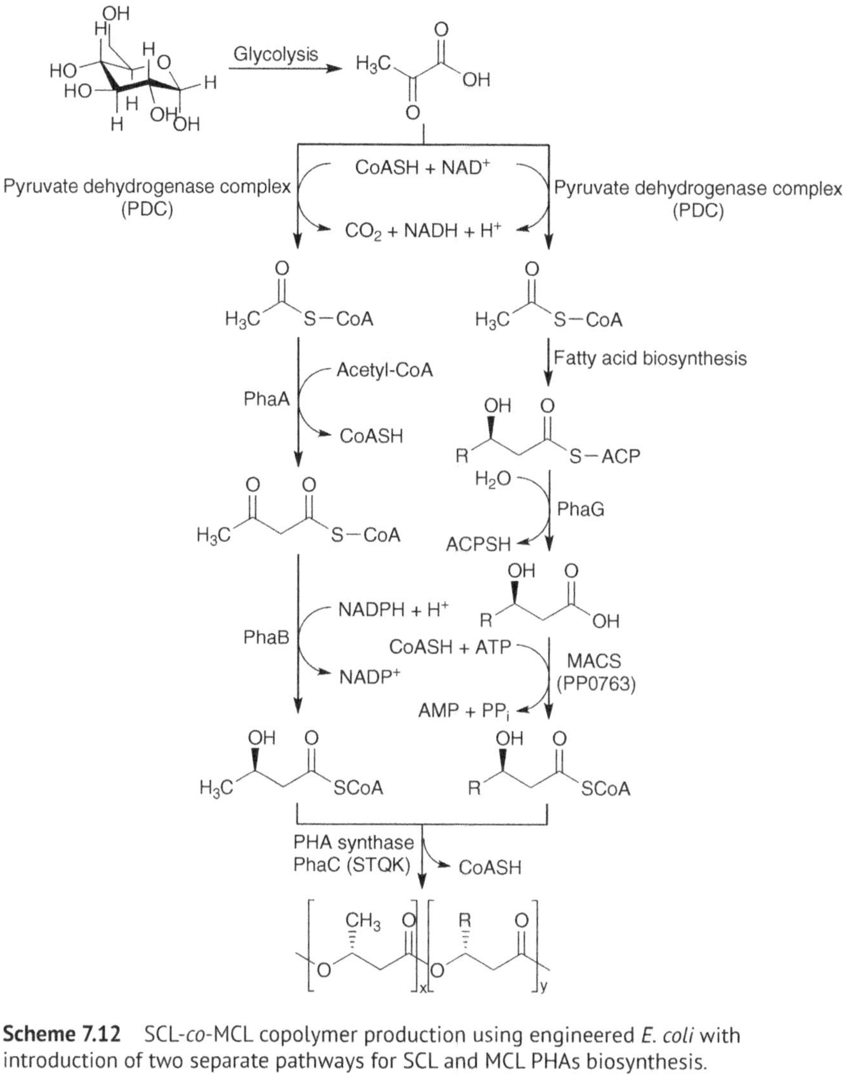

Scheme 7.12 SCL-*co*-MCL copolymer production using engineered *E. coli* with introduction of two separate pathways for SCL and MCL PHAs biosynthesis.

Microorganisms harbor structurally diverse natural products, including polyketides, peptides, and terpenoids. Understanding the biosynthetic mechanism of the atypical polyketide extender unit is important for the development of bioactive natural products. Reveromycin (RM) derivatives contain several aliphatic extender units derived from 2-alkylmalonyl-CoA, such as butylmalonyl-CoA, isobutylmalonyl-CoA, pentylmalonyl-CoA, and hexylmalonyl-CoA. While the aliphatic extender unit of Reveromycin A (RM-A), produced by *Streptomyces* sp.

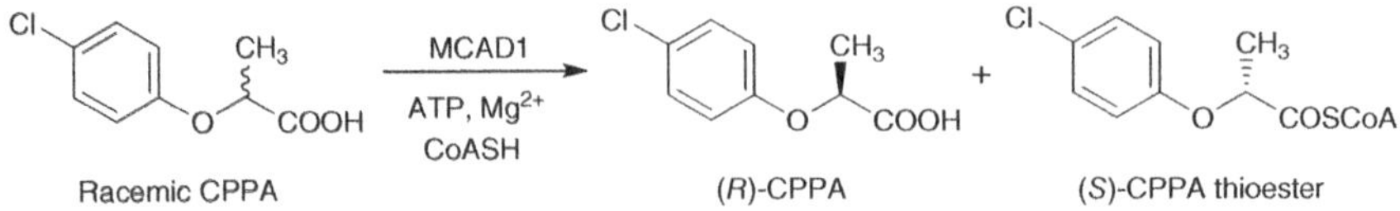

Scheme 7.13 Enantioselective thioesterification of racemic CPPA catalyzed by MCAD1.

SN-593, is derived from butylmalonyl-CoA. RM-A inhibits bone resorption by inducing apoptosis in osteoclasts, and this activity is also linked to the inhibition of osteolytic bone metastasis induced by lung and prostate cancer cells [90–95]. The molecular basis of 2-alkylmalonyl-CoA formation for RM-A biosynthesis was studied by analyzing the *revR* and *revS* genes, which form a transcriptional unit with the *revT* gene, a crotonyl-CoA carboxylase/reductase homolog (RevT) [96]. Incubation of purified RevT from expressed *Streptomyces lividans* TK23 in the presence of (*E*)-2-hexenoyl-CoA, $NaHCO_3$, and NADPH leads to the production of butylmalonyl-CoA (Scheme 7.14a). RevT also catalyzed (*E*)-2-octenoyl-CoA for the formation of hexylmalonyl-CoA with the catalytic efficiency. The *revR* gene disruption demonstrated that RevR was involved in the selective production of butylmalonyl-CoA. In the *de novo* FA biosynthesis, RevR is likely to select butyryl-CoA as a priming substrate to yield 3-oxo-hexanoyl-ACP. Next, the product may be converted into (*E*)-2-hexenoyl-ACP by β-keto processing enzymes, followed by an unknown transacylase for the conversion of (*E*)-2-hexenoyl-ACP into (*E*)-2-hexenoyl-CoA. Although the adenylate-forming enzyme RevS was classified in the fatty acyl-AMP ligase clade based on phylogenetic analysis, the enzyme catalyzed medium-chain fatty acyl-CoA ligase (FACL) but not the fatty acyl-AMP ligase activity. Moreover, the *in vitro* coupling reaction using purified RevS and RevT showed efficient conversion of hexenoic acid into butylmalonyl-CoA (Scheme 7.14b). Because RevS efficiently catalyzes C8–C10 medium-chain FACL activity, these acyl-CoA precursors have to be truncated via *in vivo* β-oxidation and converted into (*E*)-2-hexenoyl-CoA for subsequent RevT reaction to form butylmalonyl-CoA, which is catalyzed by the polyketide synthase (PKS) to produce a unique polyketide (RM-A) extender unit (Scheme 7.14c). These results provide the role of RevS, a novel medium-chain FACL, in catalyzing MCFAs to produce a unique polyketide extender unit.

The triple bond in an alkyne molecule is an important functional group widely used in chemical synthesis, material science, and pharmaceutical science. Based on previous efforts on characterizing the first carrier protein-dependent pathway for terminal alkyne formation, additional homologous gene cassettes presumably involved in alkyne biosynthesis using both *in vitro* biochemical assays and the *E. coli*-PKS reporting system for *in vivo* analysis were screened for further studies. Then, a new terminal alkyne biosynthetic pathway comprised of TtuA, -B, and -C

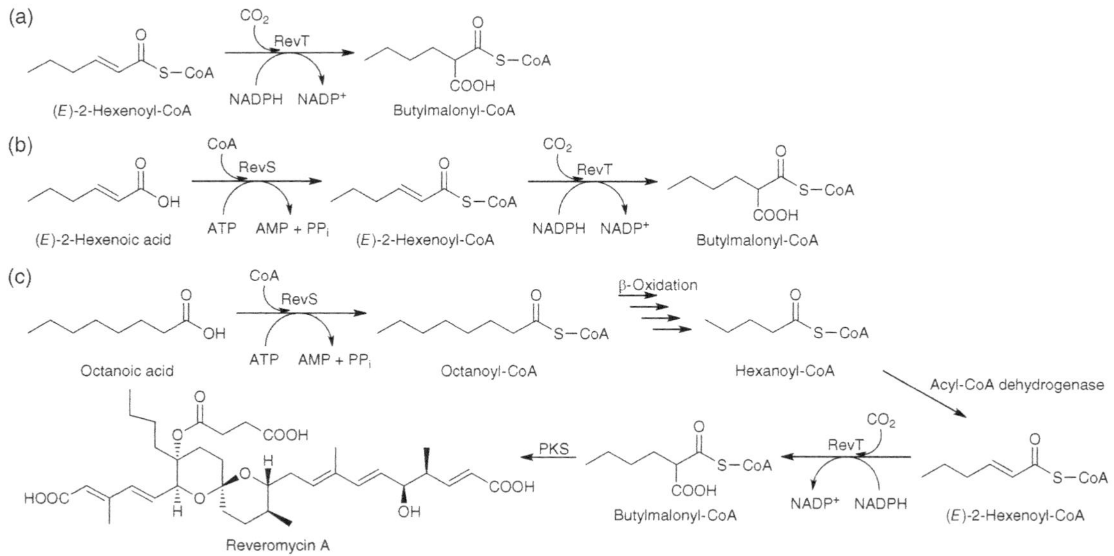

Scheme 7.14 (a) Production of butylmalonyl-CoA from (*E*)-2-hexenoyl-CoA catalyzed by RevT. (b) Successive conversion of (*E*)-2-hexenoic acid into butylmalonyl-CoA catalyzed by the coupling of RevS and RevT. (c) Biosynthetic pathway of octanoic acid incorporation into Reveromycin A.

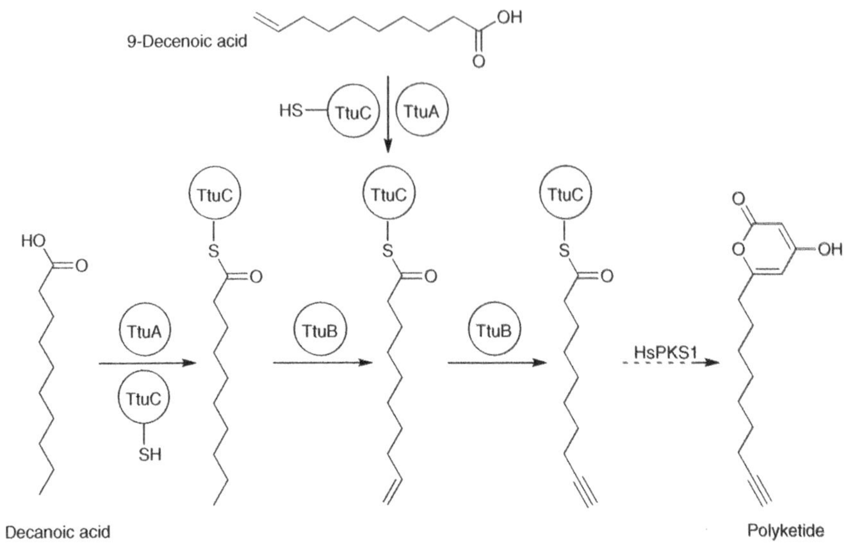

Scheme 7.15 Biosynthesis of novel terminal-alkyne-bearing polyketide starting from decanoic acid using co-expressed *ttuABC* and a PKS gene in *E. coli*. *Source:* Zhu et al. [97].

from *Teredinibacter turnerae* T7901 was discovered and characterized. While the acyl-CoA ligase homolog (TtuA) demonstrated promiscuity in the activation and loading of medium-chain FAs onto the carrier protein (TtuC), the desaturase homolog (TtuB) showed stringent substrate specificity toward decanoic acid and a bifunctional desaturase/acetylenase that efficiently catalyzed two sequential O_2-dependent dehydrogenation reactions in *E. coli*. A novel terminal-alkyne-bearing polyketide was further produced by co-expressing *ttuABC* and a PKS gene in *E. coli* (Scheme 7.15) [97].

7.2.3 Long-chain-fatty-acid-CoA Ligase

The systematic name of long-chain-fatty-acid-CoA ligase, also simply called *long-chain acyl-CoA synthetase* (LACS, EC 6.2.1.3), is a long-chain FA:CoA (AMP-forming). This category of enzyme catalyzes a two-step ATP-dependent reaction of long-chain saturated and unsaturated FAs and coenzyme A (CoA) through a fatty acyl-AMP intermediate to generate their acyl-CoA thioesters the same as that shown in Scheme 7.9 but with 12–22 carbon atoms, which are then substrates for β-oxidation as well as TAG, phospholipids, and cholesterol esters synthesis [19, 98]. The enzymes belonging to the acyl-CoA synthetase family are present in all organisms from bacteria to humans including plants and are located in various cell compartments such as mitochondria and peroxisomes, which participate in a wide variety of cellular pathways including lipid synthesis, lipid degradation, energy production, post-translational modification of proteins, membrane biogenesis, vesicular trafficking, and regulation of gene expression [99, 100].

The first LACS in guinea pig liver microsomes was reported in 1953 with activity for straight-chain SFAs of C5–C22 carbon atoms and mono-, di-, and trienoic C18 acids [101]. Since then, purification and characterization of LACS has been carried out from microsomes, mitochondria, and peroxisomes of rat liver [102–109]; from human blood platelets, erythrocyte membrane, and liver [110–114]; from calf brain [112]; and also from parasite *Cryptosporidium parvum* [115] and bacteria such as *Bacillus megaterium* and *Candida lipolytica* [116–118]. The estimated molecular weight from different cells was commonly in the range 76.0–86.1 kDa [105, 108, 115, 119], except only one with an apparent molecular weight of 250 kDa constituted by four protein components having molecular weight of 135, 115, 65, and 55 Da [106].

The first three-dimensional crystal structure of a LACS domain swap homodimer from an extreme thermophile *Thermus thermophiles* HB8 overexpressed in *E. coli* was determined at 2.0 Å resolution. In addition, the crystal structures of the enzyme complexes with ATP analogue adenosine 5′-(β,γ-imido) triphosphate (AMP-PNP) and the myristoyl-AMP complex were also determined. The enzymes consist of a large N-terminal and a small C-terminal domain, with the catalytic site positioned between the two domains. The binding of substrate may affect the relative positions of the C- and N-terminal domains. The C-terminal domain of LACS is assumed to be in an open conformation when a substrate is absent and in a closed conformation when a substrate is bound. A domain swapped dimer is formed by LACS, with monomer interacting at the N-terminal domains. A large electrostatically positive concave is located at the back of the structure in the central valley of the homodimer. The L motif, a six amino acid peptide linker, connects the large N-terminal domain and a small C-terminal domain of each LACS monomer. The N-terminal domain is composed of two subdomains: a distorted antiparallel β-barrel and two β-sheets surrounded by α-helices forming an αβαβα sandwich. The small C-terminal globular domain consists of two-stranded β-sheet and a three-stranded antiparallel β-sheet flanked by three α-helices [98, 120]. Upon ATP binding, the FA-binding tunnel gated by an aromatic residue opens to the ATP-binding site. The gated FA-binding tunnel appears only to allow one-way movement of the FA during overall catalysis. The protein incorporates a hydrophobic branch from the FA-binding tunnel that is responsible for substrate specificity. Based on these high resolution crystal structures, a unidirectional Bi Uni Uni Bi Ping-Pong mechanism for the two-step acetylation by LACS was proposed [98]. The mutational analysis of signature motif in a long-chain fatty acyl-CoA synthetase consisted of seven amino acid residues, which is essential for catalytic activity and functions in part to promote FA chain length specificity and thus may compose part of the FA-binding site within the enzyme [121].

LACS plays a central role in the physiological regulation of a variety of cellular functions by catalyzing the formation of long-chain fatty acyl-CoA from FA, ATP, and CoA via a two-step process proceeded through the hydrolysis of pyrophosphate. Long-chain fatty acyl-CoAs are bioactive compounds that have effects in protein transport, enzyme activation, protein acylation, cell signaling, transcriptional control, and organ fusion in addition to serving as substrates for β-oxidation, TAG formation, and phospholipid biosynthesis [99, 100, 119, 121–128]. LACS also has clinical significance in cancer therapy [129]. It has been reported that long-chain fatty acyl-CoA ligase 3 (FACL3), which preferentially utilizes myristic acid, eicosapentaenoic acid, and arachidonic acid as substrates to generate long-chain fatty acyl-CoAs, was upregulated by vitamin D_3 in both expression and activity levels and contributed to the growth inhibitory effect of vitamin D_3 in human prostate cancer LNCaP cells [130]. Furthermore, it was found that fatty acid synthase (FAS), the second and a key enzyme for the FA synthesis, which is overexpressed and associated with prostate cancer development, was downregulated by vitamin D_3 in LNCaP cells [131]. Then, the feedback inhibition of FAS expression by long-chain fatty acyl-CoA synthesized by FACL3 causes the downregulation of FAS mRNA by vitamin D_3. Therefore, FACL3 contributes to the antiproliferative effect of vitamin D_3 in the human prostate cancer LNCaP cells [132].

Jasmonic acid (JA), a plant hormone, and its cyclic precursors are collectively referred to as jasmonates constituting a family of bioactive oxylipins, which are derived from oxygen-containing FAs and regulate a variety of defense responses and developmental processes in plants. In plants, the biosynthetic pathway of JA is initiated in the chloroplast starting from α-linolenic acid (C18:3) or hexadecatrienoic acid (C16:3), which is released from membrane phospholipids by phospholipase and converted to 12-oxo-phytodienoic acid (OPDA) and dinor-OPDA (dnOPDA), respectively, by consecutive action of the plastic enzymes 13-lipoxygenase, allene oxide synthase, and allene oxide cyclase. Then, OPDA and dnOPDA can be released from the chloroplast and function directly as signal molecules in plant defense responses. For JA synthesis, OPDA and dnOPDA are translocated into peroxisomes by the peroxisomal ATP-binding cassette transporter COMATOSE but may also include passive mechanisms (anion trapping). Within the peroxisomes, both are converted by oxophytodienoic acid reductase 3 (OPR3) to 3-oxo-2-(2′-[*Z*]-pentenyl)-cyclopentane-1-octanoic acid (OPC-8) and 3-oxo-2-(2′-[*Z*]-pentenyl)-cyclopentane-1-hexanoic acid (OPC-6), respectively. Before proceeding the β-oxidative shortening of the carboxyl side chain for final JA biosynthesis, OPDA, OPC-8, and OPC-6 must be converted to the corresponding CoA esters by different acyl-CoA synthetases (*At*5g63380, *At*1g20510, or *At*1g20500) from *A. thaliana* (Scheme 7.16) [133, 134]. Note that both *At*5g63380 and *At*1g20501 exhibit LACS activity. Of the three acyl-CoAs, OPDA-CoA is further converted into OPC-8-CoA catalyzed by OPR3. Then, an even number of carbon

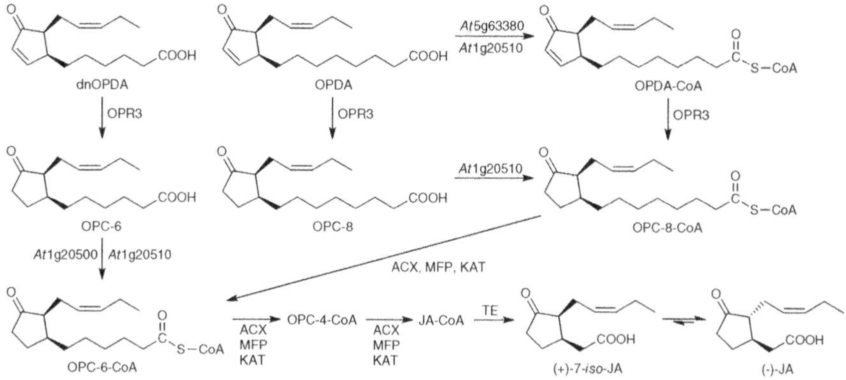

Scheme 7.16 Pathways of JA biosynthesis in plant peroxisomes. *Sources:* Based on Kienow et al. [133] and Schneider et al. [134].

is removed from the carboxyl side chains of OPC-8-CoA and OPC-6-CoA by several cycles of β-oxidation, which is catalyzed by three core enzymes, acyl-CoA oxidase (ACX), multifunctional protein (MFP), and 3-ketoacyl-CoA thiolase (KAT) to yield the penultimate intermediate jasmonyl-CoA (JA-CoA). Subsequently, (+)-7-*iso*-JA is released from JA-CoA by the action of TE and spontaneously epimerized to (−)-JA.

Owing to the limited amount, uneven distribution, and environmental pollution of fossil fuels, sustainable biofuels have garnered considerable interest for decades. Liquid biofuels, such as biodiesel and bio-alcohols, can be generated from renewable resources including edible plants, green wastes, and microalgae. Biodiesel is produced by the transesterification of TAGs, yielding mono-alkyl esters of long-chain FAs with short-chain alcohols, for example, FA methyl esters (FAMEs) and FA ethyl esters (FAEEs). In addition to being renewable and domestic origin, advantages of biodiesel over petrochemical diesel include biodegradability, higher flash point, less toxicity, and inherent lubricity. Biodiesel from microalgae is referred to as third-generation biodiesel, while the first-generation biodiesel was produced from edible resources such as edible oils and lipids and the second-generation biodiesel was obtained by the pretreatment of the biomass of non-edible oils. Microalgae lipids comprise polar and nonpolar lipids. *De novo* synthesis of FAs in microalgae occurs mainly in the chloroplast, in which CO_2 is the initial carbon source, that leads to the production of saturated palmitic acid (C16:0) and stearic acid (C18:0), which can further produce unsaturated FAs (UFAs) such as oleic acid (C18:1), linoleic acid (C18:2), and α-linolenic acid (C18:3). The biosynthesis of microalgae TAG has been suggested to take place through the direct glycerol pathway to form the storage lipid droplet [53, 54].

Although microalgae biofuel has become the most promising renewable energy, the high production cost is still a limitation. Although efforts have been made in enhancement of lipid productivity, the major cost problem in harvesting and oil extraction is still intractable. It has been reported that LACSs are involved in FAs secretion due to the ability of FAs transportation. Disruption of LACSs resulting in FAs exportation into the media was first discovered in *E. coli*. A similar phenomenon was also observed in *S. cerevisiae* when native LACSs genes, *faa1* and *faa4* that account for most of the LACSs activity, were both removed [100]. LACSs identified in eukaryotic algae were very few. However, the cDNA of two LACSs genes, *cracs1* and *cracs2*, was cloned from *C. reinhardtii*, which showed different substrate adaptation in the yeast complementation experiments. *Cracs1* could only utilize FAs C14:0, C16:1, and C18:1; *cracs2* could utilize FAs C12:0, C14:0, C16:0, C18:0, C16:1, and C18:1. Knockdown of *cracs1* and *cracs2* in *C. reinhardtii* resulted in accumulation of intracellular lipids. The total intracellular lipid content of transgenic algae q-15 and p-13 (knockdown of *cracs1* and *cracs2*, respectively) was 45 and 55%, respectively, higher than that of *C. reinhardtii* cc849. FAs secretion was observed in both transgenic algae that reach 8.19 and 9.66 mg/10^9 cells in q-15 and p-13, respectively [135]. Therefore, FAs secretion is possible by knocking down LACSs cDNA from microalgae that gives a new strategy for low-cost oil extraction.

Vegetable oil is mainly accumulated during the seed maturation phase and supplies carbon and energy for seed germination and seedling growth. Biosynthesis of TAG in plants is generally understood and considered to be a highly conserved process, but the molecular and biochemical details are mostly limited to oilseeds. In plant tissues, biosynthesis of TAG primarily involves synthesis of FAs in the plastid and their transfer to the endoplasmic reticulum (ER), where sequential esterification to a glycerol-3-phosphate (G3P) backbone in an acyl-CoA-dependent or -independent manner takes place to generate either TAG or phospholipid [136, 137]. Whereas in the cytosol and plastid, hexose is converted into acetyl-CoA through the glycolytic pathway, of which some are then used as the carbon source for FA synthesis. The general scheme for the conversion of sucrose to TAG involving degradation of sucrose; generation of pyruvate in the plastid, which involves glycolysis, pentose phosphate pathway, and plastid transporters; FA synthesis in the plastid; and TAG assembly in the ER is shown in Scheme 7.17a. In the plastid, the conversion of pyruvate to FAs involves at least 14 enzymes and/or protein complexes (Scheme 7.17b) [138].

Avocado (*Persea americana*) belongs to the family *Lauraceae*, one of the largest basal angiosperm families and is an advantageous system to study the evolution of mechanisms underlying the synthesis of storage reserves such as starch or lipids in fruit tissues other than seed. The avocado fruit, like oil palm and olive, is one of a few examples in which the mesocarp, a nonseed tissue, stores copious amounts

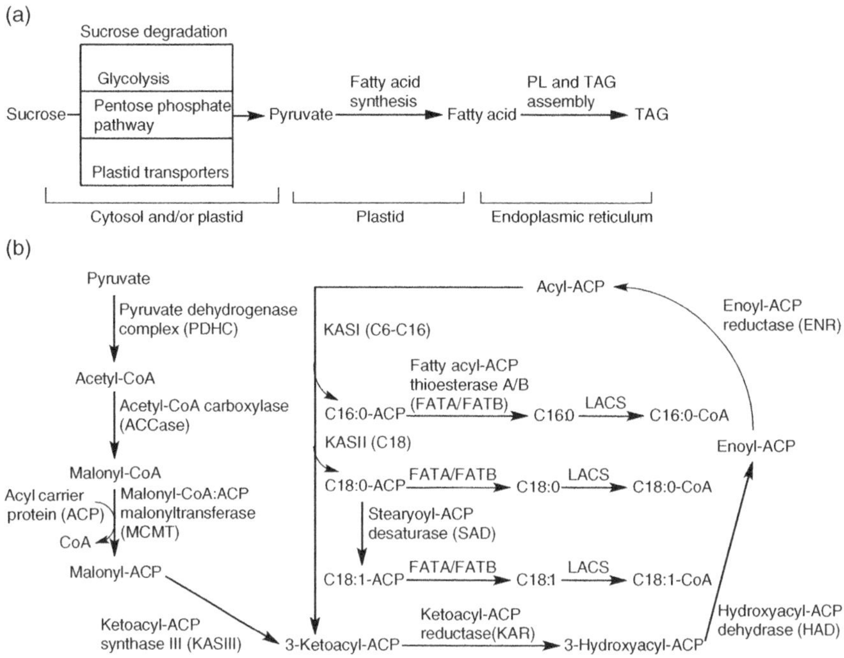

Scheme 7.17 (a) Schematic pathway for the conversion of sucrose to triacylglycerol. (b) Fatty acid synthesis pathway in the plastid. *Source:* Kilaru et al. [138].

of TAG (~70% TAG per dry weight). Comparison of the transcript levels of the orthologues of the gene family members in oil-rich tissues of avocado, oil palm, rapeseed, and castor, while indicating some similarities across diverse species and tissue, also revealed several exceptions for avocado (Scheme 7.18). The accumulation of TAG in oleic acid was associated with higher transcript levels for a putative stearoyl-ACP desaturase and ER-associated acyl-CoA synthetases, during fruit development. Gene expression levels for enzymes involved in terminal steps to TAG biosynthesis in the ER further indicated that both acyl-CoA-dependent and -independent mechanisms might play a role in TAG assembly, depending on the developmental stage of the fruit. Furthermore, in addition to the expression of an orthologue of WRINKLED1 (WRI), a regulator of FA biosynthesis, high transcript levels for WRI2-like and WRI3-like suggest a role for additional transcription factors in nonseed oil accumulation. Plastid pyruvate necessary for FA synthesis is likely driven by the upregulation of genes involved in glycolysis and transport of its intermediates. Together, a comparative transcriptome analyses for storage oil biosynthesis in diverse plants and tissues suggested that several distinct and conserved features in this basal angiosperm species might contribute toward its rich TAG content [138].

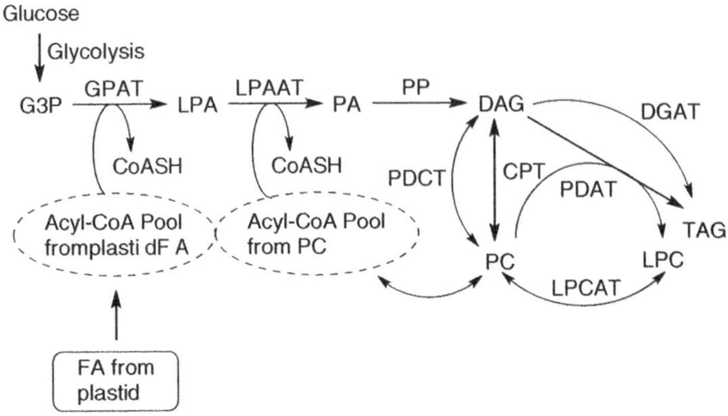

Scheme 7.18 Acyl-CoA-dependent or -independent TAG synthesis pathway derived from glucose in ER. *Legend*: CPT, cytidine-5'-diphosphocholine; DAG, diacylglycerol; DGAT, diacylglycerol acyltransferase; G3P, glycerol-3-phosphate; GPAT, glycerol-3-phosphate acyltransferase; LPA, lysophosphatidic acid; LPAAT, lysophosphatidic acid acyltransferase; LPC, lysophosphatidylcholine; LPCAT, lysophosphatidylcholine acyltransferase; PA, phosphatic acid; PC, phosphatidylcholine; PDAT, phospholipid diacylglycerol acyltransferase; PDCT, phosphatidylcholine:diacylglycerol cholinephosphotransferase (PC:DAG cholinephosphotransferase); PP, phosphatidate phosphatase; TAG, triacylglycerol.

Oilseed rape (*Brassica napus*), one of the four major oil plants in the world, plays an important role in producing vegetable proteins and edible oils for human. In addition, seed oils have long been recognized as important raw materials for biofuels. TAGs are the main component of plant seed oil. In developing oilseeds, the location for TAG biosynthesis is ER, whereas carbohydrate and FA metabolism are also involved in this pathway. While in the cytosol and plastid, hexose is degraded and converted into acetyl-CoA through Embden–Meyerhof–Parnas pathway (EMP pathway), of which some are then used as the carbon source for FA synthesis. FAs can be converted to long-chain acyl-CoAs, which are precursors for TAG biosynthesis. In *Arabidopsis*, the LACS gene family contains nine members, among which LACS1 and LACS9 have overlapping functions in TAG biosynthesis. *LACS2* gene identified in rapeseed *B. napus* (*BnLACS2*) is an orthologue of the *Arabidopsis LACS2* gene and is highly expressed in developing seed. The *Bn*LACS2-GFP (green fluorescent protein) fusion protein was mainly localized to the ER. Overexpression of the *BnLACS2* gene resulted in significantly higher oil contents in transgenic rapeseed plants compared to wild type; however, *BnLACS2*-RNAi transgenic rapeseed plants had decreased oil contents. Moreover, *BnLACS2* transcription increased the expression of several genes involved in glycolysis, as well as FA and lipid biosynthesis in developing seeds [139].

With the emergence of second-generation biofuel technologies and the need to avoid the food-fuel conflict, exploring better ways to produce biodiesel is of high interest in research. Biodiesel production using microbial lipids, known as single-cell oils (SCO), has attracted attention of researchers. They can be converted into FAME and FAEE or into hydroprocessed esters and fatty acids (HEFA). Microbial oils have many advantages over plant oils, such as short life cycle; less labor required; less affected by venue, season, and climate; and easier to scale up. The single biggest advantage is that lignocellulosic waste material can be used to reduce the available area for growing food crops or microalgae. In addition, lipid composition of microbes can easily and quickly be modified by mutation, hybridization, and/or genetic engineering techniques combined with classical fermentation [140–144]. A Gram-positive, facultative anaerobic, nonpathogenic soil bacterium, *Corynebacterium glutamicum*, was metabolically engineered to produce TAGs by constructing a *de novo* TAG biosynthesis pathway (Scheme 7.19) [145]. First, a TAG biosynthesis plasmid pZ8_TAG4 was constructed with the heterologous expression of four genes. Three of these genes: *atf1* and *atf2* derived from *Rhodococcus opacus* PD630 encode DGAT and *pgpB* cloned from *E. coli* encodes phosphatidic acid phosphatase (PAP). One gene, *tadA,* involved in the lipid body maturation was also derived from *R. opacus* PD630. Second, four strategies were employed to increase TAGs accumulation via metabolic engineering: (i) increasing precursor supply by heterologous expression of *tesA* (encoding TE to form free FA to reduce the feedback inhibition by acyl-ACP) and *fadD* (encoding LACS to enhance acyl-CoA supply), (ii) reducing TAG degradation and precursor consumption by the deletion of four cellular lipase genes (*cg0109*, *cg0110*, *cg1676*, and *cg1320*) and the diacylglycerol kinase (DKG) gene (*cg2849*), (iii) enhancing FA biosynthesis by deleting *fasR* (*cg2737*, encoding the *TetR*-type transcriptional regulator of genes for the FA biosynthesis), and (iv) eliminating the observed by-product formation of organic acids by blocking the acetic acid (*pqo*) and lactic acid production (*ldh*) pathways. The final strain (*CgTesRtcEfasEbp*/pZ8_TAG4) achieved a 7.5% yield of total FAs ($2.38 \pm 0.05\,g\,L^{-1}$ intracellular FAs and $0.64 \pm 0.09\,g\,L^{-1}$ extracellular FAs) from 4% glucose in shaker-flask experiments after process optimization, which corresponds to maximum intracellular FAs content of $17.8 \pm 0.5\%$ of the dry cell [145].

7.3 Ligases for Carbon–Nitrogen Bonds Formation

7.3.1 Glutamine Synthetase

Glutamine synthetase (GS, EC 6.3.1.2, systematic name as L-glutamate:ammonia ligase [ADP-forming]) is a crucial enzyme to all living organisms that plays an essential role in the metabolism of nitrogen by catalyzing the ATP-dependent condensation of ammonia and glutamate to yield glutamine, ADP, and inorganic phosphate in

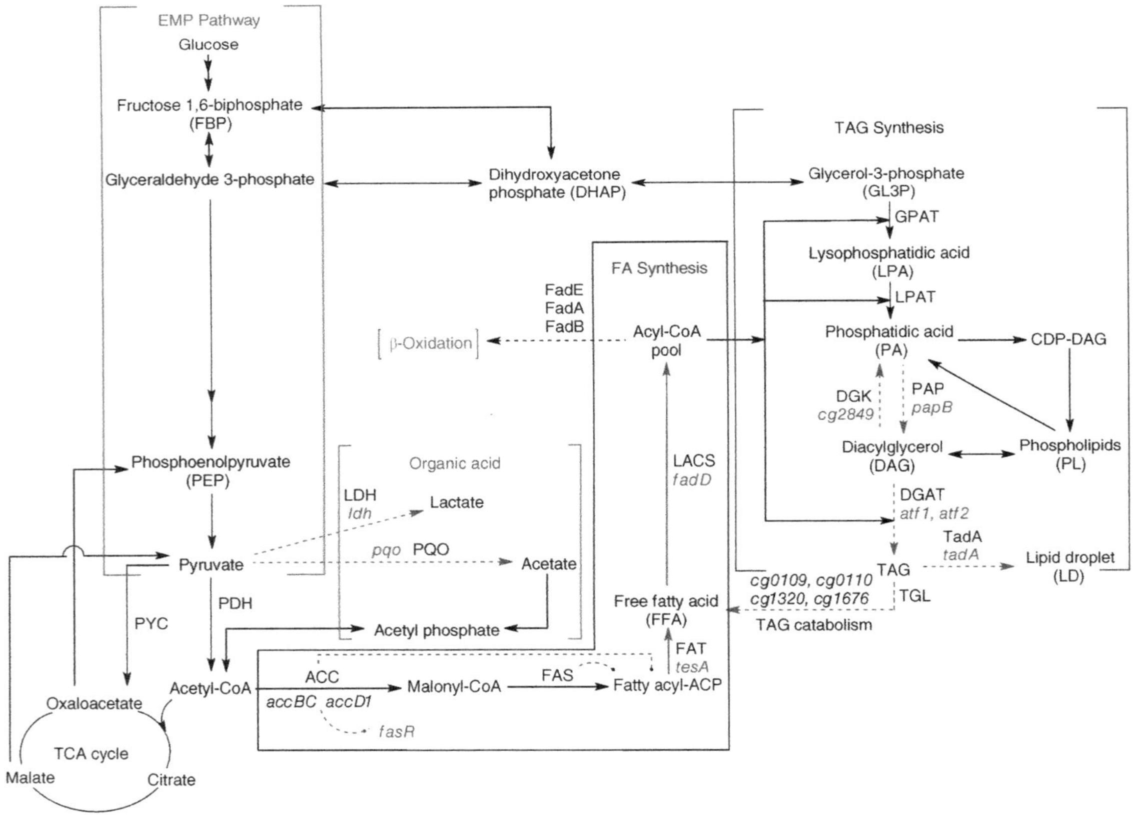

Scheme 7.19 The *de novo* TAG biosynthesis pathway constructed by modifications introduced in *C. glutamicum* ATCC 13032 for TAG production. Depicted are two main pathways for FA synthesis and TAG synthesis as well as four connected pathways (EMP, TCA, β-oxidation of FA, and organic acid production). *Legend*: ACC, acetyl-CoA carboxylase; CDP-DAG, CDP-diacylglycerol; DGAT, diacylglycerol acyl-transferase; DGK, diacylglycerol kinase; FAS, fatty acid synthetase; FAT, acyl-ACP-thioesterase; GPAT, glycerol-3-phosphate acyltransferase; LDH, lactate dehydrogenase; LPAT, lysophosphatidate acyl-transferase; PAP, phosphatidic acid phosphatase; PDH, pyruvate dehydrogenase; PQO, pyruvate/quinone oxidoreductase; PYC, pyruvate carboxylase; TGL, triacylglycerol lipases. *Source:* Plassmeier et al. [145].

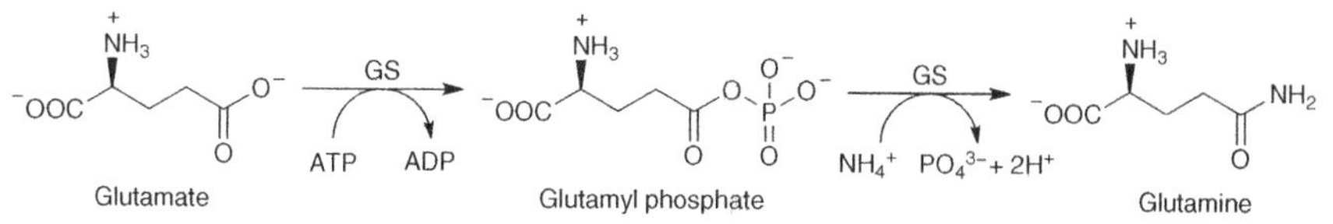

Scheme 7.20 Condensation of glutamate and ammonia to generate glutamine catalyzed by glutamine synthetase. *Sources:* Based on Dos Santos Moreira et al. [146], Moreira et al. [147], and Ginsburg et al. [148].

the presence of divalent cations such as Mg^{2+} and Mn^{2+} (Scheme 7.20) [146–148]. GS is the sole enzyme to incorporate atmospheric nitrogen into organic molecules through ammonia, which is produced by nitrate reduction, amino acid degradation, and photorespiration [149]. The first traceable paper on GS was dated back to 1953, and then more than 10000 research reports on GS were published. The plenty of research on GS reveal the wide variety of physiological roles, mostly related to its involvement in metabolic pathways, which manipulate nitrogen and carbon levels within the cell through many kinds of regulation mechanisms.

The molecular phylogenetic analyses suggest that GS genes are one of the oldest existing and functioning genes in the history of gene evolution [150]. The structure of GS has been studied extensively. The early technique used for the purified enzyme was electron microscopy [151, 152]. As the progress of protein X-ray crystallography advances, the structure of GS can be determined by X-ray crystallography initially at 3.5 Å resolution and further refined to the resolution of 2.5 Å [153–157]. There are three different classes of GS: GSI, GSII, and GSIII [146, 158]. They are mainly categorized by the difference in number of residues per monomer, while other features including the quaternary structures and the distribution among organisms are also used to classify GS enzymes. Class I enzymes (GSI) are categorized by having 450–470 residues per monomer with its oligomer of 12 identical subunits formed by two hexameric rings. GSI is typically found in prokaryotes with a few present in mammals, plants, and fungi [150, 159]. Class II enzymes (GSII) are characterized by having 350–420 residues per monomer, which is the smallest monomer among the three classes. The monomers of GSII are decamer of identical subunits formed through the combination of two pentameric rings. This class of GS is normally found in eukaryotes with some discovered in soil bacterium [150, 158, 160–162]. Class III enzymes (GSIII) are much larger than the GSI or GSII enzymes containing around 700 residues per monomer. GSIII also is a double hexameric ring associated with dodecameric homo-oligomer, but the type of association of the two hexameric rings is different from that of GSI and GSII [146, 163]. GSIII has only been found in organisms such as anaerobic bacteria, photosynthetic blue-green algae, radioresistant bacterium, the amoeba, and some protozoans [146, 163, 164]. However, it also happens that *Synechocystis* sp. encodes both GSI and GSIII in its genome [165].

GS can be composed of either 10 or 12 identical subunits separated into two face-to-face rings held by hydrogen and hydrophobic interactions [152, 153, 160–164, 166, 167]. Each active site of GS is located between the N-terminal of one monomer and the C-terminal of the adjacent monomer that means 10 active sites for the GSII decamers and 12 in the GSI and GSIII dodecamers. The C-terminal is the major contributor to the active-site cavity. The shape of each active site looks like an hourglass or a bifunnel that shows a narrow middle part near the Mg^{2+} ion binding site, and the larger part regions can access to the solvent. Note that there are three Mg^{2+} ions in each active site that are numbered according to their binding order commonly referred to as n1, n2, and n3. Two cavities are formed with one smaller and one larger. The smaller cavity is the binding site for ammonium ion and glutamate, where glutamate is generated and released. Thus, this cavity is named as "amino acids binding site," and inhibitors resembling amino acids can bind to this site. Two negatively charged regions are present on one side of the pocket to interact with the amine group and the ammonium ion. The ammonium ion binding region is constituted by Glu305, Tyr162, Asp63′, and Ser66′, while the other binding region is constituted by Asn248, Gly249, Glu134, Glu136, Glu196, Glu203, and Glu338 (the apostrophe indicates that the residue belongs to the adjacent subunit) that interacts with the backbone amine. On the other hand, the positively charged residues (Arg229, Arg319, Arg340, His253, and Asn255) are present to interact with the carboxyl groups located in the backbone and side chain of glutamate/glutamine [146, 161]. The larger cavity is the place for ATP, ADP, and AMP binding, which is named as "nucleotide binding site." Since the amino acids binding site is far more conserved than the nucleotide binding site, the nucleotide binding site is the preferential pocket to design novel GS inhibitors with selectivity toward different classes of GS enzymes [146].

The mechanism of GS catalyzed glutamate and ammonia condensation with the aid of ATP was deduced from X-ray crystallography structure data of GS-inhibitors complexes and from kinetic studies. Even though different mechanisms were proposed for the reaction, the most widely accepted hypothesis for the mechanism was the two-step reaction mechanism [153, 168–173]. In the first step, ATP binds GS and coordinates with n2 ion, and then glutamate binds next near to n1 ion and attacks the γ-phosphorus atom of ATP to form the intermediate γ-glutamyl phosphate and ADP; subsequently, after a rearrangement of the active site, ammonium ion binds and loses a proton to a basic residue (probably an Asp) as well as attacks the γ-glutamyl phosphate intermediate, resulting in a tetrahedral configuration; finally, the phosphate breaks down from the intermediate to produce a charged glutamine and loses a proton to another basic residue (probably a Glu) [146]. However, the results of three computational studies indicated three similar, but not equivalent, reaction mechanisms, in which the most differences are associated with protonation states and proton transfers. Nevertheless, all three studies pointed to an S_N2 nucleophilic substitution that the Cγ of the substrate maintained its sp^2 hybridization rather than going through a tetrahedral transition state of sp^3 hybridization. The tetrahedral intermediate does not exist but only as a transition state [147, 174, 175].

E. coli, an enteric bacterium, and many other organisms have two primary pathways for glutamate synthesis (Scheme 7.21) [176]. One involves glutamate dehydrogenase (GDH), which catalyzes the reductive amination of 2-oxoglutarate (2-OG) to form glutamate. The other pathway is the glutamate synthetase-glutamate synthase (glutamine:2-oxoglutarate aminotransferase, GOGAT, EC 2.6.1.53) pathway, comprising the consecutive reaction of GS and glutamate synthase, which is known to be essential for synthesis at low ammonium concentrations and for regulation of the glutamine pool. The occasion that GDH is used for glutamate synthesis is when the cell is limited for energy (and carbon) but ammonium and phosphate are present in excess. Therefore, the energy cost of glutamate biosynthesis may be less when energy is limited than when energy is unlimited. This pathway is also utilized by cyanobacteria for the assimilation of N_2 and NH_4^+ to produce glutamine and glutamate. In five diverse cyanobacteria, the principal initial product of metabolism of N_2 and NH_4^+ is glutamate by GS. However, glutamine declines and glutamate increases with increasing times of assimilation of N_2 and NH_4^+ in all of the N_2-fixation cyanobacteria due to the subsequent catalytic reaction by GOGAT. While the major glutamate formed by *Anacystis nidulans* incubated with NH_4^+ may be formed by GDH [177].

Trypanosoma cruzi, the etiological agent of Chagas disease, can consume glucose and/or amino acids depending on their availability. The consumption of

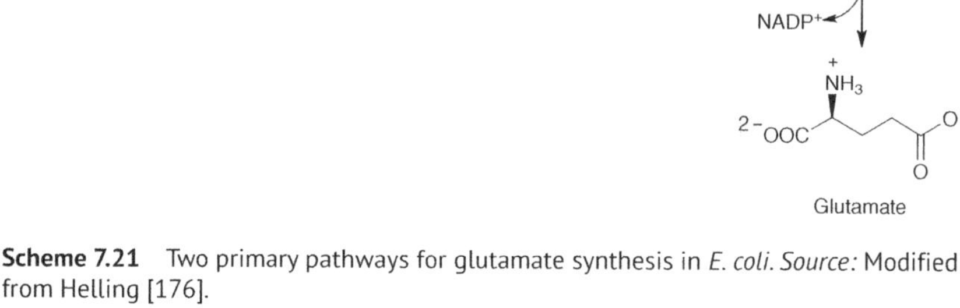

Scheme 7.21 Two primary pathways for glutamate synthesis in *E. coli*. *Source:* Modified from Helling [176].

amino acid usually produces excess toxic ammonium, which must be eliminated to convert it into a nontoxic substance. The enzyme GS serves the role for managing the excess ammonium by converting it into glutamine. The formation of glutamine from glutamate and ammonium catalyzed by GS also is essential for its invasion life cycle, particularly during the stage transiently enclosed in a tight vacuole, where it consumes amino acid and differentiates into the amastigote; in addition, extracellular Gln has a role in the survival of the parasite under severe metabolic stress conditions and in the maintenance of intracellular ATP levels [178, 179].

In nonlegume plants, nitrogen assimilation begins with uptake of inorganic N_2, NO_3^-, or NH_4^+ from soil. Subsequently, NO_3^- is reduced to NH_4^+ by the sequential reaction catalyzed by nitrate reductase and nitrite reductase. Then, NH_4^+ is converted to glutamate and glutamine by the enzymes GS, GOGAT, and GDH. Glutamate and glutamine are nitrogen donors in the biosynthesis of essentially all other amino acids and important nitrogen-containing compounds such as nucleic acids, chlorophyll, hormones, and secondary metabolite products. The amino acids glutamine, glutamate, aspartate, and asparagine serve as the principal nitrogen-transport amino acids in most crop and higher plants including *A. thaliana*. The enzymes exploited for biosynthesis of the amide amino acids glutamine and asparagine were GS and asparagine synthetase (AS), respectively, which are subject to tight regulation in response to environmental factors such as light and to metabolic factors such as sucrose and amino acids [149]. For plants, cytosolic GS1 is important for primary NH_4^+ assimilation in roots and for reassimilation of NH_4^+ generated during protein turnover in leaves, whereas the dominating role of GS2 is in reassimilation of photorespiration NH_4^+ in the chloroplasts and assimilation of NH_4^+ derived from NO_3^- reduction in plastids. GS1 activity may be downregulated via a chain of processes elicited by metabolic imbalances and environmental constraints. A pivotal role of GS1 may be related to the maintenance of essential N flows and internal N sensing during critical stages of plant development [180]. It has been noted that light absorbed for plant photosynthesis can itself be dangerous to plants. Transgenic tobacco plants enriched or reduced in plastid GS2 demonstrated that plants with twice the normal amount of GS2 had an improved capacity for photorespiration and an increased tolerance to high-intensity light, whereas those with a reduced amount of GS2 had a diminished capacity for photorespiration and C3 compounds were photoinhibited more severely by high-intensity light. Therefore, photorespiration can act as a defense mechanism [181].

The metabolism of glutamine in the leaf and subtended fruit of the aging pea (*Pisum sativum* L.) was found in relation to senescence and seed development. Experimental results suggest that the activity of GS in the supply organs (leaf, pod) furnishes the translocated amide necessary for the N nutrition of the cotyledon. The subsequent activity of GOGAT could provide a mechanism for the transfer of imported amide N to alpha amino N subsequently used in protein synthesis [182]. Nitric oxide (NO) is an important regulator in the *Rhizobium*-legume symbiosis. In root nodules of *Medicago truncatula*, the NO-mediated GS post-translational inactivation is connected to nitrogenase inhibition induced by NO that boosts glutamate channeling to the root nodules for the synthesis of glutathione (GSH), which plays a major role in the antioxidant defenses by participating in the ascorbate/GSH cycle [183].

In the vertebrate nervous system, GS in the brain participates in the metabolic regulation of neurotransmitter glutamate, the detoxification of brain ammonia, the assimilation of ammonia, recyclization of neurotransmitters, and termination of neurotransmitter signals. GS is located mainly in astrocytes, both *in vivo* and *in vitro*. Astrocytes protect neurons against excitotoxicity by taking up excess ammonia and glutamate and converting it into glutamine through GS [166, 184]. Hyper blood ammonia is a possible etiological agent for Alzheimer's disease, the most common age-related dementia, which is due to the loss of astrocytic GS activity and induces the impairment of glutamate-glutamine cycle and astrocytosis to cause an increase in brain ammonia concentration [185]. In the study of etiological mechanism of autism, a heterogeneous neurological disorder characterized by impairments in communication and social interactions, repetitive behaviors, and sensory abnormalities, significant alterations in glutamate and glutamine cycle enzymes as represented by GS and kidney-type glutaminase (GLS1), respectively, were demonstrated. While the glutamic acid decarboxylase autoantibody levels were remarkably increased, no significant difference was observed compared to the healthy control participants. Therefore, GS and GLS1 are promising indicators of a neuronal excitation and inhibition system imbalance and the combined measured parameters are good predictive biomarkers of autism [186].

In organisms, the inorganic nitrogen is assimilated through the GS/GOGAT cycle, in which glutamine and glutamate are further metabolized to form amides and amino acids [187, 188]. In plants, the amino acids glutamine and asparagine play crucial roles in their growth and development because they are the primary nitrogen transport compounds within the plant. However, growth conditions may determine the relative levels of glutamine and asparagine synthesized. In light-grown plants, glutamine is synthesized and transported to serve as the nitrogen donor in essentially all metabolic reactions, especially in chlorophyll biosynthesis. Yet, in dark-grown plants with limited photosynthetic carbon, asparagine, which has a higher N/C ratio than glutamine, is synthesized preferentially to transport nitrogen (Scheme 7.22). The key enzymes involved in this pathway are GS, AS (EC 6.3.5.4), and GOGAT [189]. Asparagine plays primary role in nitrogen recycling, storage, and transport in developing and germinating seeds, as well as in vegetative and senescence organs. The asparagine amide group can be liberated by the

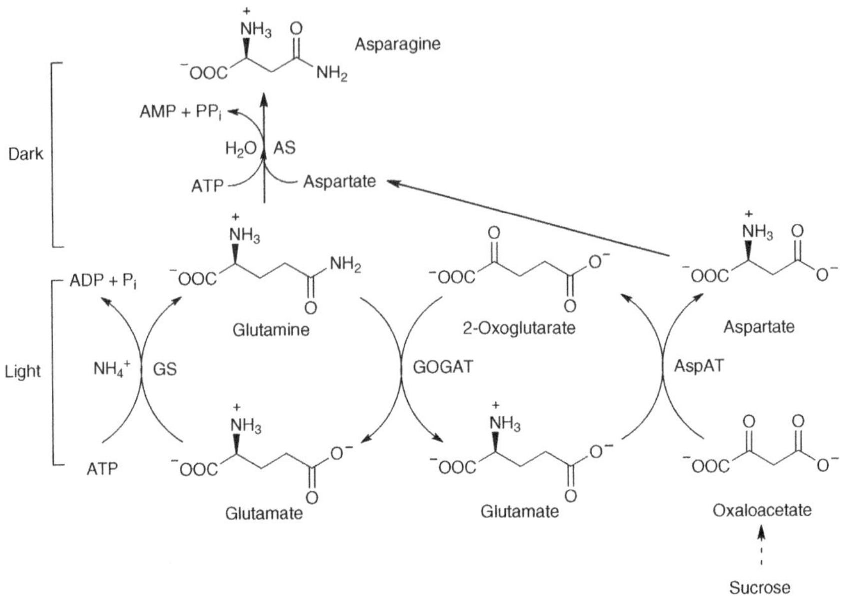

Scheme 7.22 Biosynthetic pathway of nitrogen assimilation into glutamine, asparagine, and aspartate using GS/GOGAT cycle.

reaction catalyzed by asparaginase (ASPG) to generate aspartate, and also the amino group of asparagine can be released by asparagine aminotransferase (AsnAT) for use in the biosynthesis of amino acids. In addition, the GS/GOGAT cycle can be utilized for the synthesis of aspartate through the amino group transfer reaction between glutamate and oxaloacetate catalyzed by aspartate aminotransferase (AspAT) [190].

Theanine, γ-glutamylethylamide, is an amino acid found in green tea leaves (*Camellia sinensis*), which has increased in demand not only as a taste-enhancing additive but also as a supplement to improve and/or maintain human health. GS has been isolated from the cell extract of *Pseudomonas taetrolens* Y-30, which exhibits a molecular mass 660 kDa by gel filtration chromatography and 55 kDa by sodium dodecyl sulfate-polyacrylamide gel electrophoresis (SDS-PAGE). Under the standard reaction condition for glutamine formation (with Mg^{2+}), this GS showed 7 and 1% reactivity toward methylamine and ethylamine, respectively, to that of ammonia. However, under an optimum reaction condition with Mn^{2+}, the isolated GS catalyzed the formation of theanine from glutamic acid and ethylamine with a reactivity of 7% to the reactivity of ammonia in a mixture containing the alcoholic fermentation system of baker's yeast for ATP regeneration

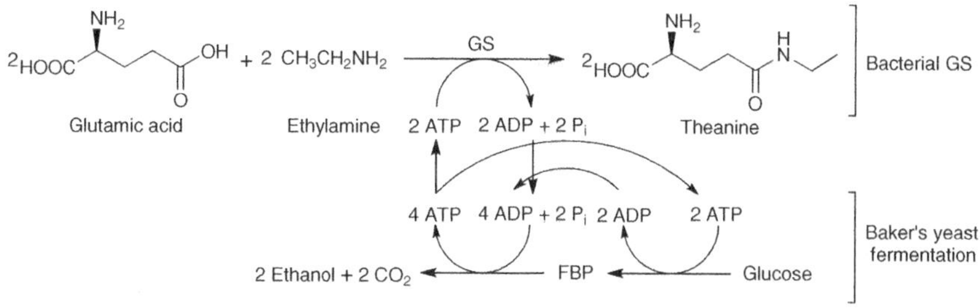

Scheme 7.23 Theanine production by coupling the reaction of GS with glucose fermentation of baker's yeast cell as an ATP-regeneration system. *Sources:* Based on Yamamoto et al. [191, 192].

(Scheme 7.23) [191, 192]. Later study showed that high concentrations of ethylamine enhanced theanine formation, whereas it inhibited yeast fermentation of sugar, and the two contrary effects of ethylamine caused a high yield of theanine based on glucose consumed. Therefore, in an improved reaction mixture, approximately 170 mM theanine was produced in 48 hours reaction period [192]. Furthermore, the enzyme productivity of recombinant GS by cloning and expression of GS gene from *P. taetrolens* Y-30 in *E. coli* was 30-fold higher than that in *P. taetrolens* Y-30. Recombinant GS had the same properties as those of unadenylated intrinsic GS and formed theanine in the mixture of coupled fermentation with energy transfer [193].

7.3.2 L-Alanine-L-anticapsin Ligase

In enzymology, L-alanine-L-anticapsin ligase, characterized from the bacterium *Bacillus subtilis*, belongs to L-amino-acid ligases (LALs, created as EC 6.3.2.28 and transferred in 2015 to EC 6.3.2.49), which is involved in the biosynthesis of the nonribosomally synthesized dipeptide antibiotic bacilysin from L-alanine and L--anticapsin in an ATP-dependent way that requires Mg^{2+} or Mn^{2+} for activity (Scheme 7.24). It is also named as BacD and is a member of ATP-grasp enzyme superfamily (ATP-dependent carboxylate-amine ligases) that catalyzes the formation of an α-peptide bond between two L-amino acids in an ATP-dependent manner [194, 195].

Bacilysin, a nonribosomally synthesized dipeptide antibiotic, was first discovered as early as 1946 from a strain of *B. subtilis* that causes partial lysis of growing cultures of *Staphylococcus aureus* [196, 197]. The exact molecular structure of bacilysin was assigned as L-alanyl-β-(2,3-epoxycyclohexanonyl)-L-alanine in 1965 with the C-terminal amino acid identical to anticapsin produced by *Streptomyces griseoplanus*, which inhibits the formation of hyaluronic acid

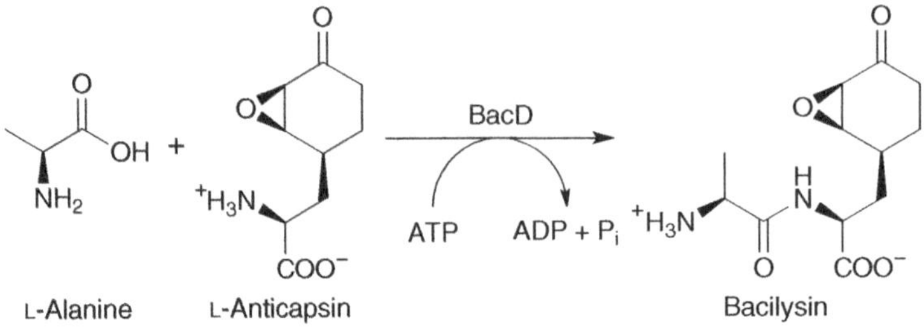

Scheme 7.24 Biosynthesis of the antibiotic bacilysin from L-alanine and the non-proteinogenic amino acid L-anticapsin catalyzed by BacD.

capsule of group A Streptococci [198–201]. The antibacterial action of bacilysin relies on its hydrolysis by intracellular peptidases to anticapsin in the host cells, which inhibits glucosamine 6-phosphate (GlcN6P) synthase [202–204]. Bacilysin had long been thought to be synthesized by a multifunctional NRPS in the genus *Bacillus*; however, the first cell-free synthesis of bacilysin was in 1987 using a reaction mixture containing L-alanine, L-anticapsin, ATP, and Mg^{2+}, as well as *B. subtilis* A14 cell-free extract [205]. Enzyme for the biosynthesis of bacilysin was partially purified from *B. subtilis* PY79 and characterized as enzyme fraction of ca. 125 kDa [206]. Later, the genes *ywfBCDEF* encoding the biosynthesis of bacilysin and its antibiotic constituent anticapsin were isolated from several strains of *B. subtilis* and *Bacillus amyloliquefaciens* as well as *Bacillus pumilus*. The *ywfBCDEF* genes of *B. subtilis* 168 were confirmed to have the core functions of bacilysin biosynthesis and accordingly renamed as *bacABCDE* [207]. In addition, *ywfE* (BacD) of *B. subtilis* was found to code for the activity forming 44 kinds of α-dipeptides from a wide variety of L-amino acids, while no tripeptide or longer peptide was detected. Therefore, *ywfE* encoded enzyme BacD was recognized as a member of the superfamily of LAL that is responsible for the ligation of L-Ala to the amino group of L-anticapsin as an ultimate step of bacilysin biosynthesis [207, 208]. The major function of bacilysin produced by *B. subtilis* is its antimicrobial activity. However, it has been found that the production of bacilysin also exhibits functions in sporulation, germination, and outgrowth of their producers [196, 197, 209]. In addition, addition of the rare earth element scandium to the growth medium can stimulate the production of bacilysin [210].

The first X-ray crystal structures of BacD from *B. subtilis* were determined in native and selenomethionine-derivatized (SeMet) crystals formed in the presence of ADP, $MgCl_2$, and the dipeptide L-Ala-L-Gln at 1.9 and 2.8 Å resolution, respectively. The molecular mass estimated from SDS-PAGE analysis is approximately 55 kDa; furthermore, the enzyme forms an active dimer state (~105 kDa) in solution [211]. The crystal structure of BacD in complex with ADP and an intermediate analogue, phosphorylate phosphinate L-alanyl-L-phenylalanine, refined to 2.5 Å were also determined [212]. BacD shows surprisingly broad substrate specificity, particularly, including the clinical drug Ala-Gln, which demonstrates its application potential. In general, BacD prefers nonbulky and neutral amino acids (e.g. Ala, Ser, Gly) as the N-terminal substrate of the final products, while bulky and neutral amino acids (e.g. Phe, Met, Leu) are more preferred as the C-terminal substrate [208]. The crystal structure information of an enzyme provides us to survey its catalytic mechanism. It appears that the catalytic mechanisms of most known LALs are similar. The proposed reaction mechanism of BacD is summarized as that the C-terminal amino acid binds to the inner side of the substrate-binding pocket, and the N-terminal amino acid (L-alanine) then binds to the lid of the pocket, which first interacts with ATP-bound state by forming complex with two Mg^{2+} ions. The ATP molecule is located in the ATP-grasp domain consisting of two α/β subdomains. The nucleophilic carboxyl oxygen atom attacks the phosphorus atom in the γ phosphate group. Once hydrolyzed, the γ phosphate groups of ATP are transferred to the N-terminal amino acid to form the acyl phosphate intermediate. Then, the amino nitrogen of C-terminal amino acid in free-base state attacks the carbonyl carbon atom of the acyl phosphate intermediate. Finally, the phosphate group is released with depriving proton from the amide group of the dipeptide product [194, 212–214].

Another implication of the crystal structure information and enzymatic characterization is the manipulation of substrate specificity for the enzyme. Previously characterized auxotrophic mutants of *B. subtilis* were analyzed with respect to their ability to synthesize bacilysin, which demonstrated that L-anticapsin moiety of bacilysin is formed from prephenate, an intermediate in the aromatic amino acid biosynthetic pathway (Scheme 7.25) [215]. Subsequently, the *bac* operon, also

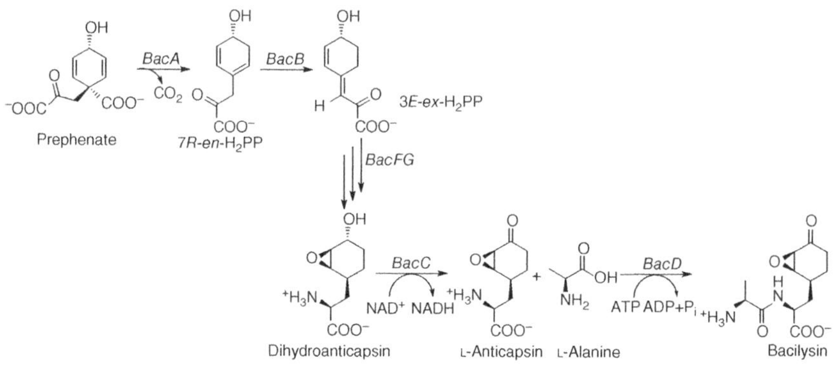

Scheme 7.25 Hypothesized biosynthetic pathway from prephenate to bacilysin. *Source:* Modified from Hilton et al. [215].

referred to as the *ywfBCDEF* gene cluster, was confirmed to have the core functions of bacilysin biosynthesis and accordingly renamed as *bacABCDE*. Extensive genetic analysis from several strains of *B. subtilis* suggests that the bacABC gene cluster encodes all the proteins for the synthesis of the epoxyhexanone ring of L-anticapsin [207, 216, 217]. In one study, cell-free extracts from *B. subtilis* PY79 were prepared and employed for *in vitro* reaction by substituting L-anticapsin for furfurylamine, norephedrine, glycine, and benzylamine, respectively. The enzyme was shown to react furfurylamine with L-alanine only to produce 2-amino-*N*-(2-furylmethyl)propanamide (AFPA) (Scheme 7.26) [218].

Engineered LAL enzymes have been considered to catalyze the biosynthesis of special dipeptides of industrial importance. The first successful production of the L-Ala-L-Gln dipeptide was reported by direct fermentation using one recombinant *E. coli* harboring the wild-type *B. subtilis* LAL expressed under a stationary-phase-specific promoter. Metabolic manipulations to the host were also carried out, including the knockout of the dipeptide-degrading protease genes and the increase of the pool of substrate amino acids by manipulating glutamine synthesis and providing high levels of alanine dehydrogenase [219]. The extracellular concentrations of L-Ala-L-Gln produced were more than 100 mM in fed-batch cultivation on glucose-ammonium salt medium without added alanine and glutamine. Except the side product L-Ala-L-Ala dipeptide, no other dipeptides or longer oligopeptides were formed. Based on the structure data of BacD, point mutant of residues that are considered to participate in the recognition of L-Ala and measured their ATPase activity was generated. The results demonstrated that Trp332 of BacD is crucial for N-terminal substrate specificity while Arg328 is significant for enantioselective recognition of L-amino acids [220]. This work provides guidelines to change the substrate specificity of LALs to catalytically synthesized desirable dipeptides useful for industrial production by site-directed mutagenesis. Therefore, engineered LAL BL00235, an L-amino acid ligase from *Bacillus licheniformis* NBRC12200, was used for the production of salt-taste-enhancing dipeptide Met-Gly without the by-product Met-Met. This has been accomplished by replacing Pro85 residue with bulky aromatic amino acids Phe, Tyr, and Trp by site-directed mutagenesis [221].

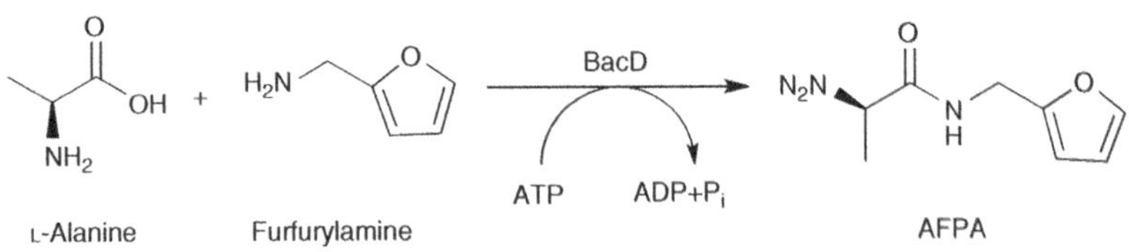

Scheme 7.26 Enzymatic production of 2-amino-*N*-(2-furylmethyl)propanamide (AFPA) catalyzed by BacD. *Source:* Modified from Tüzün et al. [218].

A critical limiting factor for the application of LALs to synthesize desirable dipeptides useful for industrial production is the sufficient supply of high-cost ATP, which is required for the formation of an amide bond catalyzed by LALs. Therefore, an economic and efficient ATP regeneration and recycling system is critical for the industrial application of LALs instead of the cost-inefficient direct addition of ATP. Several ATP regeneration systems have been developed; polyphosphate kinase (PPK) might be the most widely used biocatalyst for the regeneration of ATP, which catalyzed the degradation of polyphosphate coupled with the phosphorylation of ADP to ATP [222, 223]. Therefore, a recombinant *E. coli* strain was constructed that was capable of heterogeneously and simultaneously expressing wild-type LAL from *B. subtilis* and PPK from *Rhodobacter spaeroides*. Then, the pretreated cells were directly applied for the catalytic synthesis of L-dipeptides, in which the highest titer was for Ala-Met with the final yield of $28\,g\,L^{-1}$ and a molar conversion rate of 63.95% in a 27 hour reaction [194].

7.3.3 Carbapenam-3-carboxylate Synthase

Naturally occurring β-lactam compounds can be divided into four different subgroups based on the origin of their four-membered β-lactam ring: the penicillins/cephalosporins, the clavams, the carbapenem, and the monocyclic β-lactams. Carbapenam-3-carboxylate synthase (EC 6.3.3.6) has synonyms such as CarA, carbapenam synthetase (CPS), or carbapenam-3-carboxylate ligase, which is a cyclo-ligase involved in the biosynthesis of the carbapenem β-lactam antibiotics (5*R*)-carbapen-2-em-3-carboxylate via its catalytic ability for the ATP-dependent cyclization of (2*S*,5*S*)-5-carboxymethyl proline (5-CMP) to yield the (3*S*,5*S*)-carbapenam-3-carboxylate (Scheme 7.27) in the bacterium *Pectobacterium carotovorum* [224, 225]. Carbapenem-3-carboxylic acid, the simplest of the carbapenem compounds, is produced by several species of Enterobacteriaceae. The more

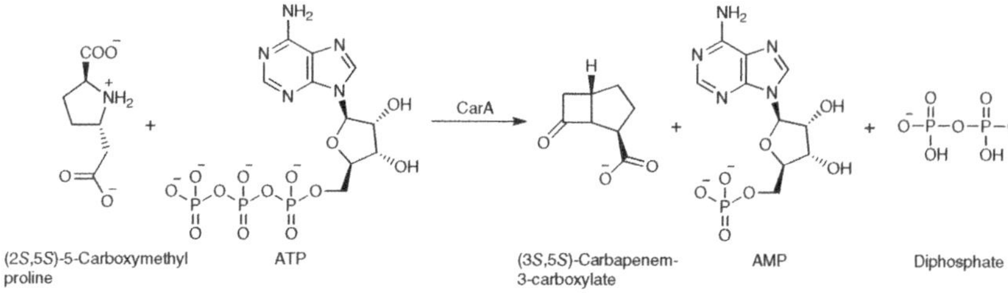

Scheme 7.27 ATP-dependent formation of carbapenam-3-carboxylate catalyzed by carbapenam-3-carboxylate synthase (CarA) from 5-carboxymethylproline.

complex carbapenem compounds such as olivanic acid and thienamycin are produced by *Streptomyces* spp.

The gene clusters for carbapenem-3-carboxylate biosynthesis in *P. carotovorum* (*Erwinia carotovora*) have been isolated and analyzed to reveal the biosynthetic pathway and enzymes. The analysis results demonstrate that the products of the first five genes of the operon are involved in the synthesis of the carbapenem molecule. Three of these, *carABC*, are absolutely required for the biosynthesis of (5*R*)-carbapen-2-em-3-carboxylate [226, 227]. Whereas gene *carA* encodes the enzyme CarA (CPS), which catalyzes the ATP-dependent formation of bicyclic (3*S*,5*S*)-carbapenam-3-carboxylic acid from the preexisting saturated five-membered ring (2*S*,5*S*)-5-carboxymethyl proline as shown in Scheme 7.26, it was purified in recombinant form from *P. carotovorum* [228, 229]. The crystal structure of CarA was reported as a tetramer that is characterized by (i) the N-terminal domain consisting of two antiparallel β-sheets forming a sandwich, flanked on each side by (relatively) short α-helices; (ii) containing C-terminal domain comprising (11–14) α-helices surrounding a five-stranded parallel β-sheet; and (iii) the active site including the ATP-binding site and substrate-binding pocket located in a cleft in the *C*-terminal domain formed by four β-strands and five α-helices [230].

From the crystal structure, the catalytic mechanism of CarA was also investigated, which involves the ordered binding of ATP/Mg^{2+} and 5-CMP to apoenzyme. The substrate-carboxyl group is first activated, *via* ATP-mediated adenylation, to form an acyl-adenylate intermediate followed by intramolecular nucleophilic attack by secondary amine to give the β-lactam product, likely *via* a tetrahedral intermediate. Pyrophosphate is the last product to be released [229–231]. Further kinetic studies on the mechanism proposed that the rate-limiting step is the β-lactam ring formation coupled to a protein conformational change and Lys443 residue in the active site is an essential residue for substrate binding and intermediate stabilization [232]. Moreover, the kinetics of CarA double-site mutants revealed that a conserved Tyr-Glu catalytic dyad in the active site functions in the deprotonation of the internal nucleophile in β-lactam formation [233].

Members of the carbapenem family and their derivatives are important among the β-lactam antibiotics for both their broad spectrum of antibiotic activity and their relative resistance to most clinically encountered β-lactamases. More than 60 naturally occurring carbapenem β-lactam antibiotics have been isolated, among these, (5*R*)-carbapenem-3-carboxylic acid is the simplest one. It has been demonstrated that the biosynthesis of (5*R*)-carbapenem-3-carboxylic acid can be efficiently carried out through only three enzymes, CarA, CarB, and CarC (carbapenem synthase), by *P. carotovorum* and *Serratia marcescens* (Scheme 7.28) [228, 229, 234–236]. In this reaction process, 5-carboxymethyl proline (**2**) was produced enzymatically by CarB from pyrroline 5-carboxylic acid (**1**) followed by

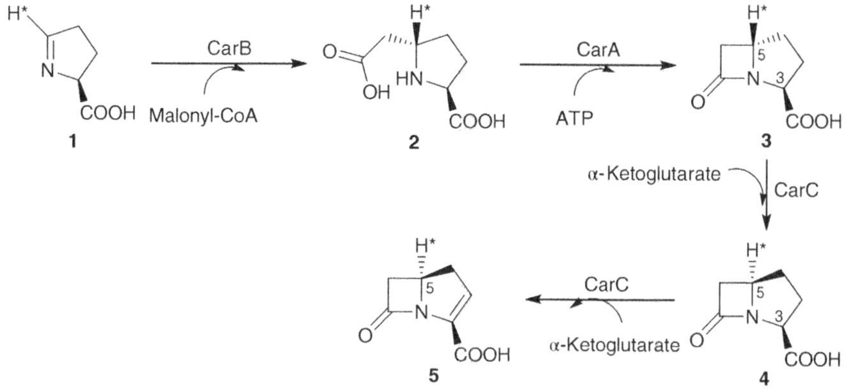

Scheme 7.28 Biosynthetic pathway of (5*R*)-carbapen-2-em-3-carboxylic acid catalyzed by CarA, CarB, and CarC. *Sources:* Based on Gerratana et al. [228], Bonder et al. [229], Stapon et al. [234], Li et al. [235], and Stapon et al. [236].

intramolecular cyclization to yield (3*S*,5*S*)-carbapenam-3-carboxylic acid (**3**) by CarA. Subsequently, a stereochemical inversion at C5 of **3** to both (3*S*,5*R*)-carbapenam-3-carboxylic acid (**4**) and (5*R*)-carbapen-2-em-3-carboxylic acid (**5**), which proceeds by loss of hydrogen labeled at that stereocenter catalyzed by CarC. Note that **3** always presents in low amounts in the final reaction mixture, which functions as an intermediate in the production of **5** [234, 235, 237]. In addition, the epimerization of **3** to **4** and subsequent desaturation of **4** to **5** are both α-ketoglutarate dependent [229, 234, 238].

Studies of the substrate specificities of CarA have shown that CarA catalyzes the *in vitro* ring closure of at least three of the four possible stereoisomers of 5-CMP to give the corresponding carbapenams (Scheme 7.29) [239], *t*-carboxyethyl proline to the corresponding γ-lactam, and some α-amino diacids such as glutaric acid, α-aminoadipic acid, and α-aminopimelic acid to the corresponding lactams [228]. Instead of five-membered *N*-heterocyclic ring carbapenams, β-lactam bicyclic systems containing six- and seven-membered *N*-heterocyclic ring analogues of 5-CMP, i.e. (2*S*,6*S*)-6-carboxymethyl-pipecolic acid (5-CMPi) and (2*S*,7*S*)-7-(carboxymethyl)-azepane-2-carboxylic acid, respectively, can also be synthesized by CarA [240]. Note that the functionalized five-, six-, and seven-membered *N*-heterocycles of 5-CMP analogues were prepared by wild-type and variant carboxymethylproline synthases (CarBs), members of crotonase superfamily enzymes, from corresponding amino acid aldehydes and (alkylated) malonyl-CoA derivatives. Functionalized 5-CMP derivatives methylated at C2, C3, C4, or C5 of the proline ring from appropriately substituted amino acid aldehydes and malonyl-CoA can be prepared by wild-type and engineered CarBs as

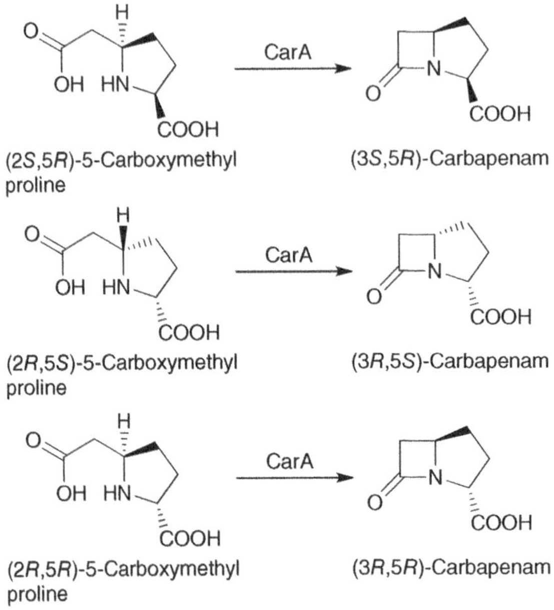

Scheme 7.29 CarA catalyzed conversion of three different stereoisomers of 5-carboxymethyl proline into corresponding carbapenams. *Source:* Sleeman et al. [239].

well. Subsequently, these substituted 5-CMP derivatives were further converted into the corresponding bicyclic β-lactams using CarA [241]. The stereoselectivity of CarA was again demonstrated herein, for example, in a 2 : 1 mixture of (7*R*)-7-methyl-5-CMPi and (7*S*)-7-methyl-5-CMPi; CarA selectively catalyzed the conversion of the former to give (4*S*,6*S*,7*R*)-7-methyl-carbacepham [240]. Further, in a 1 : 1 reaction mixture of (4*R*)-4-methyl-5-CMP and (4*S*)-4-methyl-5-CMP, CarA preferentially catalyzed (1 : 2) conversion of the latter to give (1*S*,3*S*,5*S*)-1-methyl-carbapenam [241].

The amenability of engineered enzymes for carbapenem biosynthesis shown above further promoted the stereoselective production of dimethyl-substituted carbapenams via engineered CarBs, which functionalized 5-CMP derivatives were methylated at two positions (i.e. C2/C6, C3/C6, or C5/C6), including products with a quaternary center, from appropriately substituted amino acid aldehydes and epimeric methylmalonyl-CoA. These disubstituted 5-CMPs were then served as substrates and converted by CarA into corresponding methylated bicyclic β-lactams, which obviously improved the hydrolytic stability as compared to the unsubstituted carbapenams (Scheme 7.30) [242]. Furthermore, engineered CarBs, solely, and in tandem with an alkylmalonyl-CoA-forming enzyme

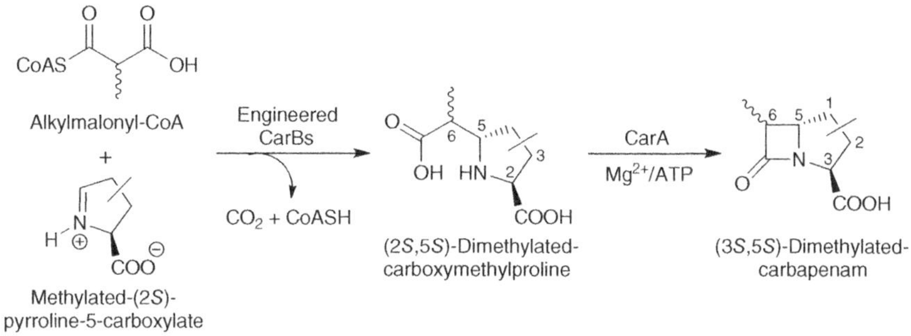

Scheme 7.30 Conversion of disubstituted-5-CMP derivatives into disubstituted-3-carboxylate derivatives by CarA through engineered CarBs. *Source:* Hamed et al. [242].

(i.e. malonyl-CoA synthetase or crotonyl-CoA carboxylase reductase) is used to produce 4,6-disubstituted-5-CMP stereoisomers, i.e. products with three contiguous chiral centers. Then, some of the prepared 4,6-disubstituted-5-CMP derivatives are further converted to bicyclic β-lactams by CarA [243].

7.4 Ligases for Carbon–Carbon Bonds Formation

7.4.1 Pyruvate Carboxylase

Pyruvate carboxylase (PYC, EC 6.4.1.1) is a member of the biotin-containing enzyme family that catalyzes the HCO_3^- and MgATP-dependent carboxylation of pyruvate to form oxaloacetate (Scheme 7.31) [244, 245]. The systematic name of PYC is pyruvate : carbon dioxide ligase (ADP-forming), which is also known as pyruvic carboxylase. PYC was first described in 1959 in the course of studies on the intracellular distribution of enzymes involved in the OAA shuttle and its relationship to gluconeogenesis in chicken liver [246, 247]. Subsequently, PYC has been found in a variety of prokaryotes and eukaryotes including fungi, bacteria, plants, and animals [246, 248]. PYC found in prokaryotes includes *Pseudomonas*,

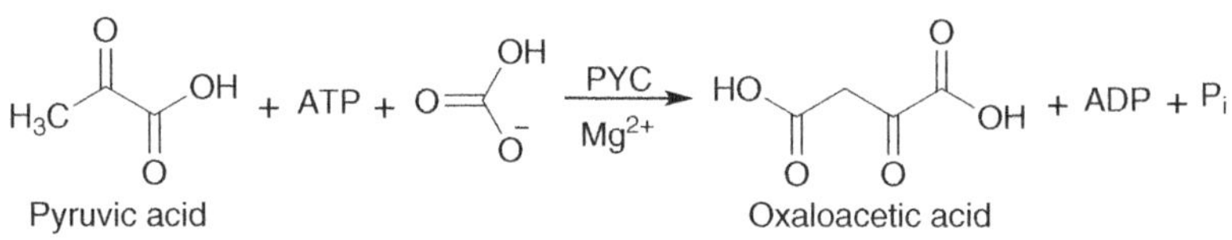

Scheme 7.31 Pyruvate carboxylase catalyzed MgATP-dependent carboxylation of HCO_3^- and pyruvate to form oxaloacetate. *Sources:* Based on Attwood [244] and Scrutton [245].

Bacillus, *Rhizobium*, *Mycobacterium*, *Staphylococcus*, and archaea but not in the enteric bacteria that can synthesize oxaloacetate directly from phosphoenolpyruvate (PEP) using phosphoenolpyruvate carboxylase (PEPC) [249–253]. In yeast, such as *S. cerevisiae*, two principal metabolic pathways leading to the production of oxaloacetate are the PYC-catalyzed reaction and the glyoxylate cycle. During growth on acetate, yeast PYC-catalyzed OAA formation is repressed but the glyoxylate cycle is active, while if yeast is grown on glucose of minimal medium, the enzymes of glyoxylate cycle are repressed [254–257]. In *E. coli*, the function of PYC is replaced by PEPC. Thus, the expression of PYC can improve recombinant protein production in *E. coli* [258]. In plants, PYC may provide an alternative gluconeogenic pathway to the photosynthetic process during germination. OAA is one of the several important intermediates in the TCA cycle that are withdrawn for use in several biosynthetic pathways. Therefore, the production of OAA from pyruvate and HCO_3^- catalyzed by PYC is considered to play an anaplerotic role for the TCA cycle in numerous biological processes. In mammals, PYC has a crucial role in gluconeogenesis, lipogenesis, glucose-induced insulin secretion by pancreatic islets, and the biosynthesis of neurotransmitters in astrocytes and the mammalian cell lines [246, 259–267]. In mammals, PYC is expressed in a tissue-specific manner, with its activity found to be the highest in liver and kidneys (gluconeogenic tissues), in adipose tissue and lactating mammary gland (lipogenic tissues), and in pancreatic islets. Activity is moderate in brain, heart, and adrenal gland and least in white blood cells and skin fibroblasts [248].

PYC found in most organisms ranges from bacteria, fungi, and invertebrates to vertebrates comprising four identical α_4 subunits, each approximately 120–130 kDa. Biotin is covalently attached to a specific lysine residue located ~35 residues from the C-terminus [248]. The three functional domains of this α_4-type PYC, i.e. biotin carboxylase (BC), carboxyl transferase (CT), and biotin carboxyl carrier protein (BCCP) domains, are all located on a single polypeptide chain. The majority of α_4-type PYCs are subject to allosteric regulation. Acetyl-CoA is a powerful activator, whereas L-aspartate is an inhibitor, of α_4-type PYC. However, PYC found in archaea, such as *Methanobacterium* sp., *Methanococcus* sp., and *Methanosarcina* sp., and certain bacteria, i.e. *Aquifex aeolicus* and *Pseudomonas* spp., is in the $\alpha_4\beta_4$ form [246, 253, 268–270]. In the $\alpha_4\beta_4$ form, each subunit is made up of two polypeptide chains, the ~55 kDa non-biotinylated α polypeptide chains, which possesses the BC activity and the ~70 kDa β polypeptide chain, which carries the biotin and also contains the CT activity. The $\alpha_4\beta_4$ type of PYC has an activity that is completely independent of acetyl-CoA [253, 271]. From the two most complete X-ray crystal structures available, an asymmetric and symmetric form of the protein have been visualized. The X-ray crystal structure for *S. aureus* PYC, in complex with the activator coenzyme A, exhibits a highly symmetrical tetramer with the active sites nearly equidistant on both faces of the tetramer, which also has been confirmed by cryo-electron microscopy (EM) studies [249, 272]. In contrast, the tetrameric structure of PYC from wild-type *Rhizobium etli* complexed with the nonhydrolyzable allosteric activator analogue, ethyl-CoA, is highly asymmetrical, with the active sites on the top face of the tetramer positioned in proximity to those at the bottom face [251, 273].

In α4 PYCs, the BC domain is at the N-terminus of the polypeptide chain, where biotin is carboxylated using bicarbonate as a substrate in a reaction requiring free Mg^{2+} for the cleavage of MgATP to form carboxybiotin, MgADP, and Pi. The CT domain contains a tightly bound transition-metal ion and is the central domain where the carboxy group from carboxybiotin is transferred to pyruvate to produce oxaloacetate. The BCCP domain is located at the C-terminus of the polypeptide chain containing the lysine residue to attach the biotin [246].

Based on the structural information, the reaction mechanism catalyzed by PYC can be divided into two partial reactions as shown below [248, 274]:

$$\begin{aligned} &\text{MgATP} + \text{HCO}_3^- + \text{ENZ} - \text{biotin} \xrightarrow{\text{Mg}^{2+},\ \text{acetyl-CoA}} \text{MgADP} \\ &\quad + \text{Pi} + \text{ENZ - biotin - CO}_2^- \end{aligned} \tag{7.1}$$

$$\text{ENZ - biotin - CO}_2^- + \text{Pyruvate} \rightarrow \text{ENZ - biotin} + \text{Oxaloacetate} \tag{7.2}$$

The first partial reaction (reaction 7.1) is the ATP-dependent biotin carboxylation, which is thought to proceed in two steps. With the first step, bicarbonate is activated by ATP to form a carboxyphosphate intermediate. Then, the second step involves the carboxylation of the 1′-nitrogen of covalently attached biotin either directly by carboxyphosphate or more likely by CO_2, which is formed via decarboxylation of the carboxyphosphate intermediate. The enol form of biotin, in which the 1′-nitrogen is much more nucleophilic than the keto form, is the carboxy-group acceptor. The second partial reaction (reaction 7.2) is the formation of oxaloacetate, which involves the carboxyltransfer to pyruvate (transcarboxylation). This reaction appears to proceed via a stepwise mechanism with proton-transfer steps flanking the central carboxyl-transfer step between carboxybiotin and the enol form of pyruvate [244, 246, 274, 275].

The synthesis of OAA from C-3 glycolysis intermediates plays a critical role in glucose metabolism, because not only is it a pivotal component in TCA cycle but also it serves as an intermediate for the biosynthesis of the aspartate family of amino acids (lysine, threonine, methionine, and isoleucine). Therefore, OAA must be replenished during glucose metabolism to make the TCA cycle continuously unabated and can be utilized for the synthesis of cellular metabolites (Scheme 7.32a). During anaerobic glucose metabolism, OAA also serves as the first metabolite of the reductive arm of the TCA cycle, which leads to the formation of the dicarboxylic acids malate, fumarate, and succinate (Scheme 7.32b).

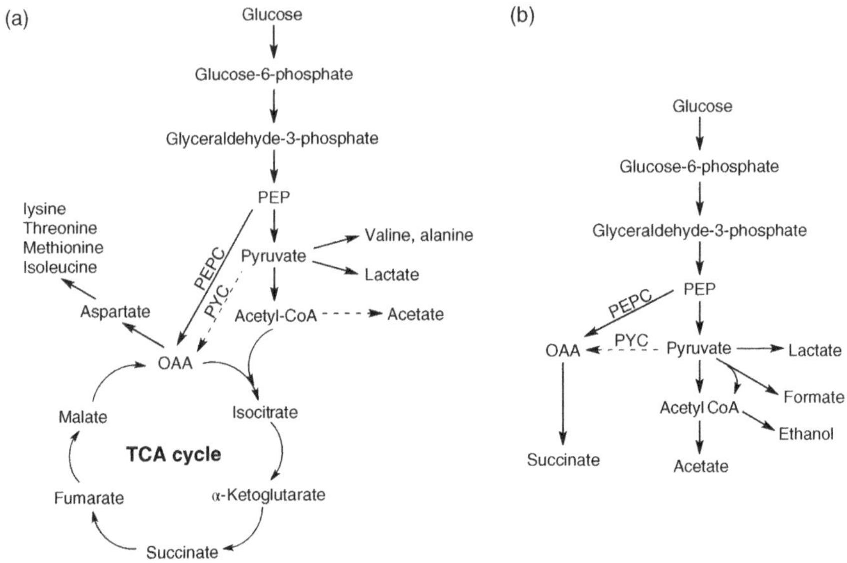

Scheme 7.32 Glucose metabolism in *E. coli* expressed with *R. etli* PYC. (a) Aerobic condition. (b) Anaerobic condition.

The primary routes used for synthesizing OAA by organisms during glucose metabolism are the ATP-dependent carboxylation of PEP using PEPC or the ATP-dependent carboxylation of pyruvate using PYC [276]. Some microorganisms, such as *R. etli* and *C. glutamicum*, are capable of synthesizing OAA during growth on glucose via either of the enzymes PYC or PEPC. However, other microorganisms, including *E. coli*, due to their lack of PYC utilize only PEPC to synthesize OAA during growth on glucose. While the expressed *R. etli* PYC not only can restore the growth of an *E. coli ppc* null mutant in minimal glucose medium, but also can increase carbon flow toward OAA in wild-type *E. coli* cells without affecting the glucose uptake rate or the growth rate. Therefore, the expression of PYC resulted in a 56% increase in biomass yield and a 43% decrease in acetate yield during aerobic glucose metabolism, as well as caused a 2.7-fold increase in succinate concentration, making it a major product by mass during anaerobic glucose metabolism [276].

Succinic acid (succinate) and its derivatives have drawn much interest because they are widely used as a precursor of numerous specialty chemicals and biodegradable plastics for applications in foods, pharmaceuticals, and cosmetics [277, 278]. Therefore, to elucidate the biochemical pathways used for succinate accumulation, both *E. coli* NZN111, which lacks activities for pyruvate-formate lyase (PFL) and lactate dehydrogenase, and AFP111, a derivative that contains an additional mutation in *ptsG* encoding an enzyme of the glucose phosphotransferase system, were engineered to introduce plasmid pTre99A-*pyc*, which expresses the *R. etli* PYC. Then, the growth, substrate consumption, product formation, and activities of seven key enzymes (acetate kinase, fumarate reductase, glucokinase, isocitrate dehydrogenase [ICDH], isocitrate lyase [ICL], PEPC, and PYC) from glucose for four strains of *E. coli* (NZN111, NZN111/pTrc99A-*pyc*, AFP111, and AFP111/pTrc99A-*pyc*) were compared under both exclusively anaerobic and dual-phase condition (an aerobic growth phase followed by an anaerobic production phase). The highest succinate mass yield was found with the strain AFP111/pTrc99A-*pyc* under dual-phase conditions with low PYC activity [279]. Furthermore, two strains AFP111 and AFP111/pTrc-pyc were aerobically fermented to catalog physiological states during aerobic growth that might influence succinate generation in the anaerobic phase, and activities of six key enzymes were also determined for these aerobic fermentation. The six transition times based on physiological states were selected for studying dual-phase fermentations. The results showed that the final succinate yield and productivity depend greatly on the physiological state of the cells at the time of transition. A final succinic acid concentration of 99.2 g L^{-1} with an overall yield of 110% and productivity of 1.3 g $L^{-1} h^{-1}$ was obtained under the best transition time [280].

A genetically engineered *E. coli*, SBS110MG, has been constructed by deleting two crucial enzymes in central anaerobic pathway, lactic dehydrogenase (LDH) and alcohol dehydrogenase (ADH), and the heterogeneous PYC has been overexpressed from *Lactococcus lactis* to achieve high succinate yield and productivity due to diverting maximum quantities of NADH for succinate synthesis (Scheme 7.33) [281]. A 1.3 yield (mol of succinate/mol of glucose) was obtained using this mutant, which exceeded the maximum theoretical yield of succinate under anaerobic conditions in terms of NADH balance. An additional mutation by eliminating the *ptsG* system was able to further increase the succinate molar yield from glucose to 1.4 mol mol^{-1}. Furthermore, a recombinant *E. coli* strain, SBS550MG, was constructed by deactivating *adhE*, *ldhA*, and *ack-pta* from the central metabolic pathway and by activating the glyoxylate pathway through the inactivation of *iclR*, which encodes a transcriptional repressor protein of the glyoxylate bypass. The activation of these genes in SBS550MG increased the succinate yield from glucose to about 1.6 mol mol^{-1} with an average anaerobic productivity rate of 10 mM h^{-1} (~0.64 mM h^{-1}-OD_{600}) [282]. These mutant strains were selected for further characterization and compared to the wild-type strain to better understand their metabolic function as a result of several genetic manipulations and to determine the significance of the fermentative and the glyoxylate pathways in the production of succinate. Metabolic fluxes including critical branch point flux split ratios were measured under batch cultivation conditions. The most favorable split ratio to obtain the highest succinate yield was the

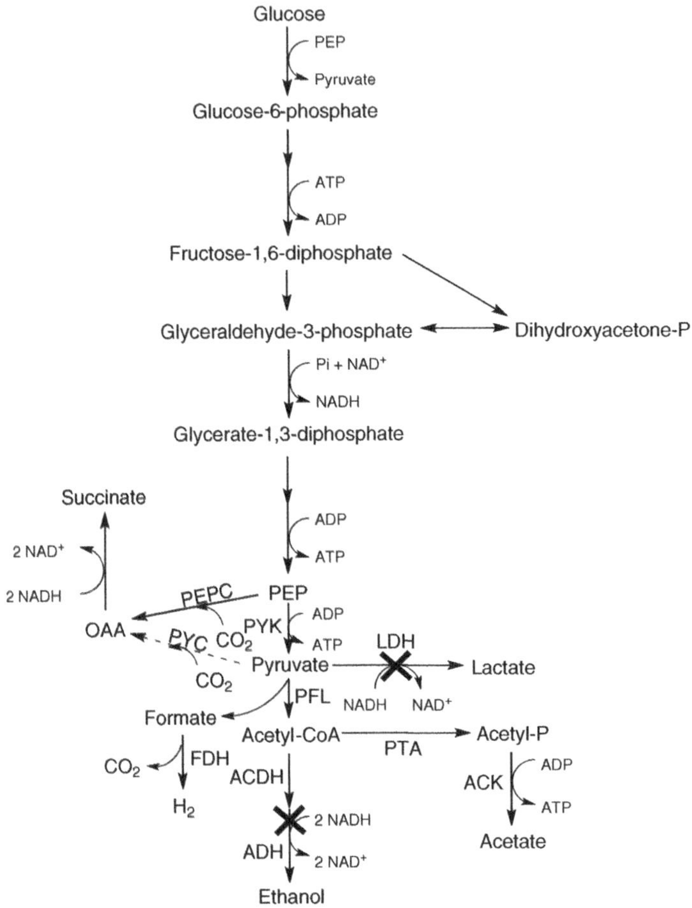

Scheme 7.33 Central anaerobic metabolic pathway of mutant *E. coli* SBS110MG by inactivating enzymes LDH and ADH and overexpression PYC from *L. lactis*. *Source:* Sánchez et al. [281].

fractional partition of OAA to glyoxylate of 0.32–0.68 to the fermentative pathway achieved in strains SBS550MG (pHL413) and SBS990MG (pHL413). The succinate yields obtained in these two strains were 1.6 and 1.7 mol mol^{-1}, respectively [283].

Engineered *E. coli* BA002 by deleting *ldhA* and *pflB* genes cannot utilize glucose anaerobically due to the inability to regenerate NAD$^+$. Overexpression of the rate-limiting enzyme of NAD(H) synthesis pathway, nicotinic acid phosphoribosyltransferase (NAPRTase) encoded by the *pncB* gene, can restore the glucose utilization and result in a significant increase in cell mass and succinate production under anaerobic conditions, while still leads to an accumulation of high concentration pyruvate. Therefore, NAPRTase and PYC from *L. lactis* were co-expressed in recombinant *E. coli* BA016, which gave a 9.8-fold higher total concentration of NAD(H) than in BA002 and a decreased NADH/NAD$^+$ ratio from 0.60 to 0.04 (Scheme 7.34a). Under anaerobic conditions, BA016 consumed 17.5 g L^{-1} glucose and generated 14.08 g L^{-1} succinate with a small quantity of pyruvate [284]. In wild-type *E. coli*, 1 mol of CO_2 was anaerobically fixated in 1 mol of succinic acid generation with the key reaction catalyzed by PEPC such that PEP is carboxylated to form OAA. Although inactivation of PFL and LDH is found to enhance the PEPC pathway for succinic acid production, it results in excessive pyruvic acid accumulation and limits the regeneration of NAD$^+$ from NADH formed in glycolysis. Therefore, co-expression of NAPRTase and PYC in a *pflB*, *ldhA*, and *pepc* deletion strain resulted in a significant increase in cell mass and succinic acid production under anaerobic conditions, that 12.08 g L^{-1} of succinic acid was produced from 14.5 g L^{-1} glucose after 72 hours (Scheme 7.34b). In addition, under optimized condition of CO_2 supply, the succinic acid productivity and the CO_2 fixation rate reached 223.88 and 83.48 mg L^{-1} h^{-1}, respectively [285].

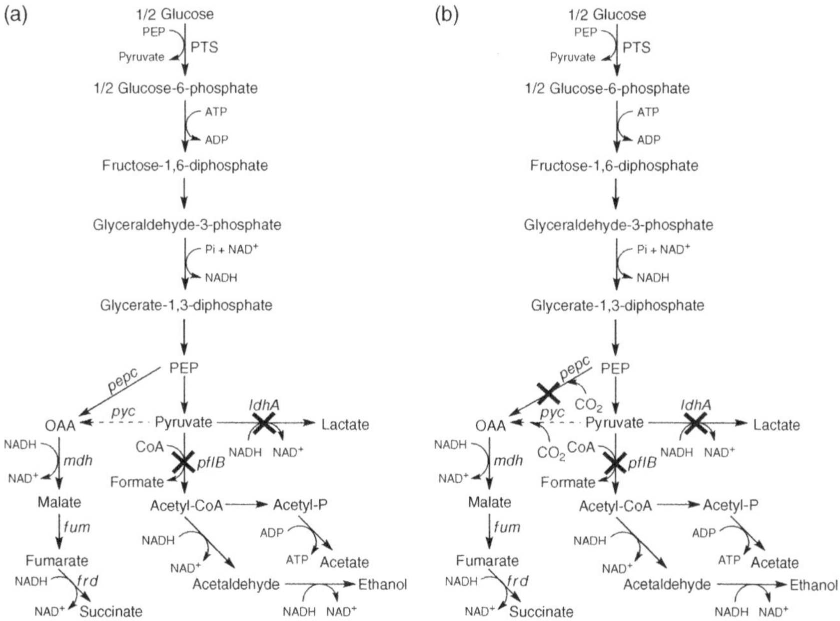

Scheme 7.34 Anaerobic mixed acid fermentation pathway for *E. coli*. (a) Coexpression of NAPRTase and PYC form *L. lactis* in *E. coli* BA016; (b) coexpression of NAPRTase and PYC in a *pflB*, *ldhA*, and *pepc* deletion *E. coli*.

Four-carbon dicarboxylic acids, in particular, malic, fumaric, and succinic acids, are industrially important chemicals that can serve as universal building blocks in the organic synthesis of a wide range of high value-added products. During anaerobic glucose utilization, the parental strain synthesized malic, fumaric, and succinic acids as the main fermentation end products, while pyruvic acid was accumulated as the main by-product resulting from the functioning of the pyruvate-oxaloacetate-malate-pyruvate futile cycle. The mutant strain by deleting individual of the *maeA* and *maeB* genes, encoding NADH- and NADPH-dependent malic enzymes, converted glucose into four-carbon dicarboxylic acids with increased efficiency, while still secreting notable amounts of pyruvic acid. The combined inactivation of both malic enzymes and the heterologous expression of the PYC in a recombinant *E. coli* strain, which exhibits inactivated mixed-acid fermentation pathways and a modified system significantly, elevated the portion of malic, fumaric, and succinic acids among the fermentation end products with a concomitant decrease in the secretion of pyruvic acid and other by-products due to the abolishment of the action of the futile cycle competing with the target biosynthetic processes [286].

Since glycerol is an abundant, low cost, and high degree of reductive carbon source, it has become an ideal feedstock for the microbial production of bio-based chemicals. A metabolically engineered *E. coli* was utilized for the microaerobic production of high value-added succinate from glycerol in minimal salts medium. Succinate production was greatly elevated through (i) blocking pathways for the synthesis of competing by-products lactate, ethanol, and acetate and (ii) expressing *L. lactis* PYC to drive the generation of succinate from the pyruvate node (as opposed to that of PEP). Thus, these metabolic engineering strategies coupled cell growth to succinate production because the synthesis of succinate remained as the primary route of NAD^+ regeneration. This feature also enabled the operation of the succinate pathway in the absence of selective pressure (e.g. antibiotics). Results demonstrated a maximum specific productivity of ~400 mg succinate $(g\,cell)^{-1}\,h^{-1}$ and yield of 0.69 g succinate $(g\ glycerol)^{-1}$, on par with the use of glucose as a feedstock [287].

The fast-growing, non-motile, Gram-positive microorganism, *C. glutamicum*, has long been employed in microbial fermentation industry for amino acids and nucleic acids. Under anaerobic condition, *C. glutamicum* cell growth is arrested, but the cells are capable of metabolizing sugars to form organic acids such as L-lactic, succinic, and acetic acids. A *C. glutamicum* strain that simultaneously deletes *ldhA* gene coding for L-lactate dehydrogenase and overexpresses the *pyc* gene encoding PYC was applied for succinic acid production (Scheme 7.35). Succinic acid was able to be efficiently generated at high cell density anaerobically with intermittent addition of sodium bicarbonate and glucose such that 1.24 M $(146\,g\,L^{-1})$ succinic acid concentration was achieved within 46 hours. The yields

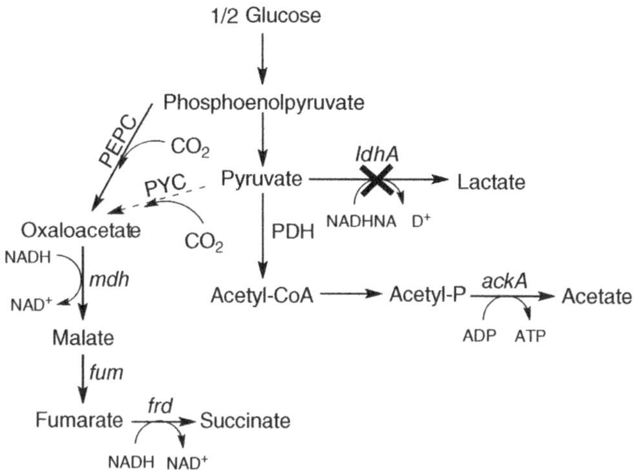

Scheme 7.35 Predominant anaerobic metabolic pathways of *C. glutamicum* for succinic acid production.

of succinic acid and acetic acid from glucose were 1.40 mol mol^{-1} (0.92 g g^{-1}) and 0.29 mol mol^{-1} (0.10 g g^{-1}), respectively. The production rate and yield of succinic acid depended on medium bicarbonate concentration rather than glucose concentration [288].

To eliminate acetate accumulation and the limitation of the yield by the availability of reducing equivalents for anaerobic conversion of OAA to succinate in the reductive TCA cycle using *C. glutamicum* with deleted *ldhA* gene for lactate synthesis, a derivative of the type strain ATCC 13032 (strain BOL-1), which lacked all known pathways for acetate and lactate synthesis, was constructed so that the pyruvate carboxylase gene *pyc* was integrated into BOL-1 to result in strain BOL-2. A fast BOL-2 catalyzed succinate production from glucose with little acetate formation and a yield of 1 mol mol^{-1} was obtained. Furthermore, the *fdh* gene from *Mycobacterium vaccae*, coding for an NAD$^+$-coupled formate dehydrogenase (FDH), was chromosomally integrated into BOL-2 to provide additional reducing equivalents derived from the co-substrate formate, which leads to strain BOL-3. A 20% higher succinate yield from glucose in the presence of formate was achieved by BOL-3 under an aerobic batch reaction. A temporary metabolic blockage of strain BOL-3 was prevented by plasmid-borne overexpression of the glyceraldehyde-3-phosphate dehydrogenase gene *gapA*. In an anaerobic fed-batch process with glucose and formate, strain BOL-3/pAN6-*gap* accumulated 1,134 mM succinate in 53 hours with an average succinate production rate of 1.59 mmol (g dry cells)$^{-1}$ h^{-1}. The succinate yield of 1.67 mol (mol glucose)$^{-1}$ accompanied with very low level of by-products is also one of the highest described currently [289].

The application of metabolically engineered *C. glutamicum* for the anaerobic production of lactic and succinic acids increased at higher temperature for long-term 48 hours reaction. The biomass-specific lactic acid production rate at 42 °C was 8% higher than that at the optimal temperature of 30 °C. In contrast, the biomass-specific succinic acid production rate was highest at 36 °C, 28% higher than that at 30 °C, although the production rate at 42 °C was still 23% higher than that at 30 °C. Among the examined five enzymes involved in the TCA cycle leading to succinic acid synthesis, only the specific activity of PYC was drastically decreased after 48 hours at 42 °C, which shows the *de novo* synthesis of PYC is required for long-term production of succinic acid [290].

Lactic acid bacterium (LAB), *Lactobacillus plantarum* NCIMB 8826, has an incomplete TCA cycle and naturally produces small amounts of succinic acid. Recombinant LABs constructed by overexpression or co-expression of enzymes, PYC, phosphoenolpyruvate carboxykinase (PEPCK), and malic enzyme, in an LDH-deficient strain of *L. plantarum* NCIMB 8826 (VL103) were investigated to examine their effects on succinic acid production. Results revealed that PYC is the key enzyme for succinic acid production by *L. plantarum* VL103, whereas PEPCK is critical for increasing biomass. The highest yield of succinic acid was achieved through co-expression of PYC and PEPCK in *L. plantarum* VL103 with a 22-fold higher amount of succinic acid than the wild type and a conversion rate of 25.3% from glucose [291].

Enterobacter aerogenes strain AJ110637, which rapidly consumes glucose under anaerobic and weakly acidic (pH 5.0) conditions, was selected as a platform for succinate production because bacterium-based succinate fermentation is considered a feasible approach to reduce total production costs. A previous metabolically engineered *E. aerogenes* strain, developed from AJ110637 by inactivating ethanol dehydrogenase and introducing *Actinobacillus succinogenes* PEPCK, generated succinate as a major product from glucose with anaerobic mixed-acid fermentation under weakly acidic conditions (pH < 6.2). The production of succinate by the Δ*adhE*/PEPCK strain was further improved by metabolic engineering to eliminate pathways that produced undesirable products and introduce two carboxylation pathways from PEP and pyruvate to oxaloacetate. The highest production of succinate was with the strain ES04/PEPCK + PYC, in which ethanol, lactate, acetate, and 2,3-butanediol pathways were inactivated and PEPCK and *C. glutamicum* PYC were co-expressed. The production yield of succinate from glucose in a pH-controlled batch culture was over 70% ($g\,g^{-1}$) without any measurable formation of ethanol, lactate, or 2,3-butanediol under weakly acidic conditions (pH 5.5) [292]. Because yeast strain *S. cerevisiae* has high tolerance toward acidity, it is also a suitable platform for succinic acid production. An engineered strain by modifying pathway for succinate production was established, which could produce up to $6.17 \pm 0.34\,g\,L^{-1}$ of succinate. The succinate titer was further improved

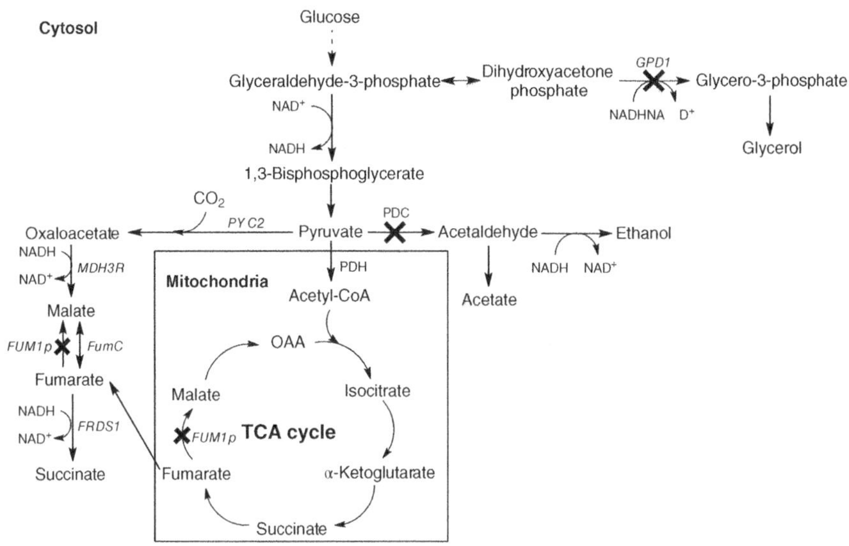

Scheme 7.36 Metabolically engineered *S. cerevisiae* employed for succinate production. PDC, pyruvate decarboxylase; PDH, pyruvate dehydrogenase.

to $8.09 \pm 0.28\,g\,L^{-1}$ by the deletion of *GPD1* and even higher to $9.98 \pm 0.23\,g\,L^{-1}$ with a yield of $0.32\,mol\,mol^{-1}$ glucose through regulation of biotin and urea levels (Scheme 7.36). In an aerobic batch fermentation with optimal supplement of CO_2, the engineered strain produced $12.97 \pm 0.42\,g\,L^{-1}$ succinate with a yield of $0.21\,mol\,mol^{-1}$ glucose at pH 3.8 [293].

C. reinhardtii is a green alga that has a network of fermentation pathways becoming active when cells acclimate to anoxia. The *C. reinhardtii hydEF-1* mutant has no hydrogenase activity and exhibits elevated accumulation of succinate and diminished production of CO_2 relative to the parental strain during dark, anaerobic metabolism. Increased accumulation of succinate indicates that the cells activate alternative pathways for pyruvate metabolism, which contribute to NAD(P)H reoxidation, and continued glycolysis and fermentation in the absence of O_2. Fermentative succinate production potentially proceeds via the formation of malate and increases in the abundance of mRNAs encoding two malate-forming enzymes, PYC and malic enzyme, are observed in the mutant relative to the parental strain following transfer of cells from oxic to anoxic conditions [294].

Malic acid, a four-carbon dicarboxylic acid, is used for the preparation of food additives in beverage and food industry as well as the synthesis of various fine chemicals. Since wild-type *S. cerevisiae* strains produce only low levels of malate, metabolic engineering is applied for achieving efficient malate production with this yeast. A putative pathway for malate production from glucose proceeds via carboxylation of pyruvate, followed by reduction of oxaloacetate to malate, which has a theoretical maximum yield of 2 mol malate per mol glucose. A previously engineered glucose-tolerant, C2-independent pyruvate decarboxylase (PDC)-negative *S. cerevisiae* strain was employed as the platform to evaluate the impact of introduction of three genetic modifications: (i) overexpression of the native PYC encoded by *PYC2*; (ii) high-level expression of an allele of the *MDH3* gene, of which the encoded malate dehydrogenase (MDH) was retargeted to the cytosol by deletion of the C-terminal peroxisomal targeting sequence; and (iii) functional expression of the *Schizosaccharomyces pombe* malate transporter gene *SpMAE1*. The results showed that single or double modifications improved malate production, while the highest malate yield and titers were obtained with simultaneous introduction of all three modifications. In batch cultures, the engineered strain produced malate at titers of up to 59 g L^{-1} at a yield of 0.42 mol mol^{-1} glucose. However, there was still substantial amounts of pyruvate accumulated in the cultures [295]. Research on the relationship between mitochondrial membrane potential and fermentation profile is also being intensely studied. A naturally occurring strain of yeast *S. cerevisiae* from sake mash was isolated that produces high levels of malic acid and demonstrates the correlation between the variations in mitochondrial membrane potential and malic acid production. Since the activities of PYC and MDH, enzymes required for malic acid synthesis, in strains that produce high levels of malic acid were elevated compared with the standard sake strain correspond to lower mitochondrial membrane potential, decreased mitochondrial membrane potential was responsible for increased malic acid synthesis was hypothesized and supported by experimental data [296].

Another yeast strain, *Torulopsis glabrata* CCTCC M202019, which is used for industrial pyruvate production, was also selected to explore the suitability of engineering this multivitamin auxotrophic yeast for increased malate production through the reductive TCA pathway (Scheme 7.37). Metabolic engineering strategies include (i) overexpression of PYC and MDH, (ii) identification of the bottleneck in malate production by model iNX804, and (iii) simultaneous overexpression of genes *RoPYC*, *RoMDH*, and *SpMAE1* from fungus *Rhizopus oryzae* and yeast *S. pombe*, respectively, which were employed to manipulate carbon flux from pyruvate to malate. Using these strategies, 8.5 g L^{-1} malate was accumulated in the engineered strain T.G-PMS, which was about 10-fold greater than that of the control strain T.G-26 under aerobic conditions [297].

Polymalic acid (PMA) is a water-soluble biopolymer composed of L-malic acid (MA) monomers and is mainly produced by the yeast-like fungus *Aureobasidium pullulans*. The free carboxyl groups in PMA make it simple to perform chemical modifications and create various derivatives or carrier-linked pro-drugs. Acid hydrolysis of PMA results in the release of widely used monomer MA in food and

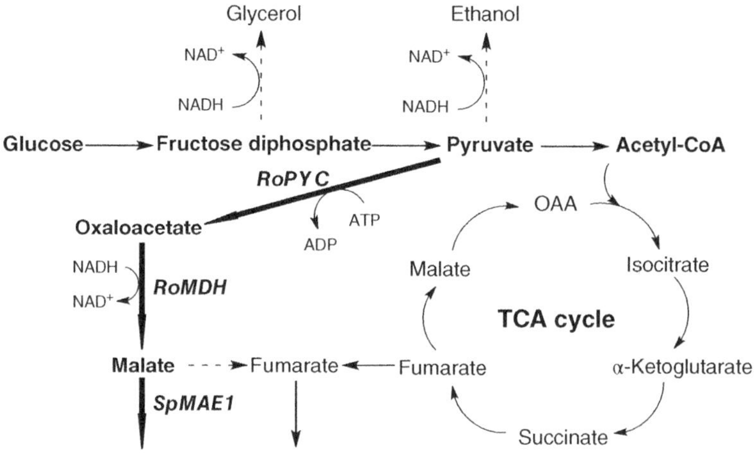

Scheme 7.37 Major metabolic pathways for the production of malate in *T. glabrata*.

chemical industry, which is a competitive process for producing bio-based platform chemicals. Comparative metabolome analysis of sucrose- and glucose-based fermentation demonstrated that pyruvate from the glycolysis pathway should be a key metabolite affecting PMA synthesis (Scheme 7.38). *In silico* simulation of a genome-scale metabolic model (iZX637) further verified that pyruvate carboxylase (*pyc*) via the reductive TCA cycle strengthened carbon flux for PMA synthesis. Therefore, a metabolic engineering of *A. pullulans* strain, FJ-PYC, was constructed by overexpressing the *pyc* gene, which increased the PMA titer by 15.1% compared with that from the wild-type strain in a 5-L stirred-tank fermenter. Sugarcane molasses can be used as an economical substrate without any pretreatment or nutrient supplementation. The fed-batch fermentation of FJ-PYC yielded the highest MA titer of $81.5\,g\,L^{-1}$ that is $94.2\,g\,L^{-1}$ of MA after hydrolysis in 140 hours with a corresponding MA yield of $0.62\,g\,g^{-1}$ and productivity of $0.67\,g\,L^{-1}\,h^{-1}$ [298].

Fumaric acid (FuA) is an organic acid with a trans-double bond and two carboxylic acid groups that has many industrial applications, mainly as a chemical feedstock for the manufacturing of synthetic resins, biodegradable polymers, and plasticizers. It is also used as an acidulant in foods and beverages and in miscellaneous industrial products, including lubricating oils, inks, lacquers, and carboxylating agents forstyrenebutadiene rubber. Metabolic engineering of *R. oryzae* by overexpressing endogenous PYC showed 56–83% increase of the PYC activity in the *pyc* transformants compared to the wild type. However, the *pyc* transformants grew poorly and had low FA yields ($<0.05\,g\,g^{-1}$ glucose) due to the formation of large cell pellets that limited oxygen supply and resulted in the accumulation

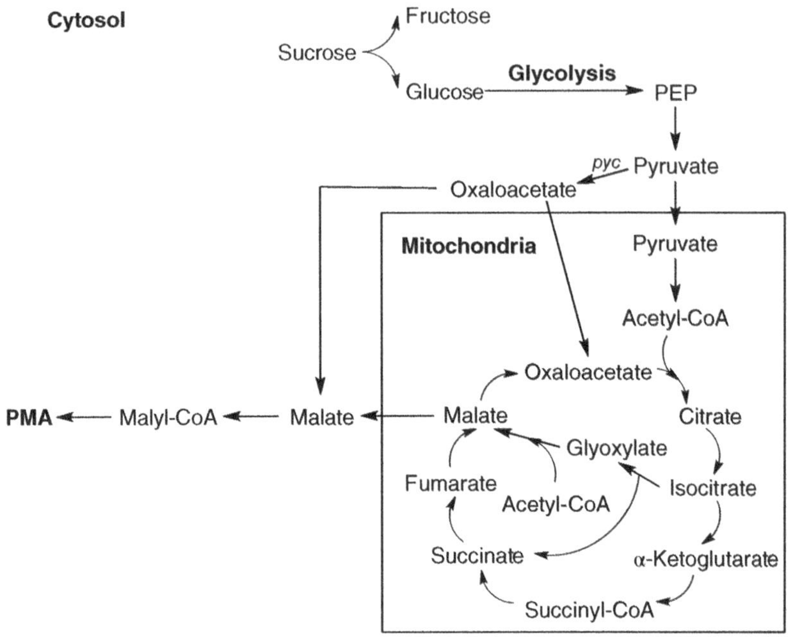

Scheme 7.38 PMA biosynthesis based on comparative metabolome analysis.

of ethanol with a high yield of 0.13–0.36 $g\,g^{-1}$ glucose [299]. FA production by direct fermentation using metabolically engineered *S. cerevisiae* with the aid of *in silico* analysis of a genome-scale metabolic model was also reported. The engineered *S. cerevisiae* strain obtained by FUM1 deletion can produce FA up to a concentration of $610 \pm 31\,mg\,L^{-1}$ without any apparent change in growth in fed-batch culture. Flux balance analysis identified PYC as one of the factors limiting higher FA production. When RoPYC gene was introduced, *S. cerevisiae* produced $1134 \pm 48\,mg\,L^{-1}$ FA. Furthermore, the final engineered *S. cerevisiae* strain by introducing the SFC1 gene encoding a succinate-fumarate transporter was able to produce $1675 \pm 52\,mg\,L^{-1}$ FA in batch culture [300]. Spatial modulation and cofactor engineering of key pathway enzymes in the reductive TCA cycle were performed for the development of a multivitamin auxotrophic *Candida glabrata* strain to efficiently produce FA (Scheme 7.39). Specifically, DNA-guided scaffold system was first constructed and optimized to modulate PYC, MDH, and fumarase (FUM), which increased the FA titer from 0.18 to $11.3\,g\,L^{-1}$. In addition, the combined tuning cofactor balance by controlling the expression strengths of ADP-dependent PEPCK and NAD^+-dependent FDH led to a large increase in FA production up to $18.5\,g\,L^{-1}$. Finally, the batch fermentation of the engineered strain T.G-4G-$S_{(1:1:1.2)}$-$P_{(M)}$-$F_{(H)}$ with a 5-L bioreactor could produce $21.6\,g\,L^{-1}$ FA [301].

Alpha-ketoglutaric acid (KGA), also known as 2-oxoglutaric acid, is a key intermediate in the TCA cycle and a major contributor to amino acid and protein metabolism. In industry, it is used as building block for the chemical synthesis of heterocycles or elastomers, as dietary supplement, and as an agent that improves athletic performance and enhances wound healing. A thiamine auxotrophic yeast *Yarrowia lipolytica* WSH-Z06 was used for α-KGA overproduction. To enhance α-KGA production and reduce the major by-product pyruvate accumulation from the substrate glycerol, the pyruvate carboxylation pathway of *Y. lipolytica* WSH-Z06 was regulated by metabolic engineering by overexpressing the pyruvate carboxylase genes *ScPYC1* from *S. cerevisiae* and *RoPYC2* from *R. oryzae*, respectively. The yields of α-KGA in *Y. lipolytica-ScPYC1* and *Y. lipolytica-RoPYC2* increased by 24.5 and 35.3%, and the yields of pyruvate decreased by 51.9 and 69.8% in shake flask fermentations, respectively. These results indicated that enhancing the pyruvate carboxylation pathway had successfully redistributed the carbon flux from pyruvate to α-KGA. Finally, a two-stage pH control fermentation strategy was applied in a 3-L fermenter, and the maximum concentration of α-KGA in *Y. lipolytica-RoPYC2* reached 62.5 $g\,L^{-1}$ with an evident decrease in pyruvate accumulation from 35.2 to 13.5 $g\,L^{-1}$, which demonstrated great potential for further industrial production [302]. Furthermore, to increase the α-KGA productivity and reduce the amounts of by-products, e.g. pyruvate, fumarate, malate, and succinate, recombinant *Y. lipolytica* strains were constructed by a gene dose-dependent overexpression of genes encoding $NADP^+$-dependent isocitrate dehydrogenase (*IDP1*) and pyruvate carboxylase (*PYC1*), respectively, or IDP1 and PYC1 genes together under α-KGA production conditions from the renewable carbon source raw glycerol. Then, a selective increase in IDP activity in *IDP1* multicopy transformants changes the produced amount of α-KGA. Simultaneous overexpression of the gene *IDP1* and *PYC1* had the strongest effect on increasing the amount of secreted α-KGA. About 19% more α-KGA in comparison with the control strain H355 was produced in bioreactor

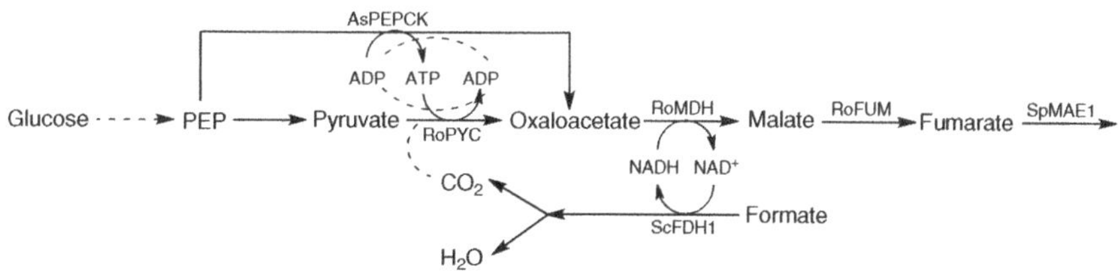

Scheme 7.39 Major metabolic pathways lead to the formation of fumarate I in *C. glabrata.* RoPYC, pyruvate carboxylase from *R. oryzae*; RoMDH, malate dehydrogenase from *R. oryzae*; RoFUM, fumarase from *R. oryzae*; SpMAE1, C4-dicarboxylic acids transporter from *S. pombe*; AsPEPCK, ADP-dependent PEP carboxykinase from *A. succinogenes*; ScFDH1, NAD^+-dependent formate dehydrogenase from *S. cerevisiae.*

with raw glycerol as carbon source resulting in the highest yield of accumulated α-KGA of about 186 g L^{-1} after 117 hours of cultivation with significantly reduced amounts of the main by-product pyruvate [303].

Propionic acid is an important chemical intermediate used in the synthesis of vitamin E, cellulose fibers, artificial fruit flavors, fragrances, and perfumes. In addition, its salts are widely used as food and feed preservatives. Since CO_2 is one of the by-product during fermentation, the heterotrophic CO_2 fixation of facultative anaerobe *Propionibacterium acidipropionici* is believed to have great potential in the industry for propionic acid production. In glycerol fermentation with *P. acidipropionici*, the volumetric productivity of propionic acid with CO_2 supplementation reached 2.94 g $L^{-1}day^{-1}$, compared to 1.56 g $L^{-1}day^{-1}$ without CO_2. The cell growth using glycerol was also significantly enhanced with CO_2. In addition, the yield and productivity of succinate, the main intermediate in Wood–Werkman cycle, increased 81 and 280%, respectively, which is consistent with the increased activities of pyruvate carboxylase and propionyl CoA transferase, two key enzymes in the Wood–Werkman cycle (Scheme 7.40). However, in glucose

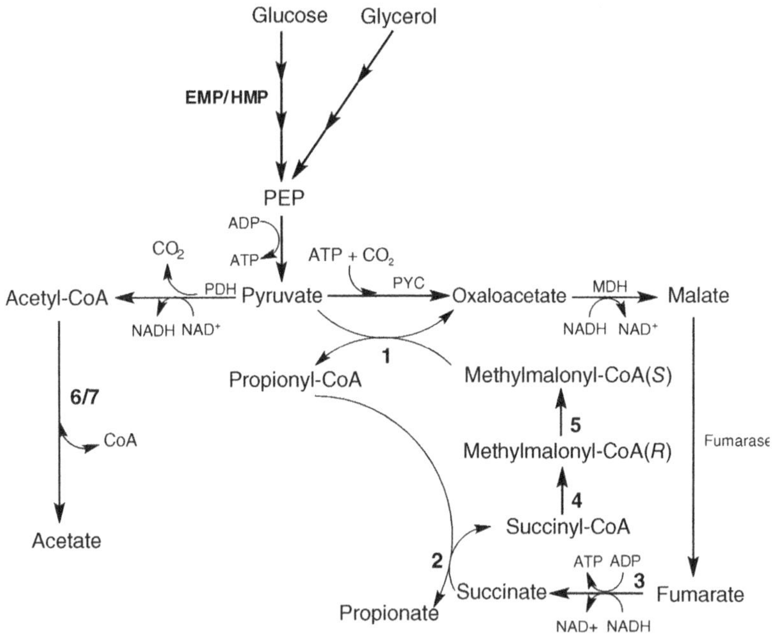

Scheme 7.40 Proposed metabolic pathways of propionic acid fermentation with glycerol and glucose by *P. acidipropionici* through CO_2 fixation via pyruvate to oxaloacetate. 1: Methylmalonyl-CoA carboxytransferase, 2: propionyl-CoA:succinate CoA transferase, 3: succinate dehydrogenase, 4: methylmalonyl-CoA mutase, 5: methylmalonyl-CoA epimerase, 6: phosphotransacetylase, 7: acetate kinase.

fermentation with *P. acidipropionici*, CO_2 had minimal effect on propionic acid production and cell growth. The carbon flux distributions using glycerol or glucose were analyzed using a stoichiometric metabolic model, and the calculated maintenance coefficient (m_{ATP}) increased 100%, which may explain the increase in the productivity of propionic acid in glycerol fermentation with CO_2 supplement [304].

Metabolic engineering of *Propionibacteriumfreudenreichii* subsp. *Shermanii* DSM 4902 was utilized for enhancing propionic acid fermentation through whole genome sequence and genetic engineering tools. The three biotin-dependent carboxylases in the dicarboxylic acid pathway controlling the carbon flux in the Wood–Werkman cycle, PYC, methylmalonyl-CoA decarboxylase (MMD), and methylmalonyl-CoA carboxytransferase (MMC) were overexpressed in *P. shermanii*, and the effects on propionic acid fermentation in serum bottles with glucose and glycerol as substrates were compared to the wild type. The mutants overexpressing MMC and MMD showed a significant increase in the metabolic fluxes toward the biosynthesis of propionic acid, against acetic and succinic acids, with a significantly increased yield up to 14% and productivity up to 17%, especially in the co-fermentation of glycerol and glucose. While the mutant overexpressing PYC grew slower, produced more succinate, and had a lower production of propionate up to 12% decrease in productivity [305].

3-Hydroxypropionic acid (3-HP), a platform chemical, is a potential chemical building block for sustainable production of superabsorbent polymers and acrylic plastics. According to the metabolic modeling, the β-alanine biosynthetic route was identified as the most economically attractive for high-level production of 3-HP. Therefore, a synthetic pathway for *de novo* biosynthesis of β-alanine and its subsequent conversion into 3-HP using a novel β-alanine-pyruvate aminotransferase discovered in *Bacillus cereus* was engineered and optimized in yeast *S. cerevisiae* (Scheme 7.41). The final strain generated 3-HP at a titer of $13.7 \pm 0.3\,g\,L^{-1}$

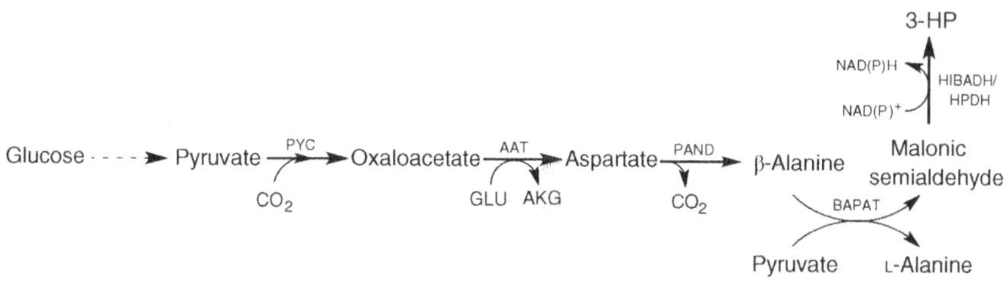

Scheme 7.41 Biosynthetic pathway toward 3-hydroxypropionic acid from glucose. PYC, pyruvate carboxylase; AAT, aspartate aminotransferase; PAND, aspartate decarboxylase; BAPAT, β-alanine-pyruvate aminotransferase; HIBADH, 3-hydroxyisobutyrate dehydrogenase; HPDH, 3-hydroxypropionate dehydrogenase.

with a 0.14 ± 0.0 C-mol C-mol^{-1} yield on glucose within 80 hours using a controlled fed-batch fermentation in mineral medium at pH 5 [306].

L-Glutamate is widely used in goods, pharmaceutical, and other industries. Particularly, glutamate is used as a food ingredient with huge market requirement in oriental countries, such as China, Japan, Korea, and Taiwan. The occurrence in *Clostridium kluyveri* of the enzymes of the upper half of the TCA cycle, citrate synthetase, aconitase, D-isocitrate dehydrogenase, and L-glutamate dehydrogenase, as well as certain other enzymes, such as PYC, oxaloacetate decarboxylase, L-malate-diphosphopyridine nucleotide (DPN) dehydrogenase, and L-malate-triphosphopyridine nucleotide (TPN) oxidodecarboxylase, made the extracts of *C. kluyveri* synthesize glutamate from citrate and precursors of citrate such as acetate, pyruvate, and oxaloacetate via the following stepwise reaction sequence [307]. *C. glutamicum* is an aerobic, Gram-positive, and biotin auxotrophic bacterium and is used for the fermentative production of glutamic acid under biotin-limited conditions. To understand the distinct roles of two important anaplerotic enzymes, PEPC and biotin-containing PYC, on glutamate production by *C. glutamicum*, the glutamate production was induced under biotin-limited conditions in the disruptants of the genes encoding PEPC (*ppc*) and PYC (*pyc*), respectively. The *pyc* disruptant retained the ability to produce high amounts of glutamate, and lactate was simultaneously produced, while the *ppc* knockout mutant could not produce glutamate. In addition, glutamate production in the *pyc* disruptant was enhanced by overexpression of *ppc* rather than disruption of the LDH gene (*ldh*), which is involved in lactate production. Metabolic flux analysis revealed that the flux for anaplerotic reactions in the *pyc* disruptant was lower than that in the wild type, concomitantly increasing the flux for lactate formation. Moreover, overexpression of *ppc* increased this flux in both the *pyc* disruptant and wild type. The experimental results revealed that singly the results demonstrate that the PEPC-catalyzed anaplerotic reaction is necessary for glutamate production induced under biotin-limited conditions [308]. During glutamate fermentation, anaplerotic reaction could be enhanced by jointly manipulating pH regulation and $NaHCO_3$ supplement by *C. glutamicum*, which lead to a 36% increase in the yield and comparably high glutamate productivity. The simulation with a novel metabolic model incorporating directed signal flow diagram and experimental results on enzymatic activities data revealed that singly regulating each individual enzyme could not increase the yield, and glutamate yield could be enhanced only when six key enzymes, PYC, PDH, ICDH, ICL, GDH, and α-oxoglutarate dehydrogenase (ODHC), work together, that is relative activities ratios of enzymatic pairs of PYC/PDH should be controlled at moderate level of 6 : 4, while those of ICDH/ICL and GDH/ODHC should be controlled at higher level of 8 : 2 simultaneously [309].

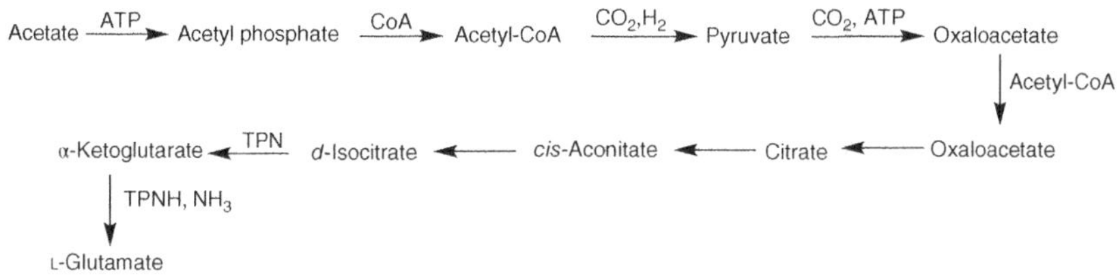

C. glutamicum was also used for the fermentative production of L-lysine in a split pathway from OAA via L-aspartate, L-aspartate semialdehyde, and D, L-diaminopimelate as the main intermediate (Scheme 7.42). The effect of improved pyruvate availability on L-lysine production was tested by deleting the *aceE* gene encoding the E1p enzyme of the pyruvate dehydrogenase complex (PDHC) in *C. glutamicum* DM1729. Strain DM1729-BB1 showed no PDHC activity, was acetate auxotrophic, and after complete consumption of the available carbon sources glucose and acetate showed a more than 50% lower substrate-specific biomass yield (0.14 vs. 0.33 mol C (mol C)$^{-1}$) and about fourfold higher biomass-specific L-lysine yield (0.13 vs. 0.09 mol C/mol C). Furthermore, overexpression of PYC or diaminopimelate dehydrogenase (DAPDH) genes in *C. glutamicum* DM1729-BB1 resulted in a further increase in the biomass-specific L-lysine yield by 6 and 56%, respectively. In addition, significant amounts of pyruvate, L-alanine, and L-valine were produced by *C. glutamicum* DM1729-BB1 and its derivatives, which indicate a further potential to improve L-lysine production [310].

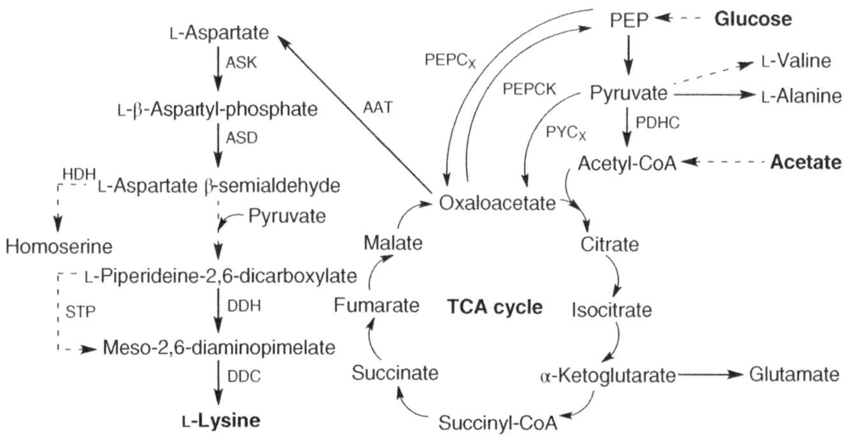

Scheme 7.42 Biosynthetic pathway of L-lysine via the D,L-diaminopimelate intermediate in *C. glutamicum*. AAT, aspartate aminotransferase; ASD, aspartate-semialdehyde dehydrogenase; ASK, aspartate kinase; DDC, diaminopimelate decarboxylase; DDH, diaminopimelate dehydrogenase; HDH, homoserine dehydrogenase; STP, succinyl transferase pathway.

The pathway for lysine and biotin biosynthesis in *C. glutamicum* AHP-3 was reconstructed by introducing biotin biosynthesis genes from *B. subtilis* on a plasmid, pBIO (Scheme 7.43). AHP-3, containing pBIO, was able to produce lysine in a medium lacking biotin and the lysine yield on glucose was similar to what is obtained when using a medium containing biotin. However, the specific growth rate decreased 20% when the strain was cultivated without biotin. To solve this problem, pimelic acid, an early biotin precursor, was added so that the growth rate could be restored fully, which demonstrates that the bottleneck is in pimeloyl-CoA (or pimeloyl-ACP) formation. PYC (*pycA*), a biotin-dependent enzyme, and biotin ligase (*birA*), needed for attaching biotin to PYC, were overexpressed by replacing the native promoters with the strong superoxide dismutase (*sod*) promoter, to examine the restoration of growth rate. Results showed that neither *pycA* nor *birA* overexpression, whether alone or in combination, had an effect on specific growth rate, but they did increase the lysine yield by 55% in the strain overexpressing both enzymes [311]. Pyruvate carboxylase variants in *C. glutamicum* that enable improved L-lysine production from glucose were identified by a lysine sensor. Two PYC variants, $PYCx^{T343A,I1012S}$ and $PYCx^{T132A}$, were identified allowing 9 and 19% increased titers upon plasmid-based expression. Chromosomal expression of $PYCx^{T132A}$ and $PYCx^{T343A}$ variants led to 6 and 14% higher L-lysine levels [312].

L-Threonine is an essential amino acid, which has an impressive history as a food additive for humans and animals. In addition, it has been widely employed in agricultural, pharmaceutical, and cosmetics industries. A metabolic modification strategy of two-stage carbon distribution and cofactor generation was proposed to tackle several limiting factors, such as the conflict between cell growth and production, by-product accumulation, and insufficient availability of cofactors ATP, NADH, and NADPH, for L-threonine production strain *E. coli* THRD. The glycolytic fluxes toward TCA cycle were increased in growth stage through

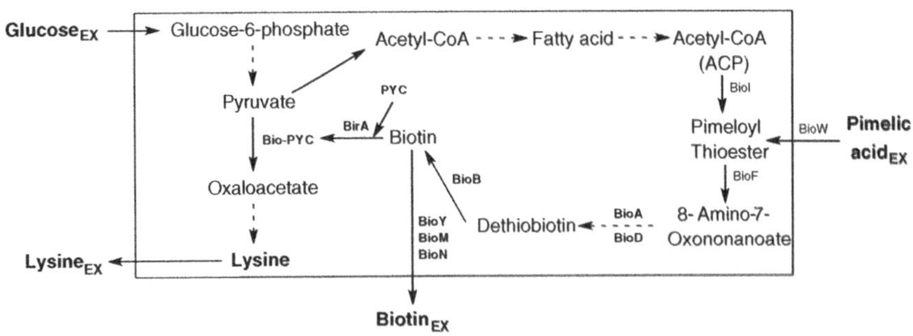

Scheme 7.43 Lysine and biotin production using PYC and biotin protein ligase overexpressed in *C. glutamicum*.

heterologous expression of PYC, PEPCK, and CS that led to improved glucose utilization and growth performance. During the production stage, the carbon flux was redirected into L-threonine synthetic pathway via a synthetic genetic circuit. Meanwhile, overexpression of GDH, FDH, and pyridine nucleotide transhydrogenase in *E. coli* THRD was developed for an L-glutamate and NADPH generation system to sustain the transaminase reaction for L-threonine. The results showed an increase in the L-threonine production using shake flask and bioreactor fermentation by 2.02- and 1.21-fold, respectively [313].

Carbon dioxide (CO_2) is an atmospheric gas contributing to the greenhouse effect that causes global warming and climate changes, which, therefore, has inspired a global effort to reduce it. Many approaches have been considered and adopted for reducing CO_2 emissions. Among these approaches, tuning inexpensive and readily available CO_2 into fuels and chemicals offers a win-win strategy to both decrease atmospheric CO_2 and efficiently exploit carbon resources. In summary, methods that have been employed for catalyzing CO_2 conversion include chemical, photochemical, electrochemical, and enzymatic. Enzymes, capable of catalyzing CO_2 fixation steps in carbon assimilation pathways, are particularly promising catalysts for the sustainable transformation of this safe and renewable feedstock into central metabolites. In this section, biotin-dependent pyruvate carboxylase, catalyzing the carboxylation of pyruvate to oxaloacetate through the use of CO_2 as feedstock, has been proved to be the efficient one [314–316].

7.4.2 Acetyl-CoA Carboxylase

Acetyl-CoA carboxylase (ACC, EC 6.4.1.2) is a biotin-containing enzyme that catalyzes the first committed step of FA synthesis, the irreversible ATP-activated carboxylation of acetyl-CoA to malonyl-CoA (Scheme 7.44) [317–319]. ACC is found in all forms of life. This enzyme has been purified and characterized from rat liver [320–323], rat skeletal muscle [324], chicken liver [325], rabbit mammary gland [326], pea plant [327], diatom [328], *Leptospira* species [329], bacteria [330, 331], archaea [332], human [333, 334], and many others. The molecular mass of ACC measured from rat liver was 230–250 kDa, as well as approximately 272 kDa (major band) and 265 kDa (minor band) from rat skeletal muscle; the chicken liver yielded a single polypeptide of 225 kDa; the apparent molecular mass obtained in pea plant was 87 kDa; the native ACC had a molecular weight of approximately 740 kDa; the purified archaeal ACC from *Acidianus brierleyi* showed a molecular mass of approximately 540 kDa; and the molecular masses of ACC in human existed between 265 and 280 kDa. Whereas ACC is a multisubunit enzyme in most prokaryotes and in the chloroplasts of most plants and algae, it is a large, multi-domain enzyme in the ER of most eukaryotes [335]. Physically

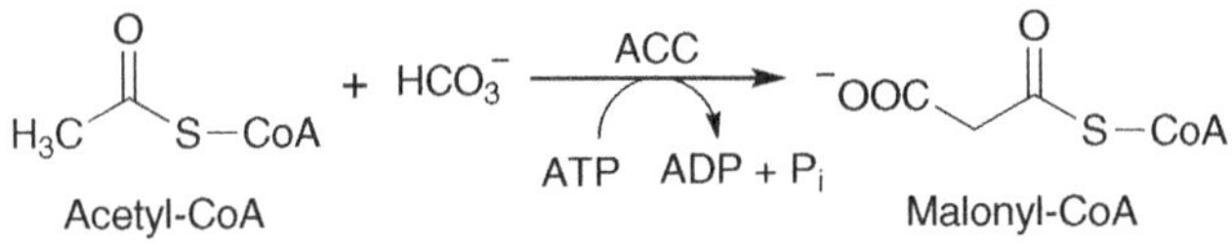

Scheme 7.44 Acetyl-CoA carboxylase catalyzed carboxylation of acetyl-CoA to malonyl-CoA. *Sources:* Based on Erb [317], Sasaki and Nagano [318], and Nikolau et al. [319].

distinct heteromeric and homomeric forms are found by the X-ray crystal structure. A heteromeric form of ACC is composed of four subunits, the biotin-linked BCCP [336, 337], the BC [338–341], and the α and β subunits of carboxyltransferase (CT) [342–344], and is usually present in prokaryotes. The homomeric form is composed of a single large polypeptide with four subunit domains and is present in eukaryotes. Most plants have both forms with the heteromeric form in plastids and the homomeric form in cytosol, while the grass family, including wheat and rice, is an exception, which has the homomeric form in both cytosol and plastids [345, 346]. Mammalian cells, including humans, express two isoforms of acetyl-CoA carboxylase, ACC1 ($Mr = 265$ kDa) and ACC2 ($Mr = 280$ kDa), and ACC1 is a cytosolic protein and ACC2 is associated with the mitochondria [333, 334, 347–349]. However, the crystal structure of full-length yeast *S. cerevisiae* ACC (*Sc*ACC) was reported to show an overall architecture of 500 kDa holoenzyme dimer, which has a structure remarkably different from that of the other biotin-dependent carboxylases [350]. The BC and CT domain dimers are located at the top and bottom of the structure, respectively. The unique central region contains five domains and is important for positioning the BC and CT domains for catalysis.

The overall carboxylation of acetyl-CoA to yield malonyl-CoA catalyzed by ACC follows a two-step mechanism as given below [318, 349, 351, 352]:

1) BCCP + HCO_3^- + Mg^{2+}-ATP → BCCP-CO_2^- + Mg^{2+}-ADP + Pi: BC
2) BCCP-CO_2^- + acetyl-CoA → BCCP + malonyl-CoA: CT

The first reaction is carried out by BC and involves the ATP-dependent carboxylation of biotin with bicarbonate serving as the source of CO_2. In the second reaction, the carboxyl group is transferred from biotin to acetyl-CoA to produce malonyl-CoA, which is catalyzed by CT. Two Mg^{2+} are coordinated by the phosphate groups on the ATP and are required for ATP binding to the enzyme. Thus, this mechanism is tightly regulated by phosphorylation [353]. Bicarbonate is deprotonated by Glu296, although in solution this proton transfer is unlikely as the pK_a of bicarbonate is 10.3; the pK_a of bicarbonate can be decreased by its interaction with positively charged side chains of Arg338 and Arg292. Furthermore, Glu296 interacts with the side chain of Glu211, that also causes an increase in the apparent pK_a. Following deprotonation of bicarbonate, the oxygen of the bicarbonate acts as a nucleophile and attacks the gamma phosphate on ATP to form the carboxyphosphate intermediate, which quickly decomposes to CO_2 and PO_4^{3-}. The PO_4^{3-} deprotonates biotin, creating an enolate, stabilized by Arg338, that subsequently attacks CO_2 resulting in the production of carboxybiotin [340]. The carboxybiotin translocates to the CT active site, where the carboxy group is transferred to acetyl-CoA. Even though little is known about the reaction mechanism of CT, it is proposed that CO_2 is released from biotin that subsequently abstracts a proton from the methyl group from acetyl-CoA. The resulting enolate attacks CO_2 to yield malonyl-CoA.

The main function of ACC is to regulate the metabolism of FAs [354]. When ACC is active, malonyl-CoA is produced that serves as a building block for new FAs and can inhibit the transfer of the fatty acyl group from acyl-CoA to carnitine, which inhibits the β-oxidation of FAs in the mitochondria [355]. In mammals, two isoforms ACC1 and ACC2 are expressed [333, 334, 347–349]. ACC1 is found in the cytoplasm of all cells but is enriched in lipogenic tissue, such as adipose tissue and lactating mammary glands, where FA synthesis is important. In oxidative tissues, such as the skeletal muscle and the heart, the express ratio of ACC2 is higher. While in liver, where both FA oxidation and synthesis are important, both ACC1 and ACC2 are highly expressed. The differences in tissue distribution reveal that ACC1 maintains the regulation of FA synthesis, whereas ACC2 is mainly responsible for FA oxidation regulation [356, 357]. In addition to the crucial roles in FA metabolism, ACC also presents a variety of clinical implications, such as the production of novel antibiotics and the development of new therapies for diabetes, obesity, and other manifestations of metabolic syndrome, which may even relate to carcinoma [335, 345, 356–362].

FAs are the essential hydrophobic component of the cellular membrane lipids of all living organisms except the archaea. In a number of bacteria, FAs also act as components of storage lipids, the most prevalent of which are PHAs [363]. Recently, the fast advancement of industrial and social activities has raised the global energy demand, which causes the problems of petroleum prices rising and global warming. Therefore, much effort has been put to develop more sustainable and cost-effective fuels from renewable sources that could reduce CO_2 emission and replace the traditional fossil fuels. Microbial fermentation derived renewable biofuels, including higher alcohols, hydrocarbons produced by isoprenoids biosynthetic pathway, and FAs as well as lipids synthesized from microbes, are being considered good alternatives [364–367]. ACC catalyzes the first committed step of the FA synthetic pathway. Although ACC has often been considered the rate-determining step in FA biosynthetic pathway for a long time, until recently this proposal has been tested *in vivo* using *E. coli*. Bacterial and plant ACC belong to type II FAS system, which is composed of a collection of individual enzymes, and

ACC reaction consists of two discrete half-reactions with four enzyme protein subunits, BC, CT complex (α and β), and BCCP (Scheme 7.45) [368].

After cloning the genes encoding the four subunits of *E. coli* in a single plasmid under the control of a bacteriophage T7 promoter, the four ACC subunits were overproduced in equimolar amounts by the induction of gene expression. Overproduction of the proteins resulted in greatly increased ACC activity with a concomitant increase in the intracellular level of malonyl-CoA and a sixfold increase in the rate of FA synthesis [369]. In another work, genes encoding ACC (*accA*, *accB*, *accC*), malonyl-CoA-ACP transacylase (*fabD*), and acyl-ACP TE (EC 3.1.2.14 gene), which are all key enzymes in the synthesis of FAs, were cloned and overexpressed in *E. coli* MG1655. The enzyme encoded by the *fabD* gene converts malonyl-CoA to malonyl-ACP, and the encoded EC 3.1.2.14 gene converts fatty acyl-ACP chains to long-chain FAs. All the genes except for the EC 3.1.2.14 gene were homologous to *E. coli* genes and were used to overexpress components of the FAS pathway through metabolic engineering. All recombinant *E. coli* MG1655

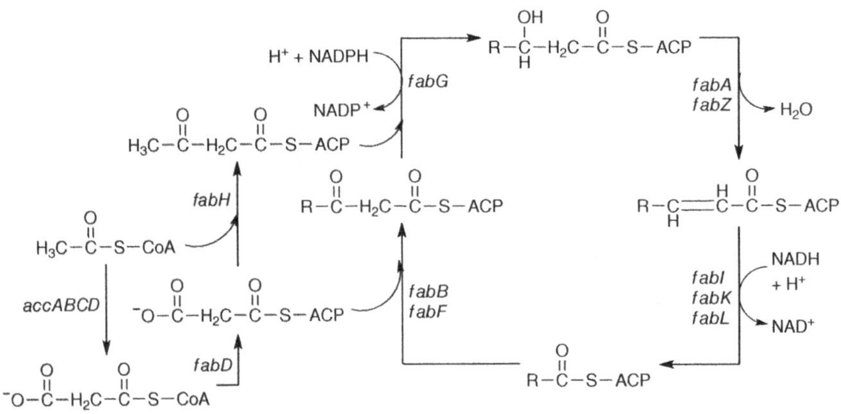

Scheme 7.45 Type II fatty acid synthetic pathway. The ACC is a heterotetramer, and the subunits are encoded by the *accA*, *accB*, *accC*, and *accD* genes. The malonate group is transferred from CoA to ACP by malonyl-CoA:ACP transacylase (*fadD*). Cycles of fatty acid elongation are initiated by the condensation of acetyl-CoA with malonyl-ACP catalyzed by β-ketoacyl-ACP synthase III (*fabH*). The second step in the elongation cycle is carried out by β-ketoacyl-ACP reductase (*fabG*). The β-hydroxyacyl-ACP intermediate is dehydrated by either *fabA* or *fabZ* to form *trans*-2-enoyl-ACP. *FabA* is also responsible for isomerization of the double bond at the 10-carbon stage to form UFAs. The final step in the elongation is catalyzed by enoyl-ACP reductase, which has three forms (*fabI*, *fabK*, and *fabL*). *FabI* and *fabL* are NAD(P)H-dependent reductases. *FabK* is an NADH-dependent flavoprotein that also oxidizes the cofactor in the absence of substrate. Subsequent rounds of elongation are initiated by the elongation condensing enzymes (*fabB* and *fabF*) whose substrate specificities govern the structure and distribution of fatty acid products. *Source:* Modified from Rock and Jackowski [368].

strains containing various gene combinations were developed using the pTrc99A express vector, and the *in vitro* metabolites and FAs produced by the recombinants were analyzed to observe the changes in metabolism. The amounts of hexadecanoic acid produced from recombinant strains were 1.23–2.41 times higher than that of the wild-type strain [370].

Modular pathway engineering strategies were developed to optimize a multigene pathway for FA production in *E. coli*. Specifically, a total of 15 essential genes of the FA metabolic pathway were selected and arranged into three modules: the upstream acetyl-CoA formation module, the intermediary acetyl-CoA activation module, and downstream FAS module, and then the pathway bottlenecks were assessed and eliminated at both transcriptional and translational levels (Scheme 7.46). Combinatorial optimization of transcriptional levels of these three modules led to the identification of conditions that balance the supply of acetyl-CoA and consumption of malonyl-CoA/ACP. Refining protein translation efficiency by customizing ribosome binding sites (RBS) for both the upstream acetyl-CoA formation and FAS modules enables further production improvement. Under controlled conditions, fed-batch cultivation of the engineered strain produced up to 8.6 $g\,L^{-1}$ FAs with a productivity of 0.124 $g\ L^{-1}h^{-1}$ [371].

The calorific value of biodiesel increases along with FA chain length, but it also decreases the low temperature fluidity. However, the *cis* double bonds in the long FA chain may decrease the viscosity. The report from the US Department of Energy reveals that the perfect biodiesel would be made from monosaturated fatty acids (MUFAs), because high levels of polyunsaturated fatty acids (PUFAs) would negatively impact the oxidative stability and increase exhaust emissions, which do not suit diesel engines. Biodiesel derived from vegetable oils or microbial lipids always contains significant amounts of PUFAs and SFAs alkyl esters, which hamper its practical applications [372]. Therefore, the fatty acyl-ACP TE (*At*FatA) and FA desaturase (SSI2) from *A. thaliana* were heterologously expressed in *E. coli* BL21 star(DE3) to specifically release free UFAs and convert SFAs to UFAs. In addition, the endogenous *fadD* gene (encoding acyl-CoA synthetase) was disrupted to block FA catabolism while the native ACC was overexpressed to increase the malonyl-CoA pool and boost FA biosynthesis (Scheme 7.47). Under shake flask conditions, the engineered *E. coli* BL21ΔfadD/pE-AtFatAssi2&pA-acc produced 82.6 $mg\,L^{-1}$ FFAs and the yield on glucose reached about 3.3% of the theoretical yield. Two types of MUFAs, palmitoleate (16:1Δ9) and *cis*-vaccenate (18:1Δ11) made up more than 75% of the FFA profiles. Fed-batch fermentation of this strain further enhanced FFAs production to a titer of 1.27 $g\,L^{-1}$ without affecting FA compositions [373].

Since FAs biosynthesis is initiated by the conversion of acetyl-CoA to malonyl-A,which requires acetyl-CoA as key substrates, the non-oleaginous yeast *S. cerevisiae* was metabolically engineered for cytosol acetyl-CoA enhancement for FA

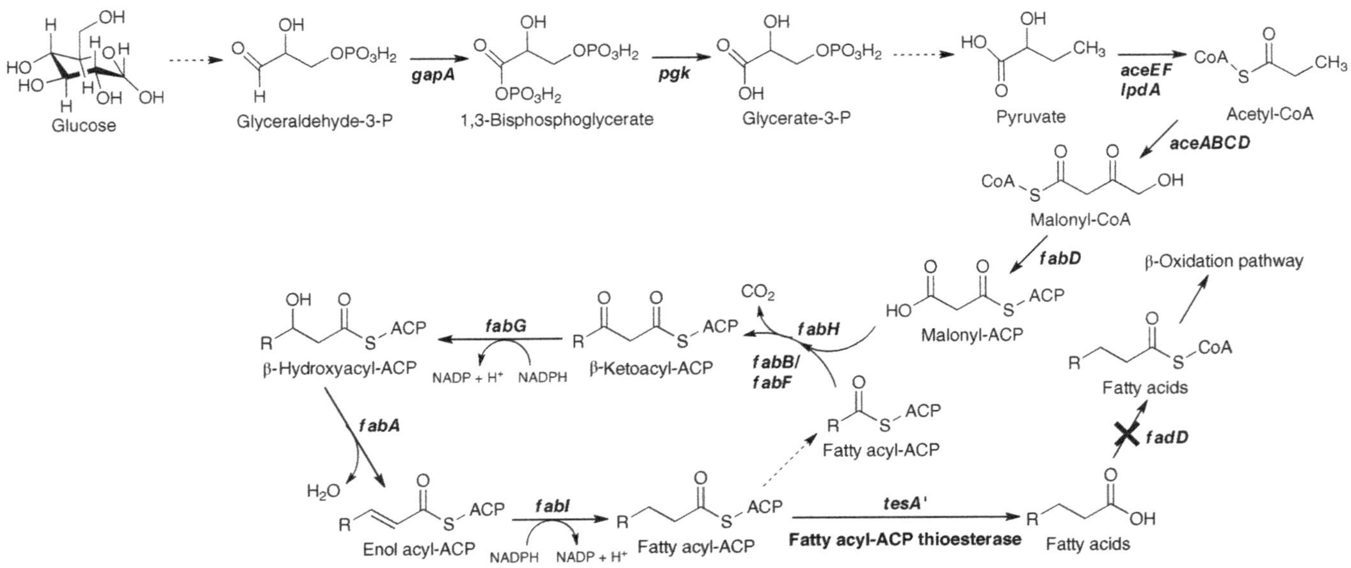

Scheme 7.46 Fatty acid metabolic pathway of engineered *E. coli*. Overexpression genes and knockout genes are shown in boldface. Dashed arrows represent multiple reaction steps. *accABCD*, acetyl-CoA carboxylase; *aceEF* and *lpdA*, pyruvate dehydrogenase multi-enzyme complex; *fabA*, β-hydroxy acyl-ACP dehydrase; *fabB*, β-ketoacyl-ACP synthase I; *fabD*, fatty acyl-CoA synthetase; *fabF*, β-ketoacyl-ACP synthase II; *fabG*, β-ketoacyl-ACP reductase; *fabH*, β-ketoacyl-ACP synthase III; *fabI*, enol acyl-ACP reductase; *fadD*, malonyl-CoA-ACP transacylase; *gapA*, glyceraldehyde-3-phosphate; *pgk*, phosphoglycerate kinase; *tesA'*, and truncated multifunctional fatty acyl-CoA thioesterase.

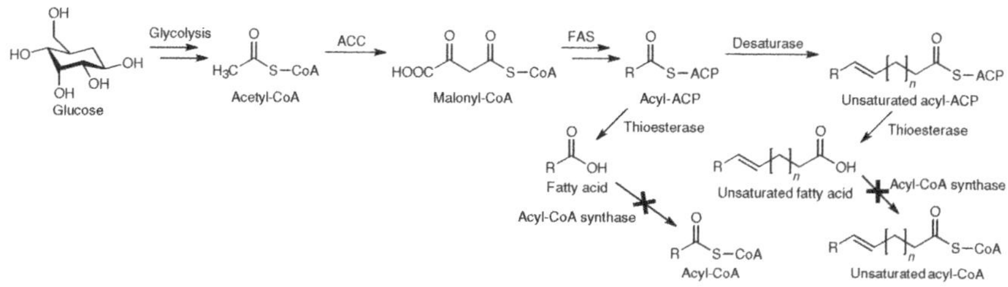

Scheme 7.47 Engineered *E. coli* for free monounsaturated fatty acid production. FAS, fatty acid synthase.

synthesis. Gene *idh1* and *idh2* involved in the citrate turnover in TCA cycle were disrupted so that the citrate level accumulated was increased four and five times in mutant yeast strains. Furthermore, a heterologous ATP-citrate lyase (ACL) was overexpressed in the wild-type and *idh1,2* disrupted yeast strains to convert the accumulated citrate to cytosol acetyl-CoA. The engineered wild-type strain expressing *acl* mainly produced SFAs with C14:0, C16:0, and C18:0 at levels about 20, 14, and 27%, respectively. In addition, the *idh1,2* disrupted and expressing *acl* strains mainly accumulated UFAs, which showed 80% increase in C16:1 and 60% increase in C18:1 with Δ*idh1* strain expressing *acl*, 60% increase in C16:1 and 45% increase in C18:1 with Δ*idh2* strain expressing *acl*, and 92% increase in C16:1 and 77% increase in C18:1 with Δ*idh1,2* strain expressing *acl* [374]. In another study, a gene coding ACC (*LsACC1*) was isolated and characterized from an oleaginous yeast, *Lipomyces starkeyi*, and the expression level was upregulated with the fast accumulation of lipids. The *LsACC1* was co-overexpressed with the endogenous glycerol-3-phosphate dehydrogenase gene (*GPD1*), which regulates lipid biosynthesis by supplying another substrate glycerol-3-phosphate for storage lipid assembly, in yeast *S. cerevisiae*. Additionally, the *S. cerevisiae* ACC (*ScACC1*) was transferred with *GPD1* and its function was compared with *LsACC1*. The overexpressed *LsACC1* and *GPD1* resulted in a 63% increase of the lipid (TAG) production in *S. cerevisiae* [375].

Y. lipolytica is another emerging nonconventional oleaginous yeast. A metabolic engineering strategy for FFA production in *Y. lipolytica* was developed, which was based on the elimination of glycerol metabolism as the main final acceptor of acyl-CoA in neutral lipid formation. FFA production was monitored under lipogenic conditions with glycerol as a limiting factor, which presents at a sufficient level for membrane synthesis but deficient for TAG accumulation. First, a glycerol auxotroph that was also unable to utilize glycerol was constructed by deleting the cytosolic glycerol-3-phosphate dehydrogenase gene (*gpd1*) and the gene encoding flavin-dependent and membrane-bound mitochondrial isozyme (*gut2*) that uncouples the glycerol-3-phosphate shuttle. In this genetic background, growth was repressed by glycerol. Moreover, oleate toxicity was observed in the *Δgpd1Δgut2Δpex10* triple mutant strain, which is deficient additionally in peroxisome biogenesis due to the deletion of peroxin gene (*per10*). Two consecutive rounds of selection of spontaneous mutants were performed: first, by relieving from growth repression by glycerol, and second, by elevating resistance to oleate. Finally, FA synthesis was upregulated by ACC gene (*acc1*) overexpression, which resulted in the production of 1436.7 $mg L^{-1}$ of FFAs. The addition of dodecane further promoted FFA secretion and enhanced the level of FFAs up to 2033.8 $mg L^{-1}$ during test tube cultivation [376, 377].

Oleaginous yeast, *Y. lipolytica*, was engineered by first establishing an expression platform for high expression using intron-containing translation elongation factor-1α (TEF) promoter, which is capable of increasing gene expression 17-fold over the intron-less TEF promoter. Then, this platform was used for the overexpression of diacylglycerol acyltransferase (DGA1) in the final step of the TAG synthesis pathway, which yielded a fourfold increase in lipid production over control, to a lipid content of 33.8% of dry cell weight. In addition, the overexpression of ACC (ACC1) increased lipid content twofold over control, or 17.9% lipid content. Therefore, the two genes were combined in a tandem gene construction by simultaneous co-expression of ACC1 and DAG1 in the lipid biosynthetic pathway (Scheme 7.48), which further increased lipid content to 41.4%. The effects of ACC1 + DGA1 transformant on lipid production were explored using a 2-L bioreactor fermentation, which resulted in 61.7% lipid content after 120 hours. The overall yield and productivity were 0.195 $g g^{-1}$ and 0.143 $g\ L^{-1} h^{-1}$, respectively, while the maximum yield and productivity were 0.270 $g g^{-1}$ and 0.253 $g\ L^{-1} h^{-1}$ during the lipid accumulation phase of the fermentation [378]. The results demonstrate the effects of metabolic engineering of two important steps of the lipid synthesis pathway, which allows effective flux inversion toward lipid synthesis and creates driving force for TAG synthesis.

The allosteric inhibition of lipid synthesis by SFAs during the conversion of sugars to oils for some microorganisms limits its conversion yield and rates. A novel metabolic regulation enzyme delta-9-stearoy-CoA desaturase (SCD) of the lipid synthesis in *Y. lipolytica* through the reverse engineering of mammalian fat-storing tissue was identified and simultaneously overexpressed together with acetyl-CoA carboxylase (*ACC1*) and diacylglyceride acyl-transferase (*DAG1*) in *Y. lipolytica* (Scheme 7.49). This engineered strain characterized highly desirable phenotypes of fast cell growth and lipid overproduction such as high carbon to lipid conversion yield (84.7% of theoretical maximal yield), high lipid titers (~55 $g L^{-1}$), and enhanced tolerance to glucose and cellulose-derived sugars. The engineered strain also featured a threefold growth advantage over the wild-type strain, which resulted in a maximal lipid productivity of ~1 $g\ L^{-1} h^{-1}$ during the stationary phase. While the engineered yeast required cytoskeleton remodeling in

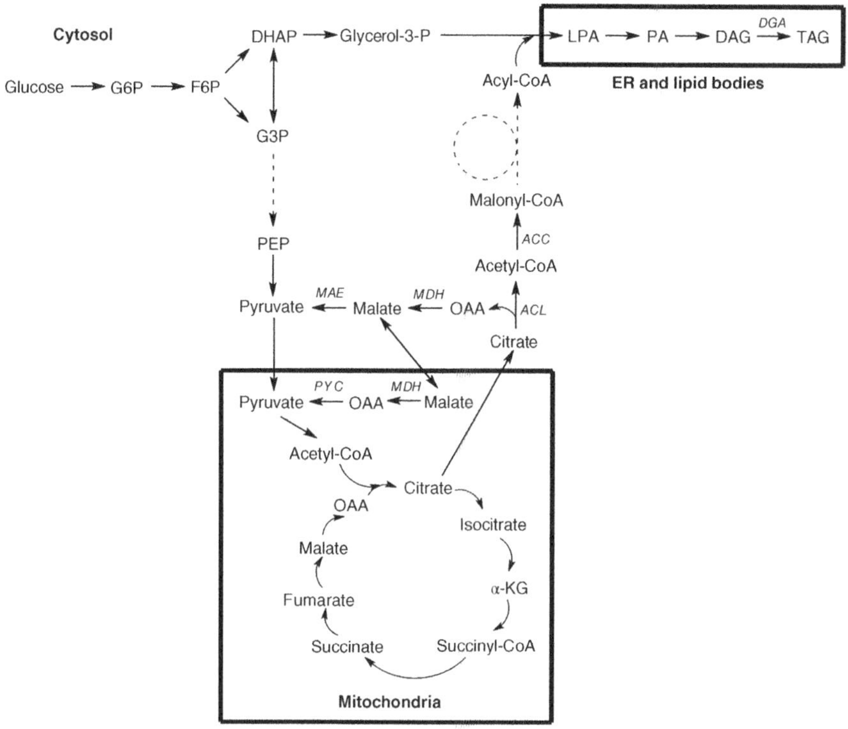

Scheme 7.48 Simplified principal metabolic pathways for lipid synthesis in *Y. lipolytica.* ACL, ATP-citrate lyase; DAG, diacylglycerol; DGA, diacylglycerol acyltransferase; DHAP, dihydroxyacetone phosphate; ER, endoplasmic reticulum; F6P, fructose-6-phosphate; G3P, glyceraldehyde 3-phosphate; G6P, glucose-6-phosphate; α-KG, alpha-ketoglutarate; LPA, lysophosphatidic acid; MAE, malic enzyme; MDH, malate dehydrogenase; OAA, oxaloacetate; PA, phosphatidic acid; PEP, phosphoenolpyruvate; PYC, pyruvate carboxylase.

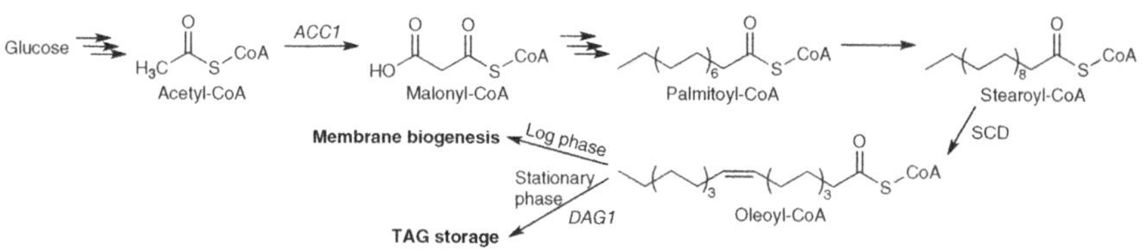

Scheme 7.49 Proposed mechanism of SCD-mediated deregulation of fatty acid pathway in oleaginous yeast *Y. lipolytica.*

the obesity phenotype. Overall, the engineered *Y. lipolytica* was demonstrated to be a superior lipid producer, which is capable of efficiently transforming glucose or renewable carbohydrates into TAGs at high yield and productivity [379].

Except for FAs and biofuels, ACC catalyzed acetyl-CoA to malonyl-CoA reaction was used for producing value-added chemicals including fatty alcohols, polyketides, flavonoids, and organic acids through the metabolically engineered microbial host using sophisticated genetic tools [363–365, 380–385]. Since fatty alcohols have been widely used in producing detergents, emulsifiers, lubricants, and cosmetics, yeast was engineered to produce 1-hexadecanol by expressing a fatty acyl-CoA reductase (FAR) from barn owl (*Tyto alba*). In order to improve the 1-hexadecanol production, the *ACC1* gene encoding ACC was overexpressed and β-oxidation enzymes POX1 and POT1 were knocked out. Moreover, the SNF1 protein that downregulates the *ACC1* expression and upregulates β-oxidation and RPD3, the negative regulator of the *INO1* gene in phospholipid metabolism, were both knocked out as well. To enhance the supply of cytosolic acetyl-CoA, thus malonyl-CoA, two pathways were engineered independently: (i) a pathway that can efficiently provide acetyl-CoA in the cytoplasm by expressing *ADH2* (alcohol dehydrogenase), *ALD6* (aldehyde dehydrogenase), and a codon-optimized ACS; and (ii) expression of ATP-dependent citrate lyase (ACL) from either *A. thaliana* or *Y. lipolytica* (Scheme 7.50). After metabolic engineering manipulation, the 1-hexadecanol titer was improved from 45 to 330 mg L^{-1} in 20 g L^{-1} glucose

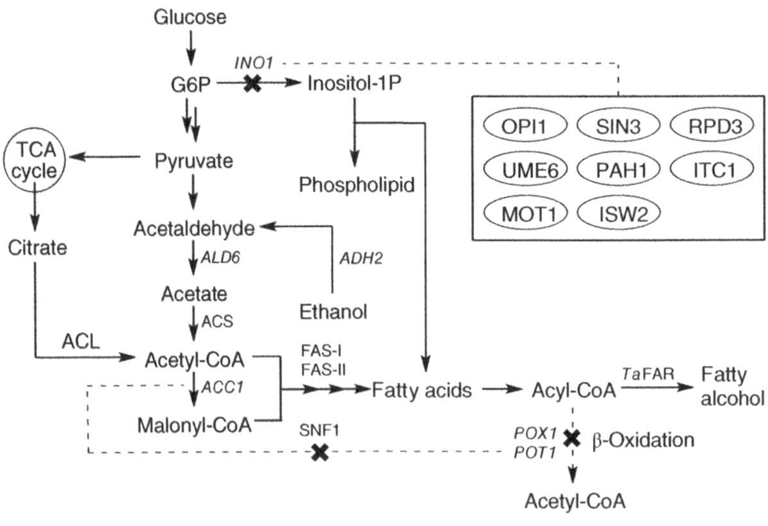

Scheme 7.50 Lipid metabolism of engineered *S. cerevisiae* for fatty alcohol production.

minimal medium. Through fed-batch fermentation using resting cells, over $1.1\,g\,L^{-1}$ 1-hexadecanol can be produced in glucose minimal medium [386].

Malonyl-CoA, the product of ACC catalyzed acetyl-CoA carboxylation reaction, is the major building block for many natural products such as polyketides and flavonoids, which have significant applications in medicine, veterinary medicine, and agriculture. In order to maintain a large amount of cellular malonyl-CoA, a variety of metabolic engineering strategies have been utilized to redirect the carbon flux inside *E. coli* to pathways responsible for the generation of malonyl-CoA. A threefold increase of cellular malonyl-CoA can be obtained by overexpression of acetyl-CoA carboxylase (*acc*). A synergistic effect with increased acetyl-CoA availability by deletion of competing pathways leading to the by-products acetate and ethanol as well as overexpression of an acetate assimilation enzyme led to an *E. coli* strain with 15-fold more cellular malonic-CoA. This engineered *E. coli* strain was applied for the production of an important polyketide, phloroglucinol, by further heterologous expression in *E. coli* of a single gene *phlD* encoding the type III PKS from *Pseudomonas fluorescens* Pf-5 (Scheme 7.51). The results showed near fourfold higher titer compared with wild-type strain, despite the toxicity of phloroglucinol to cell growth [387–389]. The metabolic engineering strategies described here should be highly useful for improved production of natural products where the cellular malonyl-CoA is rate limiting.

Mithramycin (MTM) is a glycosylated polyketide antitumor compound produced by *Streptomyces argillaceus*, which is constituted by a tricyclic aglycone with two aliphatic side chains, a disaccharide and a trisaccharide chain. The

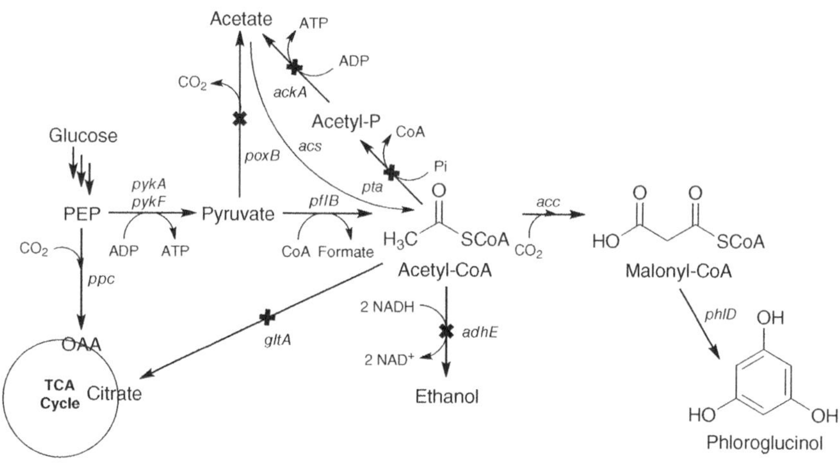

Scheme 7.51 Engineered *E. coli* for production of phloroglucinol from glucose.

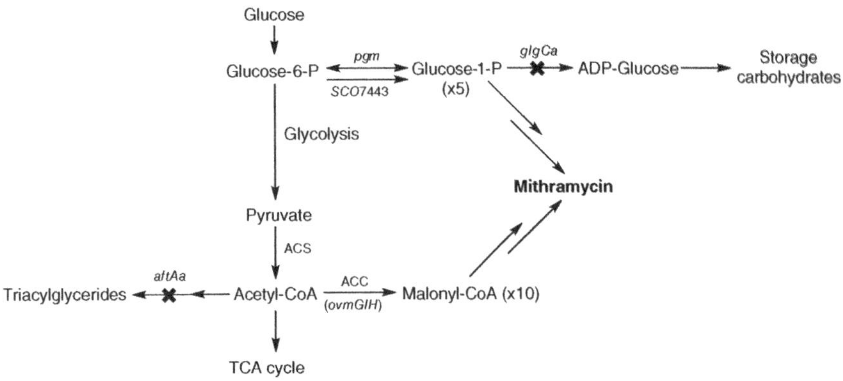

Scheme 7.52 Metabolically engineered *S. argillaceus* for production of mithramycin.

biosynthesis of MTM is initiated by the condensation of 10 malonyl-CoA units to render a carbon chain that is subsequently modified to originate a tetracyclic intermediate. The intermediate is sequentially glycosylated by five deoxysugars that in turn are synthesized from glucose-1-phosphate. Further oxidation and reduction steps render the final compound. Therefore, the supply of glucose-1-phosphate and malonyl-CoA was enhanced by overexpressing the *Streptomyces coelicolor* phosphoglucomutase gene *pgm* and the ACC *ovmGIH* genes from the oviedomycin biosynthesis gene cluster in *S. argillaceus*. Furthermore, the cloned ADP-glucose pyrophosphorylase gene *glgCa* and the acyl-CoA:diacylglycerol acyltransferase gene *aftAa* in *S. argillaceus* were inactivated to further increase the intracellular concentration of glucose-1-phosphate/glucose-6-phosphate and malonyl-CoA/acetyl-CoA (Scheme 7.52). The recombinant strain obtained by combining all metabolic engineering strategies together shows the result, which produced approximately 3.3 and 4.7 times more MTM than wild-type strain in liquid and solid media, respectively. These strategies were also applied to increase production of four MTM derivatives with improved pharmacological properties [390].

Nonconventional oleaginous yeast *Y. lipolytica* was explored for production of the simplest polyketide triacetic acid lactone (TAAL) based on its potential for high flux through the key polyketide precursors, acetyl-CoA and malonyl-CoA. Three distinct metabolic engineering strategies for acetyl-CoA formation, that are the citrate route for lipid production, the PDHC (*PDA1, PDE2, PDE3, PDB1*) pathway, and the previously uncharacterized pyruvate bypass pathway, were investigated (Scheme 7.53a). The pyruvate bypass pathway converts pyruvate to acetaldehyde through PDC, then to acetate through acetaldehyde dehydrogenase (ALD), and finally to acetyl-CoA via ACS. Therefore, overexpression of four genes (*ACS1*, *ALD5*, *PDC2*, and *ACC1*) in *Y. lipolytica* significantly improved the amounts of acetyl-CoA and malonyl-CoA, which further reacted to achieve TAAL using PKS. After optimization, the engineered strain achieved over 35% of the theoretical yield to TAAL in unoptimized tube culture and a maximum observed titer of $35.9 \pm 3.9\,\mathrm{g\,L^{-1}}$ in a 3-L bioreactor scale with glucose and spiked acetate. Moreover, a higher TAAL production (43% of theoretical conversion yield under nitrogen starvation in a test tube) can be obtained by overexpression of a β-oxidation related peroxisomal matrix protein (*PEX10*). Finally, TAAL was used for chemically modifying commodity poly(epichlorohydrin) to form poly(epichlorohydrin)-co(epoxy triacetic acid lactone) (PETAL) that upgrades its glass transition temperature [391]. Another metabolic engineering strategy employed for production of TAAL using *Y. lipolytica* and acetate as alternative carbon sources was through acetate uptake pathway as an acetyl-CoA shortcut to achieve metabolic optimality in producing TAAL. To efficiently utilize acetate for TAAL production, the metabolic bottlenecks in *Y. lipolytica*, including generation of the driving force for acetyl-CoA, malonyl-CoA, and NADH, were identified. Therefore, *Y. lipolytica* was engineered with the overexpression of endogenous acetyl-CoA carboxylase (*ACC1*), malic enzyme (*MAE1*), and a bacteria-derived cytosolic pyruvate dehydrogenase (PDH), which results in a TAAL titer of up to $4.76\,\mathrm{g\,L^{-1}}$ from industrial glacier acetic acid in shake flasks, showing 8.5 times improvement over the wild strain (Scheme 7.53b). The acetate-to-TAAL conversion rtatio ($0.149\,\mathrm{g\,g^{-1}}$) reaches 31.9% of the theoretical maximum yield, which indicates the carbon flux of this metabolic shortcut exceeds the carbon flux afforded by the native glycolytic pathways [392].

A new biosynthetic system based on Crabtree-negative yeast *Pichia pastoris*, which can grow exceptionally on ethanol and convert ethanol directly in cytosol to acetyl-CoA in three steps, was constructed to improve the production of acetyl-CoA-derived drugs. A glucose-repressed and ethanol-induced transcriptional signal amplification device (ESAD) with 20-fold signal increase was constructed to make ethanol the sole and fast-growing substrate, acetyl-CoA precursor, and strong biosynthetic pathway inducer. The developed ESAD was then used to produce dihydromonacolin L (DML), the first step intermediate of commercial hypolipidemic drug intermediate, monacolin J (MJ). DML is a fungal polyketide that is synthesized from nine malonyl-CoA and a methyl donor *S*-adenosyl methionine (SAM) with biosynthetic steps involving four enzymes, including lovastatin nonaketide synthase (*LovB*), enoyl reductase (*LovC*), and TE (*LovG*) from *Aspergillus terreus*, as well as phosphopantetheinyl transferase (*NpgA*) from *Aspergillus nidulans* (Scheme 7.54). While malonyl-CoA is directly derived from acetyl-CoA by constitutively overexpressing a *P. pastoris ACC1* in the *ADH2* and *ACS1* co-expressing strain. Then, a downstream strain for DML-to-MJ module controlled by a synthetic constitutive transcriptional signal amplification device

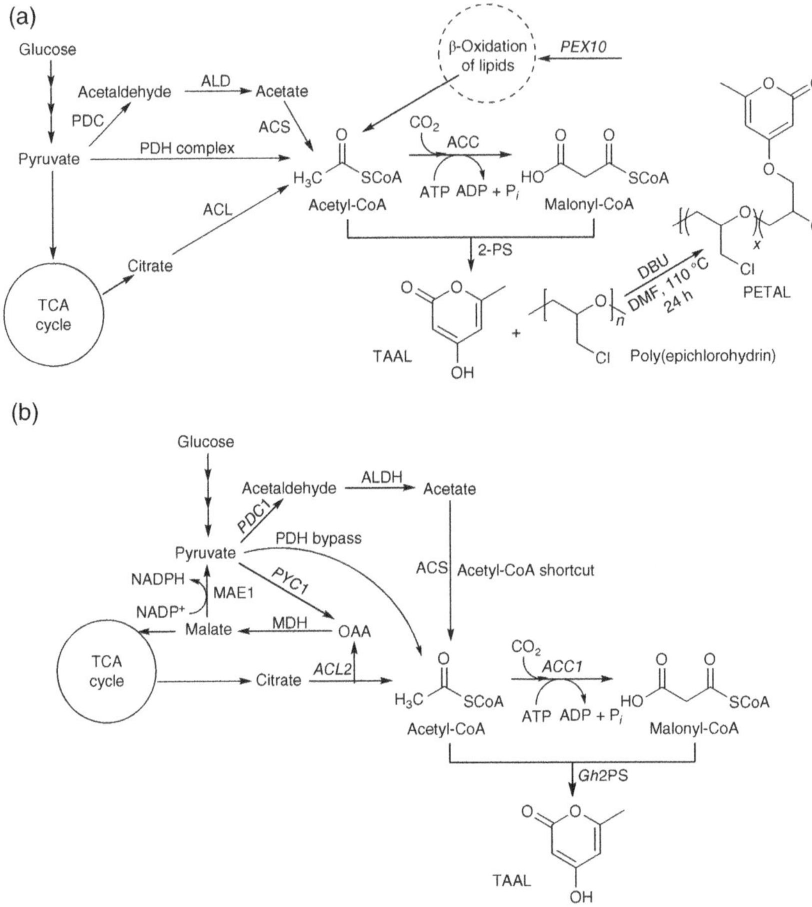

Scheme 7.53 Metabolic pathways of engineered *Y. lipolytica* for TAAL synthesis. (a) Three anabolic pathways targeted for overexpression: the citrate route, the pyruvate dehydrogenase complex, and the pyruvate bypass pathway, plus the potential β-oxidation upregulation target. (b) TAAL production from acetic acid by taking advantage of the acetyl-CoA shortcut.

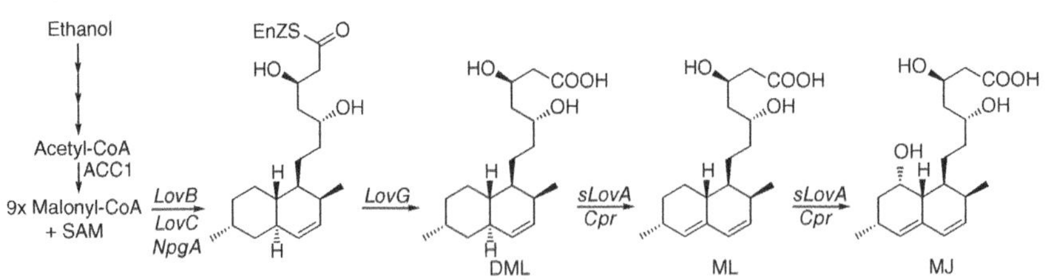

Scheme 7.54 Engineered ethanol-driven biosynthetic system for production of monacolin J in *P. pastoris*.

(CSAD) through co-expressing a cytochrome P450 monooxygenase (*sLovA*) and a cytochrome P450 oxidoreductase (*Cpr*) in a separate *P. pastoris* cell was developed, which produced a high MJ titer of $2.2\,g\,L^{-1}$ on ethanol. A 5-L stirred-tank co-culture fermentation system with the upstream strain and downstream strain allows a one-step high production of MJ (up to $3.2\pm0.1\,g\,L^{-1}$), which is far surpassing than the other reported MJ titers in heterologous hosts [393].

Phenylpropanoid-derived natural products, such as lignins, salicylates, coumarins, hydroxycinnamic amides, and flavonoids, are produced exclusively by plants through the so-called phenylpropanoid pathway. While flavonoid-derived compounds have drawn much attention by their usage as health-promoting components in human diet, due to their cancer chemopreventive, antioxidant, and anti-asthmatic activities. A plasmid in *E. coli*, which carried an artificial biosynthetic gene cluster, including *PAL* encoding a phenylalanine ammonia lyase from a yeast, *ScCCL* encoding a cinnamate/coumarate:CoA ligase from the actinomycete *S. coelicolor* A3(2), *CHS* encoding a chalcone synthase from a licorice plant, and *CHI* encoding a chalcone isomerase from the *Pueraria* plant, was constructed for the production of plant-specific flavanones. The recombinant *E. coli* cells produced (2*S*)-naringenin and (2*S*)-pinocembrin from tyrosine and phenylalanine, respectively (Scheme 7.55). The flavanone yields were greatly improved after the two subunit genes of ACC from *C. glutamicum* were expressed under the control of the T7 promoter and the ribosome-binding sequence in the recombinant *E. coli* cells, probably due to an increase of malonyl-CoA that was used for flavanone synthesis by the enhanced expression of acetyl-CoA. The yields of naringenin and pinocembrin were about $60\,mg\,L^{-1}$ when the recombinant *E. coli* cells were incubated in 3 mM tyrosine and phenylalanine [394].

Several strategies were employed for rational modifications of the acetate-acetyl-CoA-malonyl-CoA central metabolism in *E. coli* for improving malonyl-CoA precursor availability for robust synthesis of flavanones. The metabolic engineering approach targeted the multisubunit complex ACC-biotin ligase (BirA) and enzymes in acetate assimilation pathways in *E. coli* (Scheme 7.56). The metabolically engineered *E. coli* was able to increase the production of pinocembrin, naringenin, and eriodictyol in 36 hours up to 1379, 183, and 373%, respectively, overproduction with the strains expressing only the flavonoid pathway, which corresponded to 429, 119, and $52\,mg\,L^{-1}$, respectively [395].

An integrated computational and experimental approach was applied for improving the intracellular availability of malonyl-CoA in engineered *E. coli* to increase the production of plant-specific secondary metabolite naringenin. A customized version of the recently developed OptForce methodology was used to predict a minimal set of genetic interventions that guarantee a prespecified yield of malonyl-CoA in *E. coli*. Then a recombinant *E. coli* strain was constructed that exhibits a fourfold increase in the levels of intracellular malonyl-CoA compared

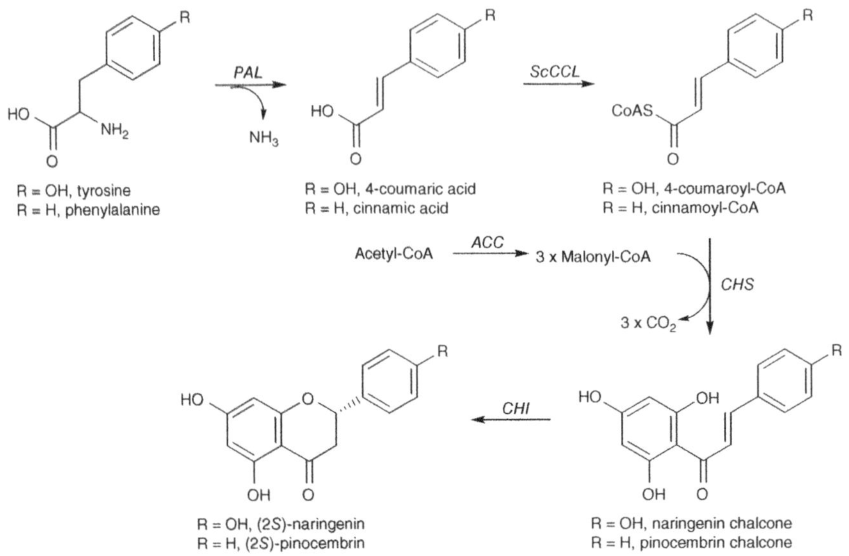

Scheme 7.55 Biosynthesis of (2*S*)-flavanones by engineered *E. coli*.

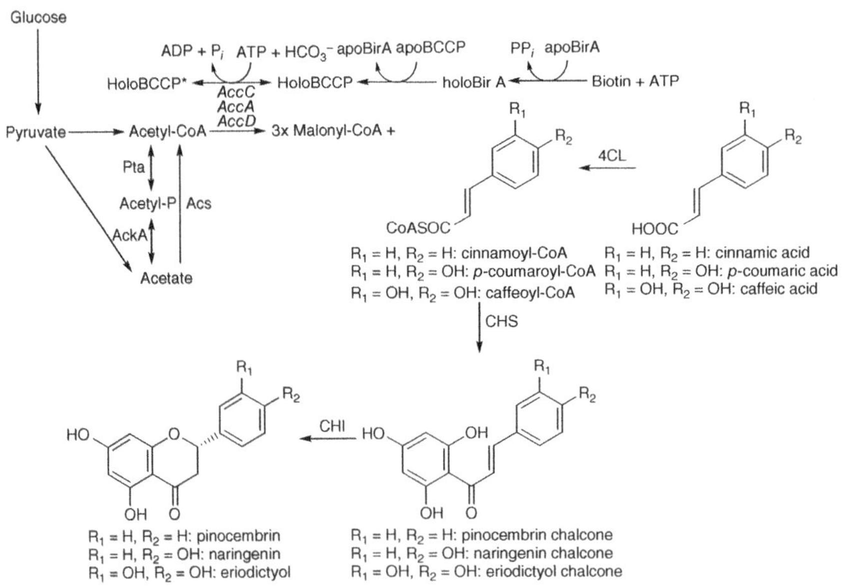

Scheme 7.56 Engineering central metabolic pathways in *E. coli* for plant flavanones production.

to the wild-type strain, which produced a naringenin titer of 474 mg L^{-1}, a highest yield ever achieved in a lab-scale fermentation process. This result was due to the combined effect of the synergistic genetic interventions that correlates genetic modification to cell phenotype, specifically the identified knockout targets (Δ*fumC* and Δ*sucC*) and overexpression targets (ACC, PGK, GAPD, and PDH) can cooperatively force carbon flux toward malonyl-CoA [396].

Eriodictyol is a flavonoid with anti-inflammatory and antioxidant activities. In order to efficiently synthesize eriodictyol, a truncated plant P450 flavonoid, flavonoid 3′-hydroxylase (tF3′H), was functionally expressed as a fusion protein with a truncated cytochrome P450 reductase (tCPR) in *E. coli*. This allowed the engineered *E. coli* to produce eriodictyol from L-tyrosine by simultaneously co-expressing the fusion protein with tyrosine ammonia lyase (TAL), 4-coumarate-CoA ligase (4CL), chalcone synthase (CHS), and chalcone isomerase (CHI). Moreover, the availability of malonyl-CoA was enhanced by metabolic engineering so as to achieve a new metabolic balance and rebalance the relative expression of genes to enhance eriodictyol accumulation. Using these strategies, the production of eriodictyol from L-tyrosine was made 203% higher than that in the control strain, which reached an amount of 107 mg L^{-1} [397].

7-*O*-Methyl aromadendrin (7-OMA) is an aglycone moiety of one of the important flavonoid-glycosides found in several plants, such as *Populus alba* and *Eucalyptus maculate*, with various medicinal applications. The biosynthesis of 7-OMA in *E. coli* from its precursor, *p*-coumaric acid, was primarily through naringenin synthesis, which was performed by feeding *p*-coumaric acid and reconstructing the plant biosynthetic pathway by introducing the structural genes, including 4-coumarate-coenzyme A ligase (4CL) from *Petroselinum crispum*, CHS from *Petunia hybrida*, and CHI from *Medicago sativa*. The enhancement of malonyl-CoA availability was overexpression genes for the acyl-CoA carboxylase α and β subunits (*nfa9890* and *nfa9940*), biotin ligase (*nfa9950*), and ACS (*nfa3550*) from *Nocardia farcinica*. Then, the produced naringenin was hydroxylated at position 3 by flavanone-3-hydroxylase (F3H) from *A. thaliana* and further methylated at position 7 to form 7-OMA using 7-*O*-methyltransferase (7-*O*-MT) from *Streptomyces avermitilis* through either dihydrokaempferol (DHK) (aromadendrin) or sakuranetin (SKN) intermediate (Scheme 7.57). Upon providing with 500 μM *p*-coumaric acid, the engineered *E. coli* strain produced 2.7 mg L^{-1} (8.9 μM) 7-OMA in 24 hours, whereas the strain expressing only the flavanone modification enzymes yielded 30 mg L^{-1} (99.2 μM) 7-OMA from 500 μM naringenin in 24 hours [398].

5-Deoxyflavanones are a distinct class of flavanone compounds in leguminous plants that play important roles in their symbiotic interactions. A four-step biosynthesis of 5-deoxyflavanones using engineered recombinant *E. coli* was constructed by overexpression of *E. coli* with four heterologous plant genes:

4-coumarate:coenzyme A ligase (4CL) from *P. crispum*, CHSs from *M. sativa* and *P. hybrida*, and chalcone reductase (CHR) and CHI from *M. sativa* (Scheme 7.58). Shake flask experiments demonstrated that *P. hybrida* rather than *M. sativa* CHS produced the highest amount of liquiritigenin from precursor *p*-coumaric acid in glucose minimal medium. Introduction of the recombinant pathway in *S. cerevisiae* also led to the formation of both 5-hydroxyflavanone and 5-deoxyflavanone, however, with yields of the later lower than that achieved in *E. coli*. Other phenylpropanoid acid precursor, such as cinnamic acid and caffeic acid, could be synthesized through the recombinant pathway as well produce corresponding 5-deoxyflavanone compounds [399].

Since it is usually lethal to microorganisms, to increase the intracellular availability of malonyl-CoA by inactivation of the malonyl-CoA consumption pathway

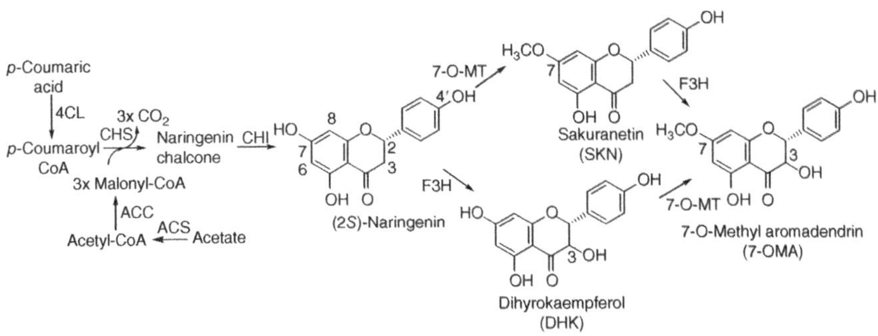

Scheme 7.57 Engineered *E. coli* for biosynthesis of 7-OMA starting from *p*-coumaric acid.

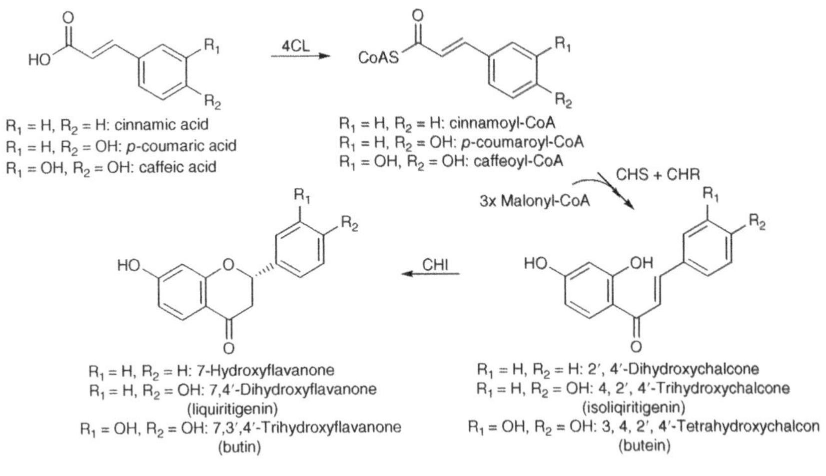

Scheme 7.58 Biosynthesis of 5-deoxyflavanones in engineered *E. coli*.

is not applicable. A novel metabolic engineering strategy is using synthetic antisense RNAs (asRNAs) to conditionally downregulate FA biosynthesis and achieve malonyl-CoA enrichment in *E. coli*. The optimized asRNA constructs with a loop-stem structure exhibit high interference efficiency up to 80%, which leads to a 4.5-fold increase in intracellular malonyl-CoA concentration when *fabD* gene expression is inhibited. The application of this strategy allows an increased production of natural products 4-hydroxycoumarin, resveratrol, and naringenin by 2.53-, 1.70-, and 1.53-fold in *E. coli*, respectively (Scheme 7.59). Besides, down-regulation of other fab genes including *fabH*, *fabB*, and *fabF* also makes remarkable increases in 4-hydroxycoumarin production [400].

Nargenicin A1 is an effective antibacterial compound produced by *Nocardia* sp. CS682 and exhibits significant activity against various Gram-positive bacteria, while possessing lower cytotoxicity than macrolide antibiotics such as erythromycin and spiramycin. In order to improve the production of nargenicin A1, a set of multi-monocistronic vectors, pNV18L1 (without transcriptional terminator (TT)) and pNV18 (with TT derived from *aac*(3)IV) containing hybrid promoter (derived from *ermE** and promoter region of *neor*), RBS, and restriction sites for cloning, so that each cloned gene was under its own promoter and RBS. The multi-monocistronic vector pNV18L2 showed better efficiency in reporter gene assay. Therefore, multiple genes involved in the biogenesis of pyrrole moiety (*ngnN2*, *ngnN3*, *ngnN4*, and *ngnN5* from *Nocardia* sp. CS682), glucose utilization (*glf* and *glk* from *Zymomonas mobilis*), and malonyl-CoA synthetase (*accA2* and *accBE* from *S. coelicolor* A3 (2)) were cloned in pNV18L2. Further statistical optimization of specific precursors (proline and glucose) and their feeding time led to about 84.9 $mg\,L^{-1}$ nargenicin A1 from *Nocardia* sp. GAP, which is approximately 24-fold higher than *Nocardia* sp. CS682 (without feeding). In addition, a dedicated hydroxylase *pikC* from the pikromycin gene cluster of *Streptomyces venezuelae* was expressed to generate *Nocardia* sp. PikC and employed to yield a novel derivative nargenicin A1 acid, which showed a reduction in antibacterial potential [401].

Lovastatin is a naturally derived drug commonly prescribed to patients suffering from heart-related diseases, which is produced as a secondary metabolite during fermentation of sugars by *A. terreus*. To increase the production of lovastatin in wild-type *A. terreus* strain, homologous recombination techniques were applied to enhance the flow of precursors toward the lovastatin biosynthesis pathway. Thus, a new strain was generated to overexpress ACC by replacing the native ACC promoter with a strong constitutive *Padh*A promoter from *A. nidulans*. A low-value substrate, glycerol, and mixture of lactose and glycerol were used independently as the carbon feedstock to determine the degree of response by the *A. terreus* strains toward the production of acetyl-CoA and malonyl-CoA. The concentration levels of malonyl-CoA and acetyl-CoA in the new strain were raised by 240 and

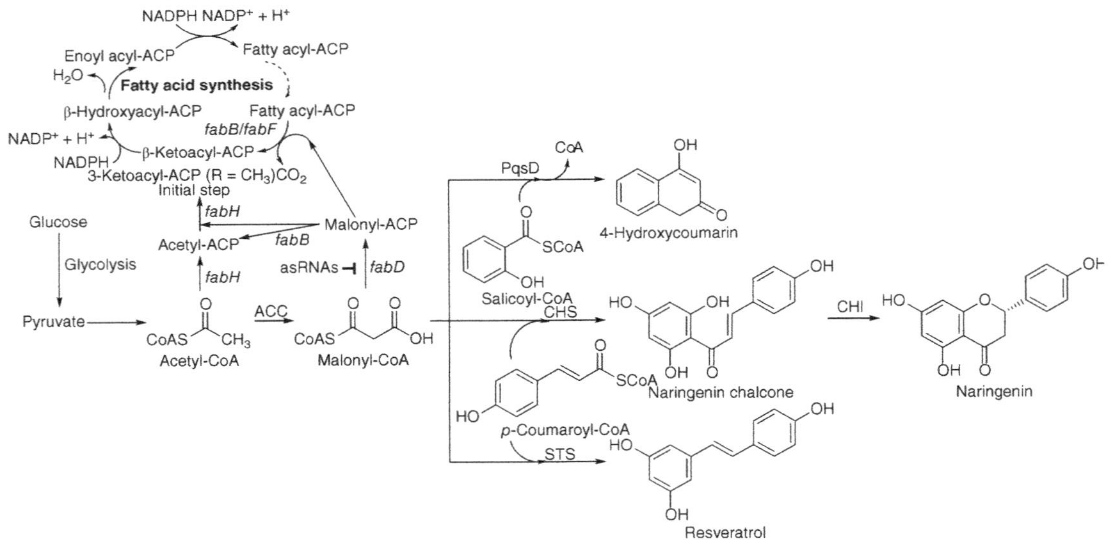

Scheme 7.59 Engineered *E. coli* for biosynthesis of 4-hydroxycoumarin, naringenin, and resveratrol.

14%, respectively, compared to the wild-type strain, which resulted in an increased production of lovastatin by 40% and a decreased production of by-product, (+)-geodin, by 31% [402].

Except for bacteria, the application of metabolic engineering on food-grade microorganisms, such as *S. cerevisiae*, is a potential alternative for sustainable production of plant secondary metabolites flavonoids. To improve the naringenin production yield due to insufficient enzyme activity, gene source screening was used as a tool to identify the best gene source for enzymes such as 4-coumarate: coenzyme A ligase (4CL) and CHS. Therefore, the 4CL gene from *M. truncatula* and the CHS gene from *Vitis vinifera* were expressed in *S. cerevisiae*, which provided 28-fold higher yield of naringenin as compared to the control strain. The combinations acquired similar performance in the Y-28 strains that resulted in a highest production of 28.68 mg L^{-1} [403]. In another study, an efficient production of (2*S*)-naringenin in *S. cerevisiae* from *p*-coumaric acid was via a biosynthetic pathway by integrating synthesis pathway genes into the yeast genome to achieve a (2*S*)-naringenin production strain. The gene dosage experiments indicate that the genes negatively regulating the shikimate pathway and inefficient CHS activity were factors limiting (2*S*)-naringenin production. Using fed-batch process optimization of the engineered strain, the (2*S*)-naringenin titer achieved was 648.63 mg L^{-1} from 2.5 g L^{-1} *p*-coumaric acid [404].

Engineered *S. cerevisiae* has been utilized for *de novo* biosynthesis of resveratrol, a natural antioxidant for food supplement and cosmetics ingredient, directly from cheap carbon sources, glucose or ethanol, via tyrosine intermediate. The biosynthetic pathway, consisting of TAL from *Herpetosiphon aurantiacus*, 4-coumaryl-CoA ligase from *A. thaliana*, and resveratrol synthase from *V. vinifera*, was introduced first and obtained 2.73 ± 0.05 mg L^{-1} resveratrol from glucose. Then, feedback-insensitive alleles of *ARO4* encoding 3-deoxy-D-arabino-heptulosonate-7-phosphate and *ARO7* encoding chorismite mutase were overexpressed to result in the production of 4.85 ± 0.31 mg L^{-1} resveratrol from glucose as the sole carbon source. In addition, the supply of the precursor malonyl-CoA was improved by overexpressing a post-translational deregulated version of the acetyl-CoA carboxylase encoding gene *ACC1*, which further increased the resveratrol concentration to 6.39 ± 0.03 mg L^{-1}. Followed by these strategies, the strain was improved by performing multiple-integration of pathway genes to increase the resveratrol production to 235.57 ± 7.00 mg L^{-1}. The final strain was fed-batch fermented to produce resveratrol using glucose or ethanol as carbon source, which resulted in a resveratrol titer of 415.65 and 531.41 mg L^{-1}, respectively [405].

Kaempferol found in many edible plants is a flavonol with broad bioactivity of antioxidant, anticancer, antidiabetic, antimicrobial, cardio-protective, and anti-asthma. The budding *S. cerevisiae* was employed for *de novo* biosynthesis of kaempferol. The biosynthetic pathway of kaempferol was constructed by first introducing high-efficient flavonol synthases from *Populus deltoids*, followed by the construction of a recombinant *S. cerevisiae* for *de novo* synthesis of kaempferol, which generated ~6.97 mg L^{-1} kaempferol from glucose. The kaempferol production was further promoted by overexpressing the acetyl-CoA biosynthetic pathway and *p*-coumarate was supplied as substrate (Scheme 7.60), which resulted in an improved kaempferol production by about 23 and 120%, respectively. Fed-batch fermentation was finally developed for better kaempferol production, which resulted in a kaempferol production of 66.29 mg L^{-1} in 40 hours [406].

Since the oleaginous yeast *Y. lipolytica* has been previously demonstrated as an optimal host for producing the simple polyketide, TAL, the capacities of this host were expanded by employing a β-oxidation-related strategy to improve more complex polyketides production directly from defined media. Therefore, the 4-coumaroyl-CoA derived polyketides, naringenin, resveratrol, and bisdemethoxycurcumin, as well as the diketide intermediate, (*E*)-5-(4-hydroxyphenyl)-3-oxopent-4-enoic acid, were established. Based on this strategy, *de novo* synthesis of naringenin was able to be produced in high level through import of both a heterologous pathway and a mutant *Y. lipolytica* allele, that generated an average maximum naringenin titer of 898 mg L^{-1}, the highest titer ever reported in any host. The results indicate that *Y. lipolytica* would be an ideal polyketide production host for more complex 4-coumaroyl-CoA derived products [407].

The oleaginous yeast *Y. lipolytica* was also employed to produce flavonoid and hydroxylated flavonoid. To overcome the limitations caused by the limited number of auxotrophic markers in *Y. lipolytica*, which restrict the ability to perform iterative genetic modifications and manipulate long gene clusters, the high recombination efficiency of the Cre-loxP system and the high integration rate of 26s rDNA were combined to develop a versatile framework to iteratively integrate multicopy metabolic pathways in *Y. lipolytica*. Both the flavonoid precursor (cytosolic acetyl-CoA and malonyl-CoA) pathways and the plant-derived cytochrome P450 enzymes were functionally integrated to improve flavonoid and hydroxylated flavonoid production. Thus, the engineered strains generated 71.2 mg L^{-1} naringenin, 54.2 mg L^{-1} eriodictyol, and 48.1 mg L^{-1} taxifolin [408]. Because oleaginous yeast produces a large amount of lipid bodies, and the abundant membrane structure and its lipophilic environment provide an ideal environment for the regioselectivity and stereoselectivity of many plant-derived P450 enzymes, modular method was used for constructing, characterizing, and optimizing the flavonoid pathways in *Y. lipolytica* (Scheme 7.61). In addition, a variety of precursor biosynthetic routes were evaluated and the metabolic potential pathways in *Y. lipolytica* were unleashed. Specifically, CHS and CPR were identified to be the bottlenecks of hydroxylated flavonoid production. Furthermore, the precursor pathway limitations by expressing genes associated with chorismate and malonyl-CoA supply were removed. Through optimization of the pH and

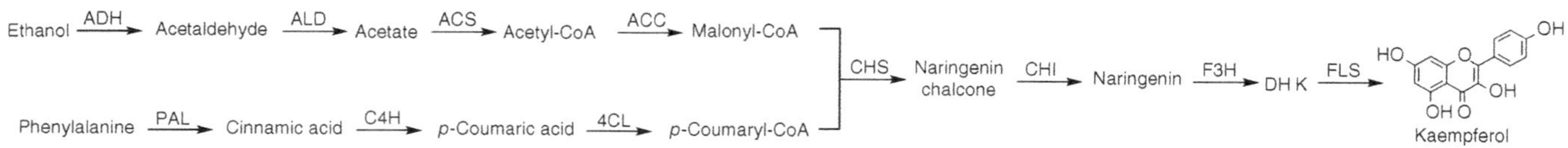

Scheme 7.60 Metabolically engineered *S. cerevisiae* for high kaempferol production.

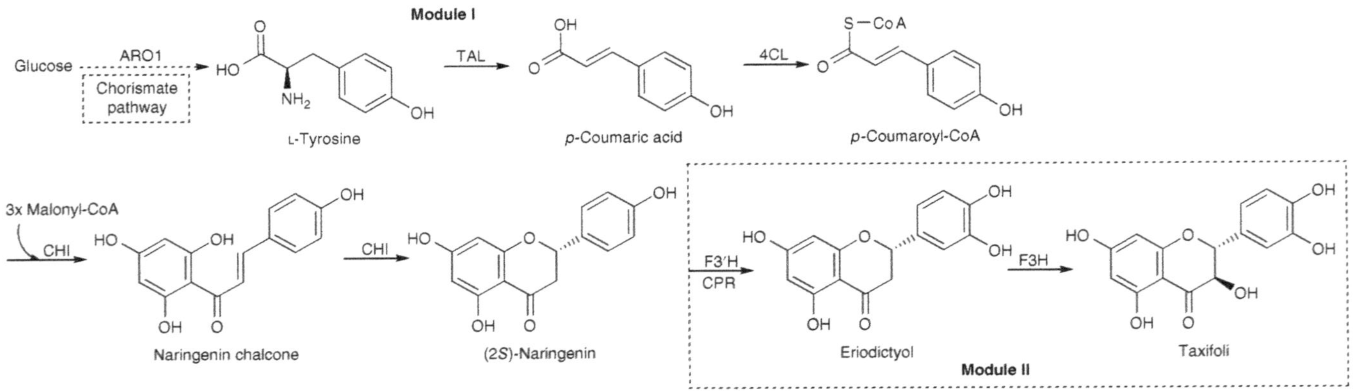

Scheme 7.61 Modular strategy to optimize biosynthetic pathways in *S. cerevisiae* for naringenin, eriodictyol, and taxifolin production.

carbon-nitrogen ratio (C/N), the engineered strain produced 252.4 mg L^{-1} naringenin, 134.2 mg L^{-1} eriodictyol, and 110.5 mg L^{-1} taxifolin from glucose in shake flask experiments [409].

Organic acids constitute a key group among top building block chemicals. With the recent desire to develop more sustainable bioprocesses, microbial production of organic acids has been expanded rapidly as alternative to petrochemical method. 3-HP is an important organic acid because it can serve as a platform for the production of 3-carbon intermediates. For example, it can easily be converted to acrylic acid upon dehydration, which is a major petrochemical product and is in increasing demand every year due to its various applications in superabsorbent, adhesive, surface coating, and paint. A variety of biological routes were employed for the production of 3-HP, including the glycerol pathway, the β-alanine pathway, and the malonyl-CoA pathway [410, 411]. The malonyl-CoA pathway involving ACC will be discussed in the following texts.

Fermentative production of 3-HP is centered on bacterial systems in most cases, in which *E. coli* is the most common one. Recombinant *E. coli* strains heterologously expressing the *mcr* gene of *Chloroflexus aurantiacus* DSM 635, encoding a bifunctional NADPH-dependent malonyl-CoA reductase (MCR), were developed to produce 3-HP from its immediate precursor malonyl-CoA (Scheme 7.62). This recombinant *E. coli* generated 0.71 mM of 3-HP in 24 hours in shake flask cultivation under aerobic conditions with glucose as the sole source of carbon. As ACC and biotinilase, encoded by the genes *accADBCb* of *E. coli* K-12 were overexpressed along with MCR, the 3-HP titer was improved by twofold to 1.6 mM L^{-1}. 3-HP production was further increased to 2.14 mM by additional expression of the gene *pntAB*, encoding nicotinamide nucleotide transhydrogenase that converts NADH to NADPH [412].

Cell growth in stationary phase potentially enhances target metabolite production due to the decreasing demand of carbon consumption for cell growth. It has been found that starvation of essential nutrients, such as nitrogen, sulfur, magnesium, and phosphorus, leads cells into stationary phase, and the overall metabolic behavior changes depend on the type of nutrient starvation in *E. coli*. Therefore, the optimum nutrient starvation type for producing malonyl-CoA-derived metabolites such as 3-HP and naringenin in *E.coli* was surveyed. Results showed that

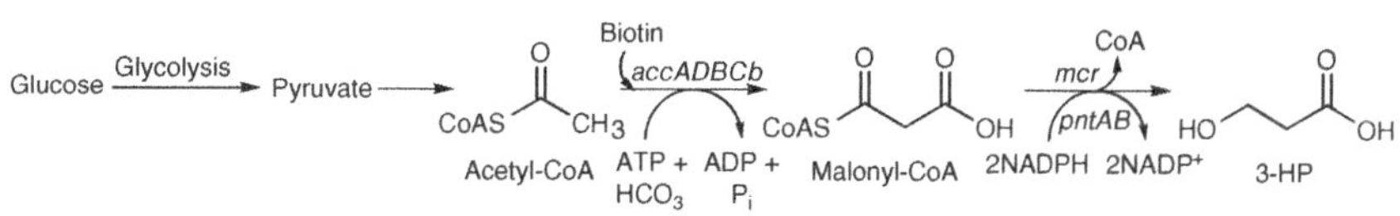

Scheme 7.62 3-Hydroxypropionic acid production from glucose through malonyl-CoA pathway with the recombinant *E. coli*.

high production titer (2.0 or 2.3 mM) and high specific production rate (0.14 or 0.28 mmol $(g\,CDW)^{-1}\,h^{-1}$) were observed for 3-HP production under sulfur or magnesium starvation, whereas almost no 3-HP was found under nitrogen or phosphorus starvation. Since malonyl-CoA is the precursor of 3-HP, the intracellular malonyl-CoA concentration was significantly increased during 3-HP production by the metabolic profiling analysis. Strong positive correlation ($r = 0.95$) between intracellular concentrations of ATP and malonyl-CoA also revealed that the ATP level is important for malonyl-CoA synthesis due to the requirement of ATP for ACC. Magnesium starvation led to high production titer and specific productivity for naringenin production as well [413].

Almost all major groups of prokaryotes possess the capability of using carbon dioxide as the sole carbon source for cell growth (autotrophy). The studies of the pathway of autotrophic CO_2 fixation in the phototrophic bacterium *C. aurantiacus* and in the aerobic thermoacidophilic archaeon *Metallosphaera sedula* revealed that none of the key enzymes of the reductive pentose phosphate cycle, the reductive TCA cycle, and the reductive acetyl-CoA pathway were detectable, but the biotin-dependent ACC and propionyl-CoA carboxylase as well as PEPC were contained in the cells. The results indicate the autotrophic growth rate via the 3-hydroxypropionate cycle, which involves the CO_2-, MgATP-, and NADPH-dependent catalytic conversion of acetyl-CoA to 3-HP via malonyl-CoA and the transformation of this intermediate to succinate via propionyl-CoA (Scheme 7.63). The investigation was further extended to the autotrophic archaea *Sulfolobus metallicus* and *Acidianus infernus*, and the cell extracts also showed acetyl-CoA and propionyl-CoA carboxylase activities. Furthermore, large amounts of a small biotin-carrying protein were detected in these aerobic archaea, as well as *C. aurantiacus*, which is most likely part of the acetyl-CoA and propionyl-CoA carboxylases. These findings suggest that the aerobic autotrophic archaea *M. sedula*, *S. metallicus*, and *A. infernus* use a yet-to-be-defined 3-hydroxypropionate cycle for their autotrophic growth, in which ACC and propionyl-CoA carboxylase are proposed to be the main CO_2 fixation enzymes, and PEPC may have an anaplerotic function. The results also support the occurrence of 3-HP cycle in the phototrophic bacterium *C. aurantiacus* [414].

Extremely thermophilic archaea *M. sedula* uses the 3-hydroxypropionate/4-hydroxybutyrate (3-HP/4-HB) cycle for fixing CO_2. Several reasons show that the 3-HP/4-HB cycle is a promising candidate for microbial production of chemical from CO_2: (i) it functions at high temperature to allow its use in extremely thermophilic host with concomitant minimal risk of contamination and reduced cooling costs; (ii) the 3-HP/4-HB cycle can function in either an aerobic or anaerobic host; (iii) the 3-HP/4-HB cycle can drive rapid autotrophic growth with a doubling time of less than 5 hours, indicating fast pathway kinetics. A reaction kinetic model was developed for recombinant *M. sedula* to examine the biological and

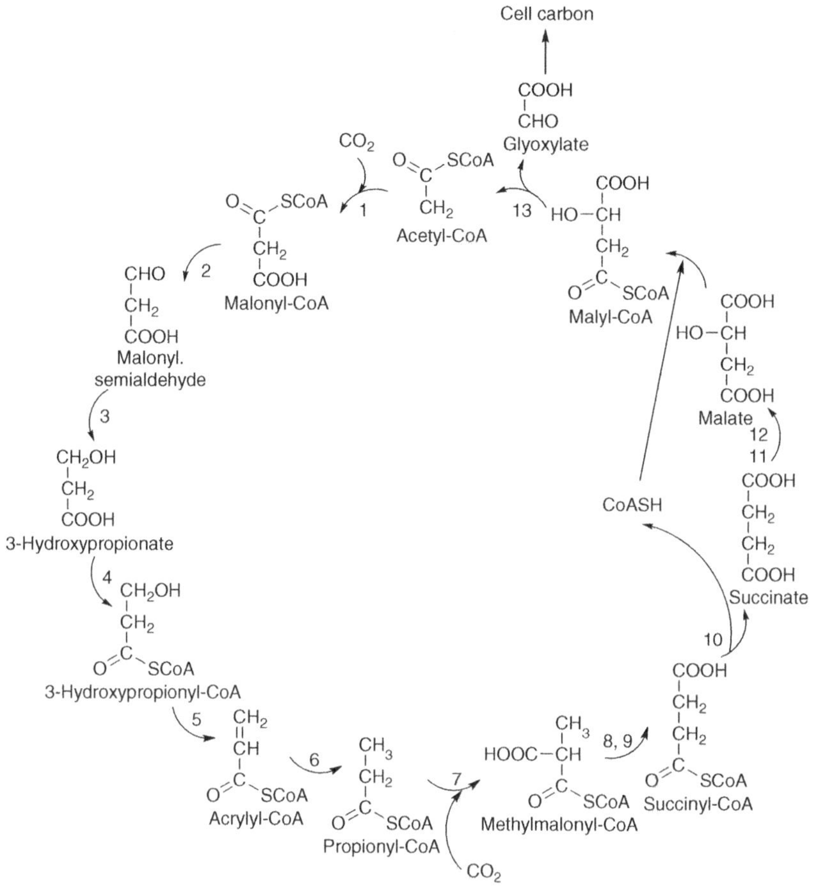

Scheme 7.63 Proposed 3-HP cycle of autotrophic CO_2 fixation in the phototrophic bacterium *C. aurantiacus*. Enzymes: **1**, acetyl-CoA carboxylase; **2**, malonate-semialdehyde dehydrogenase; **3**, 3-hydroxypropionate dehydrogenase; **4**, 3-hydroxypropionate-CoA ligase; **5**, 3-hydroxypropionyl-CoA dehydratase; **6**, acrylyl-CoA reductase; **7**, propionyl-CoA carboxylase; **8**, methylmalonyl-CoA epimerase; **9**, methylmalonyl-CoA mutase; **10**, succinyl-CoA:malate-CoA transferase; **11**, succinate dehydrogenase; **12**, fumarase; **13**, malyl-CoA lyase.

biotechnological attributes of the 3-HP/4-HB cycle for assessing metabolic engineering strategies of incorporating CO_2 into biotechnologically important chemical intermediates and products: acetyl-CoA, succinate, and 3-HP [415].

Because cyanobacteria are able to fix CO_2 directly, they have been engineered as an autotrophic microbial cell factory to produce 3-HP. Four approaches were employed to construct a biosynthetic pathway of 3-HP in cyanobacterium

Synechocystis sp. PCC 6803 that include (i) using different promoters and cultivation conditions to increase expression of MCR gene, (ii) overexpressing ACC and biotinilase to enhance supply of the precursor malonyl-CoA, (iii) overexpressing the NAD(P) transhydrogenase gene to improve NADPH supply, and (iv) inactivating the competing pathways of poly-3-hydroxybutyrate (PHB) and acetate biosynthesis to direct more carbon flux into 3-HP (Scheme 7.64). The engineered cyanobacteria system was then optimized for the production of 3-HP, which led to 837.18 mg L^{-1} (348.8 mg g^{-1} dry cell weight) 3-HP directly from CO_2 after 6 days cultivation. However, overexpression of the ribulose-1,5-bisphosphate carboxylase/oxygenase (Rubisco) gene from *Anabaena* sp. PCC 7120 and *Synechococcus* sp. PCC 7942 led to no increase of 3-HP production, which indicates CO_2 fixation may not be a rate-limiting step for 3-HP biosynthesis in *Snechocystis* [416].

Although metabolic engineering efforts on strengthening driving forces in pathway design may favor product formation in the organisms, imbalanced cellular intermediate distribution caused by excessive driving force may be detrimental to product synthesis. Therefore, an ATP-hydrolysis based driving force module provided with a reversible outlet for excessive carbon flux was engineered into cyanobacterium *Synechococcus elongatus* PCC 7942 for producing 3-hydroxybutyrate (3-HB), a valuable chemical feedstock for the synthesis of biodegradable plastics and antibiotics (Scheme 7.65). This engineering strategy effectively increases the product formation and reduces the uncontrolled accumulation of intermediate metabolites, which likely lead to metabolic imbalance and thus to cell growth inhibition. The engineered strain resulted in a significant increase of the cumulative 3-HB titer to 1.2 g L^{-1} compared to the original strain [417].

A type II methanotroph, *Methylosinus trichosporium* OB3b, was engineered by reconstructing malonyl-CoA pathway through heterologous expression of *C. aurantiacus* MCR to improve 3-HP production. To increase the malonyl-CoA pool, the supply of malonyl-CoA precursors was first engineered by overexpressing endogenous ACC, which substantially enhances the production of 3-HP. Then, overexpression of biotin protein ligase (BPL) and malic enzyme ($NADP^+$-ME) increases the 3-HP titer by ~22.7 and ~34.5%, respectively, in ACC-overexpression cells. The second strategy used to increase the malonyl-CoA pool was the reconstruction of acetyl-CoA carboxylation bypass route by co-expression of methylmalonyl-CoA carboxytransferase of *P. freudenreichii* and PEPC, which provides the MMC precursor and further improved the 3-HP titer. The highest 3-HP titer of 49 mg L^{-1} was obtained with the OB3b-MCRMP strain overexpressing MCR, MMC, and PEPC, which was 2.4-fold increase of titer compared with that in the strain only overexpressing MCR. The 3-L fermentation using the OB3b-MCRMP strain along with gaseous substrates, methane, CO_2, and air mixture resulted in 60.59 mg L^{-1} of 3-HP in 42 hours with a 6.36-fold improvement of volumetric productivity compared with that of the flask cultures [418].

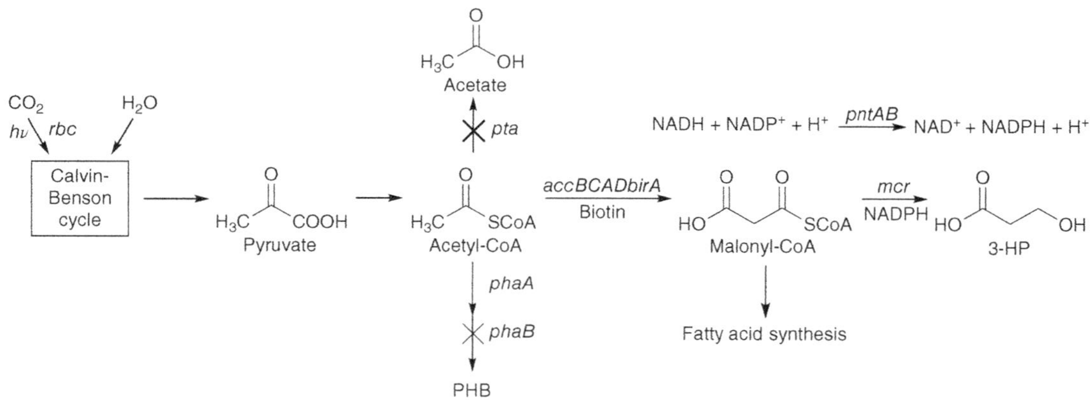

Scheme 7.64 Biosynthetic pathway of 3-HP in engineered *Synechocystis*.

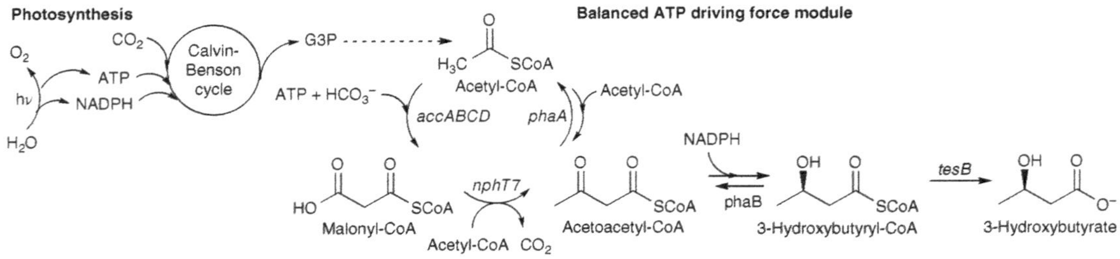

Scheme 7.65 Engineered *S. elongatus* with a balanced driving force module for 3-hydroxybutyrate production.

The use of bacterial hosts for large-scale 3-HP production exhibits certain drawbacks such as low tolerance to low extracellular pH, which means that it is not possible to produce the acid form directly. The result is the need for large volume base titrant to maintain the medium pH and the formation of gypsum as a by-product in the subsequent process of converting the salt to the acid form. Therefore, food yeast *S. cerevisiae* is an interesting alternative host as it tolerates relatively low pH, it is easy to engineer, and it has a detailed understanding of its metabolism. The malonyl-CoA pathway to 3-HP production in the budding yeast *S. cerevisiae* was reconstructed through evaluation of different MCR because this allows low pH fermentation to lead a cheaper overall process. The host strain was further engineered by increasing availability of the precursor malonyl-CoA and by coupling the production with increased NADPH supply (Scheme 7.66), which led to substantial improvement of 3-HP production by fivefold, up to a final titer of 463 mg L^{-1} [419].

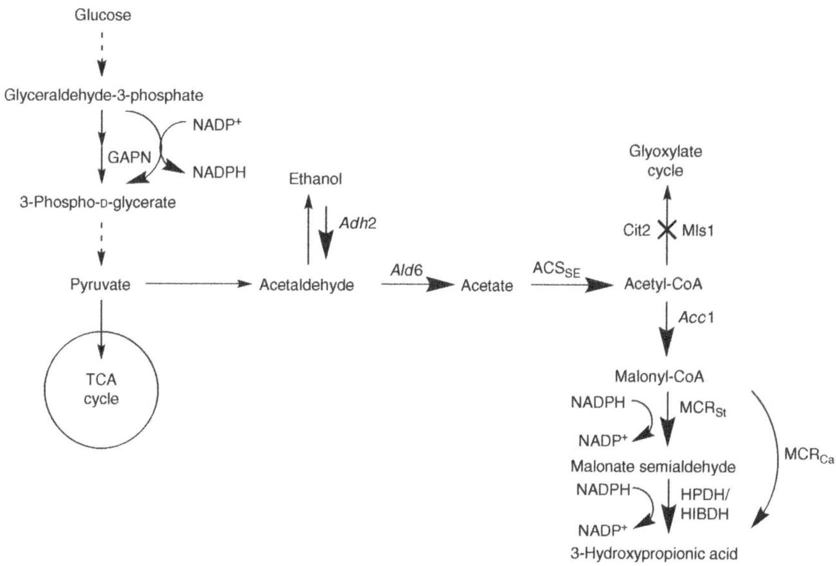

Scheme 7.66 3-HP biosynthetic pathway in engineered *S. cerevisiae*. Bold arrows show modifications. GAPN, nonphosphorylating glyceraldehy-3-phosphate dehydrogenase from *Streptococcus mutans*; *Adh2*, alcohol dehydrogenase; *Ald6*, aldehyde dehydrogenase; ACS_{SE}, modified acetyl-CoA synthase from *S. enterica*; *Cit2*, citrate synthase 2; *Mls1*, malate synthase 1; *Acc1*, acetyl-CoA carboxylase 1; MCR_{Ca}, malonyl-CoA reductase from *C. aurantiacus*; MCR_{St}, malonyl-CoA reductase from *Sulfolobus tokodaii*; HPDH, 3-hydroxypropionate dehydrogenase from *M. sedula*; HIBDH, 3-hydroxyisobutyrate dehydrogenase from *P. aeruginosa*. All overexpression of the multiple genes were conducted using plasmids.

Since the control of *Acc1* activity in *S. cerevisiae* mainly occurs at two levels, i.e. regulation of transcription and repression by Snf1 protein kinase at the protein level, a strategy for improving the activity of ACC in *S. cerevisiae* was conducted by abolishing post-translational regulation of *Acc1* via site-directed mutagenesis. The introduction of two site mutations in *Acc1*, Ser659 and Ser1157, was found to result in an enhancement of the activity of *Acc1* and an increase of the total FA content. Because Snf1 regulation of *Acc1* is particularly active under glucose-limited conditions, the effect of the two site mutations was evaluated in chemostat cultures. The results demonstrated that the modifications of *Acc1* could enhance the supply of malonyl-CoA and thus successfully increase the production of two industrially important products derived from malonyl-CoA, FAEEs, and 3-HP [420].

Because the fission yeast *S. pombe* has higher tolerance to 3-HP, this microorganism was selected to carry out the production of 3-HP from glucose via the malonyl-CoA pathway. *S. pombe* transformants were constructed by overexpressing two genes, one encoding the *S. pombe* ACC (*Cut6p*) and the other encoding the malonyl-CoA reductase derived from *C. aurantiacus* (*Ca*MCR). Moreover, an *S. pombe* protease-deficient strain was employed to prevent the degradation of these expressed proteins and acetate was supplemented to the medium to increase the cytosolic concentration of acetyl-CoA. Both strategies improved 3-HP production. A highly expressing *S. pombe hsp9* promoter was also exploited for overexpression of *Cut6p* and *Ca*MCR on 3-HP production. By culturing this engineered strain in a high-density shake flask culture, 7.6 g L^{-1} 3-HP at 31 hours with the maximum production rate of 1.183 g $L^{-1} h^{-1}$ was achieved [421].

References

1 Gupta, K. and Chundawat, T.S. (**2019**). *Asian J. Chem.* 31: 2698–2706.

2 Wandrey, C., Liese, A., and Kihumbu, D. (**2000**). *Org. Process Res. Dev.* 4: 286–290.

3 The Enzyme List (**2019**). Class 6 – Ligases – ExplorEnz. Nomenclature Committee of the International Union of Biochemistry and Biology (NC–IUBMB). www.enzyme-database.org/downloads/ec6.pdf (accessed 19 July 2021).

4 Raushel, F.M. (**2017**). *Biochemistry* 5: 1175–1176.

5 Robison, S.L., Christenson, J.K., Richman, J.E. et al. (**2019**). *ChemBioChem* 20: 1701–1711.

6 Frias, J.A., Goblirsch, B.R., Wackett, L.P., and Wilmot, C.M. (**2010**). *Acta Crystallogr.* F66: 1108–1110.

7 Christenson, J.K., Jensen, M.R., Goblirsch, B.R. et al. (**2017**). *J. Bacteriol.* 199: e00890–e00816.

8 Kancharla, P., Bonnett, S.A., and Reynolds, K.A. (**2016**). *ChemBioChem* 17: 1426–1429.

9 Christenson, J.K., Richman, J.E., Jensen, M.R. et al. (**2017**). *Biochemistry* 56: 348–451.

10 Conti, E., Franks, N.P., and Brick, P. (**1996**). *Structure* 4: 287–298.

11 Sukovich, D.J., Seffernick, J.L., Richman, J.E. et al. (**2010**). *Appl. Environ. Microbiol.* 76: 3842–3849.

12 Sukovich, D.J., Seffernick, J.L., Richman, J.E. et al. (**2010**). *Appl. Environ. Microbiol.* 76: 3850–3862.

13 Frias, J.A., Richman, J.E., Erickson, J.S., and Wackett, L.P. (**2011**). *J. Biol. Chem.* 286: 10930–10938.

14 Albro, P.W. and Dittmer, J.C. (**1969**). *Biochemistry* 8: 394–405.

15 Frias, J.A., Richman, J.E., and Wackett, L.P. (**2009**). *Appl. Environ. Microbiol.* 75: 1774–1777.

16 Bonnet, S.A., Papireddy, K., Higgins, S. et al. (**2011**). *Biochemistry* 50: 9633–9640.

17 Weibel, E.K., Hadvary, P., Hochuli, E. et al. (**1987**). *J. Antibiot.* 40: 1081–1085.

18 Wyatt, M.A., Ahilan, Y., Argyropoulos, P. et al. (**2013**). *J. Antibiot.* 66: 421–430.

19 Knights, K.M. (**1998**). *Clin. Exp. Pharmacol. Physiol.* 25: 776–782.

20 Liedvogel, B. and Stumpf, P.K. (**1982**). *Plant Physiol.* 69: 897–903.

21 Glasky, A.J. and Rafelson, M.E. Jr. (**1959**). *J. Biol. Chem.* 234: 2118–2122.

22 Ikeda, Y., Yamamoto, J., Okamura, M. et al. (**2001**). *J. Biol. Chem.* 276: 34259–34269.

23 Shockey, J.M., Fulda, M.S., and Browse, J. (**2003**). *Plant Physiol.* 132: 1065–1076.

24 Beinert, H., Green, D.E., Hele, P. et al. (**1953**). *J. Biol. Chem.* 203: 35–45.

25 Imesch, E. and Rous, S. (**1984**). *Int. J. Biochem.* 16: 875–881.

26 Yamashita, H., Fukuura, A., Nakamura, T. et al. (**2002**). *J. Nutr. Sci. Vitaminol.* 48: 359–364.

27 Ishikawa, M., Fujino, T., Sakashita, H. et al. (**1995**). *Tohoku J. Exp. Med.* 175: 55–67.

28 Fujino, T., Kondo, J., Ishikawa, M. et al. (**20001**). *J. Biol. Chem.* 276: 11420–11426.

29 Lee, H.Y., Na, K.B., Koo, H.M., and Kim, Y.S. (**2001**). *J. Biochem.* 130: 807–813.

30 Jogl, G. and Tong, L. (**2004**). *Biochemistry* 43: 1425–1431.

31 Ingram-Smith, C., Woods, B.I., and Smith, K.S. (**2006**). *Biochemistry* 45: 11482–11490.

32 Reger, A.S., Carney, J.M., and Gulick, A.M. (**2007**). *Biochemistry* 46: 6536–6546.

33 Li, R., Gu, J., Chen, P. et al. (**2011**). *Acta Biochem. Biophys. Sin.* 43: 891–899.

34 Noy, T., Xu, H., and Blanchard, J.S. (**2014**). *Arch. Biochem. Biophys.* 550-551: 42–49.

35 Ingram-Smith, C., Thurman, J.L. Jr., Zimowski, K., and Smith, K.S. (**2012**). *Archaea* 2012: 509579.

36 Gallego-Jara, J., Tero, G.L., Conesa, A.É. et al. (**2019**). *Biochim. Biophys. Acta Gen. Subj.* 1863: 1040–1049.

37 Soumya, N., Kumar, I.S., Shivaprasad, S. et al. (**2015**). *Int. J. Biol. Macromol.* 75: 364–372.

38 Soumya, N., Tandan, H., Damre, M.V. et al. (**2016**). *Gene* 580: 125–133.

39 Russ, L., Harhangi, R., Schellekens, J. et al. (**2012**). *Arch. Microbiol.* 194: 943–948.

40 Sofeo, N., Hart, J.H., Butler, B. et al. (**2019**). *ACS Synth. Biol.* 8: 1325–1336.

41 Mouterde, L.M.M. and Stewart, J.D. (**2019**). *Enzyme Microb. Technol.* 128: 67–71.

42 Wolfe, A.J., Conley, M.P., and Berg, H.C. (**1988**). *Proc. Natl. Acad. Sci. U. S. A.* 85: 6711–6715.

43 Takahashi, H., McCaffery, J.M., Irizarry, R.A., and Boeke, J.D. (**2006**). *Mol. Cells* 23: 207–217.

44 Mirzaei, H. and Longo, V.D. (**2014**). *Cell Metab.* 19: 555–557.

45 Mews, P., Donahue, G., Drake, A.M. et al. (**2017**). *Nature* 546: 381–386.

46 Bae, J.M., Kim, J.H., Oh, H.J. et al. (**2017**). *Mod. Pathol.* 30: 267–277.

47 Liang, M.-H., Qv, X.-Y., Jin, H.-H., and Jiang, J.-G. (**2016**). *Sci. Rep.* 6: 23445.

48 Patel, S.S., Conlon, H.D., and Walt, D.R. (**1986**). *J. Org. Chem.* 51: 2842–2844.

49 Jossek, R. and Steinbüchel, A. (**1998**). *FEMS Microbiol. Lett.* 168: 319–324.

50 Satoh, Y., Tajima, K., Tannai, H., and Munekata, M. (**2003**). *J. Biosci. Bioeng.* 95: 335–341.

51 Xiao, Y., Ruan, Z., Liu, Z. et al. (**2013**). *Biochem. Eng. J.* 76: 60–69.

52 Steen, E.J., Kang, Y., Bokinsky, G. et al. (**2010**). *Nature* 463: 559–562.

53 Arif, M., Bai, Y., Usman, M. et al. (**2020**). *Bioresour. Technol.* 298: 122299.

54 Ramanan, R., Kim, B.-H., Cho, D.-H. et al. (**2013**). *FEBS Lett.* 587: 370–377.

55 Cerqueira, N.M.F.S.A., Oliveira, E.F., Gesto, D.S. et al. (**2016**). *Biochemistry* 55: 5483–5506.

56 Dubey, N.C., Tripathi, B.P., Müller, M. et al. (**2015**). *ACS Appl. Mater. Interfaces* 7: 1500–1507.

57 Gulick, A.M. (**2009**). *ACS Chem. Biol.* 4: 811–827.

58 Mahler, H.R. and Wakil, S.J. (**1953**). *J. Biol. Chem.* 204: 453–468.

59 Webster, L.T. Jr., Gerowin, L.D., and Rakita, L. (**1965**). *J. Biol. Chem.* 240: 29–33.

60 Fujino, T., Takei, Y.A., Sone, H. et al. (**2001**). *J. Biol. Chem.* 276: 35961–35966.

61 Vessey, D.A., Lau, E., Kelly, M., and Warren, R.S. (**2003**). *J. Biochem. Mol. Toxicol.* 17: 1–6.

62 Kochan, G., Pilka, E.S., von Delft, F. et al. (**2009**). *J. Mol. Biol.* 388: 997–1008.

63 Shah, M.B., Ingram-Smith, C., Cooper, L.L. et al. (**2009**). *Proteins* 77: 685–698.

64 Meng, Y., Ingram-Smith, C., Cooper, L.L., and Smith, K.S. (**2010**). *J. Bacteriol.* 192: 5982–5990.

65 Du, J., Wang, X., Nie, Q. et al. (**2017**). *J. Biotechnol.* 259: 160–167.

66 Bar-Tana, J. and Rose, G. (**1968**). *Biochem. J.* 109: 275–282.

67 Oka, Y., Kobayakawa, K., Nishizumi, H. et al. (**2003**). *Eur. J. Biochem.* 270: 1995–2004.

68 Morgan-Kiss, R.M. and Cronan, J.E. (**2004**). *J. Biol. Chem.* 279: 37324–37333.

69 Lindner, I., Rubin, D., Helwig, U. et al. (**2006**). *Mol. Nutr. Food Res.* 50: 270–274.

70 Ruan, H.-Y., Yang, C., Tao, X.-M. et al. (**2017**). *Am. J. Cancer Res.* 7: 543–553.

71 Badenhorst, C.P.S., van der Sluis, R., Erasmus, E., and van Dijk, A.A. (**2013**). *Expert Opin. Drug Metab. Toxicol.* 9: 1139–1153.

72 Kasuya, F., Yamaoka, Y., Igarashi, K., and Fukui, M. (**1998**). *Biochem. Pharmacol.* 55: 1769–1775.

73 Kasuya, F., Igarashi, K., and Fukui, M. (**1999**). *Chem. Biol. Interact.* 118: 233–246.

74 Kasuya, F., Tatsuki, T., Ohta, M. et al. (**2006**). *Protein Expr. Purif.* 47: 405–414.

75 Ford, T.J. and Way, J.C. (**2015**). *PeerJ* 3: e1040.

76 Juijberts, G.N.M., Eggink, G., de Waard, P. et al. (**1992**). *Appl. Environ. Microbiol.* 58: 536–544.

77 Preusting, H., Nijenhuis, A., and Witholt, B. (**1990**). *Macromolecules* 23: 4220–4224.

78 Park, S.J., Park, J.P., Lee, S.Y., and Doi, Y. (**2003**). *Enzyme Microb. Technol.* 33: 62–70.

79 Rehm, B.H.A., Krüger, N., and Steinbüchel, A. (**1998**). *J. Biol. Chem.* 273: 24044–24051.

80 Wang, Q., Tappel, R.C., Zhu, C., and Nomura, C.T. (**2012**). *Appl. Environ. Microbiol.* 78: 519–527.

81 Sato, S., Ishii, N., Hamada, Y. et al. (**2012**). *Polym. Degrad. Stab.* 97: 329–336.

82 Sudesh, K., Abe, H., and Doi, Y. (**2000**). *Prog. Polym. Sci.* 25: 1503–1555.

83 Rai, R., Keshavarz, T., Roether, J.A. et al. (**2011**). *Mater. Sci. Eng. R Rep.* 72: 29–47.

84 Tappel, R.C., Pan, W., Bergey, N.S. et al. (**2014**). *ACS Sustainable Chem. Eng.* 2: 1879–1887.

85 Rand, J.M., Pisithkul, T., Clark, R.L. et al. (**2017**). *Nat. Microbiol.* 2: 1624–1634.

86 Feller, D.R., Kamanna, V.S., Newman, H.A.I. et al. (**1987**). *J. Med. Chem.* 30: 1265–1267.

87 Bettoni, G., Loiodice, F., Tortorella, V. et al. (**1987**). *J. Med. Chem.* 30: 1267–1270.

88 Colton, I.J., Ahmed, S.N., and Kazlauskas, R.J. (**1995**). *J. Org. Chem.* 60: 212–217.

89 Kato, K.-I., Yoshida, H., Takeo, M. et al. (**2010**). *Biosci. Biotechnol. Biochem.* 74: 2405–2412.

90 Dias, D.A., Urban, S., and Roessner, U. (**2012**). *Metabolites* 2: 303–336.

91 Osada, H., Koshino, H., and Isono, K. (**1991**). *J. Antibiot.* 44: 259–261.

92 Woo, J.-T., Kawatani, M., Kato, M. et al. (**2006**). *Proc. Natl. Acad. Sci. U. S. A.* 103: 4729–4734.

93 Muguruma, H., Yano, S., Kakiuchi, S. et al. (**2005**). *Clin. Cancer Res.* 11: 8822–8828.

94 Hiraoka, K., Zenmyo, M., Watari, K. et al. (**2008**). *Cancer Sci.* 99: 1595–1602.

95 Yano, A., Tsutsumi, S., Soga, S. et al. (**2008**). *Proc. Natl. Acad. Sci. U. S. A.* 105: 15541–15546.

96 Miyazawa, T., Takashi, S., Kawata, A. et al. (**2015**). *J. Biol. Chem.* 290: 26994–27011.

97 Zhu, X., Su, M., Manickam, K., and Zhang, W. (**2015**). *ACS Chem. Biol.* 10: 2785–2793.

98 Hisanaga, Y., Ago, H., Nakagawa, N. et al. (**2004**). *J. Biol. Chem.* 279: 31717–31726.

99 Watkins, P.A. and Ellis, J.M. (**2012**). *Biochim. Biophys. Acta* 1822: 1411–1420.

100 Black, P.N. and DiRusso, C.C. (**2007**). *Biochim. Biophys. Acta* 1771: 286–298.
101 Kornberg, A. and Pricer, W.E. Jr. (**1953**). *J. Biol. Chem.* 204: 329–343.
102 Bar-Tana, J., Rose, G., and Shapiro, B. (**1971**). *Biochem. J.* 122: 353–362.
103 Bar-Tana, J., Rose, G., and Shapiro, B. (**1973**). *Biochem. J.* 135: 411–416.
104 Polokoff, M.A. and Bell, R.M. (**1977**). *J. Biol. Chem.* 252: 1167–1171.
105 Tanaka, T., Hosaka, K., Hoshimaru, M., and Numa, S. (**1979**). *Eur. J. Biochem.* 98: 165–172.
106 Philipp, D.P. and Parsons, P. (**1979**). *J. Biol. Chem.* 254: 10776–10784.
107 Miyazawa, S., Hashimoto, T., and Yokota, S. (**1985**). *J. Biochem.* 9: 723–733.
108 Suzuki, H., Kawarabayasi, Y., Kondo, J. et al. (**1990**). *J. Biol. Chem.* 265: 8681–8685.
109 Kono, M., Hori, C., Hashimoto, T. et al. (**1996**). *Eur. J. Biochem.* 238: 104–111.
110 Berge, R.K., Villset, S.E., and Farstad, M. (**1980**). *Scand. J. Clin. Lab. Invest.* 40: 271–278.
111 Bakken, A.M. and Farstad, M. (**1989**). *Biochem. J.* 261: 71–76.
112 Bronfman, M., Inestrosa, N.C., Nervi, F.O., and Leighton, F. (**1984**). *Biochem. J.* 224: 709–720.
113 Laposata, M., Reich, E.L., and Majerus, P.W. (**1985**). *J. Biol. Chem.* 260: 11016–11020.
114 Davidson, B.C. and Cantrill, R.C. (**1985**). *FEBS Lett.* 193: 69–74.
115 Camero, L., Shulaw, W.P., and Xiao, L. (**2003**). *J. Eukaryot. Microbiol.* 50: 534–538.
116 Massaro, E.J. and Lennarz, W.J. (**1965**). *Biochemistry* 4: 85–90.
117 Mishina, M., Kamiryo, T., Tashiro, S., and Numa, S. (**1978**). *Eur. J. Biochem.* 82: 347–354.
118 Hosaka, K., Mishina, M., Tanaka, T. et al. (**1979**). *Eur. J. Biochem.* 93: 197–203.
119 Fujino, T., Man-Jong, K., Minekura, H. et al. (**1997**). *J. Biochem.* 122: 212–216.
120 Liu, Y. and Eisenberg, D. (**2002**). *Protein Sci.* 11: 1285–1299.
121 Black, P.N., Zhang, Q., Weimar, J.D., and DiRusso, C.C. (**1997**). *J. Biol. Chem.* 272: 4896–4903.
122 Glick, B.S. and Rothman, J.E. (**1987**). *Nature* 326: 309–312.
123 McLaughlin, S. and Aderem, A. (**1995**). *Trends Biochem. Sci.* 20: 272–276.
124 Lobo, S., Wiczer, B.M., and Bernlohr, D.A. (**2009**). *J. Biol. Chem.* 284: 18347–18356.
125 Weng, H., Molina, I., Shockey, J., and Browse, J. (**2010**). *Planta* 231: 1089–1100.
126 Goodridge, A.G. (**1973**). *J. Biol. Chem.* 248: 4318–4326.
127 Sanchez-Lazo, L., Brisard, D., Elis, S. et al. (**2014**). *Mol. Endocrinol.* 28: 1502–1521.
128 Kuwata, H. and Hara, S. (**2019**). *Prostaglandins Other Lipid Mediat.* 144: 106363.
129 Qiao, S. and Tuohimaa, P. (**2004**). *Biochem. Biophys. Res. Commun.* 319: 358–368.
130 Sebastiano, M.R. and Konstantinidou, G. (**2019**). *Int. J. Mol. Sci.* 20: 3624.

131 Qiao, S., Pennanen, P., Nazarova, N. et al. (**2003**). *J. Steroid Biochem. Mol. Biol.* 85: 1–8.

132 Qiao, S. and Tuohimaa, P. (**2004**). *FEBS Lett.* 577: 451–454.

133 Kienow, L., Schneider, K., Bartsch, M. et al. (**2008**). *J. Exp. Bot.* 59: 403–419.

134 Schneider, K., Kienow, L., Schmelzer, E. et al. (**2005**). *J. Biol. Chem.* 280: 13962–13972.

135 Jia, B., Song, Y., Wu, M. et al. (**2016**). *Biotechnol. Biofuels* 9: 184.

136 Lung, S.-C. and Weselake, R.J. (**2006**). *Lipids* 41: 1073–1088.

137 Dahlqvist, A., Stahl, U., Lenman, M. et al. (**2000**). *Proc. Natl. Acad. Sci. U. S. A.* 97: 6487–6492.

138 Kilaru, A., Cao, X., Dabbs, P.B. et al. (**2015**). *BMC Plant Biol.* 15: 203.

139 Ding, L.-N., Gu, S.-L., Zhu, F.-G. et al. (**2020**). *BMC Plant Biol.* 20: 21.

140 Li, Q., Du, W., and Liu, D. (**2008**). *Appl. Microbiol. Biotechnol.* 80: 749–756.

141 Pearlson, M., Wollersheim, C., and Hileman, J. (**2013**). *Biofuels Bioprod. Biorefin.* 7: 89–96.

142 Sarin, R., Sharma, M., and Khan, A.A. (**2009**). *Bioresour. Technol.* 100: 4187–4192.

143 Liang, M.-H. and Jiang, J.-G. (**2013**). *Prog. Lipid Res.* 52: 395–408.

144 Subramaniam, R., Dufreche, S., Zappi, M., and Bajpai, R. (**2010**). *J. Ind. Microbiol. Biotechnol.* 37: 1271–1287.

145 Plassmeier, J., Li, Y., Rueckert, C., and Sinskey, A.J. (**2016**). *Metab. Eng.* 33: 86–97.

146 Dos Santos Moreira, C.D., Ramos, M.J.R.N., and Fernandes, P.M.A.A. (**2019**). *Wiley Interdiscip. Rev. Comput. Mol. Sci.* 9: e1399.

147 Moreira, C., Ramos, M.J., and Fernandes, P.A. (**2016**). *Chem. A Eur. J.* 22: 9218–9225.

148 Ginsburg, A., Yeh, J., Hennig, S.B., and Denton, M.D. (**1970**). *Biochemistry* 9: 633–649.

149 Oliveira, I.C., Brenner, E., Chiu, J. et al. (**2001**). *Braz. J. Med. Biol. Res.* 34: 567–575.

150 Kumada, Y., Benson, D.R., Hillemann, D. et al. (**1993**). *Proc. Natl. Acad. Sci. U. S. A.* 90: 3009–3013.

151 Valentine, R.C., Shapiro, B.M., and Stadtman, E.R. (**1968**). *Biochemistry* 7: 2143–2152.

152 Harth, G., Clemens, D.L., and Horwitz, M.A. (**1994**). *Proc. Natl. Acad. Sci. U. S. A.* 91: 9342–9346.

153 Eisenberg, D., Gill, H.S., Pfluegl, G.M.U., and Rotstein, S.H. (**2000**). *Biochim. Biophys. Acta* 1477: 122–145.

154 Almassy, R.J., Janson, C.A., Hamlin, R. et al. (**1986**). *Nature* 323: 304–309.

155 Yamashita, M.M., Almassy, R.J., Janson, C.A. et al. (**1989**). *J. Biol. Chem.* 264: 17681–17690.

156 Liaw, S.H., Pan, C., and Eisenberg, D. (**1993**). *Proc. Natl. Acad. Sci. U. S. A.* 90: 4996–5000.

157 Pellegrini, M., Grønbench-Jensen, N., Kelly, J.A. et al. (**1997**). *Proteins* 29: 426–432.

158 Kumada, Y., Takano, E., Nagaoka, K., and Thompson, C.J. (**1990**). *J. Bacteriol.* 172: 5343–5351.

159 Wyatt, K., White, H.E., Wang, L. et al. (**2006**). *Structure* 14: 1823–1834.

160 Unno, H., Uchida, T., Sugawara, H. et al. (**2006**). *J. Biol. Chem.* 281: 29287–29296.

161 Krajewski, W.W., Collins, R., Holmberg-Schiavone, L. et al. (**2008**). *J. Mol. Biol.* 375: 217–226.

162 Torreira, E., Seabra, A.R., Marriott, H. et al. (**2014**). *Acta Crystallogr.* D70: 981–993.

163 Van Rooyen, J.M., Abratt, V.R., and Sewell, B.T. (**2006**). *J. Mol. Biol.* 361: 796–810.

164 Van Rooyen, J.M., Abratt, V.R., Belrhali, H., and Sewell, T. (**2011**). *Structure* 19: 471–483.

165 Reyes, J.C., Muro-Pastor, M.I., and Florencio, F.J. (**1997**). *J. Bacteriol.* 179: 2678–2689.

166 Liaw, S.-H., Kuo, I., and Eisenberg, D. (**1995**). *Protein Sci.* 4: 2358–2365.

167 Mary, J. and Révert, B. (**1999**). *J. Mol. Biol.* 286: 121–134.

168 Weisbrood, R.E. and Meister, A. (**1973**). *J. Biol. Chem.* 248: 3997–4002.

169 Liaw, S.-H., Villafranca, J.J., and Eisenberg, D. (**1993**). *Biochemistry* 32: 7999–8003.

170 Liaw, S.-H. and Eisenberg, D. (**1994**). *Biochemistry* 33: 675–681.

171 Rhee, S.G. and Chock, P.B. (**1976**). *Proc. Natl. Acad. Sci. U. S. A.* 73: 476–480.

172 Welder, F.C. and Horn, B.R. (**1976**). *J. Biol. Chem.* 251: 7530–7538.

173 Gill, H.S. and Eisenberg, D. (**2001**). *Biochemistry* 40: 1903–1912.

174 Issoglio, F.M., Campolo, N., Zeida, A. et al. (**2016**). *Biochemistry* 55: 5907–5916.

175 Moreira, C., Ramos, M.J., and Fernandes, P.A. (**2017**). *J. Phys. Chem.* 121: 6313–6320.

176 Helling, R.B. (**1994**). *J. Bacteriol.* 176: 4664–4668.

177 Meeks, J.C., Wolk, C.P., Lockau, W. et al. (**1978**). *J. Bacteriol.* 134: 125–130.

178 Crispim, M., Damasceno, F.S., Hernández, A. et al. (**2018**). *PLoS Negl. Trop. Dis.* 12: e0006170.

179 Damascene, F.S., Barisón, M.J., Crispim, M. et al. (**2018**). *Mol. Biochem. Parasitol.* 224: 17–25.

180 Thomsen, H.C., Ericksson, D., Møller, I.S., and Schjoerring, J.K. (**2014**). *Trends Plant Sci.* 19: 656–663.

181 Kozaki, A. and Takeba, G. (**1996**). *Nature* 384: 557–560.

182 Storey, R. and Beevers, R. (**1978**). *Plant Physiol.* 61: 494–500.

183 Silva, L. and Carvalho, H. (**2013**). *Front. Plant Sci.* 4: 372.

184 Suárez, I., Bodega, G., and Fernández, B. (**2002**). *Neurochem. Int.* 41: 123–142.

185 Jin, Y.Y., Singh, P., Chung, H.-J., and Hong, S.-T. (**2018**). *Nutrients* 10: 564.

186 Hamed, N.O., Al-Ayadhi, L., Osman, M.A. et al. (**2018**). *Psychiatry Clin. Neurosci.* 72: 362–373.

187 Lu, Y.-B., Yang, L.-T., Li, Y. et al. (**2014**). *Tree Physiol.* 34: 608–618.

188 Canas, R.A., Yesbergenova-Cuny, Z., Belanger, L. et al. (**2020**). *Plants* 9: 130.

189 McGrath, R.B. and Coruzzi, G.M. (**1991**). *Plant J.* 1: 275–280.

190 Gaufichon, L., Rothstein, S.J., and Suzuki, A. (**2016**). *Plant Cell Physiol.* 57: 675–689.

191 Yamamoto, S., Uchimura, K., Wakayama, M., and Tachiki, T. (**2004**). *Biosci. Biotechnol. Biochem.* 68: 1888–1897.

192 Yamamoto, S., Wakayama, M., and Tachiki, T. (**2005**). *Biosci. Biotechnol. Biochem.* 69: 784–789.

193 Yamamoto, S., Wakayama, M., and Tachiki, T. (**2006**). *Biosci. Biotechnol. Biochem.* 70: 500–507.

194 Wang, T., Zhang, Y.-F., Ning, L.-X. et al. (**2020**). *Enzyme Microb. Technol.* 136: 109537.

195 Fawaz, M.V., Topper, M.E., and Firestine, S.M. (**2011**). *Bioorg. Chem.* 39: 185–191.

196 Özcengiz, G. and Öğülür, İ. (**2015**). *N. Biotechnol.* 32: 612–619.

197 Stein, T. (**2005**). *Mol. Microbiol.* 56: 845–857.

198 Walker, J.E. and Abraham, E.P. (**1970**). *Biochem. J.* 118: 557–561.

199 Walker, J.E. and Abraham, E.P. (**1970**). *Biochem. J.* 118: 563–570.

200 Neuss, N., Molloy, B.B., Shah, R., and DeLaHiguera, N. (**1970**). *Biochem. J.* 118: 571–575.

201 Shah, R., Neuss, N., Gorman, M., and Boeck, L.D. (**1970**). *J. Antibiot.* 23: 613–619.

202 Kenig, M., Vandamme, E., and Abraham, E.P. (**1976**). *J. Gen. Microbiol.* 94: 46–54.

203 Chmara, H., Zahner, H., Borowski, E., and Milewski, S. (**1984**). *J. Antibiot.* 37: 652–658.

204 Chmara, H., Zahner, H., and Borowski, E. (**1984**). *J. Antibiot.* 37: 1038–1043.

205 Sakajoh, M., Solomon, N.A., and Demain, A.L. (**1987**). *J. Ind. Microbiol.* 2: 201–208.

206 Yazgan, A., Özcengiz, G., Özcengiz, E. et al. (**2001**). *Enzyme Microb. Technol.* 29: 400–406.

207 Steinborn, G., Hajirezaei, M.-R., and Hofemeister, J. (**2005**). *Arch. Microbiol.* 183: 71–79.

208 Tabata, K., Ikeda, H., and Hashimoto, S.-I. (**2005**). *J. Bacteriol.* 187: 5195–5202.

209 Katz, E. and Demain, A.L. (**1977**). *Bacteriol. Rev.* 41: 449–474.

210 Inaoka, T. and Ochi, K. (**2011**). *Appl. Environ. Microbiol.* 77: 8181–8183.

211 Tsuda, T., Suzuki, T., and Kojima, S. (**2012**). *Acta Crystallogr.* F68: 203–206.

212 Shomura, Y., Hinokuchi, E., Ikeda, H. et al. (**2012**). *Protein Sci.* 21: 707–716.

213 Wang, T., Zhang, Y.-R., Liu, X.-H. et al. (**2019**). *Biomolecules* 9: 733.

214 Ogasawara, Y. and Dairi, T. (**2017**). *Chem. A Eur. J.*: 10714–10724.

215 Hilton, M.D., Alaeddinoglu, N.G., and Demain, A.L. (**1988**). *J. Bacteriol.* 170: 482–484.

216 Rajavel, M., Mitra, A., and Gopal, B. (**2009**). *J. Biol. Chem.* 284: 31882–31892.

217 Parker, J.B. and Walsh, C.T. (**2013**). *Biochemistry* 52: 889–901.

218 Tüzün, I., Karataş, A.Y., Şeşenoğlu, Ö. et al. (**2003**). *Enzyme Microb. Technol.* 33: 725–728.

219 Tabata, K. and Hashimoto, S.-I. (**2007**). *Appl. Environ. Microbiol.* 73: 6378–6385.

220 Tsuda, T., Asami, M., Koguchi, Y., and Kojima, S. (**2014**). *Biochemistry* 53: 2650–2660.

221 Kino, H. and Kino, K. (**2015**). *Biosci. Biotechnol. Biochem.* 79: 1827–1832.

222 Mordhorst, S., Singh, J., Mohr, M.K.F. et al. (**2019**). *ChemBioChem* 20: 1019–1022.

223 Andexer, J.N. and Richter, M. (**2015**). *ChemBioChem* 16: 380–386.

224 Tahlan, K. and Jensen, S.E. (**2013**). *J. Antibiot.* 66: 401–410.

225 Hamed, R.B., Gomez-Gastellanos, J.R., Henry, L. et al. (**2013**). *Nat. Prod. Rep.* 30: 21–107.

226 McGowan, S.J., Sebaihia, M., Porter, L.E. et al. (**1996**). *Mol. Microbiol.* 22: 415–426.

227 McGowan, S.J., Sebaihia, M., O'Leary, S. et al. (**1997**). *Mol. Microbiol.* 26: 545–556.

228 Gerratana, B., Stapon, A., and Townsend, C.A. (**2003**). *Biochemistry* 42: 7836–7847.

229 Bonder, M.J., Li, R., Phelan, R.M. et al. (**2011**). *ChemBioChem* 12: 2159–2169.

230 Miller, M.T., Gerratana, B., Stapon, A. et al. (**2003**). *J. Biol. Chem.* 278: 40996–41002.

231 Kershaw, N.J., Caines, M.E.C., Sleeman, M.C., and Schofield, C.J. (**2005**). *Chem. Commun.*: 4251–4263.

232 Arnett, S.O., Gerratana, B., and Townsend, C.A. (**2007**). *Biochemistry* 46: 9337–9345.

233 Raber, M.L., Arnett, S.O., and Townsend, C.A. (**2009**). *Biochemistry* 48: 4959–4971.

234 Stapon, A., Li, R., and Twonsend, C.A. (**2003**). *J. Am. Chem. Soc.* 125: 15746–15747.

235 Li, R., Stapon, A., Blanchfield, J.T., and Townsend, C.A. (**2000**). *J. Am. Chem. Soc.* 122: 9296–9297.

236 Stapon, A., Li, R., and Townsend, C.A. (**2003**). *J. Am. Chem. Soc.* 125: 8486–8493.

237 Bycroft, B.W. and Maslen, C. (**1988**). *J. Antibiot.* 41: 1231–1242.
238 Chang, W.-C., Guo, Y., Wang, C. et al. (**2014**). *Science* 343: 1140–1144.
239 Sleeman, M.C., Smith, P., Kellam, B. et al. (**2004**). *ChemBioChem* 5: 879–882.
240 Hamed, R.B., Mecinović, J., Ducho, C. et al. (**2010**). *Chem. Commun.* 46: 1413–1415.
241 Hamed, R.B., Henry, L., Gomez-Castellanos, J.R. et al. (**2012**). *J. Am. Chem. Soc.* 134: 471–479.
242 Hamed, R.B., Henry, L., Claridge, T.D.W., and Schofield, C.J. (**2017**). *ACS Catal.* 7: 1279–1285.
243 Hamed, R.B., Gomez-Castellanos, J.R., Henry, L. et al. (**2019**). *Commun. Chem.* 2: 7.
244 Attwood, P.V. (**1995**). *Int. J. Biochem. Cell Biol.* 27: 231–245.
245 Scrutton, M.C. (**1978**). *FEBS Lett.* 89: 1–9.
246 Jitrapakdee, S., Maurice, M.S., Rayment, I. et al. (**2008**). *Biochem. J.* 413: 369–387.
247 Utter, M.F. and Keech, D.B. (**1960**). *J. Biol. Chem.* 235: PC17–PC18.
248 Jitrapakdee, S. and Wallace, J.C. (**1999**). *Biochem. J.* 340: 1–16.
249 Yu, L.P.C., Xiang, S., Lasso, G. et al. (**2009**). *Structure* 17: 823–832.
250 Cazzulo, J.J., Sundarm, T.K., Dilks, S.N., and Kornberg, H.L. (**1971**). *Biochem. J.* 122: 653–661.
251 Lietzan, A.D., Menefee, A.L., Zeczycki, T.N. et al. (**2011**). *Biochemistry* 50: 9708–9723.
252 Mukhopadhyay, B. and Purwantini, E. (**2000**). *Biochim. Biophys. Acta* 1475: 191–206.
253 Mukhopadhyay, B., Stoddard, S.F., and Wolfe, R.S. (**1998**). *J. Biol. Chem.* 273: 5155–5166.
254 Pronk, J.T., Steensma, H.Y., and Van Dijken, J.P. (**1996**). *Yeast* 12: 1607–1633.
255 Barnett, J.A. and Kornberg, H.L. (**1960**). *J. Gen. Microbiol.* 23: 65–82.
256 Gancedo, J.M. (**1992**). *Eur. J. Biochem.* 206: 297–313.
257 Frick, O. and Wittmann, C. (**2005**). *Microb. Cell Fact.* 4: 30.
258 March, J.C., Eiteman, M.A., and Altman, E. (**2002**). *Appl. Environ. Microbiol.* 68: 5620–5624.
259 Wimhurst, J. and Manchester, K.L. (**1973**). *FEBS Lett.* 29: 201–203.
260 Jassens, P.A., Jenkinson, L.A., Paton, B.C., and Whitelaw, E. (**1977**). *Aust. J. Biol. Sci.* 30: 183–195.
261 Poitout, V. and Robertson, R.P. (**2002**). *Endocrinology* 143: 339–342.
262 Osmani, S.A. and Scrutton, M.C. (**1983**). *Eur. J. Biochem.* 133: 551–560.
263 Gul, B. and Dils, R. (**1969**). *Biochem. J.* 111: 263–271.
264 Boucher, A., Lu, D., Burgess, S.C. et al. (**2004**). *J. Biol. Chem.* 279: 27263–27271.
265 Laybutt, D.R., Glandt, M., Xu, G. et al. (**2003**). *J. Biol. Chem.* 278: 2997–3005.
266 Schousboe, A., Bak, L.K., and Waagepetersen, H.S. (**2013**). *Front. Endocrinol.* 4: 102.
267 Hertz, L., Dringen, R., Schousboe, A., and Robinson, S.R. (**1999**). *J. Neurosci. Res.* 57: 417–428.

268 Mukhopadhyay, B., Purwantini, E., Kreder, C.L., and Wolfe, R.S. (**2001**). *J. Bacteriol.* 183: 3804–3810.

269 Kondo, S., Nakajima, Y., Sugio, S. et al. (**2004**). *Acta Crystallogr. D Biol. Crystallogr.* 60: 486–492.

270 Lai, H., Kraszewski, J.L., Purwantini, E., and Mukhopadhyay, B. (**2006**). *Appl. Environ. Microbiol.* 72: 7785–7792.

271 Islam, M.N., Sueda, S., and Kondo, H. (**2005**). *Protein Eng. Des. Sel.* 18: 71–78.

272 Xiang, S. and Tong, L. (**2008**). *Nat. Struct. Biol.* 15: 295–302.

273 St. Maurice, M., Reinhardt, L., Surinya, K.H. et al. (**2007**). *Science* 317: 1076–1079.

274 Attwood, P.V., Tripton, P.A., and Cleland, W.W. (**1986**). *Biochemistry* 25: 8197–8205.

275 Attwood, P.V. and Wallace, J.C. (**2002**). *Acc. Chem. Res.* 35: 113–120.

276 Gokarn, R.R., Evans, J.D., Walker, J.R. et al. (**2001**). *Appl. Microbiol. Biotechnol.* 56: 188–195.

277 Zeikus, J.G., Jain, M.K., and Elankovan, P. (**1999**). *Appl. Microbiol. Biotechnol.* 51: 545–552.

278 Lee, S.Y., Hong, S.H., Lee, S.H., and Park, S.J. (**2004**). *Macromol. Biosci.* 4: 157–164.

279 Vemuri, G.N., Eiteman, M.A., and Altman, E. (**2002**). *Appl. Environ. Microbiol.* 68: 1715–1727.

280 Vemuri, G.N., Eiteman, M.A., and Altman, E. (**2002**). *J. Ind. Microbiol. Biotechnol.* 28: 325–332.

281 Sánchez, A.M., Bennett, G.N., and San, K.-Y. (**2005**). *Biotechnol. Prog.* 21: 358–365.

282 Sánchez, A.M., Bennett, G.N., and San, K.-Y. (**2005**). *Metab. Eng.* 7: 229–239.

283 Sánchez, A.M., Bennett, G.N., and San, K.-Y. (**2006**). *Metab. Eng.* 8: 209–226.

284 Ma, J., Gou, D., Liang, L. et al. (**2013**). *Appl. Microbiol. Biotechnol.* 97: 6739–6747.

285 Liu, R., Liang, L., Wu, M. et al. (**2013**). *Biochem. Eng. J.* 79: 77–83.

286 Skorokhodova, A.Y., Gulevich, A.Y., and Debabov, V.G. (**2018**). *Appl. Biochem. Microbiol.* 54: 849–854.

287 Blankschien, M.D., Clomberg, J.M., and Gonzalez, R. (**2010**). *Metab. Eng.* 12: 409–419.

288 Okino, S., Noburyu, R., Suda, M. et al. (**2008**). *Appl. Microbiol. Biotechnol.* 81: 459–464.

289 Litsanov, B., Brocker, M., and Bott, M. (**2012**). *Appl. Environ. Microbiol.* 78: 3325–3337.

290 Hiroto, U., Kazuaki, N., Kenji, T. et al. (**2020**). *Appl. Microbiol. Biotechnol.* 104: 4313–4320.

291 Tsuji, A., Okada, S., Hols, P., and Satoh, E. (**2013**). *Enzyme Microb. Technol.* 53: 97–103.

292 Tajima, Y., Yamamoto, Y., Fukui, K. et al. (**2015**). *Appl. Environ. Microbiol.* 81: 929–937.

293 Yan, D., Wang, C., Zhou, J. et al. (**2014**). *Bioresour. Technol.* 156: 232–239.

294 Dubini, A., Mus, F., Seibert, M. et al. (**2009**). *J. Biol. Chem.* 284: 7201–7213.

295 Zelle, R.M., de Hulster, E., van Winden, W.A. et al. (**2008**). *Appl. Environ. Microbiol.* 74: 2766–2777.

296 Oba, T., Kusumoto, K., Kichise, Y. et al. (**2014**). *FEMS Yeast Res.* 14: 789–796.

297 Chen, X., Xu, G., Xu, N. et al. (**2013**). *Metab. Eng.* 19: 10–16.

298 Feng, J., Yang, J., Yang, W. et al. (**2018**). *Biotechnol. Biofuels* 11: 94.

299 Zhang, B., Skory, C.D., and Yang, S.-T. (**2012**). *Metab. Eng.* 14: 512–520.

300 Xu, G., Zou, W., Chen, X. et al. (**2012**). *PLoS One* 7: e52086.

301 Chen, X., Li, Y., Tong, T., and Liu, L. (**2019**). *Biotechnol. Bioeng.* 116: 622–630.

302 Yin, X., Madzak, C., Du, G. et al. (**2012**). *Appl. Microbiol. Biotechnol.* 96: 1527–1537.

303 Yovkova, V., Otto, C., Aurich, A. et al. (**2014**). *Appl. Microbiol. Biotechnol.* 98: 2003–2013.

304 Zhang, A., Sun, J., Wang, Z. et al. (**2015**). *Bioresour. Technol.* 175: 474–381.

305 Wang, Z., Lin, M., Wang, L. et al. (**2015**). *Process Biochem.* 50: 194–204.

306 Borodina, I., Kildegaard, K.R., Jensen, N.B. et al. (**2015**). *Metab. Eng.* 27: 57–64.

307 Stern, J.R. and Bambers, G. (**1966**). *Biochemistry* 5: 1113–1118.

308 Sato, H., Orishimo, K., Shirai, T. et al. (**2008**). *J. Biosci. Bioeng.* 106: 51–58.

309 Cao, Y., Mpofu, E., and Shi, Z. (**2013**). *Biochem. Eng. J.* 77: 136–146.

310 Blombach, B., Schreiner, M.E., Moch, M. et al. (**2007**). *Appl. Microbiol. Biotechnol.* 76: 615–623.

311 Wang, Z., Moslehi-Jenabian, S., Solem, C., and Jensen, P.R. (**2015**). *Eng. Life Sci.* 15: 73–82.

312 Kortmann, M., Mack, C., Baumgart, M., and Bott, M. (**2019**). *ACS Synth. Biol.* 8: 274–281.

313 Liu, J., Li, H., Xiong, H. et al. (**2019**). *Biotechnol. Bioeng.* 116: 110–120.

314 Long, N.V.D., Lee, J., Koo, K.-K. et al. (**2017**). *Energies* 10: 473.

315 Alissandratos, A. and Easton, C.J. (**2015**). *Beilstein J. Org. Chem.* 11: 2370–2387.

316 Von Borzyskowski, L.S., Rosenthal, R.G., and Erb, T.J. (**2013**). *J. Biotechnol.* 168: 243–251.

317 Erb, T.J. (**2011**). *Appl. Environ. Microbiol.* 77: 8466–8477.

318 Sasaki, Y. and Nagano, Y. (**2004**). *Biosci. Biotechnol. Biochem.* 68: 1175–1184.

319 Nikolau, B.J., Ohlrogge, J.B., and Wurtele, E.S. (**2003**). *Arch. Biochem. Biophys.* 414: 211–222.

320 Tanabe, T., Wada, K., Okazaki, T., and Numa, S. (**1975**). *Eur. J. Biochem.* 57: 15–24.

321 Witters, L.A. and Vogt, B. (**1981**). *J. Lipid Res.* 22: 364–369.

322 Zammit, V.A. and Corstorphine, C.G. (**1982**). *Biochem. J.* 204: 757–764.

323 Thampy, K.G. and Wakil, S.J. (**1988**). *J. Biol. Chem.* 263: 6447–6453.
324 Trumble, G.E., Smith, M.A., and Winder, W.W. (**1995**). *Eur. J. Biochem.* 231: 192–198.
325 Beaty, N.B. and Lane, M.D. (**1982**). *J. Biol. Chem.* 257: 924–929.
326 Manning, R., Dils, R., and Mayer, R.J. (**1976**). *Biochem. J.* 153: 463–468.
327 Sasaki, Y., Hakamada, K., Suama, Y. et al. (**1993**). *J. Biol. Chem.* 268: 25118–25123.
328 Roessler, P.G. (**1990**). *Plant Physiol.* 92: 73–78.
329 Peng, N., Zhong, Y., Zhang, Q. et al. (**2012**). *Acta Biochem. Biophys. Sin.* 44: 692–702.
330 Ehebauer, M.T., Zimmermann, M., Jakobi, A.J. et al. (**2015**). *PLoS Pathog.* 11: e1004623.
331 Livieri, A.L., Navone, L., Mercellin, E. et al. (**2019**). *Sci. Rep.* 9: 6725.
332 Chuakrut, S., Arai, H., Ishii, M., and Igarashi, Y. (**2003**). *J. Bacteriol.* 185: 938–947.
333 Abu-Elheiga, L., Jayakumar, A., Baldini, A. et al. (**1995**). *Proc. Natl. Acad. Sci. U. S. A.* 92: 4011–4015.
334 Widmer, J., Fassihi, K.S., Schlichter, S.C. et al. (**1996**). *Biochem. J.* 316: 915–922.
335 Tong, L. (**2005**). *Cell. Mol. Life Sci.* 62: 1784–1803.
336 Athappilly, F.A. and Hendrickson, W.A. (**1995**). *Structure* 3: 1407–1419.
337 Cronan, J.E. Jr. (**2001**). *J. Biol. Chem.* 276: 37355–37364.
338 Waldrop, G.L., Rayment, I., and Holden, H.M. (**1994**). *Biochemistry* 33: 10249–10256.
339 Shorrosh, B.S., Roesler, K.R., Shintani, D. et al. (**1995**). *Plant Physiol.* 108: 805–812.
340 Chou, C.-Y., Yu, L.P.C., and Tong, L. (**2009**). *J. Biol. Chem.* 284: 11690–11697.
341 Cho, Y.S., Lee, J.I., Shin, D. et al. (**2008**). *Proteins* 70: 268–272.
342 Zhang, H., Tweel, B., Li, J., and Tong, L. (**2004**). *Structure* 12: 1683–1691.
343 Meads, G. Jr., Benson, B.K., Grove, A., and Waldrop, G.L. (**2010**). *Nucleic Acids Res.* 38: 1217–1227.
344 Reddy, M.C.M., Breda, A., Bruning, J.B. et al. (**2014**). *Antimicrob. Agents Chemother.* 58: 6122–6132.
345 Konishi, T. and Sasaki, Y. (**1994**). *Proc. Natl. Acad. Sci. U. S. A.* 91: 3598–3601.
346 Konishi, T., Shinohara, K., Yamada, K., and Sasaki, Y. (**1996**). *Plant Cell Physiol.* 37: 117–122.
347 Brownsey, R.W., Zhande, R., and Boone, A.N. (**1997**). *Biochem. Soc. Trans.* 25: 1232–1238.
348 Abu-Elheiga, L., Brinkley, W.R., Zhong, L. et al. (**2000**). *Proc. Natl. Acad. Sci. U. S. A.* 97: 1444–1449.
349 Lee, C.-K., Cheong, H.-K., Ryu, K.-S. et al. (**2008**). *Proteins* 72: 613–624.
350 Wei, J. and Tong, L. (**2015**). *Nature* 526: 723–727.

351 Waite, M. and Wakil, S.J. (**1963**). *J. Biol. Chem.* 238: 77–80.
352 Janiyani, K., Bordelon, T., Waldrop, G.L., and Cronan, J.E. Jr. (**2001**). *J. Biol. Chem.* 276: 29864–29870.
353 Wei, J., Zhang, Y., Yu, T.-Y. et al. (**2016**). *Cell Discov.* 2: 16044.
354 Kirvoruchko, A., Zhang, Y., Siewers, V. et al. (**2015**). *Metab. Eng.* 28: 28–42.
355 Saddik, M., Gamble, J., Witters, L.A., and Lopaschuk, G.D. (**1993**). *J. Biol. Chem.* 268: 25836–25845.
356 Abu-Elheiga, L., Matzuk, M.M., Abo-Hashema, K.A.H., and Wakil, S.J. (**2001**). *Science* 291: 2613–2616.
357 Abu-Elheiga, L., Matzuk, M.M., Kordari, P. et al. (**2005**). *Proc. Natl. Acad. Sci. U. S. A.* 102: 12011–12016.
358 Lucas, C., Lucas, G., Lucas, N. et al. (**2018**). *Clin. Exp. Hepatol.* 4: 165–174.
359 Imai, N. and Cohen, D.E. (**2018**). *Hepatology* 68: 2062–2065.
360 Harriman, G., Greenwood, J., Bhat, S. et al. (**2016**). *Proc. Natl. Acad. Sci. U. S. A.* 113: E1796–E1805.
361 Lally, J.S.V., Ghoshal, S., Deperalta, D.K. et al. (**2019**). *Cell Metab.* 29: 174–182.
362 Zuther, E., Johnson, J.J., Haselkorn, R. et al. (**1999**). *Proc. Natl. Acad. Sci. U. S. A.* 96: 13387–13392.
363 Cronan, J.E. and Thomas, J. (**2009**). *Methods Enzymol.* 459: 395–433.
364 Qadeer, S., Khalid, A., Mahmood, S. et al. (**2017**). *J. Clean. Prod.* 168: 917–928.
365 Kwak, S., Jo, J.H., Yun, E.J. et al. (**2019**). *Biotechnol. Adv.* 37: 271–283.
366 Jaroensuk, J., Intasian, P., Wattanasuepsin, W. et al. (**2020**). *J. Biotechnol.* 309: 1–19.
367 Kamineni, A. and Shaw, J. (**2020**). *Curr. Opin. Biotechnol.* 62: 239–247.
368 Rock, C.O. and Jackowski, S. (**2002**). *Biochem. Biophys. Res. Commun.* 292: 1155–1166.
369 Davis, M.S., Solbiati, J., and Cronan, J.E. Jr. (**2000**). *J. Biol. Chem.* 275: 28593–28598.
370 Jeon, E., Lee, S., Won, J.-I. et al. (**2011**). *Enzyme Microb. Technol.* 49: 44–51.
371 Xu, P., Gu, Q., Wang, W. et al. (**2013**). *Nat. Commun.* 4: 1409.
372 Pfleger, B.F., Gossing, M., and Nielsen, J. (**2015**). *Metab. Eng.* 29: 1–11.
373 Cao, Y., Liu, W., Xu, X. et al. (**2014**). *Biotechnol. Biofuels* 7: 59.
374 Tang, X., Feng, H., and Chen, W.H. (**2013**). *Metab. Eng.* 16: 93–102.
375 Wang, J., Xu, R., Wang, R. et al. (**2016**). *Biosci. Biotechnol. Biochem.* 80: 1214–1222.
376 Abdel-Mawgoud, A.M., Markham, K.A., Palmer, C.M. et al. (**2018**). *Metab. Eng.* 50: 192–208.
377 Yuzbasheva, E.Y., Mostova, E.B., Andreeva, N.I. et al. (**2018**). *Biotechnol. Bioeng.* 115: 433–443.
378 Tai, M. and Stephanopoulos, G. (**2013**). *Metab. Eng.* 15: 1–9.
379 Qiao, K., Abidi, S.H.I., Liu, H. et al. (**2015**). *Metab. Eng.* 29: 56–65.

380 Zhou, S., Hao, T., Xu, S., and Deng, Y. (**2020**). *Bitechnol. Adv.* 43: 107575.

381 Matsumoto, T., Tanaka, T., and Kondo, A. (**2017**). *Bioresour. Technol.* 245: 13662–11368.

382 Marsafari, M., Samizadeh, H., Rabiei, B. et al. (**2020**). *Biotechnol. J.*: 1900432.

383 Liu, C., Ding, Y., Xian, M. et al. (**2017**). *Crit. Rev. Biotechnol.* https://doi.org/10.1080/07388551.2016.1272093.

384 Xu, P., Qiao, K., Ahn, W.S., and Stephanopoulos, G. (**2016**). *Proc. Natl. Acad. Sci. U. S. A.* 113: 10848–10858.

385 Muhammad, A., Feng, X., Rasool, A. et al. (**2020**). *Biotechnol. Adv.*: 10755.

386 Feng, X., Lian, J., and Zhao, H. (**2015**). *Metab. Eng.* 27: 10–19.

387 Zha, W., Rubin-Pitel, S.B., Shao, Z., and Zhao, H. (**2009**). *Metab. Eng.* 11: 192–198.

388 Archkar, J., Xian, M., Zhao, H., and Frost, J.W. (**2005**). *J. Am. Chem. Soc.* 127: 5332–5333.

389 Zha, W., Rubin-Pitel, S.B., and Zhao, H. (**2006**). *J. Biol. Chem.* 281: 32036–32047.

390 Zabala, D., Braña, A.F., Flórez, A.B. et al. (**2013**). *Metab. Eng.* 20: 187–197.

391 Markham, K.A., Palmer, C.M., Chwatko, M. et al. (**2018**). *Proc. Natl. Acad. Sci. U. S. A.* 115: 2096–2018.

392 Liu, H., Marsafari, M., Wang, F. et al. (**2019**). *Metab. Eng.* 56: 60–68.

393 Liu, Y., Bai, C., Liu, Q. et al. (**2019**). *Metab. Eng.* 54: 275–284.

394 Miyahisa, I., Kaneko, M., Funa, N. et al. (**2005**). *Appl. Microbiol. Biotechnol.* 68: 498–504.

395 Leonard, E., Lim, K.-H., Saw, P.-N., and Koffas, M.A.G. (**2007**). *Appl. Environ. Microbiol.* 73: 3877–3886.

396 Xu, P., Ranganathan, S., Fowler, Z.L. et al. (**2011**). *Metab. Eng.* 13: 578–587.

397 Zhu, S., Wu, J., Du, G. et al. (**2014**). *Appl. Environ. Microbiol.* 80: 3072–3080.

398 Malla, S., Koffas, M.A.G., Kazlauskas, R.J., and Kim, B.-G. (**2012**). *Appl. Environ. Microbiol.* 78: 684–694.

399 Yan, Y., Huang, L., and Koffas, M.A.G. (**2007**). *Biotechnol. J.* 2: 1250–1262.

400 Yang, Y., Lin, Y., Li, L. et al. (**2015**). *Metab. Eng.* 29: 217–226.

401 Dhakal, D., Chaudhary, A.K., Yi, J.S. et al. (**2016**). *Appl. Microbiol. Biotechnol.* 100: 9917–9931.

402 Hasan, H., Rahim, M.H.A., Campbell, L. et al. (**2018**). *N. Biotechnol.* 44: 64–71.

403 Mark, R., Lyu, X., Ng, K.R., and Chen, W.N. (**2019**). *ACS Omega* 4: 12872–12879.

404 Gao, S., Lyu, Y., Zeng, W. et al. (**2020**). *J. Agric. Food Chem.* 68: 1015–1021.

405 Li, M., Kildegaard, K.R., Chen, Y. et al. (**2015**). *Metab. Eng.* 32: 1–11.

406 Duan, L., Ding, W., Liu, X. et al. (**2017**). *Microb. Cell Fact.* 16: 165.

407 Palmer, C.M., Miller, K.K., Nguyen, A., and Alper, H.S. (**2020**). *Metab. Eng.* 57: 174–181.

408 Lv, Y., Edwards, H., Zhou, J., and Xu, P. (**2019**). *ACS Synth. Biol.* 8: 568–576.

409 Lv, Y., Marsafari, M., Koffas, M. et al. (**2019**). *ACS Synth. Biol.* 8: 2514–2523.

410 Chen, Y. and Nielsen, J. (**2016**). *Curr. Opin. Biotechnol.* 37: 165–172.
411 Chen, Y. and Nielsen, J. (**2013**). *Curr. Opin. Biotechnol.* 24: 965–972.
412 Rathnasingh, C., Raj, S.M., Lee, Y. et al. (**2012**). *J. Bacteriol.* 157: 633–640.
413 Tokuyama, K., Toya, Y., Matsuda, F. et al. (**2019**). *Metab. Eng.* 52: 215–223.
414 Menendez, C., Bauer, Z., Huber, H. et al. (**1999**). *J. Bacteriol.* 181: 1088–1098.
415 Loder, A.J., Han, Y., Hawkins, A.B. et al. (**2016**). *Metab. Eng.* 38: 446–463.
416 Wang, Y., Sun, T., Gao, X. et al. (**2016**). *Metab. Eng.* 34: 60–70.
417 Ku, J.T. and Lan, E.I. (**2018**). *Metab. Eng.* 46: 35–42.
418 Nguyen, D.T.N., Lee, O.K., Lim, C. et al. (**2020**). *Metab. Eng.* 59: 142–150.
419 Chen, Y., Bao, J., Kim, I.-K. et al. (**2014**). *Metab. Eng.* 22: 104–109.
420 Shi, S., Chen, Y., Siewers, V., and Nielsen, J. (**2014**). *MBio* 5: e01130–e01114.
421 Suyama, A., Higuchi, Y., Urushihara, M. et al. (**2017**). *J. Biosci. Bioeng.* 124: 392–399.

8

Future Perspectives

8.1 Combinatorial Enzymatic Organic Synthesis

8.1.1 Combinatorial Chemistry

In contrast to retrosynthesis or *de novo* synthesis mentioned in previous chapters, which generates only one compound at a time from the starting materials, combinatorial chemistry is a chemical synthetic method that quickly and simultaneously generates a large array of structurally diverse compounds, called a chemical library, in a single process through systematic, repetitive, and covalent linkage of various" building blocks" [1, 2]. These compound libraries with a wide range of linear or macrocyclic chemical molecules – peptides, nonpeptide oligomers, peptidomimetics, small molecules, and natural product-like organic molecules – can be made as mixtures, sets of individual compounds, or chemical structures generated by computer software [3]. The compounds in the library can then be screened by individually interacting with biological targets of interest to allow the identification of useful components of the libraries using either directly (in position-addressable libraries) or via decoding (using genetic or chemical analysis). Combinatorial chemistry has turned traditional chemistry upside down when introducing automated substance synthesis to the drug discovery process. A large number of chemical compounds generated by combinatorial library methods can be used a valuable source for the discovery of drug leads, molecular imaging agents, and capturing agents for molecular markers. Moreover, combinatorial library can provide powerful tools for basic research in various disciplines, such as to probe their effects on specific cellular function and then to study cellular pathways involved in these cellular functions, to apply for the field of chemical genetics [1, 4, 5].

The development of combinatorial chemistry and the generation of the first combinatorial library can be dated back in the early 1980s, while the word "combinatorial" appeared in the scientific word and really be taken up by the

Enzyme-Based Organic Synthesis, First Edition. Cheanyeh Cheng.

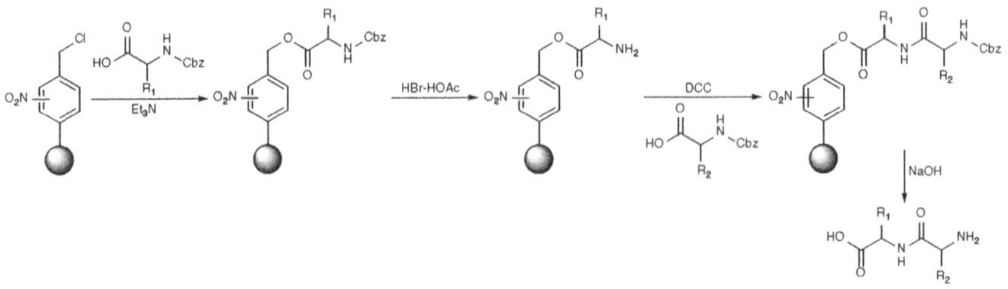

Scheme 8.1 Merrifield's solid-phase synthesis of a dipeptide; Cbz, benzyloxycarbonyl; DCC, 1,3-dicyclohexylcarbodiimide; ●, polystyrene. *Source:* Based on Rasheed and Farhat [2]; Merrifield [6]; Gordon and Balasubramanian [7].

industry since 1990s. The first report dealt with the simultaneous production of collection of chemically synthesized peptides were produced by solid phase synthesis, which can be rooted as far back as the 1960s with the researches of the solid-phase synthesis of peptides by the Nobel Laureate Bruce Merrifield at Rockefeller University (Scheme 8.1) [2, 6, 7]. In the past decades, the research and development of combinatorial chemistry to the discovery of new compounds and materials flourish. In the mid-nineties, a method combining thin film deposition and physical masking techniques has been used for the parallel synthesis of spatially addressable libraries of solid-state materials [8]. In the mid-twenties, the mRNA-display macrocyclic peptide libraries using unnatural and D-amino acids as building blocks were reported [9–11]. The introduction of the method for post-translational chemical modification of phage-displayed peptide libraries in 2009 enabled the generation of libraries of conformation constrained peptides with greater chemical diversity and resistance to proteolysis, which are potentially more useful as drugs [12]. The advancement in DNA-encoded chemical libraries (DECLs) have been used to create and decode huge diversity small-molecule organic, peptide, or macrocyclic libraries [13–19].

8.1.2 Principle of Combinatorial Chemistry

Traditionally, chemists synthesize one compound at a time, for example by reacting compound A with compound B to make the product AB, which would have been isolated after reaction workup and purified using crystallization, distillation, or chromatography. In contrast to this approach, synthesis of molecules in a combinatorial fashion can quickly lead to large numbers of molecules using the same reaction conditions and the same reaction vessels. For example, a molecule with three points of diversity (R_1, R_2, and R_3) can generate $N_{R_1} \times N_{R_2} \times N_{R_3}$ possible structures, where N_{R_1}, N_{R_2}, and N_{R_3} are the numbers of different substituents utilized. Or, given a linear amino acid sequence of n residuals X_1-X_2-X_3-X_4-. . . -X_n, the total number of different peptides obtainable equals to y^n, where n is the peptide length and y is the number of different amino acid residues in the synthesis (usually, 18). Therefore, when n equals to 3, there are 5 832 peptides; when n equals to 4, there are 104 976 peptides; when n equals to 5, there are 1 889 568 peptides; when n equals to 6, there are 34 012 224 peptides, etc. [6, 20]. While a traditional chemist can synthesize 100–200 compounds a year, advances in robotics for combinatorial synthesis lead industrial companies to routinely produce over 100 000 new and unique compounds per year, which can be high-throughput screened and tested for potential drug candidates [21].

The range of combinatorial techniques is highly diverse, and these products could be carried out either in separate reaction vessels (parallel synthesis) or simultaneously in a mixture using either solution or solid phase techniques [22–26]. Since there are specific requirements upon the reaction sequence when many reactions are performed in the same vessel, thus while solution chemistry is feasible, it is essential that reaction conditions must be carefully chosen so as to minimize byproduct formation. Yet, the application of solid-phase methodology allows for the generation of products that are physically separated from solution by their attachment to the resin, thus the removal of reagents and byproducts can be implemented by simple filtration step. In addition, the problem of the differences in reagent reactivity can be overcome by the application of the "split–mix (split and pool)" synthesis protocol (Figure 8.1) [24, 25].

The parallel synthesis method developed in 1984 by Geysen *et al.* has been used for the preparation of peptide arrays, which 96 peptides were synthesized on plastic rods (pins) coated at their ends with the solid support [22]. The pins were immersed into the solution of reagents placed in the wells of a microplate plate. Therefore, the parallel method has the advantage that which peptide or other compound formed on each pin can be exactly known, even though this method is much slower than the real combinatorial one. This method has been widely applied in many areas and automatic parallel synthesizers have been developed.

8.1.3 Combinatorial Chemistry with Enzymes

A more recent approach in combinatorial chemistry is the incorporation of biocatalysts into the array of synthetic tools to create smaller and focused libraries. Two different combinatorial schemes can be exploited to generate libraries of compounds with biocatalysts: (i) combinatorial biocatalysis, which assembles *in vitro* new compounds and derivatives using isolated enzyme and (ii) combinatorial biosynthesis, which shuffled or modified the genes of natural biosynthetic pathways to produce *in vivo* libraries of "unnatural" natural products [27, 28].

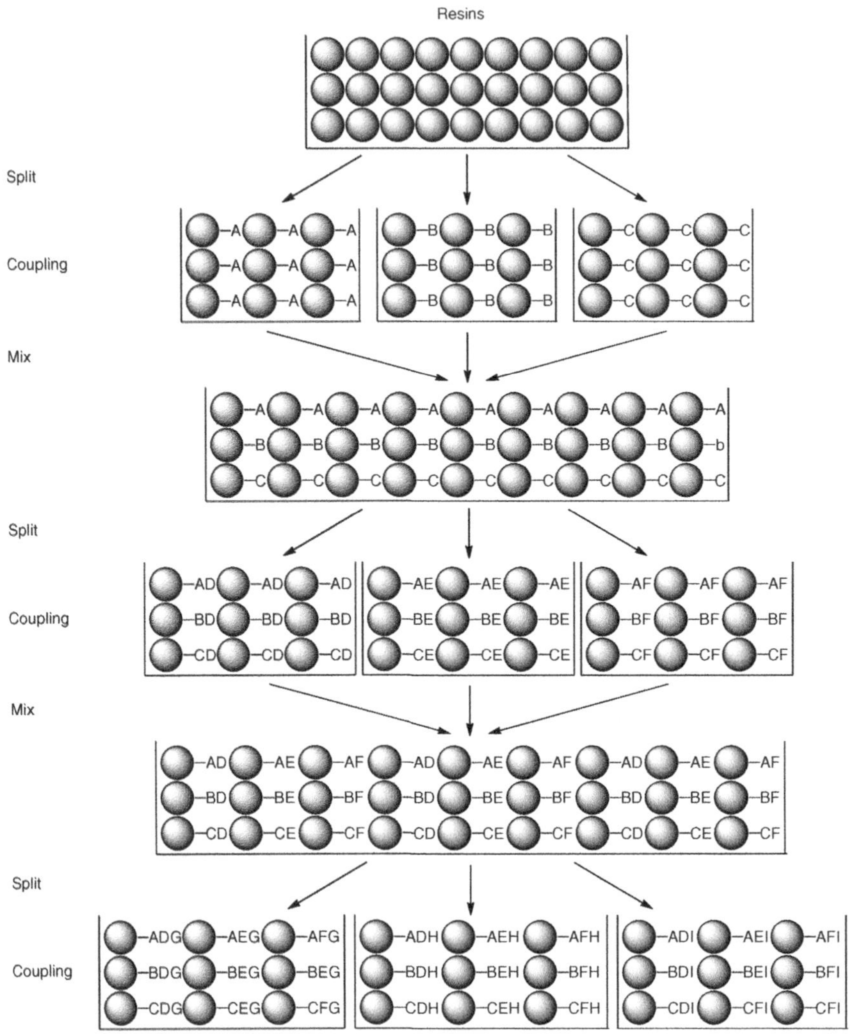

Figure 8.1 Split–mix synthesis library in three cycles and each cycle with three reagents to form 27 library members. *Source:* Based on Sebestyen et al. [24]; Furka et al. [25].

8.1.3.1 Enzymes as a Tool in Combinatorial Synthesis

Because enzymes can be applied at different stages of generating combinatorial libraries including selective deprotection of key functional groups and release of libraries from solid supports, enzymes are valuable tool in combinatorial chemical synthesis, particularly, in solid-phase and polymer-supported synthesis of compound libraries. Due to their exquisite selectivity and operating ability under mild conditions, enzymes are a unique deprotecting tool for combinatorial synthesis involving the most fragile molecules. For instance, penicillin acylase was used for selective *N*-deprotection of a highly labile *S*-palmitoylated oligopeptide in the multistep synthesis of complex lipopeptide [29]. After the removal of the 4-(phenylacetoxy)benzyloxycarbonyl protecting group, the *S*-palmitoylated oligopeptide was further employed as a building block in the following synthetic steps, such as coupling with another oligopeptide chain separately prepared by solid-phase synthesis (Scheme 8.2) [28]. Another similar enzymatic deprotection strategy has been used for the preparation of acid-labile and base-labile nucleopeptides [30]. More examples have been shown in the review of enzymatic protecting group techniques in synthetic chemistry [31].

Enzymes have been utilized in polymer-supported synthesis on which the substrate is anchored to the polymer through a functional group (linker). The linker must be stable during the synthesis and has to be easily cleavable after the synthetic sequence. Since enzymes work under mild conditions where most of compounds during the synthesis are stable, the use of enzymes has open up alternative opportunities to release compounds from polymeric supports. Many enzyme-labile linkers have been developed, which involves the use of hydrolases such as proteases, glycosidases, esterases, and amidases for their cleavages and categorized into *exo*- and *endo*-linkers [32, 33]. Whereas the *endo*-linking bond is cleaved directly by the enzyme action, the cleavage of the *exo*-bond yields an intermediate that further undergoes a spontaneous fragmentation, thereby releasing the target compound. The example of an *endo*-linker is shown in Scheme 8.3a. Following the standard solid-phase peptide synthesis, the peptidic *endo*-linker was directly cleaved by subtilisin BPN to release the target oligopeptide from the polymer support [34]. In another work, an oligosaccharide synthesized on a soluble polymer support was released into solution by α-chymotrypsin acting on an *endo*-linker containing a phenylalanine marker group [35]. *Exo*-linkers have been developed

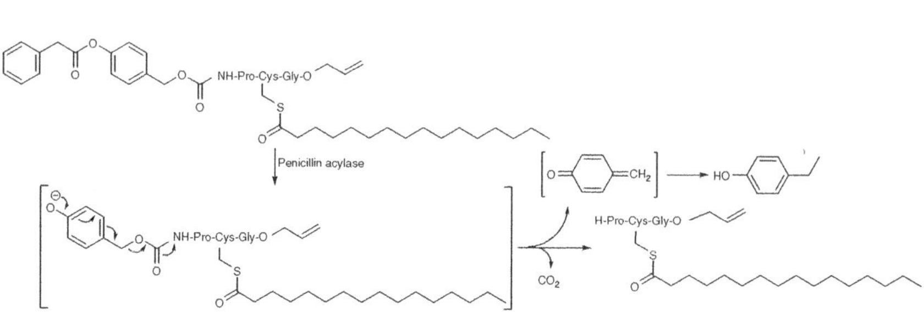

Scheme 8.2 Removal of the enzyme-labile urethane protecting group from lipopeptide by penicillin acylase. *Source:* Rich et al. [28].

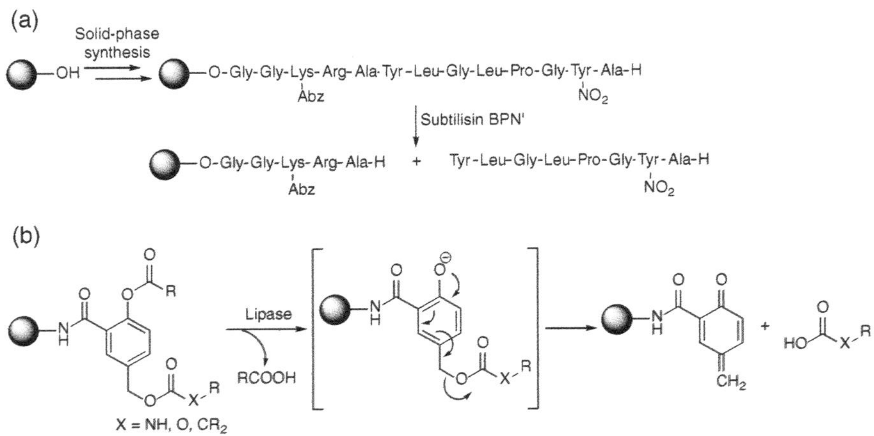

Scheme 8.3 Enzymatic cleavage of (a) *endo*- and (b) *oxo*-linker to release product compounds from the polymer support; Abz, aminobenzoyl. *Source:* Sauerbrei et al. [36].

for a variety of small molecules synthesis [36–38]. An example of the *exo*-linker comprises a 4-acyloxy-3-carboxy-benzyloxy group containing lipases or esterases cleavable acyl group, such as acetate, which is attached to the solid-phase support with an amide bond. The acyl group is cleaved by a lipase to generate a phenolate, which fragments to produce a quinone methide and releases the desired product (Scheme 8.3b) [36]. The quinone methide trapped by water or an additional nucleophile remains attached to the solid-phase support. Enzymatic cleavage of linkers has been successfully achieved both on soluble and insoluble polymer supports. However, the relatively large enzyme molecules may limit the enzyme action on solid carriers by their restrict access into the interior of polymeric beads, which results in incomplete release of synthesized compounds [33, 34].

Instead of using enzymes for cleaving chemically produced libraries from polymer supports, enzymes have also been made toward actual synthetic catalysts for polymer-supported compound libraries. For example, the pioneer work of the solid-phase synthesis of oligosaccharides was catalyzed by galactosyltransferase [39]. The synthesis of sialyl Lewis^x glycopeptide was performed on a solid support by the use of glycosyltransferase and the target product was detached from the solid support through a protease-catalyzed hydrolysis [40]. Afterward, several syntheses of oligosaccharides on solid support or on soluble polymeric support were reported [41–43]. Works on solid-phase and liquid-phase synthesis of glycopeptides, glycolipids, oligonucleotides, and oligosaccharides also have been reviewed [44]. In another approach, glycosidases were used for both glycosidation of a sugar acceptor and for the removal of the unreacted monosaccharide acceptor, and the support for this system was soluble monomethylether of polyethyleneglycol (MPEG) [45, 46]. There also has been the first report of protease-catalyzed high-yielding peptide synthesis on solid support in bulk aqueous buffer with no need for organic cosolvent or activated carboxylic acid [47]. All these and many other examples of solid- and liquid-phase biocatalytic synthesis indicate that the biocatalytic synthetic machinery is ready to be used for the preparation of compound libraries.

8.1.3.2 Combinatorial Biocatalysis

The enzymatic formation of focused libraries is based on derivatization of existing molecules, which mimics the chemistry occurring in biological systems where precursors are modified by the action of biocatalysts. There is a myriad of biocatalytic reactions available for combinatorial biocatalysis, which can be grouped into three major categories: (i) introduction of new functional groups, (ii) modification of existing functionalities, and (iii) addition onto functional groups [27]. For combinatorial biocatalysis, orthogonality is an important matter. The modification of a substrate by one enzyme "X" may prevent it from being a substrate of another enzyme "Y," while the modification of the initial substrate by enzyme

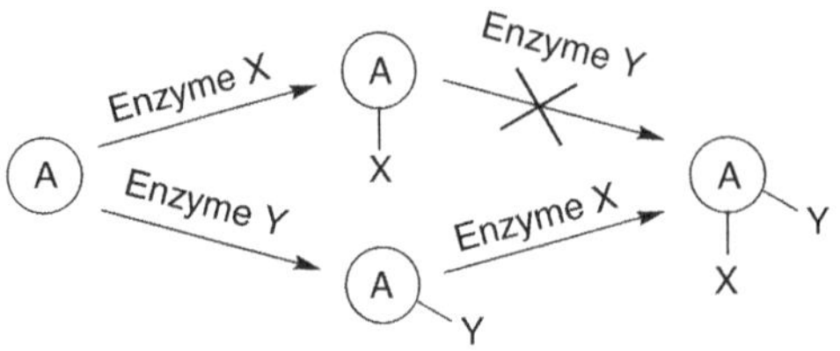

Figure 8.2 Orthogonality issues in biocatalysis. *Source:* Modified from Krstenansky et al. [48].

"Y" may not preclude it from being a substrate for the enzyme "X" (Figure 8.2) [48]. An example of this is the iterative enzymatic derivatization of bergenin to produce 7-chloro-11-(α-galactosyl)bergenin (Scheme 8.4) [48].

The enzymatic reactions are allowed to run iteratively, thus the first generation of derivatives is modified by another round of biocatalysis at different reactive sites to form a second generation of derivatives, etc. Therefore, a great number of derivatives from the original lead compound can be created after several iterations. Again, using the flavonoid bergenin as an example, modification of bergenin and its derivatives by 16 purified enzymes could be made to generate a diverse 600-member library in two synthetic rounds (Scheme 8.5) [27, 49, 50]. The progress of combinatorial biocatalysis have expanded in a growing rate with many new advances including iterative derivatization of small molecules and complex natural products, regioselectively controlled libraries, and novel one-pot library synthesis.

A 24-compound library was prepared by a one-pot reaction with four multifunctional starting materials, that is 2,4-dihydroxy-*N*-(2-hydroxyethyl)benzamide, 4-hydroxyphenethyl alcohol, 3,5-dihydroxybenzyl alcohol, and 4-hydroxybenzyl alcohol, as well as six vinyl esters using lipase from *Candida Antarctica* as the biocatalyst [51]. All the resulted esters were tested for inhibitory activity against

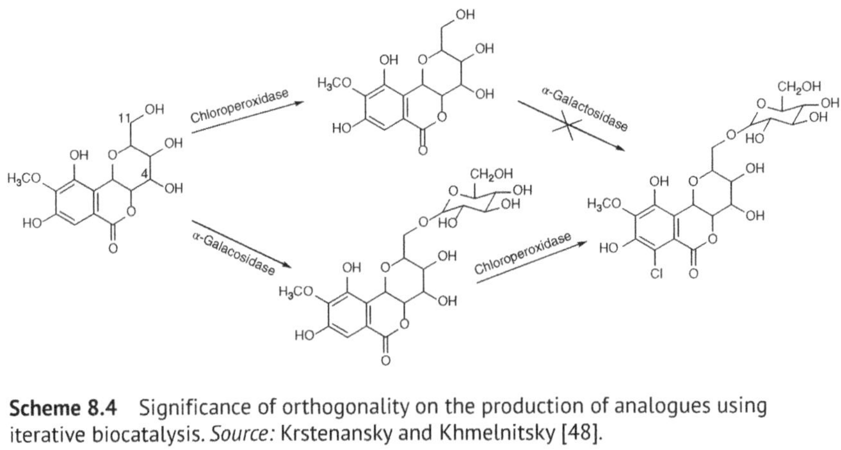

Scheme 8.4 Significance of orthogonality on the production of analogues using iterative biocatalysis. *Source:* Krstenansky and Khmelnitsky [48].

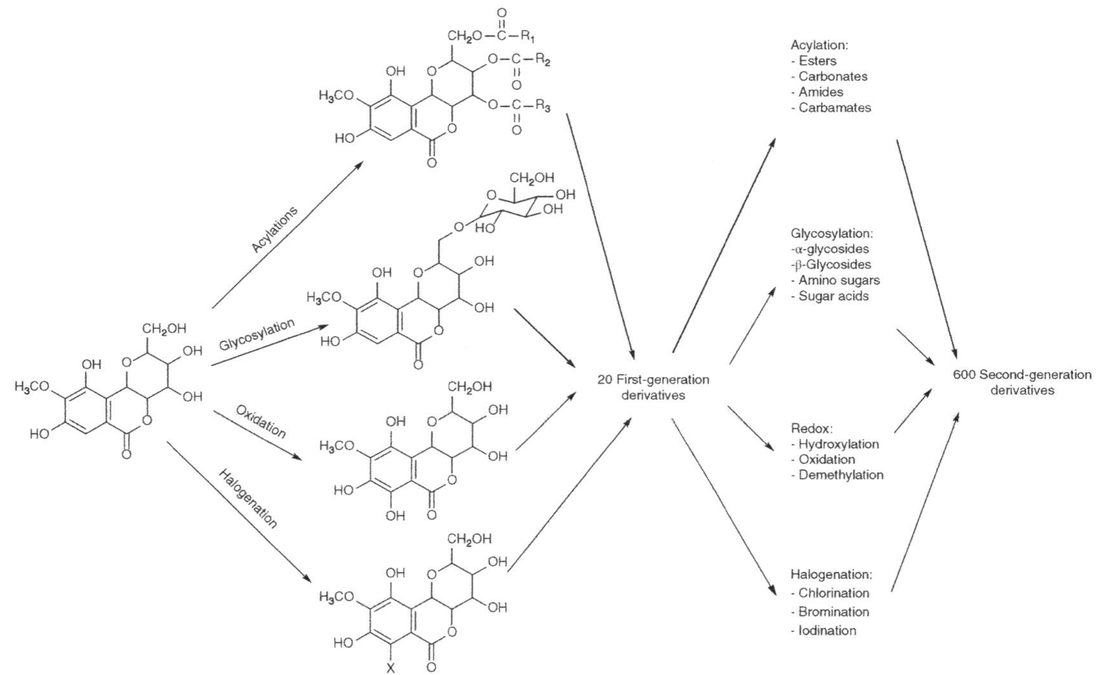

Scheme 8.5 Iterative synthesis of a 600-member library from bergenin in two synthetic rounds. *Source:* Based on García-Junceda et al. [27]; Michels et al. [49]; Altreuter and Clatk [50].

polyphenol oxidase, and a deconvolution strategy was developed to identify the active compound, which involved the resynthesis and screening of partial ester libraries prepared either with a single alcohol and mixture of vinyl esters, or single vinyl ester and mixture of alcohols. Analysis of the inhibition pattern observed with these ten partial libraries suggested that 4-hydroxybenzyl benzoate should be the most potent inhibitor.

A combinatorial library of polyesters using lipase from *Candida antarctica* was synthesized by reacting four straight-chain diesters (C_3–C_{10}) as acyl donors and 12 multifunctional alcohols as acyl acceptors. The selected alcohols included aliphatic and aromatic diols as well as carbohydrates, nucleic acids, and a steroid diol. Each of the polyesters was synthesized in different wells of a 96 deep-well plates. The molecular weights of this polyester library were as high as 20 kDa, including several novel sugar-containing polyesters. This combinatorial approach demonstrates that polymer arrays can be produced from structurally complex monomers [52].

Since the biocatalytic steps would access to structural variants that are either inaccessible or accessible with difficulty by purely synthetic means, the combination of chemical and biocatalytic reactions could provide means to quickly access much larger numbers of compounds than would be available from synthesis alone. In order to augment the structural diversity, a combination of parallel chemical synthesis and biocatalysis was employed to prepare and amplify a library of cross-conjugated cyclopentenones, which are known to incorporate in a number of marine and terrestrial natural products with antibiotic and antitumor activity. The structural space covered by the chemically synthesized combinatorial library of cross-conjugated cyclopentenones was significantly expanded by biocatalytic halohydration, reduction, and acylation using biocatalysts soybean peroxidase, yeast cells, as well as lipases and proteases, respectively (Scheme 8.6) [53].

In a different approach, the carboxylic acid and amine precursors for the chemoenzymatic preparation of a nine-member Ugi condensation library were enzymatically generated by selective acylation of 3-hydroxybutyric acid and

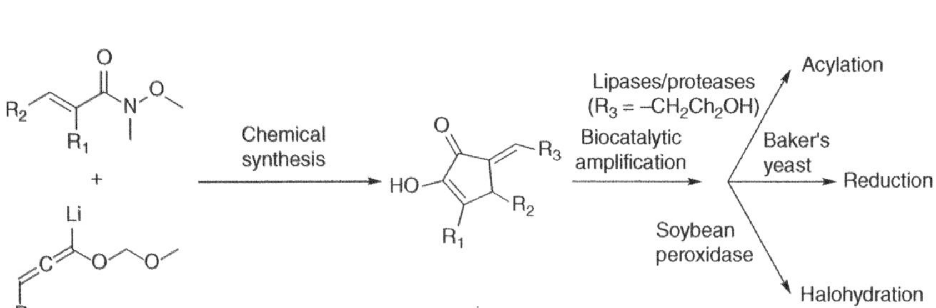

Scheme 8.6 Chemically synthesized combinatorial library of cross-conjugated cyclopentenones expanded by biocatalysis. *Source:* Modified from Jang et al. [53].

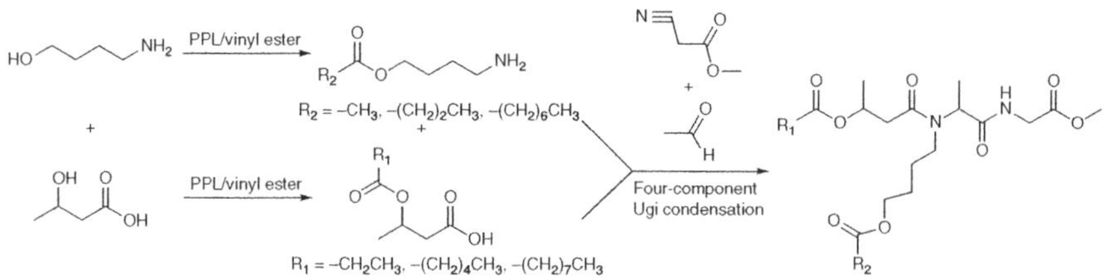

Scheme 8.7 Chemoenzymatically library synthesis by Ugi four-component condensation using enzymatically generated carboxylic acid and amine building blocks; PPL, porcine pancreatic lipase.

4-amino-1-butanol, respectively, using porcine pancreatic lipase in *tert*-amyl alcohol. These enzymatically prepared derivatives were then subjected to a four-component Ugi condensation reaction in the presence of acetaldehyde and methyl isocyanoacetate (Scheme 8.7). The range of yields for the α-(acylamino)amide Ugi products was 72–95% [54].

The glycosyltransferases, which can be used to prepare diverse libraries of oligosaccharides and glycoconjugates, show typically high promiscuity toward aglycons and almost exclusively rely on UDP- and TDP-nucleotide sugars as glycosyl donors for the construction of glycoconjugates. An efficient biocatalytic system capable of quickly generating these critical nucleotide sugars for oligosaccharides libraries has been reported, which utilizes a recombinant α-D-glucopyranosyl phosphate thymidylyltransferase to convert a wide array of α-D-hexapyranosyl phosphates to their corresponding UDP- and TDP-nucleotide sugars [55, 56].

Combinatorial enzymatic synthesis of some unnatural deoxygenated trisaccharide analogues has also been reported on the milligram scale by a series of deoxygenated uridine 5'-diphosphogalactose (UDP-Gal) derivatives and three different retaining α-galactosyltransferases (Scheme 8.8) [57, 58]. Since UDP-2-deoxy-Gal

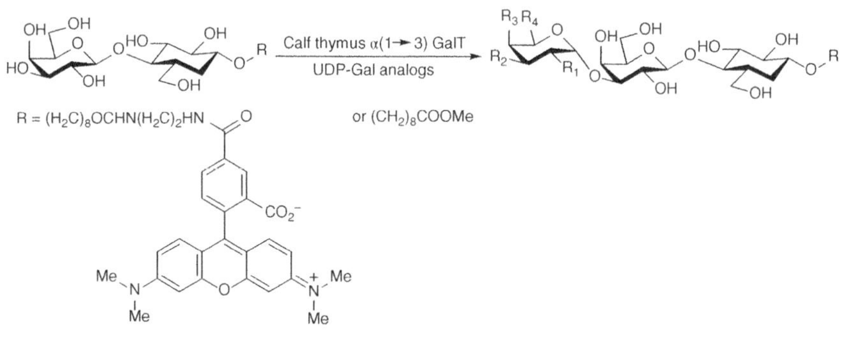

Scheme 8.8 Unnatural oligosaccharide library synthesized by using UDP-Gal analogues. *Source:* Based on Nishimura [57]; Sujino et al. [58].

and UDP-6-deoxy-Gal were found to be active as donors for all three enzymes, substrate specificity and accessibility for the assembling of oligosaccharides depend deeply on the position of the deoxygenation. This result suggests that some genetically engineered enzymes may be designed on the basis of a suited molecular modeling study on enzyme sugar-nucleotide interaction.

In contrast to combinatorial chemistry (CC), where subunits are covalently linked to yield libraries of increasingly complex molecules, dynamic combinatorial chemistry (DCC, also known as dynamic covalent chemistry) is a branch of supramolecular chemistry whereby a noncovalent assembly of subunits yields highly complex chemical systems [59]. DCC applies reversible chemical reactions to form libraries under thermodynamic control, which eventually reach the equilibrium state. However, building blocks and products in a dynamic library are still continuously interconverting one another, which enable the system to respond to an external stimulus by shifting its equilibrium composition. Therefore, artificial template molecules were used to selectively direct cyclodextrin glucanotransferase (CGTase) mediated generation a complex dynamic mixture of interconverting linear and macrocyclic α-1,4-D-glucans (cyclodextrins). CGTase catalyzes not only reversible transglycosylation but also irreversible hydrolysis. The native cyclodextrins (α, β, and γ) are formed out-of-equilibrium as part of a kinetically trapped system that operates transiently like a Dynamic Combinatorial Library (DCL) under thermodynamic control. The selectivity for the synthesis of each of the native cyclodextrins by adding different templates was 89–99%, or we can entirely alter the outcome of the synthesis to obtain larger ring cyclodextrins, δ-CD (CD9) and ε-CD (CD10). In the absence of templates, the transient DCL lasts less than a day, and cyclodextrins convert rapidly to short maltooligosaccharides [60].

Heparan sulfate, an anionic polysaccharide, is biosynthesized in the Golgi body of eukaryotic cells through sequential reactions involving up to 30 enzymes, but heparin, a widespread used clinical anticoagulant drug, is a specific highly sulfated version of glycosaminoglycan produced by mast cells [61]. A collection of recombinant *E. coli* expressed enzymes, including *N*-deacetylase *N*-sulfotransferase (NDST), *O*-sulfotransferases (OST), and C5 epimerase (Epi), is used to modify *N*-sulfo heparosan, which had been chemoenzymatically prepared from *E. coli* K5 polysaccharides, to construct a library of heparan sulfate polysaccharides. Eight heparan sulfate polysaccharides were characterized by disaccharide analysis. A heparan sulfate polysaccharide containing a repeating unit consisting of →4) β-D-glucuronic acid (1→4) α-D-6-*O*-sulfo *N*-sulfo glucosamine (1→, called "Recomparin," showed antithrombin-mediated anticoagulant activity [62].

Biocatalytic generation of potential building blocks for combinatorial libraries was described in another report that a threonine–trisaccharide conjugate was prepared in a one-pot reaction cascade involving a regioselective and stereoselective

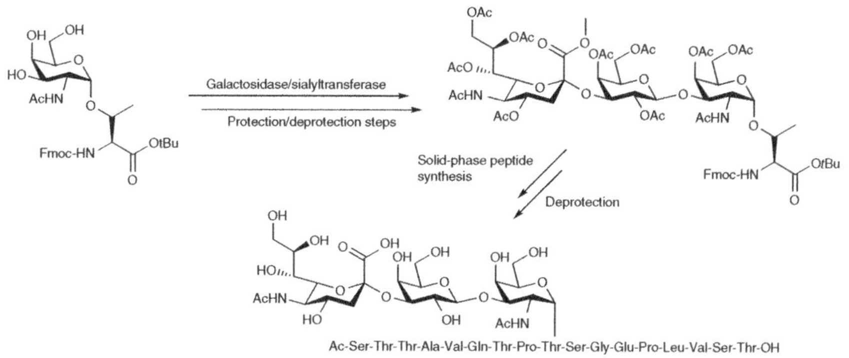

Scheme 8.9 Enzymatic synthesis of the glycosylated amino acid derivative building block in the solid-phase synthesis of a glycopeptide library; Fomc, *N*-(9-fluorenylmethoxycarbonyl). *Source:* Based on Rich et al. [28]; Bézay et al. [63].

galactose transfer from lactose, catalyzed by galactosidase, followed by enzymatic sialylation. The glycosylated amino acid derivative was then employed as a building block in the solid-phase synthesis of a glycopeptide (Scheme 8.9) [28, 63].

The chemoselective polymer blotting method has been developed for rapid and efficient synthesis of complex glycopeptides using a convenient "heterobifunctional linker" that shuttles product between solid-phase and water-soluble polymer supports. The five-step synthetic strategy follows: (i) perform the solid-phase chemical syntheses of the glycopeptides with the newly introduced heterobifunctional linker, (ii) deprotection and release of glycopeptides from the resin, (iii) chemoselective blotting of glycopeptides by a water-soluble polymer to act as a "primers," (iv) sugar elongation by glycosyltransferase, and (v) release of the target glycopeptides. The combined use of the solid-phase chemical syntheses of peptides and the enzymatic syntheses of carbohydrates on water-soluble polymers greatly contribute toward the efficient high-throughput synthesis of complicated glycopeptide libraries, that is various types of MUC1 mucin glycopeptides exhibiting a variety of sugar moieties [64].

A library of tailored multivalent neo-glycoproteins was synthesized and tested for their galectin-3 (Gal-3), a tumor biomarker involving in tumor angiogenesis and metastasis, binding properties. A set of *N*-acetyllactosamine (Galβ1,4GlcNAc; LacNAc type 2)-based oligosaccharides featuring five different terminating glycosylation epitopes was first synthesized by the combinatorial use of glycosyltransferases and chemo-enzymatic reactions, respectively. Neo-glycosylation of bovine serum albumin (BSA) was accomplished by dialkyl squarate coupling to lysine residues, which results in a library of defined multivalent neo-glycoproteins. Solid-phase binding assays with immobilized neo-glycoproteins revealed distinct affinity and specificity of the multivalent glycan epitopes for Gal-3 binding;

particularly, high selectivity was shown for neo-glycoproteins decorated with *N',N''*-dialkyllactosamine (GalNAc*β*1,4GlcNAc; LacdiNAc) epitopes. In addition, neo-glycoproteins with terminal biotinylated LacNAc glycan motif could be used as Gal-3 detection agents in a sandwich ELISA format [65].

Synthesizing metabolites with the correct stereoselective modifications usually is difficult for chemically complex parent compounds. Biocatalytic approaches using human drug-metabolizing enzymes, such as cytochrome P450 2D6 (CYP2D6), to introduce the required modifications is an alternative to purely chemical synthesis. The combinatorial libraries of ancestors for directed evolution (CLADE) approach was employed to develop a library of robust mutant CYP2D enzymes for biocatalysis. The CLADE approach increased the stability of CYP2D mutants, which were obtained by DNA shuffling using three extant CYP2D forms. The resulting CYP2D mutants exhibited divergent profiles of activity toward typical CYP2D6 substrates and an increase of the average thermostability [66].

CsyB from *Aspergillus oryzae* is a unique type III PKS (polyketide synthase) that is involved in the biosynthesis of short-chain product csypyrone B1 (Scheme 8.10a). In addition, it is the first type III PKS that performs not only the polyketide chain elongation by decarboxylative condensation but also the condensation of two β-ketoacyl units by non-decarboxylative condensation. Therefore, a highly promiscuous enzyme type III PKS was utilized to perform a combinatorial enzymatic synthesis to produce a series of unnatural long-chain β-branch pyrenes by the nondecarboxylative condensation reaction using two molecules of fatty acyl diketide-*N*-acetylcysteine (diketide-NAC) units of different chain lengths. Two molecules of non-natural long-chain (C_{16}, C_{18}) fatty acyl diketide-NACs were prepared, which were individually coupled to six short-chain (C_4-C_9) diketide-NACs to generate *in vitro* 12 novel unnatural long-chain β-branch pyrenes by one-pot synthesis (Scheme 8.10b) [67].

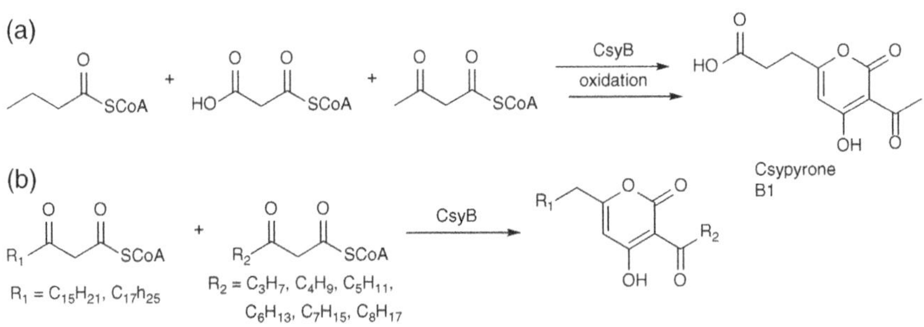

Scheme 8.10 (a) Biosynthesis of Csypyrone B1 with mutant CsyB. (b) Combinatorial enzymatic synthesis of unnatural long-chain β-branch pyrenes by mutant CsyB. *Source:* Pan et al. [67].

Integration of dynamic combinatorial chemistry and scale effect has been designed to realize the efficient synthesis of bioactive compounds. In a study, a novel microfluidic system consisting of two microreactors, coupled with online high-resolution mass spectrum detection was designed and used in the esterification. Model reaction was first optimized in microreactor A to investigate the role of scale effect in the esterification. Synthetic efficiency in the esterification reaction and combinatorial synthesis were significantly improved in the capillary with smaller diameters. Then, two libraries of long fatty acid esters were established in the flask batch mode and fixed bed flow mode, respectively. The composition of these two libraries was analyzed and compared. Based on the interaction between esters and bull serum albumin, the reaction equilibrium in microreactor B could be altered intensify the synthesis of esters with high binding activity [68].

8.1.3.3 Combinatorial Biosynthesis

Traditionally, the production of natural products and their analogs with complex structures for pharmaceutical applications was via fermentation processes. However, more research efforts on the metabolic engineering at genetic level offer the possibility to develop controllable biosynthetic pathways, which has opened the door to use the cell to exploit substrate promiscuity to generate novel "unnatural" natural products, substantially expanding the structural diversity of natural products with potential pharmaceutical application. The whole cell combinatorial biosynthesis has been extensively employed for the production of libraries such as macrolide antibiotics, carotenoids, and peptides [69–73], which many of them consist of modular enzymes that can be genetically engineered in the development of custom catalysis [74]. Since natural products have played a significant role in drug discovery, nowadays the combinatorial biosynthesis is flourishing in pharmaceutical industry. The advancement of combinatorial biosynthesis for drug discovery follows three major strategies: (i) precursor-directed biosynthesis; (ii) enzyme-level modification, which includes swapping of the entire domains, modules and subunits, site-specific mutagenesis, and directed evolution; and (iii) pathway-level recombination [75]. The followings are a few examples to highlight these strategies.

In the precursor-directed combinatorial biosynthesis, the enzymes in the biosynthetic pathways can take diverse structures of non-native building blocks to produce a variety of natural product analogs. Molecular type I polyketide synthases (mPKSs) are polyketide synthase (PKS) assembly lines that contain sequentially organized modules, each of which harbors a set of catalytic domains required for one cycle of chain extension. Thus, a chemical genetic strategy that combines precursor-directed biosynthesis and colony bioassay was developed to rapidly discover new macrolide antibiotics. A library of synthetic diketide precursors that mimics the natural diketide intermediate in the biosynthesis of

6-deoxyerythronolide B (6-dEB) was fed to cultures of engineered *E. coli* HYL3 (*E. coli* HYL3/pBP130/pBP175#8/pHL74/pHL80*) strain evolved for improved polyketide production, with plasmids expressing module 2–6 (pBP130/pBP175#8) of the mPKS 6-deoxyerythronolide B synthase (DEBS), and proteins (pHL74/pHL80*) for sugar biosynthesis and glycosyl group transfer. These precursors were converted to the corresponding glycosylated macrolides, and their bioactivities were screened against a *B. subtilis* overlaid lawn. A new class of alkynyl- and alkenyl-substituted erythromycin analogs, including an equipotent and orthogonally functional analog, with activities comparable to that of the natural product was identified (Scheme 8.11) [76].

PKSs are classified into three groups: types I, II, and III. Compared to mPKSs, type III PKSs are structurally simple PKSs, which catalyze the priming, extension, and cyclization reactions iteratively with a single active site. The sources of PKSs are plants, bacteria, and filamentous fungi and widely used to produce biologically important compounds such as chalcones, pyrenes, acridones, phloroglucinols, and stilbenes [77]. Since type III PKSs exhibit unusual broad substrate specificity and notable catalytic versability, they offer an excellent platform for precursor-directed biosynthesis of novel “unnatural” natural products [78]. A novel plant type III PKS cloned and sequenced from the Chinese club moss *Huperzia serrata* was expressed in *E. coli*, which showed unusually versatile catalytic potential to accept various starter units and produce various aromatic tetraketides, including chalcones, benzophenones, phloroglucinols, and acridones. In particular, this enzyme also accepted bulky starter substrates such as 4-methoxycinnamoyl-CoA and *N*-methylanthraniloyl-CoA, and carried out three condensations with malonyl-CoA to generate 4-methoxy-2’,4’,6’- trihydroxychalcone and 1,3-dihydroxy-*N*-methylacridone, respectively [79]. Micacocidin is a thiazoline-containing metabolite from the plant pathogenic bacterium *Ralstonia solanacearum* generated on a hybrid polyketide/nonribosomal peptide synthetase (PKS/NRPS) assembly lines. Micacocidin is of pharmaceutical interest due to its use in treating *Mycoplasma pneumoniae* infections. In the precursor-directed biosynthesis of this antibiotic, the starter-unit-activating domain, fatty acid-AMP ligase (FAAL), of the hybrid PKS/NRPS enzyme MicC activates the starter unit hexanoic acid as acyl-adenylate and forwards it to this iteratively acting polyketide synthase (Scheme 8-12) [80]. The FAAL exhibited an extended substrate tolerance, which opens the door for the modification of a micacocidin residue with non-native precursors. Thus, a total of six new analogues of micacocidin were generated with activity against *M. pneumoniae*.

Swapping of the entire domains, modules, or subunits of the enzyme has been the main classical approach for combinatorial biosynthesis. Many modular PKSs and NRPSs are suitable for the combinatorial biosynthesis by rearranging their modular organization and stepwise synthetic strategy [81–83]. One example is the

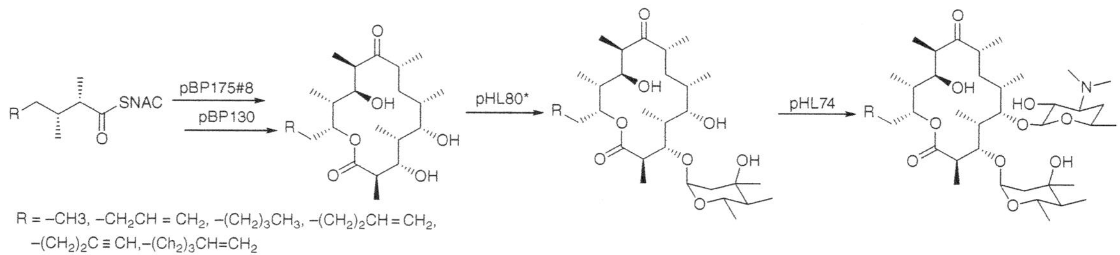

Scheme 8.11 Chemical genetic strategy for discovery new macrolide antibiotics using precursor directed biosynthesis in engineered *E. coli*. *Source:* Modified from Harvey et al. [76].

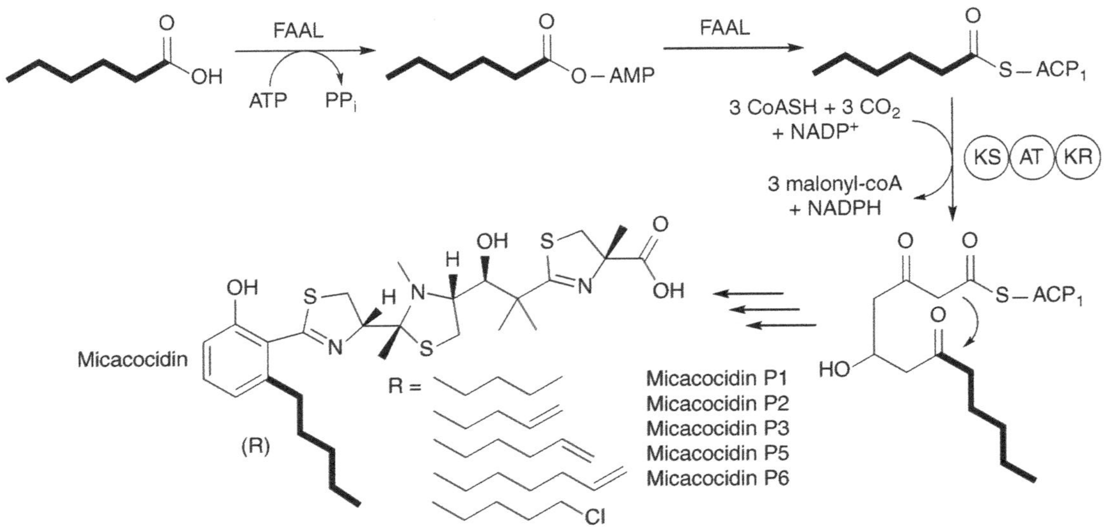

Scheme 8.12 The precursor-directed biosynthesis of micacocidin derivatives with FAAL located on MicC. Domain notation: ACP, acyl carrier protein; KS, ketosynthase; AT, acyl transferase; KR, ketoreductase. *Source:* Kreutzer et al. [80].

construction of a large combinatorial library of 61 6-dEB analogs by using 6-deoxyerythronolide B synthase (DEBS), the mPKS used for producing the macrolide ring of erythromycin. This was accomplished by substituting the AT domains and the β-carbon processing domains of DEBS with counterparts from rapamycin mPKS that encode alternative substrate specificities and β-carbon reduction/dehydration activities. The library includes analogs with one, two, and three altered carbon centers of polyketide products [84]. Daptomycin is a cyclic lipopeptide antibiotic produced by *Streptomyces roseosporus* and approved for the treatment of skin and skin structure infections caused by Gram-positive pathogens. Genetic engineering using module and subunit exchanges for NRPSs have been applied for the combinatorial biosynthesis to generate lipopeptide antibiotics daptomycin derivatives [85–87]. For instance, subunit and module exchanges between related cyclic lipopeptide A54145 and calcium-dependent antibiotic (CDA), coupled with modifications of glutamine at position 12 and variations in lipid side chain, in mutants of *S. roseosporus* were used to generate a combinatorial library of 72 lipopeptide antibiotics related to daptomycin, and most of which displayed antibacterial activity [87]. Fungal polyketide benzenediol lactones (BDLs) are important pharmacophores exhibiting a wide range biological activities. BDL scaffolds are biosynthesized by pairs of collaborating, sequentially acting iterative polyketide synthases (iPKSs) forming quasi-modular BDL synthases (BDLSs) [88–90]. Expression of random pairs of iPKS subunits from four BDL model systems in a yeast heterologous host produces a diverse combinatorial library of BDL congeners, including a novel radiolarian with a 14-membered ring exhibited more potent heat shock response-inducing activity than natural dehydrocurvularin [90].

The combinatorial biosynthesis using the classical domains, modules, or subunits swapping approaches often has difficulties with insoluble protein expression, impaired activities, and reduced product yield. These problems are most probable as a result of the disruption of protein's overall structure, which cause the loss of its function. In addition, the intermediates formed by domain swapping may result large difference in structure, which makes the intermediates inaccessible by downstream catalytic domains. Thus, modern site-specific mutagenesis, which only substitutes specific amino acid residues at the catalytic site of the enzyme, is usually less invasive and changes the enzyme function more efficiently.

After the catalytically critical active site residues in every reductive domain of the monensin PKS were identified by protein sequence alignments, a minimally invasive mutagenesis scheme was developed to inactivate the targeted reductive domains, such as ketoreductases, dehydratases, and enoylreductases of the chosen modules of monensin PKS, which leads to the formation of a library of 22 premonensin redox derivatives [91]. Organofluorines represent a rapidly expanding proportion of molecules that are used in pharmaceuticals, diagnostics, agrochemicals, and materials. The only organofluorine natural products characterized to date consist of a small set of simple molecules associated with the fluoroacetate pathway of *Streptomyces cattleya*, which have the ability to catalyze the formation of C–F bonds from aqueous fluoride. When a key S2107A ($Ser^{2107} \rightarrow Ala^{2107}$) mutation was introduced into the AT domain on the sixth module of the mPKS DEBS, the AT substrate specificity was switched from native methylmalonyl-CoA to fluoromalonyl-CoA [92]. Therefore, mutagenesis of the AT domain on either the module 2 or module 3 results in Site-selective fluorine insertion into the polyketide backbone by the modularity and the stepwise synthetic mode of mPKS, which opens the door to generating fluorinated natural product analogs [93].

Directed evolution is a powerful enzyme engineering approach that has not been widely utilized in natural product biosynthetic enzymes. Nevertheless, applying directed evolution to combinatorial biosynthesis exists several significant advantages. First, the changes by site-specific mutagenesis are more conservative, while directed evolution approaches can make more drastic alterations of substrate specificity of domains but still restoring the impaired activity due to large changes in substrate specificity. Second, in contrast to only one enzyme variant obtained with every successful domain swap, directed evolution methods significantly increase the throughput of enzyme variants that greatly benefit the combinatorial biosynthesis. Third, directed evolution can be accomplished even when the enzyme catalytic mechanism is unclear [75]. The first application of directed evolution was on NRPSs for nonribosomal peptides production. Using rounds of random mutagenesis coupled with *in vivo* screening successfully restored the activity losses of two chimeric NRPSs by the A domain swapping [94]. The success of directed evolution strategy was then employed for the combinatorial NRPS biosynthesis, which exploited saturation mutagenesis of the active site residues of A domain of the 64-kDa A-T di-domain protein AdmK within a hybrid NRPS/PKS pathway for antibiotic andrimid biosynthesis [95]. A multiplexed assay was used to screen a library of ~14,330 clones and four derivatives of antibacterial andrimid were generated including three new derivatives, which showed bioactivity toward *E. coli* imp ASR in agar overlay bioassays. Still more efforts in directed evolution of isolated A domain of DhbE were put to increase the repertoire of A domains for combinatorial biosynthesis. DhbE is part of a three-module NRPS assembly line comprising DhbE, DhbB, and DhbF responsible for the synthesis of the siderophore bacillibactin in *Bacillus subtilis* that activates 2,3-dihydroxybenzoic acid (DHB) to form the DHB-AMP acyl-adenylate and then transfer it to the aryl carrier protein (ArCP) domain of DhbB (Scheme 8.13) [96]. The substrate specificity of the A-domain mutants of DhbE of the bacillibactin NRPS complex was dramatically altered toward unnatural aromatic building blocks using a powerful high-throughput screening technique, yeast cell surface

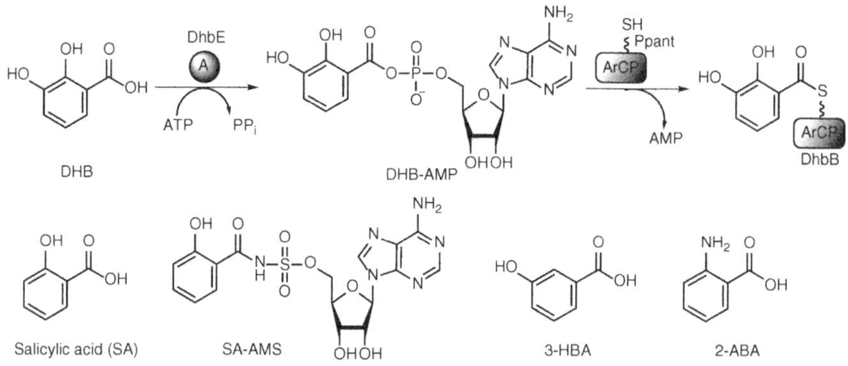

Scheme 8.13 DhbE catalyzed aryl acid activation and the generation of nonnative substrates; SA-AMP, salicylic acid-adenosine monosulfate. *Source:* Zhang et al. [96].

display. The result of iterative rounds of selection by this strategy generated a library of 5×10^6 yeast clones and acquired A-domain mutants of DhbE that identify two nonnative substrates 3-hydroxybenzoic acid (3-HBA) and 2-aminobenzoic acid (2-ABA).

In 1985, a novel antibiotic compound, mederrhodin A, was found by interchanging and combining genes from multiple species to generate combinatorial pathways [97]. Following this work, more novel natural products, especially in drug discovery, were discovered by using hybrid pathways. As an example, triterpenoid saponins, a diverse group of specialized (secondary) metabolites with a wide range of biological properties found in many plant species, can be produced with heterologously expressed combinatorial biosynthesis approaches. Recombinant yeast strains expressing the combinations of either the β-amyrin synthase (bAS), cytochrome P450 reductase (CPR), β-amyrin oxidases *CYP93E2*, and *CYP72A61v2* or the bAs, CPR, β-amyrin oxidases *CYP716A12* and *CYP72A68v2* were able to generate soyasapogenol B and gypsogenic acid, respectively [98]. The yeast strains expressing bAS, CPR, β-amyrin-30-oxidase *CYP72A63* [99], and β-amyrin oxidases *CYP93E2* or *CYP16A12* were employed to produce rare triterpenoids not occurring in *Medicago truncatula*. Furthermore, a gene, designated *CYP716Y1*, encoding a cytochrome P450 monooxygenase (P450) from *Bupleurum falcatum* was identified and exploited to catalyze the C-16α hydroxylation of saikosaponins that comprise oleanane- and ursane-type triterpene saponins. This enzyme was combined with oxidosqualene cyclase, P450, and glycosyltransferase genes available from other plant species and reconstituted the synthesis of monoglycosylated saponins in yeast (Scheme 8.14) [100]. More examples for pathway-level combinatorial biosynthesis are described in reference [75].

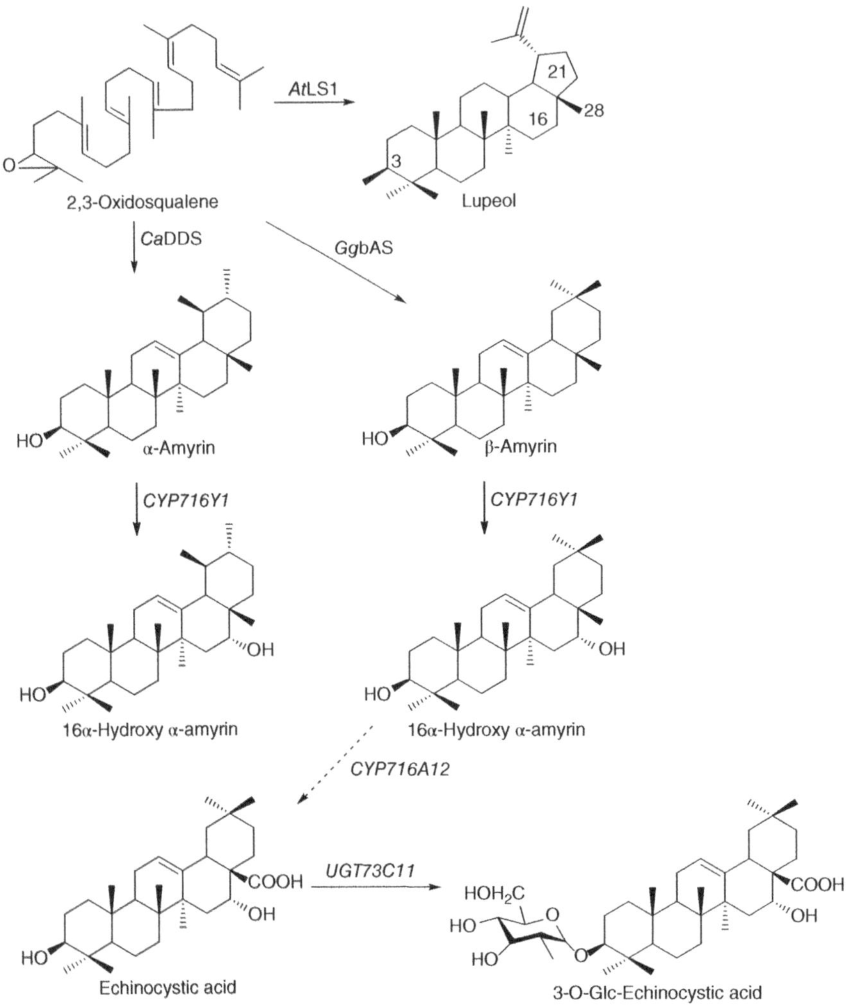

Scheme 8.14 Triterpene saponin biosynthesis pathway in engineered yeast. *Source:* Moses et al. [100].

8.2 Artificial Intelligence Assisted Enzymatic Organic Synthesis

8.2.1 Introduction to Artificial Intelligence

Intelligence means the ability to adapt to new circumstances and to solve new problems. Therefore, all but the simplest and routine human behavior is ascribed to intelligence, while even the most complicated animal instinctive behavior is never taken as an indication of intelligence. Furthermore, human intelligence is not characterized by just on trait but by the combination of many diverse abilities. In this sense, artificial intelligence (AI) refers to the simulation of human intelligence in machines (or computers) that are programmed to think like human mind and to mimic human functions, such as learning and problem solving [101].

Artificial intelligence was founded as an academic discipline in 1955 after the development of the digital computer in the 1940s, which has been demonstrated that computers can be programmed to carry out very complex tasks, for example, discovering proofs or mathematical theorems or playing chess. AI becomes an interdisciplinary science with multiple approaches in recent years. The research of AI has been divided into subfields based on technical considerations, such as particular goals, for example, "robotics" or "machine learning," the use of particular tools, for example, "logic" or artificial neural networks, or deep philosophical differences. Subfields have also been divided based on social factors, such as particular institutions or the works of specific researchers. The advancements in machine learning and deep learning are creating a paradigm shift in virtually every sector of the tech industry [102].

The goals (or problems) of AI include learning, reasoning, knowledge representation, planning, language processing, perception, and the ability to move and manipulate objects. Methods to approach the goals include statistical methods, computational intelligence, and traditional symbolic AI, and tools used in AI include versions of search and mathematical optimization, artificial neural networks, and methods based on statistics, probability, and economics. Machines are wired using a cross-disciplinary approach based on computer science, information engineering, mathematics, linguistics, psychology, philosophy, and many more. Algorithms often play an important part in the structure of AI, where simple algorithms are used in simple applications, while more complex one helps frame strong artificial intelligence. In this twenty-first century, following concurrent advances in computer power (speed and memory), large amounts of data ("Big Data"), and theoretical understanding, AI techniques have become an essential part of the technology industry.

Nowadays, AI has been applied for medical care and diagnostics, transportation, public safety, service robots, education, and entertainment, but the collection of "Big Data" and the expansion of the Internet of Things (IoT) have made a perfect environment for new AI applications and services to grow. We will foresee that with the combination of Internet AI will serve as a powerful engine for the society and economic growth [103].

8.2.2 Categorization of Artificial Intelligence

Artificial intelligence can be divided into two broad categories: narrow AI and general AI. Narrow AI, sometimes referred to as "Weak AI," is a kind of AI operates within a limited context to perform single specific tasks within a domain (e.g.

language translation). Although narrow AI may seem intelligent, they operate under far more constraints and limitations than even the most basic human intelligence. General AI, sometimes referred to as "Strong AI," is hypothetical and not domain specific but can learn and perform tasks anywhere (e.g. robots). General AI is a machine with general intelligence that carry on the tasks much like a human being, thus it can apply to solve any problem [103]. In contrast to the more complex and complicated general AI systems, narrow AI is all around us and is easily the most successful realization of artificial intelligence to date that has experienced numerous breakthroughs in the last decade to have "significant societal benefits and have contributed to the economic vitality of the nation," according to a 2016 report released by the Obama Administration. Therefore, the following subjects will focus on advances in narrow AI, especially on the development of new algorithms and models referred to machine learning.

8.2.3 Machine Learning

Venture capitalist Frank Chen provides a note:

> *Artificial intelligence is a set of algorithms and intelligence to try to mimic human intelligence. Machine learning is one of them, and deep learning is one of those machine learning techniques.*

This note indicates AI often revolves around the use of algorithms. An algorithm is a set of unambiguous instructions that a mechanical computer can execute that is used to solve a problem. Algorithms are developed by programmers to instruct computers in new tasks, which are the building blocks of the advanced digital world we meet today. Computer algorithms organize enormous amounts of data into information and services based on certain instructions and rules. Here, an important concept must be understood that in machine learning (ML), learning algorithms and not computer programmers create the rules [102–104].

In the ML approach, algorithms give the computer instructions that allow it to learn from data without new step-by-step computer programming instructions by the programmer, which means computers can be used for new complicated tasks that could not be manually programmed (e.g. photo recognition for the visually impaired or translating pictures into speech). In other words, in the basic ML process, we put training data to a learning algorithm, so the learning algorithm generates a new set of rules based on inferences from the data, which in essence is the generation of a new algorithm, formally referred to as the machine learning model. The same learning algorithm could be used to generate different models from different training data. Put it simply, ML is a process that feeds a computer data and uses statistical techniques help it "learn" how to get progressively better at a task, without having been specifically programed for that task, eliminating the need for millions of lines of written code. Therefore, the same type of learning algorithm could be used to teach the computer how to translate languages or predict the stock market. Note that a complex algorithm is often built on top of other, simpler, algorithms [102–104]. From the arguments above, the core strength of ML is the inference of new instructions from data, which in turn highlight the critical role of data, that is the more data available to train the algorithm, the more it learns.

Even an ML model may apply a mix of several technique, the methods for ML can be categorized as three general types: supervised learning, unsupervised learning, and reinforcement learning. For supervised learning, the learning algorithm is given labeled data sets and the desired output, for example, pictures of cat labeled "cat" will help algorithm identify the rules to classify pictures of cats. In unsupervised learning, unlabeled data sets are given to the learning algorithm and the algorithm is asked to identify patterns in the input data, for example, the recommendation system of an e-commerce website where the learning algorithm discovers similar items often bought together. While for reinforcement learning, the algorithm interacts with a dynamic environment that provides feedback in terms of rewards and punishments, for example, self-driving cars being rewarded to stay on the correct direction of the road [103]. In addition to the three general types of ML, deep learning is a type of ML that runs inputs through biologically inspired layered neural network architecture. The layered neural networks contain a few hidden layers through which the data are processed, allowing the machine to go "deep" in its learning as well as making connections and weighting input for the best results [104].

With the advent of computer, the use of computer in organic chemistry or computer-aided organic synthesis has also rapidly and vigorously developed [105]. The idea and research studies of computer-assisted organic syntheses can be traced back as early as 50 years ago [106]. AI development and ML move the organic synthesis toward a new era, which can be employed for not only synthesis planning, suitable reaction conditions prediction, and search for new activity but also optimizing the prediction of chemical patterns and robot-drive automation and autonomous synthesis [107–112].

8.2.4 Artificial Intelligence with Enzymatic Organic Synthesis

Biocatalysis with either isolated enzymes or whole cell microorganisms is increasingly interested in synthetic organic chemistry and being applied for diverse large-scale industries [113]. In the past few years, a plethora of methods have been developed to enable the rather rapid tailored-design of an enzyme for a targeted reaction such as asymmetric synthesis of a chiral building block by the combination of information from sequence and structure databases with modern molecular biology methods and high-throughput screening tools [114], while the application of disconnection approach guided by retrosynthetic analysis of possible intermediates and the chemical reactions involved to search ready available starting materials for the planning of a biocatalytic synthesis route is just get developed [115]. All these synthetic strategies indicate that the selection of the most appropriate existing enzymes or the discovery of novel enzymes for targeted reactions, cascade reactions, or reaction pathways is the utmost important point for successful enzymatic synthesis purpose.

Protein engineering or specifically enzyme engineering has emerged as an important tool for the development of new biocatalyst entities with tailor-designed desired properties [116]. Accompanied with the development of computational techniques and site-directed mutagenesis, or directed (molecular) evolution techniques, researchers can produce enzymes with optimized features such as activity, (enantio-, regio-, and chemo-) selectivity, stability, substrate specificity, cofactor specificity, tolerance of cosolvents, pH optimum and cofactor requirements, and many more others. However, the high intrinsic structural complexity of enzyme still makes it difficult to understand how enzymes function and control the selectivity of different reactions. Therefore, it is urged to (i) develop computational protocols that can reliably describe and predict enzyme mechanisms and controlling selectivity, (ii) design of new enzymes, (iii) develop computational tools that can efficiently scan sequence space in the search for optimal biocatalysts, and (iv) engineering thermostability and solvent tolerance [117].

In addition to the experimental techniques, enzyme computational simulations are a powerful addition to the protein design toolbox, and their practicability and effectiveness in the study of biocatalysis have been greatly increased by the development of modern theoretical algorithms concurrent to the improvement of computer power. These methods are contingent on the knowledge of the enzyme's 3D structure and of the reaction mechanism, which is very much easier to obtain nowadays. Not only do simulations allow for a more ambitious and tentative exploration of the sequence space but also a much higher number of variants and protein-ligand systems can be analyzed *in silico*. Modern computational simulations have been applied for the re-design and *de novo* generation of enzymes. If *de novo* or re-designed enzymes are not very efficient catalysts, further refinement by such as directed evolution can increase the catalytic parameters. At present, computational simulation tools such as molecular dynamics (MD) and hybrid methods combining quantic mechanics and molecular mechanics (QM-MM) have been mainly employed for rationalize the success (or failure) of designed enzymes or to reduce the universe of possibilities by ranking feasible designs [118].

Computer modeling approaches have been utilized to the quantitative understanding of enzyme catalysis. The method used for modeling an enzymatic reaction is the QM-MM method in well-defined active site. Because it is believed that the catalysis is mainly due to electrostatic preorganization of the active site, to convince this point may require the demonstration of effective rational enzyme design. The effectiveness of computer-aided enzyme design is based on the ability to predict the actual catalytic power of different design constructs in spite of the fact that the most of current advances are still done by direct evolution [119].

Computer-aided molecular techniques, which are widely used in academia and industrial settings, can assist the selection of new compounds for modulating biological targets. When applied to the discovery of compounds supposed to become future drugs, these techniques are commonly referred as computer-aided drug design (CADD) techniques. In the last decade, epigenetic modifications of histones, particularly histone deacetylases of classes I, II, and IV (HDACs), are found to be the catalytic result of a large set of enzyme families that operate covalent modifications on specific residues at the histone tails. These modifications elicit a complex and concerted process that greatly contribute to the chromatin remodeling and may drive different pathological conditions, especially cancer. The computer-aided molecular design approaches have been applied for several epigenetic targets under validation for drug discovery purposes to produce the first preclinical and clinical outcomes [120].

FuncLib is a new automated computational protein design strategy that designs a small set of stable, efficient, and functionally diverse multipoint active-site mutants using phylogenetic analysis and Rosetta design calculations and is amenable to low-throughput screening. FuncLib, in principle, can be applied to any natural enzyme using its molecular structure and a diverse set of homologous sequences. Two unrelated enzymes, a phosphotriesterase and an acetyl-CoA synthetase, were then tested and most of the designs showed significant different activity profiles from the wild-type and from one another. Several dozen designs with only 3-6 active-site mutations exhibited 10- to 4 000-fold higher efficiencies with a range of alternative substrates, including hydrolysis of the toxic organophosphate nerve agents soman and cyclosarin and synthesis of butyryl-CoA. FuncLib is implemented as a web server (http://FuncLib.weizmann.ac.il) [121].

In the past few decades, AI and ML have been gaining attention in designing new biocatalysts. Apart from the classical model-driven rational design and iterative selecting of the existing mutants in directed evolution, ML strategy is data-driven, which identifies patterns in the existing data to predict properties of the previously unseen but similar input or generates new, previously unseen but promising variants. Specifically speaking, the patterns found from the collected data can help predict protein structures, improve enzyme stability, solubility and function, predict substrate specificity, and guide rational protein design, which indicate ML strategy is the combination of rational design and directed evolution.

Results on ML-guided directed evolution, analysis in systems metabolic engineering, implementation of biocatalysis in systems in the industry, and biosystems design were also reported in literatures. Multiple ML algorithms have already been applied to enzyme engineering such as random forests used to predict protein solubility, support vector machines and decision trees to predict enzyme stability changes upon mutations, *K*-nearest-neighbor classifiers to predict enzyme function and mechanisms, and various scoring and clustering algorithms for rapid functional sequence annotation. The advances in the analysis of human genetic variation data in biomedicine and health care further increase the appeal of this approach for the design of beneficial mutations [122].

The success of an ML predictor for previously unseen data crucially depends on both the quality of the data used for training and the efficiency of the underlying algorithm. A lot of notable databases have been used in enzyme engineering. The most abundant are databases of protein sequences followed by databases of protein structures. Protein stability and solubility are the next two qualities that have routinely been measured for several decades now [122]. The applications of ML for enzyme engineering using these databases include protein structure prediction, predicting enzyme catalytic activities, precise functional prediction, protein solubility prediction, and the design of smart combinatorial libraries for directed protein evolution.

Focus on the biocatalytic synthesis reaction, a reaction database is needed for searching reactions and relevant information of reactions. A reaction database, which covers chemical and biochemical reactions, has been developed to assist small molecule pharmaceutical process development during the early stages of the synthesis route selection and process-product development by providing a data rich environment for enhanced process understanding. The database structured in terms of reaction classification of reaction types, compounds participating in the reaction, use of organic solvents and their function, information for single step and multistep reactions, target products, and reaction conditions and reaction data. Information for reactor scale-up together with information for the separation and other relevant information for each reaction and reference are also available in the database. In addition, the retrieved information obtained from the database can be evaluated in terms of sustainability using published "green" metrics. The application of the database has been highlighted for the synthesis of ibuprofen, for which data on different reaction pathways retrieved from the database were compared using "green" chemistry metrics [123].

Unlike pathway databases such as EcoCyc, MetaCyc, and Kegg, which focus primarily on intermediary metabolism, a free access database (http://umbbd.ahc.umn.edu) concentrated on biocatalysis/biodegradation was established by the University of Minnesota. The University of Minnesota Biocatalysis/Biodegradation Database (UM-BBD) is an online compilation of microbial catabolic enzymes, reactions, and pathways for primarily synthetic organic chemical compounds, which can be queried (i) for specific compounds via name, CAS number, or chemical formula; (ii) for enzymes by means of name or EC code; or (iii) for specific microorganisms, as well as can be browsed by compounds, enzymes, reactions, pathways, or microorganisms from the search page. Information in the database is directly applicable toward enhancing the understanding of biocatalysis leading to specialty chemical manufacture and the biodegradation of environmental pollutants. Catabolic routes for compounds having any of the 50 functional groups listed as of January 2002 are provided. The UM-BBD is a key resource for biodegradation pathway information and is evolving to include the prediction of specialized catabolic routes for new compounds. An example the text and graphic formats of the propylene degradation pathway from the UM-BBD database is shown in Figure 8.3 [124]. The UM-BBD showed several strengths including its focusing on specialized enzymes and genes not found in many databases of intermediary metabolism, focusing on functional group transformations, and numerous static and dynamic links to other databases; however, complete UM-BBD pathways may not necessarily be present in their entirety within a single organism is its potential weakness.

The advent of AI algorithms to automatize, improve, and generalize predictions is gaining importance in the organic synthesis that holds the potential to advance organic synthesis in a revolutionary way, and the development of AI together with "Big Data" have been extended to the applications such as targeted drug discovery, diabetes care, medicine, and public health [125–128]. Then, it comes out a question: Are robots replacing chemists? [129] I think the answer should be "No" based on the most basic rule: the created "thing" cannot exceed creator. Robots may be superior to chemists in some routine works due to their creation by a group of intelligent people and the fast processing of large amounts of data and knowledges. However, certain characteristics of human such as intuition and inspiration, which can never endow to robots, make chemists prevail robots.

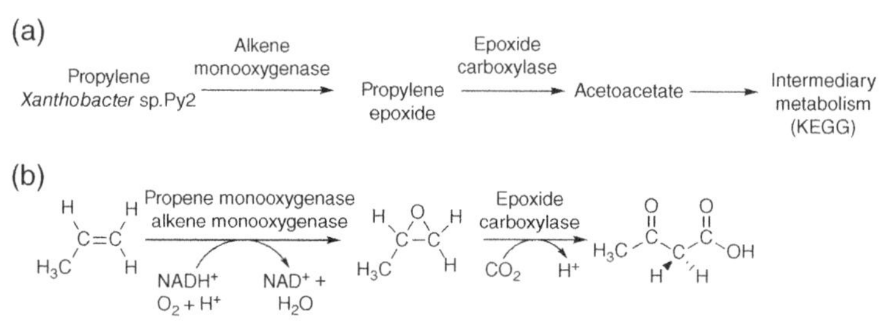

Figure 8.3 Biodegradation pathway for propylene retrieved from the UM-BBD. (a) Text-based and (b) graphical-based representations. *Source:* Burgard and Maranas [124].

References

1 Liu, R., Li, X., and Lam, K.S. (**2017**). *Curr. Opin. Chem. Biol. 38*: 117–126.
2 Rasheed, A. and Farhat, R. (**2013**). *Int. J. Pharma. Sci. Res. 4*: 2502–2516.
3 Pottel, J. and Moitessier, N.J. (2017). J. Chem. Inf. Model. 57: 454–467.
4 Darvas, F., Dorman, G., Urge, L. et al. (2001). Pure Appl. Chem. 73: 1478–1498.
5 Verdine, G.L. (1996). Nature 384: 11–13.
6 Merrifield, R.B. (1963). J. Am. Chem. Soc. 85: 2149–2154.
7 Gordon, K. and Balasubramanian, J. (1999). Chem. Technol. Biotechnol. 74: 835–851.
8 Xiang, X.-D., Sun, X., Briceño, G. et al. (1995). Science 268: 1738–1740.
9 Frankel, A., Millward, S.W., and Roberts, R.W. (**2003**). *Chem. Biol. 10*: 1043–1050.
10 Josephson, K., Hartman, M.C.T., and Szostak, J.W. (**2005**). *J. Am. Chem. Soc. 127*: 11727–11735.
11 Murakami, H., Ohta, A., Ashigai, H., and Suga, H. (**2006**). *Nat. Methods 3*: 357–359.
12 Heinis, C., Rutherford, T., Freund, S., and Winter, G. (**2009**). *Nat. Chem. Biol. 5*: 502–507.
13 Franzini, R.M., Neri, D., and Scheuermann, J. (**2014**). *Acc. Chem. Res. 47*: 1247–1255.
14 Franzini, R.M. and Randolph, C. (**2016**). *J. Med. Chem. 59*: 6629–6644.
15 Li, Y., Zimmermann, G., Scheuermann, J., and Neri, D. (**2017**). *ChemBioChem 18*: 848–852.
16 Favalli, N., Bassi, G., Scheuermann, J., and Neri, D. (**2018**). *FEBS Lett. 592*: 2168–2180.
17 Ratnayake, A.S., Flanagan, M.E., Foley, T.L. et al. (**2019**). *ACS Comb. Sci. 21*: 650–655.
18 Sannino, A., Gironda-Martínez, A., Gorre, É.M.D. et al. (**2020**). *ACS Comb. Sci. 22*: 204–212.
19 Lemke, M., Ravenscroft, H., Rueb, N.J. et al. (**2020**). *Bioorg. Med. Chem. Lett. 30*: 127464.
20 Furka, A. (**2002**). *Drug Discov. Today 7*: 1–4.
21 Noonan, J.W., Brown, J.N., Kershaw, M.T. et al. (**2003**). *Assoc. Lab. Automat. 8*: 65–71.
22 Geysen, H.M., Meloen, R.H., and Barteling, S.J. (**1984**). *Proc. Natl. Acad. Sci. U.S.A. 81*: 3998–4002.
23 Houghten, R.A. (**1985**). *Proc. Natl. Acad. Sci. U.S.A. 82*: 5131–5135.
24 Sebestyen, F., Dibo, G., Kovacs, A., and Furka, A. (**1993**). *Bioorg. Med. Chem. Lett. 3*: 413–418.
25 Furka, A., Sebestyen, F., Asgedom, M., and Dibo, G. (**1991**). *Int. J. Peptide Protein Res. 37*: 487–493.

26 Lam, K.S., Salmon, S.E., Hersh, E.M. et al. (**1991**). *Nature 354*: 82–84.
27 García-Junceda, E., García-García, J.F., Bastida, A., and Fernández-Mayoralas, A. (**2004**). *Bioorg. Med. Chem. 12*: 1817–1834.
28 Rich, J.O., Michels, P.C., and Khmelnitsky, Y.L. (**2002**). *Curr. Opin. Chem. Biol. 6*: 161–167.
29 Machauer, R. and Waldmann, H. (**2001**). *Chem Eur J 7*: 2940–2956.
30 Jeyaraj, D. and Waldmann, H. (**2001**). *Tetrahedron Lett 42*: 835–837.
31 Pathak, T. and Waldmann, H. (**1998**). *Curr Opin Chem Biol 2*: 112–120.
32 Reents, R., Jeyaraj, D.A., and Waldmann, H. (**2002**). *Drug Discov. Today 7*: 71–76.
33 Reents, R., Jeyaraj, D.A., and Waldmann, H. (**2001**). *Adv. Synth. Catal. 343*: 501–513.
34 Rademann, J., Grotli, M., Meldal, M., and Bock, K. (**1999**). *J. Am. Chem. Soc. 121*: 5459–5466.
35 Nishiguchi, S., Yamada, K., Fuji, Y. et al. (**2001**). *Chem. Commun.* 1944–1945.
36 Sauerbrei, B., Jungmann, V., and Waldmann, H. (**1998**). *Angew. Chem., Int. Ed 37*: 1143–1146.
37 Grether, U. and Waldmann, H. (**2000**). *Angew. Chem. Int. Ed. 39*: 1629–1632.
38 Grether, U. and Waldmann, H. (**2001**). *Chem. Eur. J. 7*: 959–971.
39 Zehavi, U., Sadeh, S., and Herchman, M. (**1983**). *Carbohydr. Res. 124*: 23–34.
40 Schuster, M., Wang, P., Paulson, J.C., and Wong, C.-H. (**1994**). *J. Am. Chem. Soc. 116*: 1135–1136.
41 Blixt, O. and Norberg, T. (**1998**). *J. Org. Chem. 63*: 2705–2710.
42 Lubineau, A., Malleron, A., and Le Narvor, C. (**2000**). *Tetrahedron Lett. 41*: 8887–8891.
43 Brinkmann, N., Malissard, M., Ramuz, M. et al. (**2001**). *Bioorg. Med. Chem. Lett. 11*: 2503–2506.
44 Zehavi, U. (**1999**). *React. Funct. Polym. 41*: 59–68.
45 Corrales, G., Fernández-Mayoralas, A., García-Junceda, E., and Rodríguez, Y. (**2000**). *Biocatal. Biotransform. 18*: 271–281.
46 Schmidt, D. and Thiem, J. (**2000**). *Chem. Commun.* 1919–1920.
47 Ulijn, R.V., Baragañ, B., Halling, P.J., and Flitsch, S.L. (**2002**). *J. Am. Chem. Soc. 124*: 10988–10989.
48 Krstenansky, J.L. and Khmelnitsky, Y. (**1999**). *Bioorg. Med. Chem. 7*: 2157–2162.
49 Michels, P.C., Khmelnitsky, Y.L., Dordick, J.S., and Clark, D.S. (**1998**). *Trends Biotechnol. 16*: 210–215.
50 Altreuter, D.H. and Clatk, D.S. (**1999**). *Curr. Opin. Biotechnol. 10*: 130–136.
51 De la Gôutte, J.T., Khan, J.A., and Vulfson, E.N. (**2001**). *Biotechnol. Bioeng. 75*: 93–99.
52 Kim, D.-Y. and Dordick, J.S. (**2001**). *Biotechnol. Bioeng. 76*: 200–206.
53 Jang, W.B., Hu, H., Lieberman, M.M. et al. (**2001**). *J. Comb. Chem. 3*: 346–353.
54 Liu, X.-C., Clark, D.S., and Dordick, J.S. (**2000**). *Biotechnol. Bioeng. 69*: 457–460.

55 Jiang, J., Biggins, J.B., and Thorson, J.S. (**2000**). *J. Am. Chem. Soc. 122*: 6803–6804.

56 Thorson, J.S., Hosted, T.J. Jr., iang, J. et al. (**2001**). *Curr. Org. Chem. 5*: 139–167.

57 Nishimura, S.-I. (**2001**). *Curr. Opin. Chem. Biol. 5*: 325–335.

58 Sujino, K., Uchiyama, T., Hindsgaul, O. et al. (**2000**). *J. Am. Chem. Soc. 122*: 1261–1269.

59 Frei, P., Hevey, R., and Ernst, B. (**2019**). *Chem. Eur. J. 25*: 60–73.

60 Larsen, D. and Beeren, S.R. (**2019**). *Chem. Sci. 10*: 9981–9987.

61 Bhaskar, U., Sterner, E., Hickey, A.M. et al. (**2012**). *Appl. Microbiol. Biotechnol. 93*: 1–16.

62 Linhardt, R.J. and Kim, J.-H. (**2007**). *Chem. Biol. 14*: 972–973.

63 Bézay, N., Dudziak, G., Liese, A., and Kunz, H. (**2001**). *Angew. Chem. Int. Ed. 40*: 2292–2295.

64 Fumoto, M., Hinou, H., Ohta, T. et al. (**2005**). *J. Am. Chem. Soc. 127*: 11804–11818.

65 Laaf, D., Bojarová, P., Pelantová, H. et al. (**2017**). *Bioconjugate Chem. 28*: 2832–2840.

66 Gumulya, Y., Huang, W., D'Cunha, S.A. et al. (**2019**). *ChemCatChem 11*: 841–850.

67 Pan, L., Yang, L., Huang, Y. et al. (**2019**). *ACS Omega 4*: 21078–21082.

68 He, W., Gao, Y., Zhu, G. et al. (**2020**). *Chem. Eng. J. 381*: 122721.

69 Xue, Y. and Sherman, D.H. (**2001**). *Metab. Eng. 3*: 15–26.

70 Méndez, C. and Salas, J.A. (**2001**). *Trends Biotechnol. 19*: 449–456.

71 Schmidt-Dannert, C. (**2000**). *Curr. Opin. Biotechnol. 11*: 255–261.

72 Mootz, H.D., Schwarzer, D., and Marahiel, M.A. (**2000**). *Proc. Natl. Acad. Sci. U.S.A. 97*: 5848–5853.

73 Doekel, S. and Marahiel, M.A. (**2000**). *Chem. Biol. 7*: 373–384.

74 Khosla, C. and Harbury, P.B. (**2001**). *Nature 409*: 247–252.

75 Sun, H., Liu, Z., Zhao, H., and An, E.L. (**2015**). *Drug Des. Dev. Ther. 9*: 823–833.

76 Harvey, C.J.B., Puglisi, J.D., Pande, V.S. et al. (**2012**). *J. Am. Chem. Soc. 134*: 12259–12265.

77 Yu, D., Xu, F., Zeng, J., and Zhan, J. (**2012**). *IUBMB Life 64*: 285–295.

78 Wakamoto, T., Mori, T., Morita, H., and Abe, I. (**2011**). *J. Am. Chem. Soc. 133*: 4746–4749.

79 Wanibuchi, K., Zhang, P., Abe, T. et al. (**2007**). *FEBS J. 274*: 1073–1082.

80 Kreutzer, M.F., Kage, H., Herrmann, J. et al. (**2014**). *Org. Biomol. Chem. 12*: 113–118.

81 Dutta, S., Whicher, J.R., Hansen, D.A. et al. (**2014**). *Nature 510*: 512–517.

82 Whicher, J.R., Dutta, S., Hansen, D.A. et al. (**2014**). *Nature 510*: 560–564.

83 Condurso, H.L. and Bruner, S.D. (**2012**). *Nat. Prod. Rep. 29*: 1099–1110.

84 McDaniel, R., Thamchaipenet, A., Gustafsson, C. et al. (**1999**). *Proc. Natl. Acad. Sci. U.S.A. 96*: 1846–1851.

85 Nguyen, K.T., He, X., Alexander, D.C. et al. (**2010**). *Antimicrob. Agents Chemother. 54*: 1404–1413.

86 Doekel, S., Coëffet-Le Gal, M.-F., Gu, J.-Q. et al. (**2008**). *Microbiology 154*: 2872–2880.

87 Baltz, R.H., Brian, P., Miao, V., and Wrigley, S.K. (**2006**). *J. Ind. Microbiol. Biotechnol. 33*: 66–74.

88 Chooi, Y.-H. and Tnag, Y. (**2012**). *J. Org. Chem. 77*: 9933–9953.

89 Xu, Y., Espinosa-Artiles, P., Schubert, V. et al. (**2013**). *Appl. Environ. Microbiol. 79*: 2038–2047.

90 Xu, Y., Zhou, T., Zhang, S. et al. (**2014**). *Proc. Natl. Acad. Sci. U.S.A. 111*: 12354–12359.

91 Kushnir, S., Sundermann, U., Yahiaoui, S. et al. (**2012**). *Angew. Chem. Int. Ed. 51*: 10664–10669.

92 Walker, M.C., Thuronyi, B.W., Charkoudian, L.K. et al. (**2013**). *Science 341*: 1089–1094.

93 Wang, J., Sánchez-Roselló, M., Aceña, J.L. et al. (**2014**). *Chem. Rev. 114*: 2432–2506.

94 Fischbach, M.A., Lai, J.R., Roche, E.D. et al. (**2007**). *Proc. Natl. Acad. Sci. U.S.A. 104*: 11951–11956.

95 Evans, B.S., Chen, Y., Metcalf, W.W. et al. (**2011**). *Chem. Biol. 18*: 601–607.

96 Zhang, K., Nelson, K.M., Bhuripanyo, K. et al. (**2013**). *Chem. Biol. 20*: 92–101.

97 Hopwood, D.A., Malpartida, F., Kieser, H.M. et al. (**1985**). *Nature 314*: 642–644.

98 Fukushima, E.O., Seki, H., Sawai, S. et al. (**2013**). *Plant Cell Physiol. 54*: 740–749.

99 Seki, H., Sawai, S., Ohyama, K. et al. (**2011**). *Plant Cell 23*: 4112–4123.

100 Moses, T., Pollier, J., Almagro, L. et al. (**2014**). *Proc. Natl. Acad. Sci. U.S.A. 111*: 1634–1639.

101 Russell, S.J.; Norvig, P. (**2009**). *Artificial Intelligence: A Modern Approach*, 3rd ed., Upper Saddle River, New Jersey: Prentice Hall. ISBN 978-0-13-604259-4.

102 Artificial Intelligence, https://en.wikipedia.org/wiki/Artificial_intelligence.

103 Artificial Intelligence and Machine Learning: Policy paper, Internet Society, April (**2017**). https://www.internetsociety.org/resources/doc/2017/artificial-intelligence-and-machine-learning-policy-paper/.

104 34 Artificial Intelligence Companies Building a Smarter Tomorrow. https://builtin.com/artificial-intelligence/.

105 Peiretti, F. and Brunel, J.M. (**2018**). *ACS Omega 3*: 13263–13266.

106 Corey, E.J. and Wipke, W.T. (**1969**). *Science 166*: 178–192.

107 Coley, C.W., Green, W.H., and Jensen, K.F. (**2018**). *Acc. Chem. Res. 51*: 1281–1289.

108 Gao, H., Struble, T.J., Coley, C.W. et al. (**2018**). *ACS Cent. Sci. 4*: 1465–1476.

109 Granda, J.M., Donina, L., Dragone, V. et al. (**2018**). *Nature 559*: 377–381.

110 Cova, T.F.G.G. and Pais, A.C.C. (**2019**). *Front. Chem 7*: 809.

111 Empel, C. and Koenigs, R.M. (**2019**). *Angew. Chem. Int. Ed. 58*: 17114–17116.

112 Gromski, P.S., Granda, J.M., and Cronin, L. (**2020**). *Trends Chem. 2*: 4–12.

113 Sheldon, R.A., Brady, D., and Bode, M.L. (**2020**). *Chem. Sci. 11*: 2587–2605.
114 Bornscheuer, U.T. (**2017**). *Phil. Trans. R. Soc. A 376*: 0063.
115 De Souza, R.O.M.A., Miranda, L.S.M., and Bornscheuer, U.T. (**2017**). *Chem. Eur. J. 23*: 12040–12063.
116 Balke, K., Kadow, M., Mallin, H. et al. (**2012**). *Biomol. Chem. 10*: 6249–6265.
117 Enzyme Engineering: Bright Strategies from Theory and Experiments (**2016**). Flagship Workshop, Centre Européen de Calcul Atomique et Moléculaire (CECAM), Lausanne, Switzerland,
118 García-Guevara, F., Avelar, M., Ayala, M., and Segovia, L. (**2015**). *Biocatalysis 1*: 109–117.
119 Frushichea, M.P., Mills, M.J.L., Schopf, P. et al. (**2014**). *Curr. Opin. Chem. Biol. 0*: 56–62.
120 Andreoli, F. and Rio, A.D. (**2015**). *Comp. Struct. Biotechnol. J. 13*: 358–365.
121 Khersonsky, O., Lipsh, R., Avizemer, Z. et al. (**2018**). *Mol. Cell 72*: 178–186.
122 Mazurenko, S., Prokop, Z., and Damborsky, J. (**2020**). *ACS Catal 10*: 1210–1223.
123 Papadakis, E., Anantpinijwatna, A., Woodley, J.M., and Gani, R. (**2017**). *Processes 5*: 58.
124 Burgard, A.P. and Maranas, C.D. (**2002**). *Metab. Eng. 4*: 111–113.
125 Liu, B., He, H., Luo, H. et al. (**2019**). *Stroke Vasc. Neurol. 4*: 206–213.
126 Musacchio, N., Giancaterini, A., Guaita, G. et al. (**2020**). *J. Med. Internet Res. 22*: e16922.
127 Lovis, C. (**2019**). *J. Med. Internet Res.* 21: e16607.
128 Bemle, L' Benke, G. (**2018**). *Int. J. Environ. Res. Public Health 15*: 2796.
129 Maryasin, B., Marquetand, P., and Maulide, N. (**2018**). *Angew. Chem. Int. Ed. 57*: 6978–6980.

Index

b

C

d

e

f

g

i

j

k

l

O

q

r

S

u

Printed and bound by CPI Group (UK) Ltd, Croydon, CR0 4YY

21/08/2023

08102587-0002